T0295653

Facilities @ Management, Concept - Realization - Vision, A Global Perspective

Facilities @ Management, Concept - Realization - Vision, A Global Perspective

Edited by Edmond Rondeau and Michaela Hellerforth

Published by John Wiley & Sons, Inc., Hoboken, New Jersey.

Published simultaneously in Canada.

A catalogue record for this book is available from the Library of Congress

Hardback ISBN: 9781394213283; ePub ISBN: 9781394213306; ePDF ISBN: 9781394213290; oBook ISBN: 9781394213313

Author Photo: © Pexels
Cover Design: Wiley

Set in 9.5/12.5pt STIX Two Text by Integra Software Services Pvt. Ltd, Pondicherry, India

SKY10064569_011124

DEDICATED TO[1]

The hardworking, knowledgeable, and skilled facility professionals around the world who acquire, plan, design, construct, support, maintain, and manage their organizations facilities to assist their customers and organizations to succeed and excel.

The in-house customers who require and accept professional assistance and entrust their facility service providers with corporate resources and timely confidential information to meet their strategic business requirements.

The in-house customers who require and accept professional assistance and entrust their facility service providers with corporate resources and timely confidential information to meet their strategic business requirements.

The suppliers, vendors, and consultants who provide quality, timely, creative, and cost-effective services to help their facility professional clients succeed and excel.

The bosses, leaders, and organization officers who support and provide their facility professional staff with the responsibility, authority, and resources to execute their duties.

The students, teachers, educational organizations, and FM associations who sustain a growing body of knowledge and research base of facility management information and are the future of the profession.

Keeping up the good work!

1 From the Dedicated To found in the book "Facility Management," Wiley, 2006, Edmond P. Rondeau, Robert K. Brown and Paul D. Lapides.

Contents

Foreword

Among my collections of books and reference materials that have accrued over my 50 years of Facilities Management diverse work experience, involvement with global FM educational initiatives, and even a personal desire to learn more about the profession, this is one that will be among the most treasured.

As I have personally and professionally used and owned many books written by one of the authors, Edmond P. Rondeau, I can testify to the content of this being one of the most advantageous for you to devour the contents with the knowledge that your valuable time will not be wasted and there will be a great "return on your investment."

Facilities @ Management, is not only a great global outlook from the many contributors but provides a historical perspective of the FM profession as well as providing current and relevant topics and impacts to FM's such as the globalization, sustainability, technology, and integration of the FM into the professional business world.

The ever evolving, changing, and the speed of change is deeply engrained in the FM world that brings resilience and the ability to propel the organizations we work with forward much faster and in alignment than ever before.

I highly recommend this as a resource for everyone interested in, already working in, or in any way related to the FM profession with all the passion and earnest drive that I have for our profession. It will be read often by myself and not be gathering any dust on the shelf.

Jon E. Martens
FMP, SFP, CFMJ, RCFM
IFMA Fellow
IFMA Certified Instructor
The JEMCOR companies, USA

Preface

Introduction

This book was developed as an anthology to celebrate the past 40 years of international Facility Management (1982–2022) and beyond with a look to the future challenges for the FM profession.

The book was jointly initiated by Josef R. G. Mack (Germany), and Edmond P. Rondeau (USA) who have known each other for many years. They have collaborated on several projects together during the past 30+ years, including working on a book on Facility Management (FM) 20 years ago together with Walther J. F. Moslener that was published in Germany in 2001.

The co-editors of this book are Professor Dr. Michaela Hellerforth (Germany) and Edmond P. Rondeau, AIA Emeritus, RCFM, IFMA Fellow (USA). Detailed information on the co-editors can be found in Appendix C. Both have authored/co-authored numerous FM and real estate books, have spoken and taught at universities, and have been active in FM internationally for decades.

The concept for this book has a mixture of **50+** FM contributors from 16 countries around the world. These contributions have been provided from the contributors' point of view, from their memory and from documents and other credited sources. Contributors have many different backgrounds and education including FM professionals, academics, vendors, architects, engineers, interior designers, consultants, and business developers from the US, Canada, Mexico, Brazil, South Korea, Hong Kong, China, Singapore, Australia, England, Scotland, Ireland, Germany, Austria, Switzerland, and the Netherlands. All have an affiliation with one or more related associations, and many have been recognized for their service and support to FM.

The main topics you will see that the contributors addressed included one to some of the following areas from their objective point of view:

- the making of the term "facility management" in their country
- the acceptance of FM in their country
- the implementation of FM within their country
- the spreading of the term FM globally
- the similarities and differences of FM globally
- the status of academic efforts, continuing education, and realizations to promote the field of FM worldwide.
- the growth of women in FM and women FM leaders
- the transformation necessities for traditional structures
- the challenges in the future for the built environment in general, especially due to the COVID-19 pandemic and the general impact to offices in regard to People, Process, Place, Technology, and remote working by the bringing together of workplaces and office needs – all of this amplified by climate change, energy crises, sustainability, social implications, etc.

Goal

The goal of this book is to describe why from a personal perspective Facility Management developed formally in the US and also Canada, from a nongovernmental and business perspective, compared to how Facility Management developed in various countries in the Americas, Europe, Australia, and Asia.

The selection of **FACILITIES**, as a plural term, is used to show the perspective chosen by the initiators for this interdisciplinary approach of their anthology and this legacy covering the forty years of international FM and beyond. In combination with @, the icon for Information Technology as an essential bridging element to **MANAGEMENT**.

The title is designed to express the general importance that facilities have in the built environment not only for working processes, cost, and productivity, but also for research, educational, care and living space. The three nouns of the subtitle reflect the three dimensions of the book's content which displays the origin, the implementation, and the potential still inherent in FM.

In *the* ***Concept*** part principals, sponsors, and creators of the research project spanning over 40 years into the relationship between people, workplace, and buildings are identified. The abstract academic FM idea is explained, leading to the first FMI and the subsequent coining of the term as well as forming the National Facility Management Association (NFMA) in the USA in 1980.

The ***Realization*** describes the development of the concept in academia and the work environment, and the transformation of the professional association into a transnational activity, claimed in 1982 by the name-change to International Facility Management Association (IFMA). The educational and professional activities, the founding of FM associations in other countries, their realization, and activities are characterized. Development of technology, scientific placement, academic classification, implementation of the profession in different areas and companies internationally are represented.

The ***Vision*** part includes success as well as shortcomings of the realization and transformation in the academic, the institutional, and the commercial world. To create a platform of demands for the immediate future it is required for the built environment to respect social, economic, ecological, and demographic changes under sustainability demands and global climate-change.

Insight

Facility Management is still facing misunderstanding as a concept in some parts of the world, despite the role it plays for over four decades already internationally since its invention, "born and coined" by academia, raised and brought into professional and corporate hierarchies through the initiative of an association, and finally accepted as a profession with academic training, university seats, and research far beyond the USA, implemented in Europe, and spread out worldwide.

A variety of technical FM books on the subject have been available for some time, not only in English but in many languages. They have aimed to serve as introduction to the body of knowledge, the description of tools and technics developed and available, as well as functional approaches to the phenomenon that is FM, whereas the origin of the term FM and its potential as a holistic concept remained often only worth a footnote or vague annotations.

To fill this gap, the vision of this anthology strives to offer academic and sponsoring thought leaders of the beginning, drivers of the professional development throughout the past, and today's experts and protagonists in FM and adjacent fields including academics in educational institutions nationally and internationally, a platform to reveal their very personal involvement, ideas, experience, and outlook gained in their practice, simultaneously disclosing roots, perspectives, and even inherent answers to questions still pending today and the future.

This approach allows the reader insights into the process of the creation of the term up to what the development and present state of the art can offer. Furthermore, the unique display of experience and background of the contributing contemporary witnesses gives an insight into the causes leading to the development of the concept FM in the last quarter of the past 20th century providing a global perspective.

Readers of This Book

This anthology is for readers interested in the built environment and its personal, social, structural, and technical aspects; factors important to all of us throughout our entire lives. Readers are primarily students, professionals and researchers in planning, engineering, structural analysis, building, real estate, HVAC, IT, etc. Included are top experts in related social and political science, laymen wanting to understand the FM phenomenon, gain an overview of the history, implementation, and perspectives of development and its possible potential for the future.

What is special about This Book?

The publication offers a platform for recalling the original vision, experience realized, and the insight of founders and contributors to the development as well as their findings, achievements, and legacy for FM. All contributions are pinpointed and standalone but are not redundant, highlighting the contributor's view.

Being contemporary researchers, educators, experts, and professionals with different backgrounds from building, real estate, and industry from the US, Canada, Mexico, Brazil, China, Hong Kong, Singapore, South Korea, Australia, England, Scotland, Ireland, the Netherlands, Germany, Austria, and Switzerland. These contributors summarize the state of the art in international FM in their countries to provide an assessment of the existing solutions and their potential for meeting the requirements regarding facilities in the future under the challenge of sustainability, climate change, and – last but not least – growing social and political issues all over the world.

The volume of the anthology was designed to be approximately 500 pages and includes essays from the 50+ contributors. The average length of the contribution was 10 pages (plus/minus) each, which included citations, figures, charts, plots, a brief bibliography, with a short educational and professional history of the author.

Making a Difference

Despite the path of success of FM books in the decades passed, the awareness about the origin of FM, the real idea behind the concept, and the vision of the beginning in some countries has decreased, remained untold to a wider public, or even got lost, despite the academic and corporate spread internationally. Under this experience, the proposal to use the anniversary mentioned for an academic checkout into the achievements seemed not only of academic interest, but of an almost historic importance, as long as eyewitness of the founding period are still available.

However, when the idea became a project in 2021, it was not certain, who could still be found from the early days in the USA and who from the current protagonists there, being colleagues, professionals, academics, and practitioners, would be willing to contribute in order to help achieving the intended goal. It was a pleasant surprise that the first cautious inquiries brought already more interest than expected, and in the past months most contributions had been turned in already, most from North America and international regions with English as the lingua franca.

This book would only be an incomplete picture of the founders' original concept and IFMA's intentions as forerunners of an idea that became an international success. To reflect the development in the Americas, Asia, Australia, and Europe and especially in the German-speaking countries, it turned out that there the situation to win interested and for the reader interesting contributors was rather complex. Not because of language hurdles, which nowadays are not a real issue anymore, but the variety of different strands of development in academia, business, and associations.

To respect multiple requests of contributors, learning late of the initiative, an extensions of the submission period had become necessary, to give enough time for qualified adaptations and the transcription of texts.

Mission Statement

To differ from text and technical books, we choose the form of an anthology, to allow contemporary witnesses and contributors to reveal the roots, the ideas, the foresight, and the support of sponsors necessary to create a first FM Institute as well as an association like IFMA. Their roles and contribution in the dissemination of the term and the creation and forming of a profession are still widely unknown.

The interlocking of the development of research activities in academia regarding work processes within corporate offices and structures, is not only of historic interest. It can be taken as a very interesting case study for a dynamic reaction to adapt to changes in a completely new setting – then caused by the intrusion of Electronic Data Processing (now called IT) in the work processes, taking place in offices, housed in buildings. A situation not too different from the challenges demographic, social, and environmental changes that are now confronting the developed post-industrial nations and their built environment under resource, climate, and sustainability aspects.

Only this holistic approach we believe can do justice to the industry, the academic development of curricula, and the economic importance FM has gained in these four decades passed. And furthermore, this can unveil some of the inherent potential still dormant in the concept of Facility Management, to assist in solving problems now and in the future, especially regarding resource, climate, and sustainability challenges.

We trust that you will enjoy this book, the legacy, and the perspectives that our contributors have provided. The editors are donating the royalties of this book to one or more educational focused Foundations for the benefit of FM university students.

EPR
MH
JRGM

Acknowledgments

The co-editors/initiators acknowledge and especially thank the 50+ contributors internationally whose experience, writing, speaking, teaching, and sharing have contributed to many of the Facility Management (FM) concepts, and experiences related in this F@M book about the 40 years of international facility management (1982–2022) and into the future. And to the many others globally who have helped in the development of the facility management profession which has led to the organizing of over 40 FM associations globally.

We also acknowledge and thank those who were invited to contribute to this unique anthology and legacy publication, but for various business, time, personal, health, and/or family issues had to decline the sharing of their contributions. And we sadly recognize that a number of important FM colleagues have passed away before their contributions could be requested.

We are grateful for and acknowledge the many manufacturers, vendors, and consultants who have supported the FM profession and its growth around the world for over 40 years. These organizations include furniture manufacturers, flooring, lighting, rentals, food, beverage, cleaning, FM outsourcing, real estate, architects, engineers, interior designers, contractors, IT and special software developers, etc. Many of these organizations have helped FM associations, chapters, and conferences with their sharing and supporting the FM message and education with their time, knowledge, experience, support, and sponserships.

The success of the last 40+ years of the FM profession also resides with the request for and development of FM education and research. Most and possibly most all of the contributors have taken advantage of the education and research that FM associations have developed and created. We acknowledge the establishment of formal university FM degree programs, certification and certificate courses, and continuing education courses at FM conferences, universities, and private education vendors. Also, association foundations have established and raised funds for many scholarship programs for FM students from around the world. Many of the contributors have been the beneficiaries of these scholarship programs.

The reader may have noticed that some organizations vendors, or businesses that have supported FM in the past have not provided a contribution, and we wanted to acknowledge this. After a number of attempted contacts, no representative from these organizations responded. This is part of the business cycle where the reader may have noticed that some prominent organizations at FM conferences may no longer be as visible as many years ago. We find that the management of these organizations, vendors, or businesses have changed, and new leadership provides different directions either as new strategic goals or because leadership has no or little understanding of FM and/or their history of FM purchasing their products or services.

We also thank the editors with John Wiley & Sons (Wiley), our F @ M book publisher, and especially thank Kalli Schultea, Editor, Civil Engineering and Construction, Isabella Proietti, Editorial Assistant, and Indira Kumari, Managing Editor, for their support of this book. And the editors worked with Wiley on the book cover.

A world map was identified in night mode which shows pinpoints of night activity in the various continents which reflected the built environments that FM supports providing A Global Perspective.

Finally, we thank the FMs, and contributors whose senior managers who have supported them, their staff, the profession, and the work they accomplish for their customers and their organizations to succeed and excel. We celebrate these senior managers who have provided the time and resources for these contributors to share their FM story in this anthology.

The information in this book is for educational and informational purposes only. It should not be taken as professional advice. Also, the views expressed in this book are solely those of the editors and contributors and do not necessarily reflect the views of the publisher.

MH
EPR
JRGM

1

Contributors

Concept

In *this **Concept*** part, principals, sponsors, and creators of the research project spanning over 40 years into the relationship between people, workplace, and buildings are identified. The abstract academic FM idea is explained, leading to the first FMI and the subsequent coining of the term as well as forming the National Facility Management Association (NFMA) in the USA in 1980.

The following contributors provided their anthology and legacy on the early years of facility management:

1.1 David Armstrong
1.2 William (Bill) Back
1.3 Mary Day Gauer
1.4 Melvin Schlitt
1.5 Eric Teicholz
1.6 Christine H. Tobin (formerly Neldon)
1.7 Gunter Neuman
1.8 Duncan Waddell

Facilities @ Management, Concept - Realization - Vision, A Global Perspective, First Edition. Edited by Edmond Rondeau and Michaela Hellerforth.

1

Contributors

Concept

In this Concept part [illegible] approaches and [illegible] of the research program spanning over 40 years into the relationship between people working and building [illegible]. The [illegible] was [illegible] first [illegible] of the [illegible] Association (PMA) in the USA in 1983.

The following contributors provided [illegible] and [illegible] to the [illegible] of [illegible]:

1.1 David Armstrong
1.2 William (Bill) Back
1.3 Mary [illegible]
1.4 [illegible]
1.5 [illegible]
1.6 Christina [illegible] (formerly [illegible])
1.7 [illegible]
1.8 [illegible]

[illegible] Concepts [illegible] First Edition. Edited by [illegible]
[illegible]

1.1

When and How "Facility Management" Became an Identified and Needed Profession

David L. Armstrong

Introduction

In 1968, while Associate Dean of Agriculture of Resident Instruction in the College of Agriculture and Natural Resources at Michigan State University (MSU), an unusual and career-changing situation occurred. The College was growing very rapidly and was running out of office space. The MSU architect mentioned that Herman Miller, Inc was doing something very interesting new things and that I should contact Bob Propst, President of the Herman Miller Research Corporation (HMRC). Michigan State University (MSU) became the first government contract for Herman Action Office Systems (AO2). It became a research site and a publication by HMRC titled "Facility Influence on Productivity."

The MSU experience set Bob Propst on a mission. He wanted to visit office facilities in as many companies as possible. He wanted to see firsthand what equipment and furnishings were present, how they were used, determine if a plan existed and how facilities were managed. The observations were all over the place. Since Bob Propst was running a research corporation, he wanted answers. For the next eight years, Bob Propst and I visited dozens of companies. During this same period, we were invited lecturers at Architectural Colleges, corporations and were sought after for American Management Association (AMA) Lectures. Office environments were growing at exponential speed, and the nagging questions remained, who was in charge and how were decisions being made. It was obvious, that decisions were "reactive" to a request. Very few decisions were "fact" based. Bob Propst was rather quickly concluding; "If there is not competent management associated with these environments in place, why should I as a researcher and designer continue to promote intelligent and adaptable environments that are ending up in unmanaged and misunderstood situations once on site?"

In 1978, I left Michigan State University and joined the Herman Miller Research Corporation Team in Ann Arbor, Michigan. The Information Age and the growth of Office Environment were all the rage. To understand what was happening, you have to relive the previous decades of manufacturing and assembly businesses that were everywhere. Colleges of Engineering had majors or focus programs that were labeled "Plant Engineering" or similar. These engineers were trained and focused on the factory environment, kinds and size of structure, factory line design, logistic requirements, time and motion studies and more. The associated paperwork (The Office) had an area in the corner or a "lean to shed" to the main factory. The Plant Engineering profession was often not the favored engineering career, and those programs were dropped, and the subject matter incorporated into the other engineering majors. But the Information Age emerged with intensity and unprecedented growth and the evidence of active management of office spaces was all over the place. Generally speaking, no one desired the task or the assignment of managing this new movement. Managers were told that this was an "add-on" responsibility and

Facilities @ Management, Concept - Realization - Vision, A Global Perspective, First Edition. Edited by Edmond Rondeau and Michaela Hellerforth.

you have been selected. Many positions were filled by persons about to retire and on and on. The observations of how the office facility was managed became almost humorous. Growth was rampant and a "management function" was badly needed. In most cases there was only minimal evidence of office facility management, except for a moving crew in maintenance. When new office building projects were started, more was needed. Managers were "volunteered" as kings of the new building. No one wanted this job – often they were well respected leaders about to retire. Bob Propst and I talked to many of these managers. When they inquire about staff, budget, or anything else, they were told this is the company's top-one priority and they would get the resources they need!

After several years of corporate visits, lectures and conversations, Bob and I knew there was a serious void in the non-researched, informally managed and massive confusion in managing the growth of facilities in the Information Age. During that time, we had accumulated a treasure chest of names deeply interested in this specific topic. In late 1978, a conference was put together, by Herman Miller Research Corporation in Holland, Michigan, to focus on the dilemma of managing office facilities. It was a "Think Tank" type of conference, with everyone being in an explorative mood. The conference did set off an aftermath of response.

Two attendees at the conference, Charlie Hitch, of Manufacturers Bank and George Graves, from Texas Eastern, wrote a letter to Max De Pree, CEO of Herman Miller asking for a seat on Herman Millers Board of Directors, representing users of their products. The letter was forwarded to me and Max De Pree noted on the letter that "He did not need an added Board Member or a committee or anything. If there is something here that the Company should be working on, it should start as part of the programs at Herman Miller Research at Herman Miller Corporation."

There is no question that this letter started the ball rolling and was based on dialogue from the conference in Holland, Michigan, just a few months earlier.

Bob Propst gave me the assignment to develop a program that would be part of Herman Miller Research Corporation and would focus on management of the office facility. Here is where I started:

1) The "White Collar" workforce was growing at an exponential rate. No one questioned that and it brought a bundle of decisions, problems, and a need for informed decision making.
2) Without history, nearly everything was on a "first time" basis. Whoever was asked to help manage the processes was developing their position as they worked with virtually nowhere to go for help or assistance?
3) Colleagueship was minimal. The typical professional society of fraternity did not exist.
4) How could current professionals augment their backgrounds with this "facility thing"?
5) The need for professional literature, courses, meetings, and professional dialogue were also missing.
6) What would the handle be? What should the name be?

The Process for a Plan

I asked former colleagues at Michigan State University what was it like to develop a new academic program in Parks and Recreation. In a chance meeting, I asked him, "How did you start?" He said, "Strange you would ask. To put it simply, there is no history, no data, you had to first count the 'canoes'. You need information. How big is this thing and those kinds of questions?" It was obvious; we had to do some counting and data collection. For example, what percent of the corporate operating budget should be allocated to managing the "white collar" facilities? What is a reasonable investment in facilities, equipment etc. for this workforce? What would the staff needs be and what would be the most effective reporting structure? It was obvious early on that a focused group of research professionals was needed. There was a need for an academy. We should stop right there and see if a college or university would be interested. If they were, we could direct our efforts to assist them, rather than do it ourselves. Our inquiries instilled little interest, but shifting or adding academic staff to take on the responsibility was barely a lukewarm reaction. It sounded "trade-like" to most and perhaps our presentations were not as convincing as we thought. If you remember, higher education was also experiencing a huge growth. They had their plates full and a new program, even though it had grant money attached, our idea was not very convincing. Our conclusion was going to take more resources and time to develop and an effective approach with universities than we could afford. It was better to focus on our idea independently.

The need for an academy or an institute had to be established. A dedicated and professional research presence was required. There was also a need for literature, outreach, and public relations, maybe even a professional organization. Meetings, short courses, seminars, and other career enhancing activities seemed like a requirement. The number crunching convinced the Herman Miller Research Corporation that the idea for an academy made sense. It would not be self-sufficient in the early years; therefore, funds would be required. We wanted to have control of the development in the formative years, so we decided to present the funding requirement to the Board of Herman Miller, Inc.

Bob Propst strongly supported the idea. In his presentation to the Board, he emphasized repeatedly, there was a need for the Herman Miller Research Corporation to add this venture to its program agenda. To be an "Out Front" research and development organization in the office environment, this was a critical component. He sold that idea very effectively. In our dialogue with the Board, we used the word institute. The Board was also told the entire focus of the Institute would be on the management of the white-collar workforce environment. The emphasis would be on workplace needs, efficiency, and productivity, technical requirements and of course esthetics.

The controversy arose when the Board realized that the Institute would focus on management and not hardware. The Institute would be using Herman Miller furniture but would not be promoting or selling Herman furniture in its programs and activities. In addition, the Institute would be accepting participants regardless of furniture used, competitor or not! The concept was accepted, but it was the funding that was a problem. Bob Propst and team promised if the Institute were proven successful, we would make every attempt to transfer it to a neutral college or university. The Institute was approved and launched in 1979.

Launching the Institute

Naming is never an easy process. There had to be an emphasis on management and obviously we were interested in the focus to be on facilities. Calling it the Facility Management Institute was not very creative, but it was basically unused. In the corporate world "Facility Management" was a responsibility to run away from, and that is exactly where we wanted to go. We checked all references relative to the words "Facility Management" and found no competition to the term. We were convinced we had a clean and definitive use of the term. Herman Miller Research Corporation registered the name and definition with the US Library of Congress with no dissenting comments or concerns.

During the process, we did find two uses of the term "Facility Management." IBM was using the term in their installation of new computers during this time. When a major IBM purchase was made and shipped, an IBM person was there to receive and install the product. These representatives were called facility managers. The US Army was also using the term.

On September 14, 1979, a report to the Herman Miller Research Corporations Board of Directors stated we were ready to launch. Our Model for the institute was formed and field tested. We were ready to have a soft launch. In May 1979, a task force of four people from Proctor and Gamble visited HMRC (Herman Miller Research Corporation) when they learned about plans for a Facility Management Institute.

P & G stated quite candidly that they suffered from a problem, which is characteristic of many research and development lab facilities, namely, that organizational changes occur more rapidly than facilities can respond. This was taking a fresh look at research and development itself and the relationships need between office, laboratory, and pilot production facilities.

Many of the ideas discussed were closely related to concepts developed in *The Integrated Offices Facility* research report. Additionally, our work at Michigan State University, Kellogg Community College, and American Productivity Center was reviewed to illustrate some of the principles, which were being discussed.

Discussions also touched on relationships between the personality types and task profiles of research and development workers and their attitudes toward the use of systems environment for offices. Traditionally, research scientists involved in creative activity, particularly senior scientists, have been insistent on high levels of privacy. While it is true that high level of concentration is required for some tasks, it was suggested that facilities must also encourage an appropriate amount of informal interaction and communication among scientists. The P & G task force seemed to be impressed with the perspective and ideas that HMRC had on these problems and were very open in discussing even sensitive corporate issues.

The HMRC task force invited Procter and Gamble to visit the facilities, and the issue of collaborating on a joint project was touched upon but no commitments were made. It is possible that an interesting research proposal would be well received by the task force, and that an interesting project could be developed based on this meeting.

The development of the Facility Management Institute had two equally important components: (1) HMRC corporation developed the macro-concepts. There was a plan, but the "facility managers" in the field were also reinforcing this plan. And (2), corporations had to be supportive of the concept. The Facility Management Institute had "no legs" if corporations turned a blind eye to the concept or refused to pay for their personnel to be FMI participants. Major corporations had to see the void, and actively support research and the involvement of their personnel. That is why the P & G involvement was so critical. Another major company joined the ranks early.

On July 26 and 27, 1979, the Facility Management Institute hosted a day and a half workshop and discussion session centered around three visitors from Honeywell, Inc of Minneapolis, with the addition of two others from the Detroit area. Representing Honeywell were Will Grove, Director, Corporate Real Estate and Field Administration; Joseph Riordan, Director, Corporate Productivity; and John Rousseau, Manager of Corporate Real Estate. The Detroit group included Greg Dymanski, Second Vice President, Bank Properties Department at Manufacturer's Bank plus Richard L. Deatherage, Facilities and Planning for GM truck and Coach Division in Pontiac, MI.

Initially, the meeting was to include only Honeywell people, resulting from their request for a follow-up discussion of a visit Dave Armstrong and Hal Hanson had in Minnesota on February 28, 1979, with Will Grove, John Rousseau, and two other staff members. Will Grove also attended the HMRC sponsored, "Facility Influence on Productivity Conference" held at Marigold Lodge on November 15–17, 1978. To gain experience in a teaching seminar format, an agenda similar to the future Institute sessions, was the basis of the visit.

Will Grove responded with a letter to Dave Armstrong. "Dear Dave, I thank you and others for your hospitality last month. I am convinced that you are on the right course in devoting a major effort to raising the level and recognition of the facility manager. It impressed me that your plan is to acknowledge this manager as basically a 'facilitator' of the business functions of the various department managers he serves."

Those of us on the staff of the newly formed Facility Management Institute in Ann Arbor, MI, had been provided sufficient verification to the "mission" that an all-out launch could now happen.

The FMI Facility

There was no question; the Facility Management idea was catching on fast. With the help of public relations firms, trade organizations, trade magazines and many others were pushing us for more information, ways they could become involved, invitations to speak to staff, invitations to speak at conferences and more. We were going crazy. I was invited to do a "Dear Dave" column in a leading trade publication all directed to facility management.

A flip chart talk I made to the Executive Committee of Herman Miller was presented to several individuals and groups over a period of a few weeks. The material from that talk was organized into a formal paper to be presented on October 23, 1979, at the Automated Office Conference for the American Institute of Industrial Engineers.

It was anticipated that by the end of the third quarter, fiscal year 1979/80, the Institute would have a total of 28 professional and support staff on board.

Our own facility had to be planned. If we were going to invite many groups to activities at the FMI, our facility had to incorporate the principles we were teaching. A "smart" facility had to demonstrate to all visitors the design and organization of the FMI facility. The staff of the Facility Management Institute had either experienced or knew of field experiences focusing on the office environment. In a "Think Tank" approach, a team was assembled that would illustrate the integration of "People. Place and Process" had to be real. The interior/planning of the new Facility Management Institute had four major concepts/philosophies that are being used as guidelines to create the "atmosphere tone" of the building:

1) Modular limited using a limited vocabulary of parts
2) Planning areas for "degrees" of activities/functions.
3) Planning for "degrees" of manipulation of systems parts.
4) Selecting a family of fabrics and finishes that work together on various combinations to create many visual settings.

It was truly a stunning example of what could be accomplished using "People, Place and Process" into an integrated and balanced workplace. It virtually "spoke" to you when you first experienced it. Functionality, simplicity, traffic flow, topography, attractiveness were evident. Rather than spend too much time on the facility, I have included excerpts from a major designer's handwritten comments to HMI corporate.

Tom Newhouse wrote:

"On August 1979, I spent the day in Ann Arbor consulting with the FMI 'New Facility Team' on the interior space planning problems of the very young organization.

I am pleased to say that during a very intense conceptual day, Martha Whitaker, Mike Wodka. John Adams and I resolved a concept/philosophy that should not only help the FMI to define itself as teaching/research organization, but also produce one of the more process innovative Action Office facilities installed in recent years.

The concept is innovative on several levels of open plan system application and space planning management; because it is a different approach from HMI Corporate standard, and because it is an experimental concept that will require support to make it happen...

I am committed to ongoing support of this project and recommend the Facility Management Institute be given the design freedom and necessary support to pull it off." Tom Newhouse.

Our own facility became the basis for more detailed research and resulted in a book, *The Negotiable Environment People, White-Collar Work, and the Office*, by Cecil Williams, David L. Armstrong, and Clark Malcohm, 1985.

"The Workshops"

The Facility Management Institute offered a workshop in Ann Arbor almost every quarter. Each staff member developed a session around their interest and specialty. These were typically two-day experiences. Discussions were lively. The evenings were usually dinner at an FMi staff member's home. The discussions were very energetic and long lasting. We knew instantly that these facility management professionals needed to talk to each other–they needed a forum! They wanted information. Motivating these groups was not a problem, but an exciting resource. It is impossible to reiterate all that happened at the Workshops, but attendance was great. During these sessions, we often talked about; "People. Place and Process" and the "Crazy Daisy." There was much more, but these centered the conversation for many more topics.

It was fun starting the Facility Management Institute, or was it? The Facility Management had a small staff initially. All were experienced members of the Herman Miller Research Corporation staff. However, for the new FMI staff to be redirected to the world of facility management was a new venture. "New" means new! Empty files, no publications, no experienced professionals, only a challenge to fill the void of the under or unmanaged dilemma of the emerging white-collar workplace.

Bob Propst and I had met many managers immersed in this situation. Each of our staff had an assignment to call these people and pick their brains on the subject. We had brainstorming sessions at the office. I even stuffed fortune cookies with facility management questions. At every coffee or lunch break, our staff, including administrative staff, was required to pick a fortune cookie out of the bowl and discuss the facility management topic stuffed inside. This proved more stressful than anticipated. But we had to develop a language – a new language to keep us focused and moving forward. This experience was always a group effort. The experience brought up things we needed to know, research that had to be done, and classes that needed to be developed. It was difficult to get started, but quickly became our passion. Our conversations in the field had more focus and content, confidence grew and in the institute was off and running.

People, Place, and Process

Every organization deals with people, whether it is Human Resources, People, or some other "tag." Several organizations had used a more contemporary tag: Department of People or similar. The "People" departments were getting more and more legal with time, and therefore more clearly defined. The work process, the engine of most organizations, was usually incorporated in the Operating Departments or Division. When visiting organizations, you would notice great differences in how work was organized and evaluated at the group and individual level. Work processes were obviously more decentralized than human resources, data collection and analysis varied considerably, and accountability for workplace efficiency was quite varied. It was common to see reports on revenue per employee, or the growth of employee revenue compared with previous years or similar macro data. There was evidence of "efficiency teams," workstation reorganization, and more.

Since our interest was in the management of facilities, the presence of a formal organization in facilities rarely existed. Associated facility tasks were often dispersed, almost hidden everywhere, with little or no consistency from organization to organization.

At the new Facility Management Institute, we started to diagram our observations. In a very subjective way, each of our researchers drew a diagram with three circles: One for People, one for Process, and a third for Place. The size of the circle was to measure effectiveness of the function. Thus, three circles were placed on a diagram; usually each of the three would be of different sizes. The next part of the diagram was to spatially place the three circles. This gave us some indication of how closely and interdependent the functions were to each other and how effectively they were working together.

In an ideal world, optimum efficiency and productivity would occur when the three circles would intersect. Thus, the interconnected circles were often a model that started many discussions and conversations called the People, Place, and Process diagram.

You can imagine very few interconnected. People was always the larger circle, nested close to Process, but rarely intersecting, and Place was a small little circle struggling for synergy with the other two. Nearly everyone could identify with the findings. Since we were focusing on "Place," we knew if we believed in our premise, there was work to be done. See Figures 1a, 1b, and 1c.

The incongruity of the People Place and Process findings led us to another rather interesting routine, but enlightening process. Interviewing those involved in the facility business (concentrating on white-collar issues) resulted in asking what facility issues you have been asked to oversee or help with. We started to represent each activity as a petal of a daisy. You would have small petals and larger petals. The diagram was quickly labeled the "crazy daisy." A list like the following happened:

1) Workplace design
2) Installation and moves
3) Employee moves
4) Acquisition
5) Warehousing
6) Travel services
7) Motor pool
8) Corporate aircraft

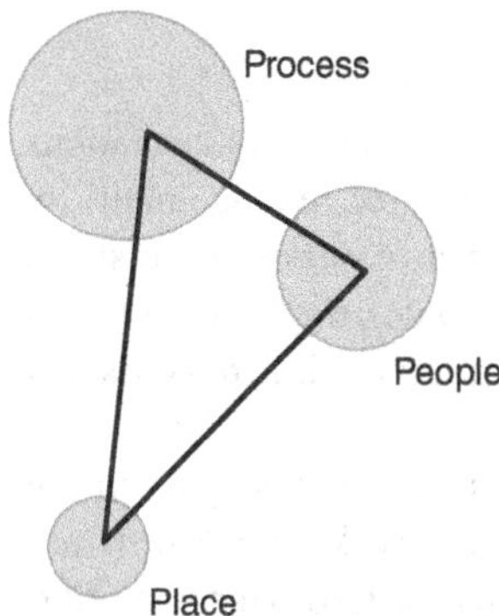

Figure 1a Dissonance Triangle: People. Place and Process are also there, but how are they positioned and how does it function? (*Source:* D. Armstrong).

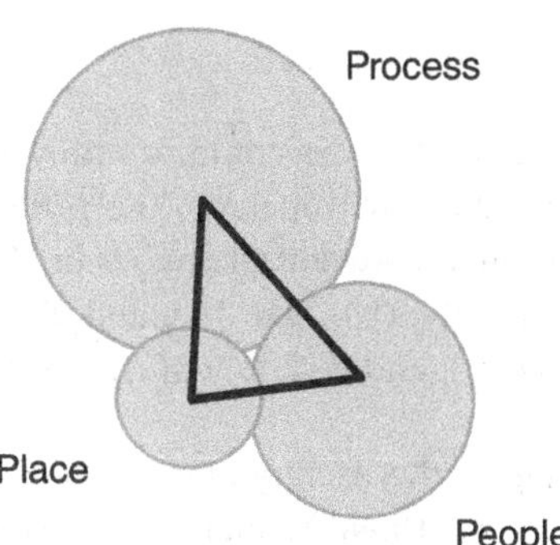

Figure 1b Dissonance Triangle: They should at least touch! (*Source:* D. Armstrong).

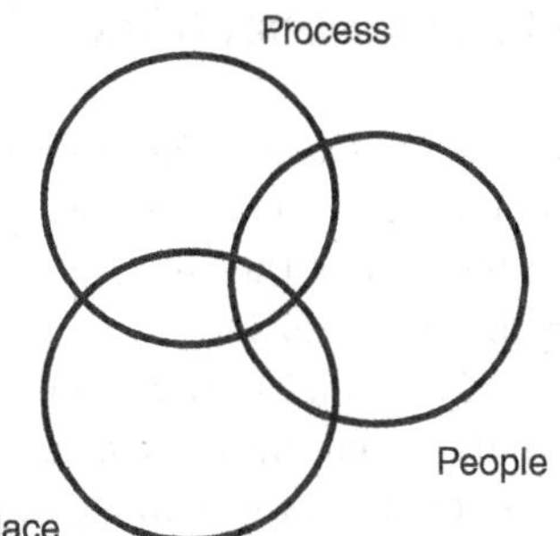

Figure 1c Synergy achieved! (*Source:* D. Armstrong).

9) Food service
10) Maintenance and cleaning
11) Building and grounds
12) Mail rooms
13) And more...

It was not a pretty flower in most cases. The "crazy daisy" had to have a center holding the pedals. No one disagreed that all of these activities existed, but how were they positioned in the organization and how effectively were they managed. Like it or not, the facility manager of every organization was at the center of this "crazy daisy" of activity. The Facility Management Institute had to help identify and strengthen the skills needed to do the job. See Figures 2a, 2b, and 2c.

The "Crazy Daisy"

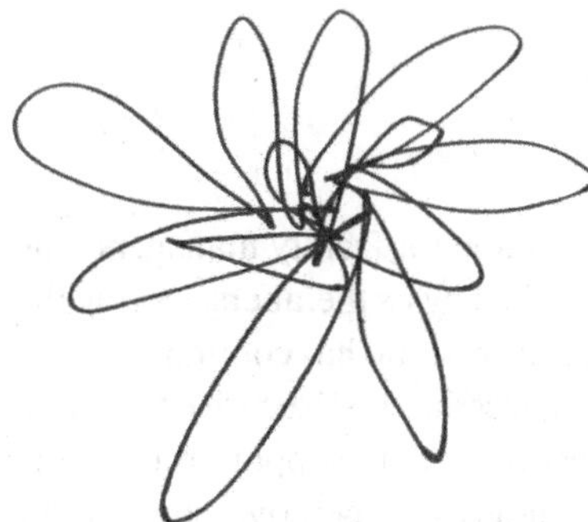

Figure 2a Facility Management often feels like a "Crazy Daisy" – a lot of mass, but no beauty! (*Source:* D. Armstrong)

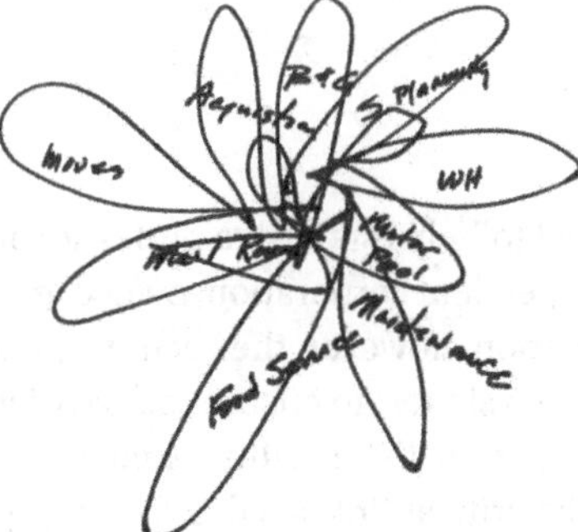

Figure 2b It is a "Crazy Daisy," but it is real! (*Source:* D. Armstrong).

Figure 2c Demands a focus on Management and Coordination. (*Source:* D. Armstrong).

The International Dimension

Besides possible association with Management Centre Europe, contacts with various other national associations, institutions, and individuals are being pursued. These will include: the British Institute of Management, the British Institute of Directors, the European Institute for Advanced Studies in Management Science, the Institute of Organization and Social Studies at Brunel University, Uxbridge, England (Prof. Alan Dale), the London School of Business Administration (Prof. Anthony J. Eccles), and a variety of American and national business associations in the major European cities.

To assist with the educational logistics activities of the Facility Management Institute in. Europe, particularly this first activity in February 1980, Steve Snowey and David Williams suggested that we work with Ann Hines, formerly of Herman Miller Ltd.'s staff in Bath. Ann was currently working on a free-lance basis for Herman Miller and had been involved in a number of UK seminar and conferencing activities. She had been contracted and was beginning initial conference site groundwork investigation (e.g., conference possibilities, European speakers, and public relations)

After the initial February 1980 program, the Facility Management Institute intended to have educational program activity on a regular basis in Europe three times a year (early fall, mid-winter, mid-spring). This regular activity began in the fall of 1980.

Research activity in Europe was focused on two areas. A project with Phillips in Eindhoven, Netherlands, had been in progress for the last six months and continued in accordance with the research proposal submitted to Phillips. Essentially the project involved an experimental office renovation and an evaluation of the applicability of systems environments to Philips facility needs. The second research area in Europe was a broad, descriptive survey of the state of the art in facility management in Europe. This study identified differences in research, educational and information needs between Europe and the US. A parallel study was conducted in the US to allow this comparative analysis.

Besides the education and research activities in Europe, there were a significant number of European contributors to the information side of the Facility Management Institute, particularly in the *Facility Management Journal*. Also, the International Association of Facility Managers was expected to be quite active in Europe and provided new contacts for Institute involvement.

In addition to involvement in planning for the above activities, Harry Phillipins from the Netherlands and on FMI staff was able to spend time with various members of the FMI staff discussing project activity underway in the US and to familiarize himself with the programs being developed. Harry participated in one of the one-day research retreats in which the issues of developing an interdisciplinary professional role were discussed. Overall, his visit was viewed as worthwhile and productive, and he continued to travel to the US two to three times a year to insure close contact and program continuity.

A Professional Organization?

The Institute launched a program designed to help establish a professional association for facility managers. The association was organized as a legally independent corporation, owned and controlled by its members. During the association's start-up and first year of operation, however, the Facility Management Institute has committed some personnel and funding (for a journal and newsletter) to ensure the association's success. The Executive Secretary of the association for the first year came from the Facility Management Institute. This support role to the association provided the Institute with the critical link needed to develop a long-term cooperative program for advancing the Facility Management discipline.

The first interest-testing meeting on the formation of an association was planned for early October 1979. It was hosted by Charles Hitch of Manufacturer's Bank of Detroit and George Graves of Texas Eastern Transmission,

Houston. Approximately 15 to 20 facility managers were invited. The Facility Management Institute, in its supportive role, developed the agenda and logistics support. Interested individuals from this group organized into a working group to consider by-laws and association structure. The formal announcement of the association was projected for early 1980.

Interest in an association grew like wildfire. FMI assigned Mel Schlitt to manage the formation of the phase of the facility management launch. After some thought, we changed the formation approach to "Bottom/Up" rather than "Top/Down." The thought was to form NFMA Chapters, National Facility Management Association, and have the chapters advise FMI on the structure needed. Mel Schlitt was a busy man. The Chapters grew very fast and needs for the central organization grew with it "Hand in Hand." Obviously NFMA as a name was short-lived, as the chapter idea was also in Canada and on it went. IFMA (the International Facility Management Association) was born in 1982.

A plan for a quarterly journal was written encompassing purpose, sponsor, editorial policy, pay, rent, reprint, and copyright policy, format (size and shape), and general content ideas. Plans were to go ahead with publication of the first journal ahead of the formal start of the association so that the journal could be used as part of promotion for membership.

Facility Management Institute projections, at that time, assumed the association would be able to attract the following membership levels in the first five years: 1st year: 150–300 (membership by invitation only). 2nd and 3rd years: 300–3,000. 4th and 5th years: 3,000–6,000.

In Appreciation

I have tried to give you a glimpse of what it was like forming the Facility Management Venture. It was a long time ago. Probably, very important things were left out, but I assure you it is an honor for me to be asked to contribute to this history.

Sometimes the easiest way to know the weather is to open the window or walk outside. Doing that, you know exactly what is happening. Just like the weather, things change and sometimes it is cloudy. There is a two-volume publication written by FMI Staff when the organization ceased operations in Ann Arbor in 1986. The conclusion was that it was a "good" thing. Also, forming FMI and IFMA were a great contribution to the community of facility professions and the organizations they serve. Interesting that skepticism was evident during the formation, saying HMI would never stand behind the concept, but they did and deserve major credit for doing so. I must express thanks to the great colleagues at FMI. It was a magical place. The synergy of the staff, the place and our clients were the greatest I have ever experienced. Rare indeed, but FMI had it!

I am also grateful to the many that continued as professionals in facility management that kept it going. It is an interesting account. But, when I look out the "Facility Management" window 40 years later, I see clearly an active IFMA organization, professional activities, and individuals with pride of their vocations, using the words "facility management" with ease and comfort. Therefore, the weather is fine. Thank you!

David L. Armstrong, PhD, IFMA Fellow
Holland, MI USA

Born in 1935 in rural Ohio. Higher education was not my family's goal, but with a teacher's encouragement, I attended Ohio State University, ending with a PhD in Agricultural Economics in 1960. During that time, I was an ROTC (Reserve Officer Training Corps) cadet and served my active duty leaving as a Captain in the US Army Reserves. I pursued an academic career as a professor and researcher at University of California Davis, Southern Illinois, and Michigan State University. I was the Associate Dean of Agriculture and Natural Resources at MSU when I ended my exciting and successful academic adventures in 1975.

Joined the Herman Miller Research Corporation in Ann Arbor, Michigan. Had eight exciting years in research and development at HMRC. Became Vice President of HMRC and founder and Director of the Facility Management Institute. Co-authored *The Negotiable Environment*, a book dedicated to the dissonance in the office workplace. It is all about "People, Place and Process" almost the fight song of the FMI.

Joined Herman Miller Corporation as Vice President of Marketing. Implemented several innovative changes in market application, including the integrated dealer called the Office Pavilion. Later becoming the Executive Vice President of Herman Miller Inc. in Sales and Marketing worldwide. Retired in 1991.

Formed Armstrong Associates, Inc. as a consulting and business development business. Was a founder, investor, and executive in several organizations including Intertrade, Inc., The Waypointe Companies, Inc., The Maginus Group, LLC, and Viability, LLC. Worked actively with Baker Scholar Program at Hope College and mentored several young entrepreneurs.

In 2008 my wife and I purchased an art gallery; Armstrong De Graaf International Fine Art, which we operated till December 2022.

Jane and I were married in 1957. Live in Holland, Michigan. Have three daughters, seven grandchildren, and one great grandchild.

1.2

"In the Beginning of FM..."

William W. 'Bill' Back Jr.

... there was, in the mid-1970s, a crisis brewing in the office work environment in the USA, which was then dominated by private offices. The workplace was becoming more open with the introduction of systems furniture. At this same time technology was proliferating creating electrical load and cabling problems, as well as security issue never before dreamed of. The then typical office space did not have the flexibility or capacity to accommodate these innovations. Because all of these innovations and issues were interrelated there was an additional concern, there was no single individual within most corporate organizations designated to deal with these issues as a total package in a coordinated and seamless manner. Within existing corporate organizations, each of the issues was usually managed separately, causing significant management conflict and subsequent operating cost increases. Recognizing that these issues were impacting productivity Herman Miller Research Corp. hosted "Facility Influence on Productivity" at Herman Miller's Marigold Lodge in Holland Michigan in December 1978. This event brought together three leaders from different business sectors and different parts of the USA:

- George Graves, Manager of Office Services, Texas Eastern Transmission Corp., Houston, Texas,
- Charles Hitch, Manufacturers National Bank, Detroit, Michigan, and
- Dave Armstrong, Michigan State University, East Lansing, Michigan.

Each of these gentlemen had a vested interest in and a deep understanding of the issues then facing the workplace. They also shared a vision of what the modern office work environment should look like and how technology should be integrated, implemented, and utilized. They knew this technology driven workplace innovation would eventually impact every business sector and company on the earth. Graves, Hitch and Armstrong realized a whole new profession would need to emerge to integrate and manage these innovations and changes so that this new work environment would be a more productive, efficient, and effective workplace. To facilitate the success and acceptance of this new profession, they began to discuss the need to develop potential organizational structures and performance standards for the profession. At this point it became obvious to them that there would need to be a standard bearer for this new profession to help it develop and become more widely recognized.

With these realizations the question arose who would be responsible for developing the performance standards and potential organizational designs. Consequently, the idea of creating a professional organization similar to BOMA, IREM or the AIA materialized. They realized that all of this couldn't be accomplished at this December 1978 meeting, there would need to be more input and participation from a much wider and varied audience. Graves, Hitch, and Armstrong vowed to spread the word and started to organize a second gathering that would have broader attendance with differing views and inputs. These three gentlemen returned home and began the task of contacting their acquaintances from other companies in their vicinity. Herman Miller Research Corp recognized the possibilities for this new profession and established, in early 1979 the Facility Management Institute (FMI), to

Facilities @ Management, Concept - Realization - Vision, A Global Perspective, First Edition. Edited by Edmond Rondeau and Michaela Hellerforth.

be located in Ann Arbor, Michigan, and enticed Dave Armstrong to be its first Director. Armstrong almost immediately began a tour of major US cities to promote Facility Management as a profession. In late 1979 Hitch and Graves hosted a second conference for Facility Management professionals at the Renaissance Center, in Detroit, Michigan, USA, which was attended by 23 facility managers (see Attachment 1 for names and companies) from across the US. Much of the discussion at this conference centered around the formation of a professional organization for Facility Managers.

(At this point I must ask the reader's indulgence as things get very personal for me from this point on.) Subsequent to the Detroit meeting there were numerous private conversations and conference calls to garner support for the development of a professional organization. On May 28 and 29, 1980, George Graves hosted an organizational meeting for this professional associations in Houston, Texas, at Two Houston Center. This meeting was attended by 10 Facility Management professionals (see Attachment 1 for names and companies), including Yours Truly.? Graves being the leader and results producer he was, took the initiative and locked the conference room door and telling everyone, "No one is leaving until we get this done!" There was a resounding "ABSOLUTELY" from all present. We were locked in that room for nine to ten hours. We determined our first task would be to create a governance document. We all agreed this document would be much easier to create if we had an existing organizations governance document as a "go-by." Mr. Andy Pedrazas, of Columbia Gulf Transmission Corporation, who was a member of the "Texas Passenger Traffic Association" (TPTA), offered to let us use the TPTA's Constitution and By-Laws as this "go-by." Mr. Pedrazas had the documents faxed, to George's office. These documents were then delivered to us in the conference room since we were not allowed to leave the meeting. The door was not unlocked even for this important delivery, the document was slipped under the door. At this point everyone in the room realized that "we 'ABSOLUTELY' were going to get this done."

The question then arose, what do we call this new organization? "We need a name!" Initially the new organization became the "FACILITIES MANAGER'S ASSOCIATION," or "FMA." Since several FMAs already existed, and we didn't want to be confused with the "Florida Moose Association," the "Fabricators & Manufacturers Association," or the "Financial Management Association," although financial management, and hopefully no Moses, would certainly be a core responsibility of a Facility Manager, there was only one real alternative that struck a chord with everyone present. The "NATIONAL FACILITIES MANAGER'S ASSOCIATION" or "NFMA." At this early stage in the life of a new Facility Management Association we were most concerned with gaining traction and growing the association's membership within the confines of the United States, hence the use of the word National. Following some additional discussion, we decided that using the words Facilities Manager in the name would be a bit limiting. We were all acquainted with individuals, some actually attending this meeting, who should be eligible for membership in the association, and who were responsible for managing facilities, but who did not bear the title "Facility Manager." Consequently, with a few minor adjustments the name "NATIONAL FACILITY MANAGEMENT ASSOCIATION" "NFMA" was unanimously adopted. The remainder of the first day was spent reviewing, editing, adapting, and modifying the Texas Passenger Traffic Association Constitution and By-Laws to fit the purpose and needs of NFMA.

The second day saw additional editing and review of the Constitution and By-Laws. A large part of the day was spent with individuals making personal commitments to support the establishment of the association by recruiting other members from among their FM friends and acquaintances. We also spent time drafting a letter that would be used as a recruitment mailer to our FM acquaintances. This mailer was also sent to the Human Resource Departments of companies we knew must have some form of FM presence and ask that the letter be passed along to their FMs. After lunch the second day, it came time to decide who was going to lead this monumental effort. With input, via conference calls, from other interested FMs who could not be physically present, it was unanimously decided that George W. Graves would be NFMA's first President, and Charles F. Hitch its first Vice President. Charles A. White, of Houston Natural Gas, volunteered to fill the office of the Treasurer. The office of the Secretary was a bit of a challenge, since everyone knew this person would be responsible for the final drafting of NFMA's first Constitution and By-Laws. With several of FMI's staff present and with their realization that "we were actually going to get this done," they committed FMI to host an association meeting, in October 1980, where the Constitution and By-Law's would be ratified by a larger and more diverse group. After much discussion George, as President, took charge and said, "Bill Back, you are the Secretary, and I know you'll get the job done." *(Besides George was my boss, how could I refuse, and I knew he would make sure that our regular workload didn't interfere with getting NFMA's Constitution drafted and finalized.)*

On October 9, 1980, the Facility Management Institute in Ann Arbor, Michigan, hosted the First Annual National Facility Management Association Conference. It was at this first conference that FMI presented a graphic definition of Facility Management for the first time (see Figure 1):

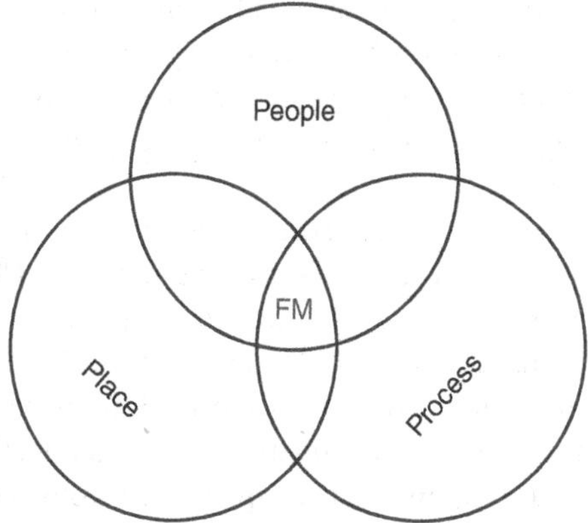

Figure 1 This graphic definition quickly became NFMA's Logo. (*Source:* adapted from NFMA).

There were "In the Beginning..." Forty-Four (44) Facility Management professionals (see Attachment 1 for names and companies) attending this conference who reviewed and ratified the Constitution and By-Laws of the National Facility Management Association originally drafted during the Houston meeting in May 1980. Of this infamous group of 44, four would later become President of the association. Charles Hitch was elected at this first national conference to succeed George Graves as the President of NFMA.

Additionally at this first conference FMI generously agreed to support this new association financially, by providing office space and allowing Melvin R. Schlitt to work part-time as NFMA's first executive director. Also, at this first conference, committees and project teams were formed to help accelerate the expansion and membership growth of the associations. Among these committees, the contingent from the Houston, Texas, area committed to create the Association's First Chapter and to host the Second Annual NFMA Conference in October 1981.

Over the next two and a half months this Houston contingent worked tirelessly to identify and commit additional FMs from the area to become members and form the Houston Chapter, of NFMA. A meeting was organized in mid-November 1980 and hosted by Judy Brady Farrar, an NFMA Founder, at the American Productivity Center in Houston, Texas. At this meeting William W. Back Jr., an NFMA Founder, was elected the first President of the first Chapter of NFMA. The Chapter voted to modify and adapted the recently ratified NFMA Constitution and By-Laws to fit the needs of a chapter. The Chapter also formed a committee to Organize the 2nd Annual NFMA Conference scheduled for October 13–16, 1981. Over the next 11 months, with a great deal of assistance from Mel Schlitt and FMI, the venue (The Galleria Plaza Hotel, in Houston) was contracted, several education seminars were organized, and Mr. Bill Caudill, of CRS (Caudill, Rowlett and Scott Architects) in Houston, was engaged as the first keynote speaker for the annual NFMA Awards Banquet.

The creation of NFMA was not the only significant development in the FM world in 1980. Dr. Franklin Becker, with Cornell University in Ithaca, New York, developed the curriculum, college courses, workshops and the first Facility Management degree program.

NFMA members recognized early on that the association would not survive if membership were limited to just those individuals who carried the title and filled the role of a Facility Manager (Professional Member). There was a need to involve the companies and vendors who provided support, products and consultation to the Professional Facility Managers. The Companies and Vendors (Allied Members) who desired to join NFMA were required to recruit three (3) professional FM members before they could join. This became the most successful recruiting tool during the early years. The remainder of 1980 was spent mostly recruiting additional members for the association. By the end of 1980 there were probably less than 100 Professional Members, the majority of which were located in either Southeastern Michigan or the Houston area.

In early 1981 the members from Southeastern Michigan gathered and formed the second chapter to the association, the Southeastern Michigan Chapter of NFMA. The balance of 1981 was spent recruiting additional members, organizing and publicizing the conference scheduled for October in Houston. Most of the publicity for the conference was handled by the part-time editor and creator of the first NFMA Newsletter, Ms. Joan Dent. Ms. Dent was a full-time staff member of FMI, whom Mel Schlitt recruited to be the second NFMA Staff member. The October 1981 NFMA Annual Conference was very successful drawing approximately 100 attendees from across the US.

So, what was the catalyst for the growth of NFMA to become an international organization as opposed to being a limited to the USA. In 1981 James W. Chamber, IFMA Fellow, was the Facility Manager for Wood Gundy in Toronto Canada. He was in the process of writing the constitution for the formation of the Canadian Facility Management Association when he heard that NFMA had formed the previous year in the United States. Recognizing that the two organizations on opposite sides of the border had the same objectives, Chambers contacted NFMA and offered to establish the first Canadian Chapter.

During one of his trips to visit both NFMA Chapters and NFMA Headquarters at FMI, we had dinner in Houston. Jim wanted to experience authentic TexMex food for the first time. We went to a local TexMex Cantina and ordered a variety of authentic TexMex. I thought Jim was going to have heart failure when he bit into his first jalapeno pepper. After some huffing and puffing, he said something like, "... just another reason to make this association international, so others can experience a jalapeno pepper." Consequentially, not because of the jalapeno pepper, the National Facility Management Association embraced its future and changed its name to the International Facility Management Association (IFMA) See Figure 2.

Those are my recollections of the founding of NFMA (see Table 1), its first couple of years and evolution to the IFMA we all respect and cherish today. I hope there are another 40 times 40 years of IFMA.

Figure 2 IFMA logo in 1982. (*Source:* IFMA).

Table 1 NFMA organizational meeting attendees & founders.

Name	Company	11/79	05/80	10/80	FOUNDER
Allen, Russell M.	Pennsylvania Power & Light			X	
Arick, Dick	Lincoln National Life Insurance			X	
Armstrong, Dave	**Facility Management Institute**	X		X	X
Autrey, Dorothy	Frito-Lay, Incorporated			X	
Back, William	**Texas Eastern Transmission Corp**		X	X	X
Beattie, Steve	Naval Construction Battalion Center			X	
Brooks, A. C.	Shell Oil Company		X		
Demanski, Greg	**Manufacturer's National Bank of Detroit**	X		X	X
Dent, Joan	Facility Management Institute			X	
Dergis, William	**University of Michigan**	X		X	X

Table 1 (Continued)

Name	Company	11/79	05/80	10/80	FOUNDER
Dethridge, Richard	GM Truck and Coach	X			
Dodson, Tim	Smith, Hinchman and Grylls			X	
Farrar, Judy Brady	**American Productivity Center**			**X**	**X**
Foley, E.P.	**Chrysler Corporation**	**X**		**X**	**X**
Fruechtemeyer, Don	**Cincinnati Milacron**	**X**		**X**	**X**
Galuardi, John	General Services Administration	X			
Gauer, Mary	**Kellogg Company**	**X**		**X**	**X**
Gleason, Jack	Chrysler Corporation	X			
Graves, George	**Texas Eastern Transmission Corporation**	**X**	**X**	**X**	**X**
Healy Jr, Daniel	Bendix Corporation Research Laboratories	X			
Hendricks, Rick	Office of Space Management	X			
Hitch, Charlie	**Manufacturer's National Bank of Detroit**	**X**		**X**	**X**
Houston, Suzanne	The Western Company		X		
Hufschmidt, Judy	Johnson Controls, Inc.			X	
Kaplan, Lucy	Herman Miller, Inc.			X	
Karamitsanis, Pete	Smith, Hinchman and Grylls			X	
Kennedy, Jim	Michigan Mutual Insurance	X			
Keranen, Thomas	Eastern Michigan University			X	
Klausmeyer, Tom	Smith, Hinchman and Grylls			X	
Koeze, Mary Beth	Facility Management Institute	X			
Larson, E. V.	Chrysler Corporation	X			
Liese, Hank	Facility Management Institute	X			
Lilly, Paul	Facility Management Institute			X	
Man, Robert	University of Michigan			X	
Mills, Wayne	**Pennzoil Company**		**X**	**X**	**X**
Mitchell, Neal	Mitchell Systems Structural Programming	X			
Mumby, Steve	Lincoln National Life Insurance			X	
Northam, Tom	Naval Ship Weapons Systems			X	
Olson, Robert	First Chicago Building Corporation			X	
Paparella, Anthony	Central Michigan University			X	
Pedrazas, Andy	**Columbia Gulf Transmission Company**	**X**	**X**		**X**
Peterlin, John	Michael Reese Hospital			X	
Peters, Jim	Michigan State University	X			
Schafer, Lawrence	Cummins Engine Company			X	
Schlitt, Melvin	**Facility Management Institute**		**X**	**X**	**X**

(Continued)

Table 1 (Continued)

Name	Company	11/79	05/80	10/80	FOUNDER
Sherman, Douglas	**University of Michigan**	X	X	X	X
Sleeper, Charles	State of Michigan			X	
Snyder, Bob	**Facility Management Institute**	X	X	X	X
Stanley, William	Americana Healthcare Corporation			X	
Steinmann, Jim	The Planning Collective Limited			X	
Teitelbaum, David	Teitelbaum Holdings, Ltd.			1	
Terpstra, Harry	Manufacturing Data Systems, Inc.			X	
Theis, Joseph	Valero Energy Corporation			X	
Trayer, George	Central Bank of Denver			X	
Vaillant, Jeff	Wells Fargo Bank			X	
Walters, Howard	C & P Telephone			X	
West, Robert	University of Michigan			X	
White, Charles	**Houston Natural Gas**	X	X	X	X
Williams, Cecil	Facility Management Institute			X	
Zager, Robert	Work in America Institute, Inc.			1	
		23	10	44	16

Note: All Highlighted names in Bold Face are Founders of NFMA.

William W. 'Bill' Back Jr., RCFM, IFMA Fellow, CxA, LEED AP
Austin, Texas, USA

William W. 'Bill' Back, Jr. was born and raised in Helena, Arkansas. He graduated from Arkansas State University in 1971 with a BS Degree in Business and was commissioned a 2nd Lieutenant in the US Army, in which he served for the next six years. In 1978 he joined the Office Services Department of Texas Eastern Transmission Corporation (TE) in Houston, Texas and for the next 14 years was mentored by NFMA Founder and First President George W. Graves.

Bill was the Office Facility Manager for 14 years, overseeing all internal office renovations and relocations, as well as coordinating TE needs with the building management staff. After leaving TE he sat for and passed the First CFM Exam offered by IFMA and then worked in Facility Management for Flour Daniel in Houston and Aguirre Corporation in Dallas. In 1995 Bill was honored to be made an IFMA Fellow and became the Director of Education for IFMA and for the next four years lead the initiative to create several FM Educational Seminars including the first CFM Exam Review course.

After leaving IFMA Bill became the South-Central District Facility Manager for Computer Associates (CA). Upon leaving CA Bill joined the Sebesta Blomberg team and over the next ten years established Sebesta's Houston and Austin, Texas offices, and earned both his LEED AP and Commissioning Authority (CxA) designation. During these years he led the building commissioning of several buildings for Houston's Methodist Hospital, MD Anderson Cancer Center, Houston, the University of Houston and the University of Texas at Austin. Bill retired from the Facility Management Profession in 2014.

1.3

The Birth of FM and IFMA

Mary Day Gauer

The term "Facility Management" as we use it today was initially coined in the 1960s by Electronic Data Systems founder Ross Perot in reference to the network management of IT systems, but was soon expanded to include elements of commercial space management.

By the time Herman Miller hosted its "Facility Influence on Productivity Conference" in November 1978 the term "facility management" was already widely in use. It was stated at this conference that the Herman Miller Research Corporation (HMRC) wanted to understand how the "human performer, the individual, his life in large organizations and how these organizations entrust a healthy, appropriate atmosphere on society" (Facility Influence on Productivity Conference Report)

The genesis of discussions about facility management is found in the history of Robert Probst, who started the Probst Co. in 1953, a Denver-based firm specializing in speculative product development. He called on Herman Miller in 1958 to discuss one of these ideas, a unique fishbone connection system for furniture components. He was asked by Hugh DePree, then president, to sign on as a consultant (later to become president of HMRC) and began conducting unprecedented studies of people in work settings. DePree asked him to also "find problems outside of the furniture industry and conceive solutions for them" (www.hermanmiller.com/designers/Probst), which led to many inventions, most of which had nothing to do with furniture, but also included a mobile office design for a quadriplegic.

The final design for Probst's Action Office System was completed in 1968 and Herman Miller began producing the product, revolutionizing how knowledge workers are accommodated in office environments. This was the world's first open-plan office system of reconfigurable components and a huge departure from what we fondly called "four on the floor," the static row upon row of desks. It was immensely successful and transformed the workplace. Probst's *The Office: A Facility Based on Change* became a bible for those engaged in understanding and managing this new phenomenon.

Inherent in the success of this invention was the need to manage the environment in order to support work relationships and better understand organizational objectives. HMRC also recognized that a revolutionary revision of the work process and how information is used would be critical. This led to a desire to understand management processes, effective communications and relationships, communication channels and how to organize all of these elements.

The idea to assess the effect of the environments on people's productivity who had been working in the new responsive furniture environments for 10 years drove the desire to survey operations and facilities managers for data leading to findings on the effect of these systems on productivity. The top priority for facility managers was the ability to fit the facility to management programs and restructuring, while the top priority for operational

Facilities @ Management, Concept - Realization - Vision, A Global Perspective, First Edition. Edited by Edmond Rondeau and Michaela Hellerforth.

managers was the ability of the work environment to support efficient work processes and human user needs. These slightly opposing goals dealt with the question of the people versus the structure, drawing attention to the need to expand the FM's understanding of the true client, the end user, along with that of organizational needs. These results were presented at the 1978 conference by Probst. A presentation was also given by Dr. Abraham Korman, a consultant in the area of management and organizational training and development which expanded the issue of productivity on the individual level. This focus on the people within the workplace would become central in the development of facility management.

The 37 attendees at the Facility Influence on Productivity Conference came from the USA, England, and France. They were managers, presidents, chairmen of the board, directors and vice presidents, people who dealt with company facilities and management policies and objectives. Several of these attendees became the nucleus of the group which founded the National Facility Management Association, which two years later became the International Facility Management Association.

After the conference, Herman Miller established an independent Facility Management Institute (FMI) in order to accomplish its objectives and to be positioned to include all professionals regardless of the furniture systems they managed. FMI held several small conferences starting in 1979 at its offices in Ann Arbor, MI. Some of the attendees from the 1978 Productivity Conference were also there and discussions started about developing a common language for the emerging field of FM as well as providing a network for these professionals. There was also discussion about the differences in the field of FM versus those of engineering, architecture, interior design, business management, and human resources. Many attendees at these meetings were members of other professional organizations, none of which met the overall needs of the people in this emerging field.

Charles Hitch, the VP and Senior Operations Officer from Manufactures National Bank in Detroit, and George Graves, Manager, Offices Services from Texas Eastern Transmission Corporation, both of whom attended the 1978 conference, began discussions with David Armstrong of FMI and the idea to create an association that would give this new profession support, education, research and a way to network with other facility managers.

An organizing meeting in Detroit in 1979 and one in Houston in 1980 resulted in the creation of the framework for the association that became the backbone of who we are today. In October, 1980 at the FMI, participants gathered to discuss the profession and the vision for the association. The constitution was ratified, officers were elected and the participants who were there that day who had also attended either or both of the organizing meetings in Detroit and Houston became the 16 founding members. The National Facility Management Association (NFMA) was born. I was fortunate to be one of those people.

The seminars at the FMI were a revelation in their focus on blending the physical aspects of the environment with that of the psychology of people. Dave Armstrong's theory during these sessions described "people, process and place" as intersecting circles, and he posited that these intersections were the foundation of the profession of FM.

While the creation of Action Office was a response to a fundamental need, the creation of the FMI was a response to discovering ways to manage environments more dynamically and responsibly to changing conditions. Cecil Williams, then associate director of research at FMI, indicated that people who are responsible for facilities needed to understand the people who work for them and within their buildings.

It wasn't simply about the envelope, heating, cooling, and cleaning but meeting the needs of the people. He also put forth that facility managers must understand their role within the corporation and the goals of the organization, or they wouldn't be successful. The idea of planning for day-to-day as well as planning to envision the future was a cornerstone for much of the early research and educational tracks within the association.

As recognition of the value of facility management grew within corporations, so did IFMA. The ability to provide an environment which made a company more competitive, managed growth, improved employee productivity and contributed directly to the bottom line began to be seen as a strategic role rather than simply one of operations, maintenance and services. IFMA provided the educational and networking opportunities which served to help it grow as well as the profession.

Gradually, it was understood that facility management reached far beyond the office environment, and professionals from all industries representing a myriad of facility types became not only IFMA members, but brought their expertise and passion to the organization.

The association recognized that it needed to provide a framework for its educational offerings as well an understanding of how competency could be measured. In the early 1990s, the first eight core competencies were developed: Long Range and Annual Planning, Financial Forecasting & Management, Telephone, Security & Administration, Interior Space Planning & Space Management, Architecture & Engineering, New Construction & Renovations, Maintenance & Operations, and Real Estate.

The first "exams" were done via essays with descriptions of work performance within the competencies, and the first tests were begun in 1994. The ninth competency was added several years later as Technology took over our world on all levels, including our building operating systems. Risk management and Sustainability have also been added to further define the performance of the profession and the professionals who work within it, and now gives us the foundation for this career path. As the profession and its focus and complexity have shifted, so have our competencies in an effort to define our current and future states as well as global performance.

The pandemic has created an opportunity for us to in many ways, return to our roots. It is interesting to note that conversations are shifting back to a focus on the individual contributor within the context of their organizations and their needs. Going back to people, process and place and looking at the intersections of management processes, effective communications and relationships and how to organize all of these elements in new and inspiring ways is the source of the next level of growth within the profession.

I was fortunate to be working for The Kellogg Company in Battle Creek, MI, during the formation of the association. My education and experience was in interior design but I found myself intrinsically identifying with FM and its philosophy.

A shift from consumer based goods to one of engineering design and production for Honeywell Aerospace allowed me to later apply my knowledge of FM to a completely different environment. Immersed in the aerospace industry for 24 years found me making a move to FM within higher education, specifically focused on Health Sciences.

IFMA provided me with the ability to make this transition and meet the challenges. Without this association, its networking, its learning opportunities, and its support, I would have not had the chance to experience the challenges and achievements during my career.

I've had the honor of volunteering on many levels within the association for many years and can only say without IFMA, I wouldn't be who I am today professionally or personally. My accomplishments are based not only on my abilities but also on the wealth of opportunities offered through this association.

Reference

Facility Influence on Productivity Conference Report. www.hermanmiller.com/designers/Probst.

Mary Day Gauer, RCFM, IFMA Fellow
Albuquerque, NM, USA

Mary Gauer retired as Group Manager with the University of New Mexico Planning, Design & Construction Department, where her focus was on education, medical research, clinical and administrative facilities for the Health Sciences Center.

She has since served as a consultant with the UNM Comprehensive Cancer Center on two large tenant improvement and life safety projects: her former department of Planning Design & Construction on projects for the Main Campus, and a project for the National Nuclear Security Administration.

Her early career was spent at The Kellogg Company as a Facility Manager and at Honeywell Aerospace as the Supervisor of Design and Construction.

Mary is a 1980 founding member of IFMA, a 1990 founding member of the New Mexico Chapter, recipient of IFMA's 1999 Distinguished Member Award, was inducted as an IFMA Fellow in 2000, and has been a speaker at numerous local, regional, and international conferences.

1.4

The Birth of Facility Management and IFMA

Melvin R. Schlitt

The birth of the profession facility management and the International Facility Management Association (IFMA) and how I was intimately involved in the launch of and served as its Executive Director for the first 10 years is an interesting story with a somewhat serendipitous set of circumstances.

My father and his four brothers started a construction company after World War II to build homes for the returning soldiers and their families. The brothers had all learned home construction skills from my grandfather, who was a home builder prior to the war. The demand for new homes was very high after the war, and lumber companies were willing to supply lumber to home builders on a contingency basis with no interest since the homes were selling as fast as they could be built. My father and his brothers kept the construction business going for several years until the demand for new homes had cooled and they all went their own ways into other trades. However, my father continued to build homes for himself and when I was just a young teenager, I began to help him with home construction and learned a lot of the skills of the trade, which gave me an interest and knowledge of construction.

After graduating from Michigan State University in 1972 with a degree in Business Administration, I landed a job with Michigan National Bank in Lansing, Michigan, and started my career there as a management trainee. As a new trainee your task is to learn the complete banking business by working in the various departments of the bank to see how they all integrate with one another. As I was rotating through the different departments, I had an opportunity to work in the Property Management Department, which had the responsibility of managing all the bank's properties including the headquarters' office towers, branch bank offices, data processing center, and security department. Services provided by the Property Management Department included: real estate leasing, contract negotiations, building maintenance and physical plant operations, security, space planning, renovations, and new construction coordination for all the internal bank departments as well as building tenants. Of all the departments that I worked in at the bank during my training period, I found Property Management to be the most interesting because they were supporting the functions of all the bank's various departments by providing them with the office space, office furniture, office equipment and the support services they needed to do their work.

A few years later, much to my delight the position as the head of the Property Management Department opened and I was offered the job in part due to my knowledge of construction that I had learned from helping my father with building our homes in my youth. In my new role as VP of Property Management, I joined the local chapter of the Building Owners and Managers Association (BOMA) and began to network with other property management professionals in the business community. However, while many of the local BOMA members dealt with building operations, especially for tenant occupied buildings, I was not finding professionals like me, who

Facilities @ Management, Concept - Realization - Vision, A Global Perspective, First Edition. Edited by Edmond Rondeau and Michaela Hellerforth.

were managing corporate office buildings and providing facility management services to support the work of the company's own employees.

What happened next set the stage for my entry into the newly developing field of facility management. The banking industry in Michigan had recently gone from a limited branching banking system based on a restricted service area to a statewide branch banking system with no restrictions on service area. To take advantage of this change in the law, our bank merged with several other independent banks in the state to form Michigan National Corporation and we began a rapid expansion of our branch banking network across the state. As part of my job, I was responsible for coordinating the design and construction of all our new branch bank offices across the state. Our company's goal was to open these new branches as quickly as possible for us to be able to apply for additional branch bank offices through the federal banking regulatory authorities in Chicago. To help facilitate this process, I asked our furniture supplier, Herman Miller (HMI), if they could design and manufacture a new variable-height panel connector to allow us to construct teller lines out of their *Action Office* panel system. After some discussion they agreed to produce the new connector. Later our new teller line system was featured in an article in the new *Facility Management Magazine* after which Bank of America used our panel system design for their branch banks throughout California.

Then one day I got a call from a job recruiter saying they were looking for a facility manager to join a new Herman Miller company called the Facility Management Institute (FMI) in Ann Arbor, Michigan, that was being formed with the mission to do research, consulting, and education into the emerging field of facility management. I was quite intrigued with the idea, and I ended up going to Ann Arbor to meet with Dave Armstrong, the Director of FMI, and some of his newly assembled team. Dave had been a professor at Michigan State University in the school of Agriculture and Packaging and he had organized a group of diverse team members for FMI including several people who had worked at Herman Miller Research with Bob Propst on the development of the *Action Office* panel system. The team comprised an architect, an interior designer, a product engineer, a graphic artist, a psychologist, a research analyst, and several support staff. It was a very interesting group of people with a lot of energy, and they all seemed to be excited about this new concept of facility management.

After thinking about the opportunity and doing some additional research on Herman Miller and its history as an innovative company in the field of office environments, I decided to join FMI as a senior associate in 1980. One of my first assignments, because of my banking background, was to set up interviews with facility managers from several major financial institutions across the US to learn about how they managed their organization's facilities. Some of these companies included: Citibank, Wells Fargo, Morgan Guaranty, Metropolitan Life Insurance, and Bank of America. Other FMI team members interviewed facility managers from different industry sectors. Our goal was to gain insight into the organizational structure of different company's facility management departments to see what services they provided and where their departments were located within the organization and at what level. Our research data revealed, among many other things, that the larger a company's number of employees and the larger amount of square feet of facilities they occupied, the higher up in the organization the FM Department was located. Also, manufacturing companies often had their FM departments as part of their engineering group.

As we conducted more research and did consulting with major corporations on facility management, we used the data we captured to develop new facility management educational courses that were held at FMI for facility managers. Dave Armstrong developed a model for facility management that he called the three Ps: People, Process, and Place. The concept was that the intersection of the three functions of People (employees), Process (work procedures/equipment), and Place (office environment) was the area of responsibility where the facility management department provided the necessary support services to facilitate the work of the organization. These FM services as depicted in a model of a daisy flower with its numerous pedals including space planning, move coordination, telecommunications, leasing, construction, maintenance, physical plant operations, and more could either be outsourced or done in-house depending on the needs of the company. Either way all these services were overseen by the facility management department. These two simple but effective FM models became the basis for our explanation of facility management.

Additionally, as part of my responsibilities at FMI, Dave asked me if I would be interested in working on starting some sort of organization to represent the emerging profession of facility management. It seemed that some of Herman Miller's customers wanted to start a Herman Miller users' group so that they could meet to discuss the use of HMI's *Action Office* systems furniture product and to tell Herman Miller what they thought could be done to improve their product and services. Dave said that Herman Miller was not interested in having a users' group and that I should try to come up with some sort of association that companies could join to share ideas on best practices for facility management. His idea was that the individual companies would join as members. I told Dave that when I was at Michigan National Bank I belonged to the local chapter of the Building and Owners Management Association (BOMA), but that it was mostly focused on building management, tenant leasing and operations, so I felt there was definitely a need for a new organization for facility management that was broader than just building management. I started doing research to see what other organizations, if any, might already exist for facility management. After combing through the *Encyclopedia of Non-Profit Associations* with over 16,000 listings, the only organization I found with any focus on the management of the physical work environment was the Office Landscape Users' Group (OLUG). However, OLUG was only focused on the practice of using plants and panels to provide visual privacy in open office settings, which was very limited in scope.

I was then introduced to George Graves from Texas Eastern Transmission Corporation in Houston and Charles Hitch from Manufacturers' National Bank in Detroit, both of whom were major Herman Miller customers and were interested in forming a Herman Miller's users' group. Our discussions instead lead to ideas about forming a new association for the emerging profession of facility management, which would be much broader and more encompassing that just a Herman Miller *Action Office* users' group. I went down to Houston several times to meet with George and his assistant Bill Back to discuss the idea further. We worked on putting together a draft set of bylaws and a constitution for the National Facility Management Association (NFMA) using a set of legal documents as a guide from another professional association George belonged to. One of the key things we all agreed on was that the membership in the association should be with the individual facility manager and not the company and that the membership should transfer with the individual if he/she should change employers. I also felt strongly that the dues should be set and collected by the Association's headquarters with a set portion going back to the individual chapters for consistency.

In October of 1980 the National Facility Management Association's first annual meeting was held at the Facility Management Institute in Ann Arbor. The constitution was ratified, officers for the first full one-year term were elected, committees were created, and the members agreed to begin identifying other facility managers, who would benefit from joining. FMI generously agreed to support the new association by providing office space and allowing me to work part-time as the National Facility Management Association's first Executive Director for one year. After the October meeting I put together a Chapter Formation Manual that contained the steps and requirements involved in starting a new NFMA Chapter, along with a set of chapter bylaws and constitution to be followed by the NFMA chapter organizers. Additionally, I developed a simple pocket size NFMA Membership Brochure that explained the association's mission and membership benefits. The membership brochure was designed so that it could be handed out or mailed in a #10 envelope along with a cover letter from the chapter organizers to member prospects. These documents proved to be invaluable in the growth of the association. It was very exciting to see the first group of membership applications and checks begin to roll into the NFMA office.

The first chapter formed was in Houston in 1980 followed by the Southeastern Michigan Chapter in 1981. Seven additional chapters were formed in 1982. As each new chapter was ready to be chartered, I would fly out to their first meeting and welcome the new chapter officers and members to NFMA. With the formation of the Toronto Chapter in 1982 the NFMA Board agreed to change the name of the Association to the International Facility Management Association (IFMA) and to collect dues from Canadian Members in Canadian dollars. That was a big step for the association, as we realized facility management was of global interest and IFMA could indeed grow internationally. I also joined the American Society of Association Executives (ASAE), which is sometimes referred to as the association for associations, to learn more about association management. I found ASAE to be very helpful with information on working with your Board of Directors, strategic planning, educational program development, association software and systems, and much more.

While IFMA got its start with the support of FMI and its parent company, Herman Miller, the IFMA Board and I felt we needed to get the support and involvement of other industry manufacturers if we were going to be able to grow the association and not be viewed as just a part of Herman Miller. To do so, Charles Hitch, the current President of NFMA, and I made an appointment to meet with executives from Steelcase in Grand Rapids in the winter of 1981 to pitch them on joining NFMA. However, the weather was bad the morning of our meeting and Charles decided not to drive over from Detroit. Not wanting to miss the opportunity to meet with the Steelcase executives, I decided to go to the meeting by myself. Upon arriving at the Steelcase World Headquarters in Grand Rapids, MI, I was greeted by three Steelcase executives including: Paul Whitting, EVP of Worldwide Sales and Marketing, their Midwest Regional Sales Director, and their National Marketing Director. I explained that Charles was not able to make the meeting due to the poor weather conditions and that I would proceed with the presentation about NFMA as its Executive Director. All went very well until the end of my presentation when Paul Whitting asked who I worked for and when I said FMI, a division of Herman Miller Research, he jumped up and said to his team to call their Security Department and have me thrown out of their building. I quickly explained to Paul that while Herman Miller had been very instrumental in the formation of NFMA, the association was an independent 501C Professional Society and that it was the Association's goal to involve other industry manufacturers and suppliers. Paul then said if NFMA was still around in one year for me to come back and they would reconsider joining at which point he turned and left the room.

In 1982 we held the IFMA Annual Conference in Dearborn, Michigan and for the first time we added an area for exhibits. Interestingly Steelcase took exhibit space in the show along with about forty other exhibitors. This is when I realized that having exhibits at the annual conference was not only a good draw for attendees but could make significant money for the Association. That year we netted $30,000 on the annual conference and exhibition. When I told Dave Armstrong about our net earnings from the event, he said you mean $3,000 don't you and I said no $30,000. He was impressed. I realized then that the Annual Conference and Exhibition could be a significant source of income for the Association in the future that could be used to fund other causes and member services.

Not wanting to miss out on following up with Steelcase for their support for NFMA now IFMA, I met again with Paul Whitting at their offices in Grand Rapids in 1982 one year after our first and somewhat stormy meeting. This time the atmosphere of the meeting was much more congenial. Not only did Steelcase agree to join as a corporate sustaining member, but they also assigned one of their marketing executives, Jim Hickey, to work with me to help form new IFMA chapters across North America through their network of national sales representatives. With Jim's help over the next few years the number of new Association chapters grew dramatically.

In 1984 the IFMA Board and I felt it was time for IFMA to get its own office space and move out of FMI. In addition to myself, I had an Administrative Assistant, an IFMA News Editor and a Membership Records Assistant working for the Association. George Trayer, who was the current IFMA President, and I thought it might be a good idea to move the association's offices to Grand Rapids to be next to the headquarters' offices of the Business and Institutional Furniture Manufacturers Association (BIFMA), since their members represented many of the key contract office furniture manufacturers. George and I had found suitable office space in Grand Rapids, and we were getting ready to sign a lease for the new space when another serendipitous event happened. Without any appointment or previous correspondence of any kind, a gentleman showed up at the IFMA offices at FMI in Ann Arbor asking if he could meet with me. He said he represented Century Development in Houston, TX and that he wanted to speak with me about the possibility of IFMA moving its offices to their Houston Design Center and Integrated Facilities Institute in Houston.

They felt IFMA's mission and type of membership would be a good fit with the Integrated Facilities Institute, which would be doing research and education into the growing complexities of designing and managing white collar work environments. They offered to pay all our moving costs to Houston, give us a fully built out office suite for five staff along with free rent for up to 10 years and pay an office manager's salary for 3 years. Their offer sounded too good to be true. However, after discussing it with the IFMA President, we agreed I should go down to Houston as their guest to check out the Houston Design Center and the Integrated Facilities Institute. From my investigation, it appeared that the offer was indeed legit and would be a good move for IFMA. For a small, but growing, association this deal really gave us a shot in the arm financially and allowed us to show the association's independence as a separate entity for everyone to see. The move to Houston in 1985 was very scary for me personally to undertake on a wing and a prayer, but for some reason I was so passionate about the future and purpose of IFMA that I just decided to go for it. Unfortunately, for Century Development, the Integrated Facilities Institute and the Houston Design Center never achieved their vision. Despite their demise, IFMA just kept on growing.

In 1985 I was also approached by the Merchandise Mart in Chicago wanting us to hold IFMA's Annual Conference and Exhibition at the Mart in conjunction with NeoCon, the National Exposition of Contract Furnishings in June of each year, featuring over 500 manufacturers of office, healthcare, educational and hospitality furnishings. NeoCon was started in 1969 at the Mart and had grown to encompass over 1 million sq. ft. of contract furnishings manufacturers' showroom space on five floors in the Mart. NeoCon had over 50,000 attendees to the show each year that included interior designers, architects, contract furniture dealers, and independent sales representatives. The Mart felt that IFMA represented the influential target audience of corporate end users and decision makers that companies like Steelcase, Haworth, Herman Miller, and other major manufacturers of contract furnishings wanted to see at NeoCon. However, after giving it much thought the IFMA Board and I felt we should keep IFMA's Annual Conference and Exposition separate in October of each year and not be part of NeoCon. As an alternative the Mart offered to help us grow our Annual Exposition if we were to hold the IFMA Annual Conference and Exposition at the Merchandise Mart Expo Center in October.

The Mart assigned two of their staff members to sell exhibit space for us at no charge during the three years, 1985–1987, that we held our annual event at the Merchandise Mart Expo Center. With the help of the Merchandise Mart, we were able to dramatically increase the size of our Annual Conference and Exposition and resulting revenues for the Association. However, in 1988 we decided to leave Chicago to go to Atlanta at the Atlanta Market Center for a change of venue. Following Atlanta, it became the practice of the Association to move our Annual Conference and Exposition around the country, which helped to increase our attendance by going to new cities as well as tapping different chapters for volunteer help.

During the period 1985–1990 we began to get inquiries from FM contacts in other countries about IFMA and their interest in forming Facility Management Societies in their countries. Dave Cotts, the current IFMA President, and I were invited in 1987 to go to Japan to speak to the Japan Facility Management Association and tour the country to speak to different FM groups. It was an interesting but exhausting trip as we went from one end of the country to the other speaking to several different audiences over three days. As a result of the interest in IFMA from other countries around the globe including Canada, Japan, Germany, Australia, the Netherlands, and others, the IFMA Board began to grapple with international expansion. Issues like the collection of dues in different currencies, different languages, chapter formation, staff support, and board seats all began to be discussed. International expansion, while alluring had its pros and cons.

During the first 10 years of IFMA we developed many programs and services for the members including the following:

- **Educational Seminars** – As the membership in the Association grew, we felt it was necessary to begin offering seminars on facility management to the members to help them succeed in their careers. Our first program was held in 1986 and was titled, "Principles of Facility Management." We soon realized, however, that to breakeven financially we needed to attract at least 100 attendees to a program, so we began holding mini regional conferences with multiple speakers, which were very successful.

- **Councils** – in 1983 the first special interest Utilities Council was formed. The purpose of a Council was to provide an opportunity for members from similar industries to be able to network together to discuss unique issues on facility management for their particular industry. Other Councils were soon formed, including the Research and Development Council in 1985, the Information Technology Council in 1986, and the Health Care Council in 1987.
- **Annual Conference and Exposition** – The Association's Annual Conference and Exposition grew rapidly and increased in sophistication each year. With the addition of an exposition at the 1982 event in Dearborn, MI, we quickly grew from holding the event in a hotel to moving to a convention center. The program grew to include keynote speakers, multiple session tracks, Council meetings, and a major awards banquet. As the event grew in size and complexity, we hired production companies to help coordinate registration, hotel and airline bookings, the meeting room presentation technology, banquet hall special effects, and entertainment. It became a major production. I remember in Dallas we had five former Miss Americas and a singing cowboy riding around the dance floor for entertainment during the Awards Banquet. We also had a cocktail party with live long horn steers for members to sit on for souvenir photos. Another year in Atlanta we had Ed Rondeau magically appear on stage in the middle of a laser light show. The exposition also became a major production with large exhibits and displays requiring three days to set up. The contract furnishings manufacturers and FM service companies turned out to exhibit in large numbers at the event with each trying to outdo the others with their displays. The Annual Conference and Exposition became a much-anticipated event for the members and a major source of revenue for the Association.
- **International Facility Management Symposium** – As more and more interest about facility management was being directed to IFMA from other countries, it was decided to hold our first International Facility Management Symposium. The event was held at a hotel in Washington D.C., because of it being such an international city, and facility managers from other industrialized countries around the world, who had expressed interest in forming FM Groups or had already done so, were invited to attend. Each country was represented by its own flag on stage, much like the UN. Attendees spoke about the status of facility management in their country as well as their facility management group/association. It was a fun and thought-provoking event and helped to set the stage for IFMA's growth internationally.
- **Research and Benchmarking** – To assist the members with locating facility management information and to be able to have benchmarks to compare each other's facility management practices, we started a Research Department with a full-time research librarian. Members would pay a listed hourly fee for the service, which became very popular and self-funding over the years. We also conducted research to establish FM benchmarks with the assistance of FMI and leading universities like Cornell and others.
- **IFMA News/Journal** – In the first year of the Association, with the help of FMI, we began to produce *IFMA News*, the association's monthly newsletter that was sent to all the members. It was a great vehicle for keeping the members informed of happenings in IFMA and to promote upcoming events. Around 1989 the *IFMA News* transitioned into the *IFMA Journal* with four color paid advertisements, submitted articles from members and experts in the field and a calendar of events. According to ASAE, when your association publishes its own magazine, it means you have arrived.
- **Job Listings Service** – I remember when one of the IFMA members approached us about running an advertisement for an open facility management position with his company in the IFMA News. After discussing the request with the IFMA Board, it was decided to move forward with publishing the job listing. The Board was concerned about how this might affect companies wanting to let their facility managers join IFMA if they could find jobs through the association with other employers. However, the Board felt that most professional associations provide job listings services for their members. It was also decided if a member were to lose their job, they

would have their membership in IFMA extended for up to one year for free or until they got a new job, whichever was sooner.

- **Leadership Conference** – As IFMA began to grow and the number of chapters and Councils increased, we realized that we needed to provide a way to educate the leaders of the chapters and Councils to develop consistency and standards of operation. Hence the Annual Leadership Conference was born. It was a two-day program held in Houston to allow IFMA staff to interact with leaders from the IFMA Chapters and Councils to exchange ideas and procedures for best practices.
- **Certification** – I believe it was Dave Cotts with the World Bank, who was the IFMA President promoting the idea of a Certification Program for facility managers. He felt that it was important to spell out the body of knowledge and experience that was necessary to be recognized as a certified facility manager and for facility management to be listed in the Library of Congress as a true profession. Dave and Ed Rondeau both published their own books on facility management around that time, which helped to set the framework for the Certification Program.
- **Foundation** – The IFMA Foundation was founded in 1990 as a separate entity from the Association. The purpose of the Foundation is to support the knowledge base centered around the management of the built environment through research projects, support of educational programs and the awarding of student scholarships. Ed Rondeau served as the first IFMA Foundation Chair.

Surprisingly as other manufacturers' support for IFMA grew, Herman Miller's support seemed to decline. Eventually, Herman Miller pulled the plug on FMI as they felt their mission had been accomplished and now the job of growing the facility management profession was left to IFMA. By 1990 IFMA had grown to over 8,000 members, 83 chapters, 14 staff, $5 million in annual revenues, and $5 million in reserves. After 10 years as a founder and the first Executive Director of the Association, I left IFMA in 1990 and joined the Merchandise Mart in Chicago as the VP of Marketing, but that is another story. I will always treasure my experiences gained with the Facility Management Institute, NFMA and IFMA, those were exciting times, and I met a lot of smart, interesting, and dedicated people along the way.

Melvin R. Schlitt, Principal
M.R. Schlitt & Associates
Harbor Springs, MI USA

Schlitt is currently the principal of M.R. Schlitt & Associates, a consulting firm specializing in the development of marketing and trade show management strategies for the residential and contract furnishings' industries worldwide.

Prior to forming M.R. Schlitt & Associates he served as a Business Development Consultant for International Market Centers, where he worked on tradeshow development and acquisitions in the areas of commercial and residential furnishings. IMC is the largest operator of premier showroom buildings for the furnishings, home décor, and gift industries, with more than 14 million square feet of world-class exhibition space in High Point, NC, Atlanta, and Las Vegas.

Prior to joining the International Market Centers, Schlitt had a 21-year history with Merchandise Mart Properties, Inc. where he was vice president of marketing and new development for MMPI's trade shows throughout the United States and Canada, including the contract furnishings, residential furnishings, casual furnishings, franchising and the gift and apparel shows. Additionally, Schlitt lead the production, budgeting, programming, registration, list management, trade show operations, and meeting planning for these events.

Before joining MMPI, Schlitt played an integral role in establishing the International Facility Management Association (IFMA), serving as the executive director from its inception in 1980 until 1990. Under his direction,

IFMA grew to becomc the leading international association for the profession of facility management, with more than 10,000 members and 100 chapters throughout the United States and Canada.

Prior to his involvement in IFMA, Schlitt taught, researched, and consulted on facility management for the Facility Management Institute, a division of Herman Miller, Inc., in Ann Arbor, Michigan. He also served as vice president of bank properties for Michigan National Bank, a major bank holding company based in Lansing, Michigan.

Schlitt is active in several industry associations, including the International Interior Design Association (IIDA), American Society of Interior Designers (ASID), International Association of Exposition Managers (IAEM), Society of Independent Show Organizers (SISO), and the American Society of Association Executives (ASAE).

Schlitt holds a Bachelor of Arts Degree in business administration from Michigan State University, East Lansing, Michigan. He currently lives in Harbor Springs, Michigan, with his wife Kathryn and their three cats: Joey, Chloe, and Milo.

1.5

Beginnings

Eric Teicholz

Introduction

Mere chance and pure luck were probably the dominant factors that determined the trajectory of my professional career. For example, it was by chance that I got involved with computer-aided design (CAD) applied to architecture in its infancy in the mid-1960s, and it was luck that I attended Harvard's Graduate School of Design (GSD) in Cambridge, MA, USA, where much of this graphic technology and architectural research was taking place, although for the most part at the Massachusetts Institute of Technology (MIT).

In any case, I stayed in Cambridge after graduating from the GSD and then continued at Harvard to teach and do research and development in CAD, at first in architecture then in facility management (FM) at a time when database technology first enabled building data to be effectively integrated with graphics (which I see as another lucky coincidence).

As an academic (until the early 1980s) I also kept good records of my professional career and my published books and research papers keeping up this habit until about 2010. I thought that perhaps one day someone might be interested in the early days of CAD as it applied to architecture and facility management. Sure enough, a few PhD architecture students recently contacted me about architectural CAD in the 1960s and the GSD library has asked for copies of my early work, research, and publications for archival purposes.

Which brings me to the content of this chapter. I was approached by Ed Rondeau to write about the early days of facility management technology. I have reviewed my early resume to see exactly when I began working in the FM field and I've decided to divide this chapter into three sections: the first to be about my work in architectural and graphic technology at Harvard which became the foundation for my FM work; the second to be about my FM work at Graphic Systems – a company I started after leaving Harvard; and the third about my work with IFMA which started pretty much at the beginning of this century and continues to this day.

The Beginning: Architecture and Computer Graphics at Harvard

I enrolled as an architectural student at Harvard's Graduate School of Design in 1962. Facility managers were primarily involved in maintenance and custodial services related to buildings at the time. FM was not part of the architectural curriculum: Design with a capital D was what was considered most important by the architectural faculty.

But what did get my attention was computers. Ivan Sutherland, a professor of electrical engineering at MIT, came to Harvard to set up a computer graphics laboratory. A local computer manufacturer, Digital Equipment

Facilities @ Management, Concept - Realization - Vision, A Global Perspective, First Edition. Edited by Edmond Rondeau and Michaela Hellerforth.

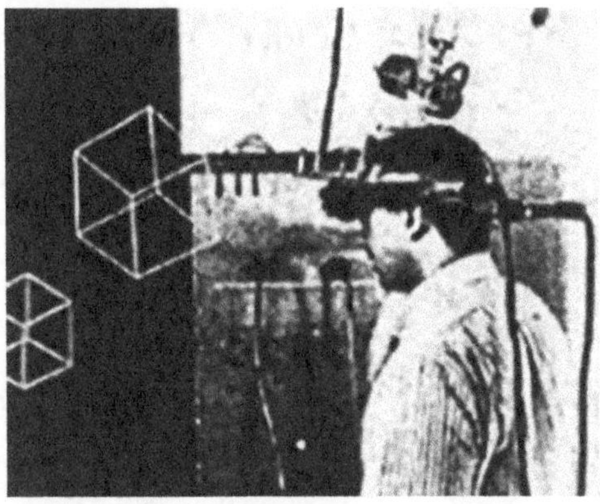

Figure 1 Sutherland 3D display. (*Source:* E. Eichholtz).

Company, donated a PDP 15 minicomputer as well as access to a mainframe PDP 10 computer. Sutherland then visited all of Harvard's graduate schools to try and recruit graduate students to develop applications on the PDP 15 related to their fields of study. I volunteered as the GSD representative.

Dr. Sutherland also brought an analog and digital 3-D device with him from MIT which had two small CRTs that fit over one's eyes and some gears that measured vertical and horizontal movement of the head (see Figure 1). The PDP 15 then would display crude (40-degree cone of vision) 3D line drawings on the CRT displays. That was enough to excite me, following which I began my initial research related to generating 3D perspective displays of buildings.

Also at the GSD at the time was a newly formed research group called the Laboratory for Computer Graphics and Spatial Analysis which was involved in research related to cartography. The Lab staff initially developed maps using a line printer which used overprinted alphanumeric characters to produce shades of gray. I joined this group and later became its Associate Director. The Lab purchased a digitizer to measure and record coordinate data. A digitizer is a device that moves a puck around a flat surface and can measure horizontal/vertical movement of the puck and output the result (plus any other codes you want to manually enter) on punch cards. In this way, I was able to use the computer to output 3D perspective drawings. It was a laborious task because the algorithm for solving the hidden line problem had not yet been invented (i.e., it was not yet possible to generate drawings that automatically eliminated lines that could not be seen from a particular location). Thus the "hidden" lines had to be eliminated manually using the digitizer to re-enter only those line segments that were visible from the location of the viewer (a location called the station point) – a laborious task at best (see Figure 2). It was time to move on to other applications.

The year is now early 1967. My research interest had turned to automating building space allocation, a task that's at the heart of architecture. This became the subject of my Master's thesis and the program I developed to deal with the problem was called Generation of Random-Access Site Plans, or GRASP. As input the user enters desired relationships between spaces. For example, in the diagram below the user would enter how she would like

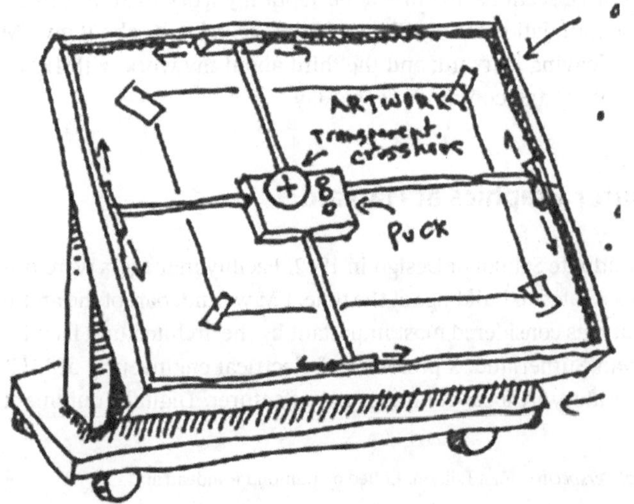

Figure 2 Digitizer. (*Source:* E. Eichholtz).

space A to relate to space B and C, with lower numbers indicating adjacency and higher numbers indicating non-adjacency. It's also possible for the input matrix to be non-symmetrical whereby the user can enter desired adjacencies for space B and C that are different from space A's adjacencies. For example, space A might have a desired adjacency for space B that is different from how space B should relate to space A (see Figure 3).

	Space A	Space B	Space C
Space A	3 2	4	1
Space B	1	3 2	4
Space C	2	3	1 2

Figure 3 Symmetric space interaction matrix. (*Source:* E. Eichholtz).

Interestingly, this problem has no solution in two-dimensional space and space allocation becomes an "over constrained" problem with a heuristic solution required. Fortunately, there was a mathematical solution to collapsing N-dimensional space (where a perfect solution exists) to the best possible solution in two-dimensional space.

The GRASP program started with the user inputting certain data such as site information (without buildings on it), architectural space configurations, an adjacency matrix, and a table of "loose ends" (walls where units can attach themselves to other units). The user can also enter other data, such as desired density, mix of units, desired distance to street, "no build" areas and so on.

The computer uses random number generators to pick a loose end and one of the architectural configurations. It then tests that configuration against the input constraints. If it passes those tests, the unit is placed on the site. The process continues until no further development can take place. If a particular solution is not found after one minute of computer time, the program tells the user it is done or that she should "relax" the input constraints (see Figure 4).

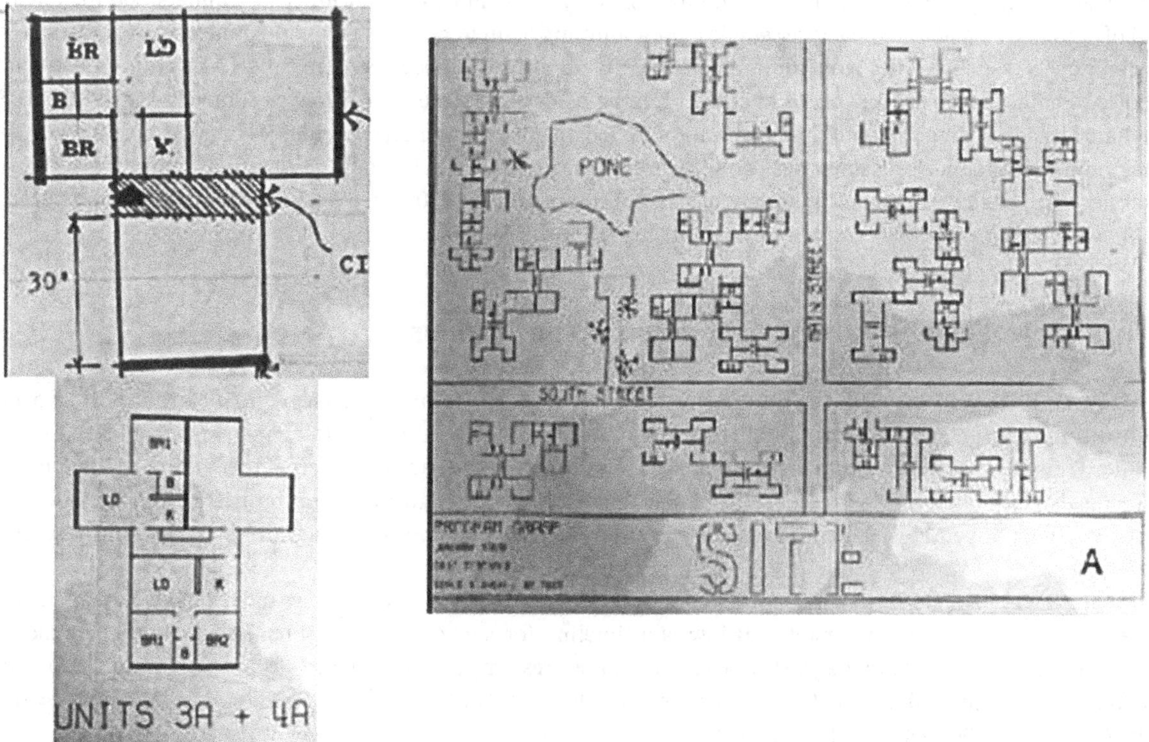

Figure 4 GRASP output. (*Source:* E. Eichholtz).

Figure 5 The Author (seated) at Perry Dean and Stewart's firm. (*Source:* E. Eichholtz).

After graduating from the GSD, I spent two years as an apprentice architect mostly designing parking lots and refreshment stands for a large cinema company based in Boston. This was enough to send me back to Harvard where I joined the GSD faculty to continue architectural research and to teach. This indeed proved extremely satisfying. I started out as a teaching assistant in 1968, eventually becoming an Associate Professor before leaving Harvard in the early 1980s. Along the way, I also had semester long guest lectureships at the University of Pennsylvania and several two-week programs at MIT as well as at universities in Finland and Egypt.

There's also one final story worth telling related to architecture and technology. I had a professor, Cliff Stewart, who was a senior partner at the architectural firm of Perry Dean and Stewart. Cliff clearly understood the potential of automation and asked me to assemble a programming team to help design an integrated set of applications for architects (see Figure 5). His firm purchased a PDP 15 and developed a system called AIDS (an unfortunate choice of names for the acronym Architectural Interactive Design System), which was completed in 1971. AIDS included basic drawing software, space allocation (based on adjacency analysis), quantity take-off/costing, bubble diagramming, and blocking software. The software was licensed to the Digital Equipment Company and sold to architectural, engineering, and construction (A/E/C) companies for $7000 per workstation or $14,000 for two or more workstations. I believe that a total of about 15 systems were sold.

The Middle: Facility Management and Graphic Systems

I spent the 1970s and early 1980s primarily teaching at Harvard, doing extensive lecturing around the world, consulting to large A/E/C firms, and becoming an editor for several newsletter, magazine, and book companies. I also served on several non-profit Boards including the Research Council of the National Academies. During this period, I also wrote several books related to architecture and the built environment. At the current time (November 2022), I have written or edited 17 books and over a hundred articles on these and FM topics.

During the 1970s and 1980s significant changes were taking place both in technology and in facility management, often with technology as the enabler. Below is a timeline for some technology breakthroughs that impacted vendor offerings in facility management. In case it is of interest, my personal resume from 1968 through 2007 can be found here: (https://docs.google.com/document/d/19ts2aquZDmOQ1-BAu-KQZk2SP60W9IXL/edit?usp=sharing&ouid=103890692844072421153&rtpof=true&sd=true)

FM Enabling Technology timeline from 1972 through 1984:

- 1973: First cell phone call – leading to FM mobile computing applications
- 1974: First commercial barcode scan – pointing towards asset management applications
- 1975: Production of the Altair 8800 – the first successful microprocessor-based computer and use of the general-purpose BASIC software interpretive language. BASIC was developed in 1964, a time when most computer companies required custom machine-level software which greatly limited their use to mostly scientists or mathematicians.
- 1976: Apple Computer Company formed by Steve Jobs, Steve Wozniak, and Ronald Rayne. Apple 1 appears.
- 1977: Apple II appears and begins to revolutionize personal computing. Apple III was used for some of my teaching work at the GSD.
- 1978: First bulletin board system (BBS) developed by the DoD's Arpanet group` (precursor of the Internet) using MODEM protocols for file transfer. Arpanet also sponsored some early GSD technology research.
- 1981: IBM manufactures its first PC. IBM also sponsored some of my early work by making me a Fellow of the Harvard Computer Center in 1969.
- 1983: Apple's Lisa is produced, one of the first PCs to use a graphical user interface (GUI). Over 10,000 Lisas were sold at an introductory price of $9,995.
- 1984: CD ROM developed. CD ROMs are optical discs, originally created by Sony and Philips, for storing large amounts of data.

Graphic Systems Inc. (GSI)

GSI was established in 1981 with me as its founder and president. GSI was an independent technology and management consulting firm specializing in corporate real estate and facility management (CRE+FM). We believed that technology was an essential means to achieving business objectives, but not an end goal in itself. During these exciting times for technology consulting in FM, PCs were becoming much more powerful; data storage was expanding; databases, networks and integration possibilities were more flexible; and vendors were proliferating while developing new FM+CRE applications and tools for integration into clients' databases.

GSI's projects involved some or all of the following consulting and implementation activities:

Business Analyses

- Definition/prioritization of business objectives
- Business process improvement opportunities
- Alignment of real estate services with fundamental corporate and business unit objectives
- Change management

Technology Planning

- Technology audits and strategy development
- Integration using the Internet/management implications and involvement
- Costs and benefits analyses
- Project planning and software/hardware phasing
- Strategic technology planning / business objective analysis / road map development

Design & Implementation

- Standards development
- Vendor analyses

- Implementation and configuration planning
- System integration and Implementation support

Support

- Data management services
- Data migration and training support
- Development of system benchmarks and key performance indicators (KPIs)

Education

- Seminars and educational programs
- Documentation

The Endgame: IFMA

I became a member of IFMA early in the 21st century. Boston had an active chapter, and many technology companies were located here as well as Archibus, one of the major FM vendors. I was active in the Boston Chapter and participated on their technology committee.

I was awarded an IFMA Fellowship (IFMA's highest award) in 2003 and became a Trustee of the IFMA Foundation in 2007, then a Board Member of IFMA's Global Board of Directors in 2013. I was excited about the possibilities of working with these groups, and I initiated or managed a number of interesting projects. Some of the more exciting ones for me included the following:

2003: FMpedia

I was head of the Foundation's Knowledge Management Committee and designed FMpedia to bring a wiki-like web-based glossary of FM terms to members of our industry. It was a dynamic, real time, open, Internet-based glossary where contributors were welcome to add new FM terms and definitions and comment on existing ones. It was meant to be as broad and international as possible. The glossary was noncommercial, supported by the IFMA Foundation and GSI, and contained no product endorsements or paid sponsors.

We later partnered with BOMI (a large building educational and training organization) and merged its 343-page *Dictionary of the Built Environment*, which added 4,500 terms to FMpedia (see Figure 6). At its peak, the glossary had more than 3,000 registered users from over 25 countries and its website received more than 12,000 "hits" per month.

2007: BIM Research for Facility Management

The use of Building Information Modeling (BIM) was just beginning to grow at this time as more architects, engineers, contractors, fabricators, and sub-contractors learned how to use data-rich 3D models to accurately portray the design and construction details of buildings and other types of structures. They have learned that BIM supports much better collaboration from the very beginning of projects so that higher quality buildings can be designed and then built with confidence that the results will reflect the model. A FM BIM book was the result of this research effort.

"It's all about the data." That expression was a good starting point for this book. In this case the data referred to the massive amounts of information needed by facility managers for their work and the systems that provide the basis for effective and efficient facility management. This book described current best practices to support the

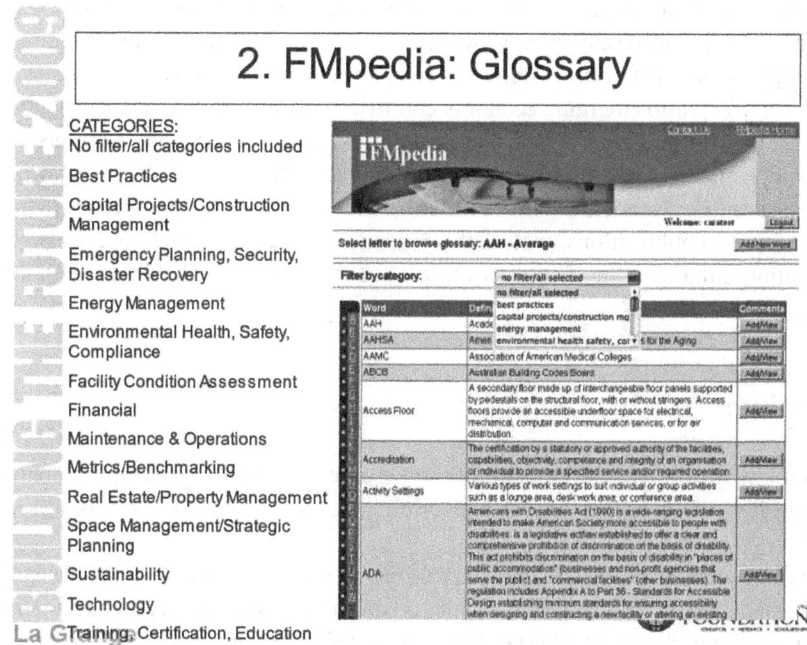

2. FMpedia: Glossary
CATEGORIES:
No filter/all categories included
Best Practices
Capital Projects/Construction Management
Emergency Planning, Security, Disaster Recovery
Energy Management
Environmental Health, Safety, Compliance
Facility Condition Assessment
Financial
Maintenance & Operations
Metrics/Benchmarking
Real Estate/Property Management
Space Management/Strategic Planning
Sustainability
Technology
Training, Certification, Education
FMpedia
BUILDING THE FUTURE 2009

Figure 6 FMpedia Glossary Categories. (*Source:* E. Eichholtz).

integration of BIM with FM systems and how to collect the data needed to support this integration. The emphasis was on what the owner and FM staff needed to know to ensure that these practices are used on new projects.

While this book did not focus on how to use BIM, it did illustrate the importance of FM participation in the early stages of a project. This in turn ensured that at the end of the project, there would be a successful start to the operation and maintenance of the facility.

2007: The "International Journal of FM"

Kathy Roper, an academic from Georgia Tech and, as president of the IFMA Board at that time, wanted to develop a juried peer-reviewed Journal for IFMA. We worked together to design and implement this project (see Figure 7). The *International Journal of Facility Management* was an open source, Internet accessible journal devoted to the science, technology, and practice of facility management. It was designed for both academics and professionals, and derived its audience from universities, researchers, practitioners and allied professional organizations throughout the world. The journal enabled the sharing of new theories, research, thoughts, experiences, and concepts for building and workplace design, management, operations, use and disposition. The global focus brought contributions from around the world, bringing this emerging and growing industry key insights for improvement and increased professionalism.

Figure 7 ijfm logo. (*Source:* IFMA).

The primary purpose that this journal served was to not only educate practitioners and students, but to provide an outlet for and to motivate researchers to begin new research, collaborate with allied professionals, advance the research agenda and professionalism of the entire field of facility management. As more research was established and published, the professionalism of the industry was enhanced, creating further need for research and researchers, providing more growth to the academic side of facility management, and in preparing more and better practitioners to further advance the industry.

In 2007 facility management was expanding into emerging economy countries, where design and construction had been their primary focus. With competitive pressures of globalization, however, these countries were increasingly embracing the concepts of facility management and expanding its impact. The need for peer-reviewed research was well established in these countries' academic communities, so acceptance for facility management was aided by additional, strong peer-reviewed publications, such as this journal, to expand international research and knowledge. Open access publication enhanced the ability for all to have the journal available and benefit from its existence.

2013: FM in the City Symposium

Marc Liciardello was IFMA's president in 2013. I was the Director of the Research Committee and on the Board at this time. Marc asked me to help design and implement a major research program related to the role of FM in the context of the city (see Figure 8).

This research resulted in a series of three workshops focused on what role FM should play in the planning, design, and management of the urban environment and, in turn, the effect of cities on the profession of facility management. Rapid urbanization was taking place throughout the world. This fact brings with it physical, environmental, cultural, and technological changes that facility managers had to deal with and formed the focus of the symposium.

Sustainability at IFMA

My involvement and passion related to sustainability started in 2007. At that time Dave Brady was the president of IFMA and there were few activities or programs related to this topic taking place within IFMA. I was a Foundation Trustee at the time. At a Trustee meeting that took place in La Grange Georgia, I gave a presentation

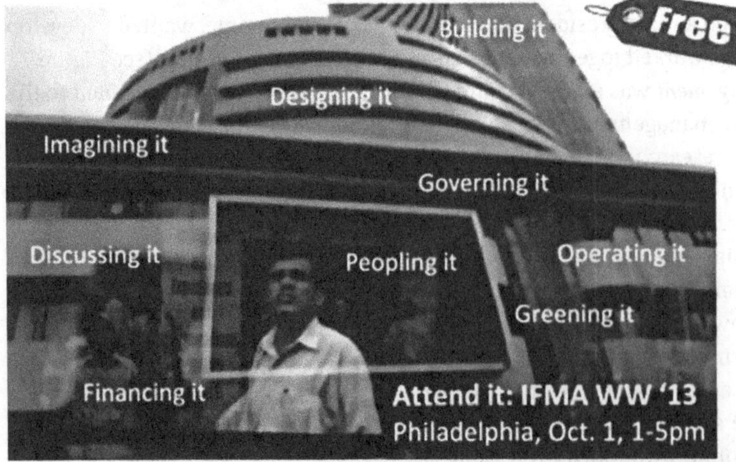

Figure 8 IFMA's FM in the City Workshop Promotion Illustration. (*Source:* IFMA).

to the Trustees entitled "It's Not Easy Being Green." Dave Brady was in attendance and asked me to help structure a more formal program related to sustainability activities within IFMA. I readily accepted his invitation.

Phase 1 of my research on this topic consisted of my taking a few months to travel around the country to visit various organizations that were doing relevant research on this topic or had developed on-going programs of interest. This effort then resulted in the formation of a Sustainability Task Force that I chaired. Established in November 2008, IFMA's Sustainability Task Force consisted of 12 industry leaders ranging from facility managers of global corporations to key players from the US Department of Energy and the US General Services Administration. Its goal was to recommend a framework and a set of supporting actions that, once implemented, would result in IFMA becoming the primary information source for all matters pertaining to sustainability initiatives in the built environment. The task force produced a report which served as a guide in the near-term development of IFMA's strategic roadmap for sustainability initiatives.

The repost we generated covered the following topics:

1) **Current Situation at IFMA** – Evaluation of the existing sustainability situation and initiatives within IFMA;
2) **Analysis of Global Sustainability Issues** – Work to gather facts regarding existing relevant sustainability initiatives associated with the built environment, regulatory issues, global issues, suppliers, resources, economic issues, etc.;
3) **Path** – development of a road map for prioritizing and implementing initial and potential objectives;
4) **Resources Required for Planning Effort** – job description for a full time IFMA sustainability hire, which IFMA committed to, and financial requirements to develop recommendations associated with the strategic plan;
5) **Schedule** – tasks, milestones, schedule associated with road map;
6) **Communication with IFMA** – risk analysis, tracking and reporting progress to the President / CEO and Executive Vice President / COO of IFMA on a regular basis.

The Sustainability Task Force morphed into an IFMA Sustainability Committee and eventually into IFMA's current Environmental Stewardship Utilities and Sustainability Community (ESUS), which I chaired until 2020. Numerous activities, programs, and partnerships were achieved during this timeframe – a few of which are described below.

2009: The How-To Guides

I initiated this project as part of my role as the Foundation's Director of Knowledge Management Committee. The general objectives of these "How-to Guides" were as follows:

To provide data associated with a wide range of subjects related to sustainability, energy savings, and the built environment.

1) To provide practical information associated with how to implement the steps being recommended;
2) To present a business case and return-on-investment (ROI) analysis, wherever possible, justifying each green initiative being discussed;
3) To provide information on how to sell management on the implementation of the sustainability technology under discussion;
4) To provide case studies of successful examples of implementing each green initiative;
5) To provide references and additional resources (e.g., Web sites, articles, glossary) where readers can go for additional information;
6) To work with other associations for the purpose of sharing and promoting sustainability content.

How-To Guide Title	Published	Pre-April 2016 Downloads	Post-April 2016	Download Total
Commissioning Existing Buildings	2012	16,001	288	16,289
A Comprehensive Guide to Waste Stream Management	2013	15,039	338	15,377
Getting Started	2009	12,949	305	13,254
Carbon Footprint	2012	10,805	824	11,629
US Government Policy Impacts	2012	9,021	229	9,250
EPA ENERGY STAR Portfolio Manager	2015	8,511	348	8,859
Global Green Cleaning	2011	7,432	1,390	8,822
Lighting Solutions (Rev 2)	2015	6,984	401	7,385
Sustainability in the Food Service Environment	2011	6,984	257	7,241
Sustainable Landscaping	2010	6,689	388	7,077
Green Building Rating Systems	2010	6,111	261	6,372
A Comprehensive Guide to Water Conservation	2010	5,523	346	5,869
No-Cost/Low-Cost Energy Savings Guide	2010	4,967	387	5,354
Turning Data Centers Green	2010	4,947	294	5,241
Engaging Occupants in Your Sustainability Program	2009		776	776
Measuring and Monitoring	2015		204	204
Total Guide Downloads		121,963	7,036	128,999

Figure 9 How-To Guide Titles and Downloads. (*Source:* adapted from IFMA).

The guides were reviewed by an editorial board, an advisory board and, in most cases, by invited external reviewers. Once the guides were completed, they were distributed via the IFMA Foundation's Web site free of charge. The Guides were produced over a six-year period and were widely used by the IFMA and guide-sponsoring communities (see Figure 9).

2011: ENERGY STAR

Measuring building performance in a consistent manner, benchmarking energy, and utility use, sharing and implementing efficiency measures and best practices helps reduce costs and consumption, reduce greenhouse gas emissions, and improve facility performance (see Figure 10). Since 2009, IFMA had worked with ENERGY STAR's

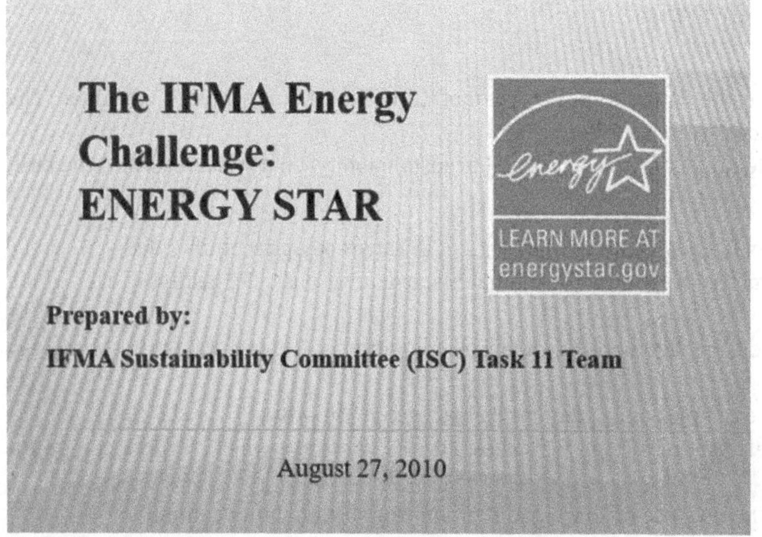

Figure 10 IFMA's ENERGY STAR Challenge. (*Source:* IFMA Energy Star/US department of energy / Public domain).

Portfolio Manager, developed by the Environmental Protection Agency (EPA) to help the FM community improve building energy performance in existing facilities.

In 2011 IFMA launched the IFMA **ENERGY STAR** Challenge. As part of the challenge, IFMA members were encouraged to benchmark their buildings in the ENERGY STAR Portfolio Manager software and share their data with IFMA through a master account. This allowed for the aggregation of data and the generation of benchmarking reports based on shared information. This capability was particularly important for non-benchmarkable facilities such as convention centers and airports whose facility managers otherwise had a difficult time finding peer data they could utilize.

There were close to 2,900 facilities that initially shared their information with IFMA through the IFMA Master Account. This information was incorporated into an IFMA O&M Benchmarks report. The initial report was generated by Laurie Gilmer, the current (2022) chair of the IFMA Board and past member of the IFMA Sustainability Committee.

The data in the report was provided to facility managers for the purpose of comparing individual facility performance to other facilities within the IFMA database. Data collected included facility type, size, geographic location (state or province), energy use per square foot, ENERGY STAR score (when available), and greenhouse gas emissions (as related to building energy use).

Current Activity: Eric Teicholz Sustainability Facility Professional Scholarships

The Eric Teicholz Sustainability Facility Professional (SFP) Scholarships (see Figure 11) were endowed in May 2020. The scholarships are open to young professionals with a demonstrated financial need who are currently practicing facility management or a related field and are interested in earning a specialty credential in sustainability. Additional criteria considered in the selection of candidates for scholarships included diversity and the impact the scholarship was expected to have on the applicant's environment.

The program was expanded in 2022 to include additional scholarships provided by the development of a new corporate sponsor program. Current 2022 corporate sponsors include JLL, SDI, Planon, and Jumbo Chains plus three IFMA chapters (San Fernando Valley, Kansas City, and South Florida).

Partners have established two new scholarship tracks: one focused on FM technology and the other related to sustainable supply chain management. The program has grown 450% since its inception.

The IFMA Foundation awarded its Chairman's Award of Excellence to me at the 2022 World Workplace event.

Some Closing Thoughts

I was honored to be selected to write this chapter on facility management and my early work in architectural technology. I have worked and written about these topics for my entire career which spans over 50 years. I tried to be as accurate as possible in my dating of some individual projects and my books, research papers and articles that I initiated or managed. As stated in the introduction to this chapter, I was lucky to still have my detailed resume from about 1968 through 2007. I want to thank all of the people I worked with at Harvard, Graphic Systems and IFMA who deserve much of the credit for the projects I chose to discuss in this chapter.

Figure 11 The Eric Teicholz Sustainability Facility Professional (SFP) Scholarship Logo. (*Source:* E. Eichholtz).

Eric Teicholz, IFMA Fellow
President, Graphic Systems, Inc.
Lexington, MA USA

Eric Teicholz is president and founder of Graphic Systems, Inc., a firm specializing in facility management and real estate automation consulting.

- Fellow of the International Facility Management Association
- Ex-officio chair of IFMA's Strategic Advisory Group, Sustainability Task Force and Sustainability Committee
- Member of IFMA's ESUS sustainability community,
- IFMA Foundation's Board of Trustees
- IFMA's Global Board of Directors
- Board on Infrastructure and the Constructed Environment (BICE) for the Federal Facilities Council at the National Academy of Sciences
- Commonwealth of Massachusetts Integrated FM and Advanced Energy Projects Committees

Eric graduated with a Master of Architecture degree from the Graduate School of Design at Harvard University where he later became an Associate Professor and the Associate Director of the university's largest R+D facility, the Laboratory for Computer Graphics and Spatial Analysis.

He also is the author/editor of 17 books related to FM, CAFM/IWMS, and GIS technology. Teicholz was awarded the Navy's Superior Public Service Award by the Secretary of the Navy under President George W. Bush for his participation on a blue-ribbon FM panel defining the future strategy for naval shore facilities.

For a complete resume of Eric Teicholz' research and publications from 1968 to 2007 please visit https://docs.google.com/document/d/19ts2aquZDmOQ1-BAu-KQZk2SP60W9IXL/edit?usp=sharing&ouid=103890692844072421153&rtpof=true&sd=true

1.6

FM in the Beginning

Christine H. Tobin (formerly Neldon)

In the late 1970s, the business world changed radically. Computers invaded the workplace, and everything exploded. Traditional furniture no longer worked, and phones became complicated. Power and cable requirements placed new demands on building systems and capacities. HVAC became a critical system and new space was needed for mainframes and their support teams.

Engineers, Architects, Building Owners, and Operations personnel were struggling for solutions and understanding. Costs were skyrocketing with no clear business leader within organizations to deal with the crisis. Vendors were struggling to come up with solutions as lead times on construction, products, and services increased dramatically. Board of Directors had no clear understanding of the issues or costs involved.

In the early 1980s, the office lease for the Atlanta office of Arthur Andersen was expiring. A "Big Eight" accounting firm owned by partners, a move was planned to the new (client-owned) Georgia-Pacific Headquarters building. A move coordinator was required, and it was considered a no-win situation. As a young, female, operations manager and the most expendable member of the staff, I was selected for the job.

At about the same time as this assignment, an invitation came in the mail from the Facility Management Institute (a Herman Miller organization). A lecture was being presented by Dave Armstrong on the new field of Facility Management (FM). Not wanting to lose my job, and understanding clearly help was needed, I attended. Listening to Dr. Armstrong describe people, process, and place, an epiphany occurred.

I was a Facility Manager.

Looking for information, I attended the organizational meeting of the Atlanta Chapter of NFMA (National Facility Management Association). It was clear that technology in the workplace had leveled the playing field. Regardless of age, sex, education, or race, knowledge was needed from anyone who had it. The Chapter and National organization became an instant sharing ground for those seeking knowledge and anyone was welcome.

My field was technology and I understood it very well. I needed knowledge about everything else! My main mentor was Ed Rondeau who was an architect and very experienced with real estate and construction. He along with others in the chapter put me in touch with the best product and service companies. They also spent time explaining various aspects of my new responsibilities. In return, I answered their questions about technology and its implementation. It became obvious that one of the primary benefits of membership was the networking and education we could obtain from each other.

Ed quickly drew me into IFMA leadership. I was offered opportunities for learning and skill development that my employer was not providing. It was understandable because their core business was accounting, and I wasn't an accountant. Also, at that time, women were not typically identified for leadership positions in business. Our

Facilities @ Management, Concept - Realization - Vision, A Global Perspective, First Edition. Edited by Edmond Rondeau and Michaela Hellerforth.

chapter leaders along with others around the United States of America (USA) started inventing the field of FM and encouraging anyone with FM related skills.

In the early years, there was no clear definition of FM and people in a variety of fields were performing various FM responsibilities. In order to encourage them to join a professional association, IFMA had to offer benefits. The first task for chapters was to recruit new members by providing education and networking.

The Atlanta Chapter IFMA Board arranged interesting topics for monthly meetings. We addressed current needs in the workplace with local speakers who had found solutions. Often we would have a panel discussion with each person describing things that worked for them. So, educational content was beginning to develop but we needed members.

It was not clear at first how to find these people. When the Atlanta Chapter chartered in May, 1982, I was the secretary for the chapter. In 1983, I became Vice President and in 1984 the Chapter President. We needed to spread the word about FM and recruit but how?

Most US cities had libraries that published books of lists. There was a list of the employers in the area by number of employees. I reasoned that large companies or organizations had someone who was acting in an FM capacity even if they didn't know it. With no internet at that time, I began to call a few of these companies every day to ask their switchboard who their operations people were. The operator would connect me through, and I would begin my FM speech. If interested and appropriate, they or someone they volunteered would be added to my list and invited to a meeting.

At the same time, an interesting phenomenon was occurring with FM vendors. They were finding that their buyers didn't understand their products. So, vendors actually became primary IFMA recruiters because for them an informed customer was actually the best customer.

In 1982, NFMA, which later became the International Facility Management Association (IFMA), required 75% of the Association to be FM practitioners and only 25% could be vendors or consultants. So in order to join, we encourage non-practitioners to recruit three professional members to maintain the percentage. I began to contact companies with FM products or solutions to explain IFMA and promote FM education.

Also, there were beginning to be publications in the field. Ann Fallucchi was Editor of the *Facility Management and Design* magazine, published out of New York City. In 1984 she joined the Board of Directors (BOD) of IFMA as the first volunteer Director at Large – Marketing. She became one of the strongest international supporters and promoters of FM and IFMA for many years. She and I were the first two female members of the IFMA Board of Directors. My new role was Regional Vice President (RVP) for the Southeast.

That year, one of my first committee assignments was to work with then IFMA President George Trayer, IFMA Director Mel Schlitt, and IFMA Secretary Ed Rondeau to develop a revised Constitution and membership categories. Being on a low budget, we worked in a hotel room at a Holiday Inn in Atlanta, Georgia. Amusingly, the National Association of Street Vendors was taking place in the hotel at the same time which made things loud and interesting! The new constitution was implemented by the IFMA BOD later that same year.

In 1984, IFMA achieved a definition of FM in the United States Library of Congress and defined the first eight core competencies. It now became much easier to explain FM to potential members. Also, education committees could begin developing specific standards and programs in each category. Colleges and Universities began to consider FM courses and degree avenues.

As a new RVP, I wanted to encourage chapter formation in other cities. As President of the Atlanta IFMA Chapter that same year, I encouraged the Atlanta BOD to fund trips for me to other southeastern US cities to promote FM. Keep in mind that at this time, IFMA staff in Houston was very limited and in early organizational stages. The international association also had very limited funds.

So, during my role as IFMA RVP from 1984 through 1987, I followed the same method as Atlanta to identify potential members and vendors in sister cities to recruit. Then, we would have a meeting and in every case a chapter formed within months. These included the USA cities of Richmond (1984), Greater Triangle Raleigh/

Durham/Chapel Hill (1986), Birmingham (1986), Greater Louisville (1987), Charlotte (1987), Columbia (1987), Jacksonville (1987), and Nashville (1987).

It is interesting to note that the organizational meeting for IFMA in Richmond was organized by locals to be held in an exclusive local business club. It actually had a policy in place that women were not allowed entry. When they realized the speaker was going to be a woman, they accelerated a policy they were contemplating of allowing women to join. Also, after the Richmond chapter visit, I wrote a Chapter Implementation Guide that was later used by other RVPs.

It is interesting to note that in 1984 there was a first at the IFMA Annual Conference in Denver. In the hotel lobby right outside the conference room, vendors were invited to pay for tables where they could promote their company. Passing by, George Trayer commented that "there may be a future in these." The Annual Expo was born that later morphed into World Workplace in Miami in 1995. This event is a major fund raiser for IFMA and also a significant source of education.

In 1987, I became the first female on the IFMA Executive Committee taking the role of Secretary. In 1988, I moved onto the role of Treasurer and established the first independent financial oversight and investment committees. In 1989, I moved into the role of Vice President. By this time, it was clear that IFMA was beginning to have a truly International impact.

In 1986, IFMA formed its first international alliance with the British Association of Facility Management. (AFM) It was followed in 1987 with an alliance with the Japan Facility Management Association. It was important for us to understand other countries and their needs and differences. The IFMA BOD was also beginning to hear that our materials were in use in other countries and opportunities for chapter formation might exist. It was decided that visits to other countries would be appropriate.

So, in 1989, I took a week-long visit to Canada to visits IFMA chapters that were forming there. I began in Winnipeg then went to Regina, Calgary, and Vancouver attending chapter meetings in each city. I also met with chapter leaders to answer questions and encourage development.

About that same time in 1989, Eric Lund introduced the concept for an FM Research and Education Foundation to the Board. In 1990 the IFMA Foundation was established with Ed Rondeau as its first Chairperson. Norman Polsky of Fixtures Furniture donated the first endowment fund. He was a major promoter of FM and IFMA for many years.

In 1990, I became the first female President of IFMA. IFMA leadership at that time came from volunteers with limited support or funding from their employers. The staff in Houston was still small and funding was limited. Members of the BOD paid their own travel and expenses personally and often had to use vacation time for meetings. This was the case for me until a single incident changed it all.

As my knowledge of FM grew, my job grew. I successfully accomplished our office move and many other FM projects. My role expanded to include other branch locations. My reputation was growing within the company but more importantly in the larger business community. I was invited to write articles about FM and speak at business meetings.

One day, the President of a large client of my employer approached the Managing Partner of my office and asked about my expertise. Taken aback, the boss struggled to remember me and what exactly FM was. Consequently, he started recognizing my expertise and leadership capabilities. Shortly after, my IFMA travel, and activities were partially funded. This was a significant step forward at that time for someone not engaged in the core business (accounting) and especially for a female.

For the first time in 1990, the IFMA BOD decided to sponsor a major overseas fact finding mission. We wanted to know the progression of FM in other countries and whether chapter or country FM associations were better. (We later found it varied by country and different solutions applied to each one.)

So in spring of 1991, Robert J. Gross (1989–1990 IFMA President) and I took separate trips overseas. Traveling with Dennis Longworth (IFMA Executive Director), we first traveled to London to meet with the British Council of Offices. This was a new organization formed by the College of Real Estate Management in Reading, England. Next, we met with the British AFM which had formed an alliance with IFMA in 1986. We then met with the Institute of Facilities Management, a division of the United Kingdom's Institute of Administrative Management (IAM). At that point in time, FM had grown in England faster than anywhere in the world except the USA and Canada but had no single association for the country.

Next we went to Helsinki, Finland where we met the Director of Facilities at Digital Equipment Corporation and toured their "Digital Office of the Future." We met with the Finnish Real Estate Federation (17,000 member companies) where there was no Finnish word for Facility Manager. The nearest translation was Economical Manager. An IFMA Finland Chapter actually chartered later in 1995.

In Stockholm, Sweden we met with the National Board of Public Buildings. They were the largest property manager with responsibility for government-owned facilities, including castles dating from the Middle Ages. We also met with a Milliken (carpet manufacturer) representative there who provided us with a list local FMs. We learned that Swedish FMs were interested in information from other countries but felt strongly that their needs should be addressed by their own research and associations.

In Copenhagen, Denmark, I spoke at the organizational meeting of the Danish Facilities Management Association. An IFMA overview and key FM concepts were presented. We toured the new Scandinavian Airlines System (SAS) headquarters where employee well-being and indoor air quality were paramount. This was some of my first exposure to these topics since they were not popular in the USA yet.

The final stop was Amsterdam for a meeting with the Netherland Facility Management Association (NEFMA), an Alliance partner with IFMA. Primarily a group of professional members practicing FM they were actively forming committees to spread FM education across Europe. They were planning a EuroFM Conference and Exhibition in 1992 to be held in Rotterdam.

As a result of the trip, we formed alliances with the Institute of Facility Management in England and Danish Facilities Management Association. We also found that issues were similar between us and the countries we visited. It encouraged us to continue outreach of IFMA's educational programs to other countries.

Also during 1991, I traveled to Japan where we negotiated for the development of the designation of Japan Certified Facility Manager by a combined group of Japanese Facility Management Groups. Coincidently, I arrived there on December 7 which was the 50th anniversary of the bombing of Pearl Harbor. I also traveled to Australia to speak at the IFMA Australia (not associated with the US IFMA) annual meeting.

Also, in my presidency, I realized the need for Councils to grow. These were groups of FMs from specialized facilities who could learn from each other. In 1983, the Utilities council had formed and was highly successful. I established the role of Director at Large – Councils and the number of councils significantly increased that year.

My vice President and the 1991–92 IFMA President was Samuel E. Johnson CFM, IFMA Fellow. Sam was an outstanding leader who was the first African American to serve on the Board. In late 1991 South Africa invited the IFMA President to visit and promote FM. They were still in apartheid, and the invitation was declined since they had not offered to fund the trip in their invitation.

During my year as Past President (1991–1992), I was asked to chair a committee to determine how to honor and encourage the continued involvement of key members who were significant leaders. In a brainstorming session, we looked at other organizations and what they did to recognize their most active and involved members. From this research, the IFMA Fellows program was born. The most prestigious title bestowed by IFMA is that of IFMA Fellow. The program was formally established by the Board of Directors on January 31, 1992. No more than five percent (0.5%) of the IFMA membership may hold the title.

Bestowing the title recognizes outstanding contributions and service rendered to the profession, IFMA and other facility management-related organizations and who embody the character and experience sought in an IFMA Fellow. The program establishes an elite core of respected leaders who are called upon to act as advisors and ambassadors for IFMA.

The first group of Fellows I had the honor to award was at the annual conference in 1992. It was an extremely impressive group involved in the profession in excess of ten years in various roles. Each was multifaceted and volunteered as teachers, writers, mentors, advisors, speakers, or whatever task needed to be accomplished.

Foremost was the first IFMA President George Graves. In May 1980, George hosted a meeting at his offices in Houston to talk about forming an association. Being a leader and results producer, he took the first action by locking the conference room door and telling everyone, "no one is to leave until we get this done!" There was a resounding "ABSOLUTELY." George and his wife JoAn were tireless organizers, supporters, and promoters of IFMA until his death in 2015.

Dave Armstrong, employed by Herman Miller's Facility Management Institute, travelled to numerous cities educating people in the industry who didn't know yet they were actually facility managers. With the famous model of "People, Process and Place," he showed where they all came together to form facility management. For years, he taught and promoted facility management to individuals and businesses.

Anne Fallucchi was Editor of *Facilities and Design Management* magazine and through thought provoking and intelligent articles deftly promoted and educated about facility management. This diminutive powerhouse from New York was the first female Fellow. At a time when the field was male dominated, she was an impressive and forthright proponent of the profession.

Jim Hickey, working for Steelcase, helped to create and promote the Annual exhibition and sponsorships for IFMA. These activities raised significant funds resulting in the ability to hire permanent IFMA staff. He educated other vendors and his customers on facility management and the value of an educated customer. Through his encouragement, other companies recruited members and became active internationally or in chapters.

Jim Chambers was IFMA's first international member. Hailing from Toronto, Canada, he formed the first international chapter there and put facility management on the map in his country. He was also instrumental in forming the first International Conference and Exhibition which was held in Toronto.

Ed Rondeau and Dave Cotts were active Board members for many years, and each served as President for IFMA. Both were strong educators and contributed heavily to the first education programs. Prolific authors, Ed and Dave co-authored separate books on facility management published in the 1990s. Their book *The Facility Manager's Guide to Finance and Budgeting* first published in 2004 is used as a text at many universities in the USA. Dave taught at George Mason University for many years and was an excellent mentor for many FM leaders. Ed was the first Chairman of the IFMA Foundation in 1990 and speaks and teaches at FM conferences, universities, and colleges in the USA and internationally. He has also authored a *Corporate Real Estate* manual used as a text at a number of universities in the USA.

After my last year on the IFMA International Board, I continued to support FM by writing, lecturing and serving on various committees. In 2021, I had the distinct honor of working on the committee to develop the new title of IFMA Fellow Emeritus. This title is for Fellows who have passed away or are no longer active (usually due to illness or retirement). At World Workplace in 2021, I was honored to present the 30th Class of Fellows and the first class of IFMA Fellow Emeritus.

Over 40 plus years in FM, the one constant I find is change. FM continues to evolve daily. There is a plethora of products and services creating a need for standards and evaluation. IFMA serves to educate and provide a networking ground for members to grow and learn. It is clear that continued collaboration and communication are keys to success.

See the next two pages for articles by Christine written as the first women President of IFMA.

IFMA News, December 1990 7

New president's acceptance speech

Neldon announces priorities for '91

I am very honored to have been elected president of IFMA. I would like to thank the membership for its vote of confidence.

I would also like to thank my employer, Arthur Andersen and Andersen Consulting, for giving me this opportunity; without their strong support, I would not be here. Additionally, I would also like to thank my family and my husband, Russ, for their tolerance and support of IFMA. Serving on IFMA's board is a demanding and time-consuming job. It can be a real challenge when balanced with a full-time job and a family.

My facility management involvement began nine years ago. I was given a new assignment in my company. My boss called it a wonderful opportunity. My peers muttered under their breath that it was corporate suicide. I didn't know what to call it because it did not have a name.

We should be proud of our accomplishments, but we cannot become complacent.

I began to go to preliminary planning meetings. The people talked about plenums, soffits, netmuxes and rosettes, and all sorts of other things I had never heard of. But they told me not to worry. They were the experts and they would take care of everything. When I heard this, alarm bells started to ring in my head. How could these people tell me what was best for my company when they did not know anything about it? I knew that I was in big trouble and needed help very quickly.

About that time, an invitation came in the mail to an organizational meeting of a group called the National Facility Management Association. I decided to go listen to Dave Armstrong of the Facility Management Institute talk about the relationship of people, process and place, and something called facility management. For the first time, someone said something that made sense.

After the meeting, several of us stayed to talk. We discovered that we had all kinds of things in common. Like other cities around the country, we decided to form a chapter of what quickly became the International Facility Management Association.

We have accomplished a lot in the past decade. We host the premiere conference and exposition in the facility management field. Our educational programs are unmatched in quality and in depth. The research we undertake is focused and supportive of the profession. We provide excellent networking opportunities for facility management professionals all over the world. We should be proud of these accomplishments, but we cannot become complacent.

We need to develop our leadership skills so that we cannot only become association leaders, but become corporate leaders as well. We need to educate our bosses and our management on the value of asset management. We need the ability to network with people in many different countries to give our companies a competitive edge in a globalized economy. We need better education and research so we can perform our jobs faster and better than others. That is why we formed IFMA 10 years ago and that is why we need it today.

There are several areas that must be priorities for 1991. We will stay firmly on schedule in implementing facility management certification. In addition, we will enhance our current education and research programs to meet the changing information needs of members. The new IFMA Foundation will be very active in securing the financial support necessary for these programs.

We will encourage new council development and strengthen and define our existing councils. A new board position for councils has been established to accomplish this goal.

1991 President Christine H. Neldon

We will continue in our world leadership role and promote facility management, both here and abroad. We will actively seek new alliances and chapters in other countries.

We will provide leadership training and encourage members to take leadership roles in IFMA and their company. IFMA members will be the recognized spokepersons for matters pertaining to facility management.

In summary, I have a brief story to tell. My company recently celebrated its 50th anniversary of serving the Atlanta metro area. We attended a big meeting and party to celebrate our success and commitment to the community. During the proceedings, a wise old leader of the firm stood before us as a group and reminded us of how we achieved our leadership position. Our motto, he said, was to think straight, talk straight and act with dispatch. That is what we at IFMA intend to do in 1991.

Thank you.

Christine H. Neldon

IFMA News, monthly IFMA newsletter, by Christine Neldon, IFMA President, "Neldon Announces Priorities for '91," page 7, December 1990. Image(s) courtesy of IFMA's FMJ.

Facility Management: Its Growing Stature

By Christine H. Neldon, President, International Facility Management Assn.; Director of Facilities, Arthur Andersen Consulting, Atlanta, Ga.

According to the Library of Congress, facility management is "the practice of coordinating the physical workplace with the people and work of the organization. It integrates the principles of business administration, architecture and the behavioral and engineering sciences..."

During the 1980s, the technological explosion caused complexities in the way facilities were managed. Facility management was clearly defined during this period, and standard practices developed. This structure is just in time for the challenges which await in the 1990s. While some problems will be new, others will be a continuation of existing ones.

There is a great deal of concern over the present economy. But long before there was talk of a recession, companies began a "learning" process. Mergers, bankruptcies, reorganizations and buyouts caused companies to become extremely cost conscious. Now there will be an even stronger focus on the bottom line and profitability.

Ten years ago, facilities needs were simpler, but as technology increased, more and more products are competing for budget dollars. In addition to furniture and telephones, the facility manager has to contend with computers, HVAC needs, complex security systems and a host of other concerns. Managements are viewing these growing costs with increased wariness. Thus, the facility management professional of the 1990s must have the necessary financial skills to demonstrate how wise asset investments will improve productivity and contribute to the bottom line.

INTERNATIONAL FACILITY MANAGEMENT ASSN. was founded in 1980. Its 1991 conference is Nov. 10–13 in San Diego, Calif. Executive director is Dennis Longworth at 1 Greenway Plaza, Houston, Texas 77046. 713/623-4362.

With a corporate need to tighten the purse-strings, and the cry for safer work areas, the facility manager gains in importance.

This decade will see the removal of many trade barriers between nations. With the restructuring of governments throughout the world, new markets will open up for companies, more international products will be available and competition will become fierce.

These changes will have a dramatic impact on facility managers. Purchasing decisions will require sophisticated information and analysis to insure the best results. In many instances, the facility manager will lead a team of experts in purchasing decisions.

The popularity of outsourcing will become more visible, and complexities in purchasing decisions will require specialized expertise. Peak and valley work loads will require supplemental help. Separate facilities management companies may be able to perform many traditional in-house facilities services at a financial savings to the organization. But there will always need to be an executive and appropriate support staff within the company to be sure such needs are effectively managed.

Another significant trend for the 1990s is the increased need and concern for environmental awareness. In a recent IFMA survey, it was found that many companies have started voluntary recycling programs.

Health and safety in the workplace is another major concern as increasing governmental involvement brings new laws. Chemicals used in the workplace, and their disposal, will be more carefully regulated.

Fewer young people will be available to enter the workplace of the '90s. Current employees may begin opting for a simpler lifestyle, less stress and fewer hours. Non-traditional sources of labor such as the elderly or the disabled will become more prevalent. Offices will need designs that meet the special needs of this new labor source.

An important new part of employee benefits packages may be enhanced facilities. Child care, health clubs, and food services may become important incentive and recruitment tools. Attractive facilities and advanced technology may be primary retention tools.

All of these issues may seem overwhelming. Actually, they present an outstanding challenge for facility managers to become leaders in their organizations. By being aware of changes, facility managers can be proactive. TO

42 The Office, January 1991

The Office magazine, Article by Christine Neldon, "Facility Management: Its Growing Stature", page 42, January 1971. Image(s) courtesy of IFMA's FMJ.

Christine H. Tobin (formerly Neldon)
RCFM, IFMA Fellow
1990–91 IFMA President (position is now known as IFMA Chairperson)
Atlanta, GA USA

Before retiring, Christine was Regional Project Manager for Travelers Corporate Real Estate in Alpharetta, Georgia. Responsibilities included planning project work, facilitating space negotiations, overseeing outsourced Allied Partners and department relationships, annual budgeting, emergency, and operational support.

Previously the Director of Facility Programs for Andersen Global Real Estate, she developed best practices, policy and procedures for facility management including annual capital and expense planning, lease management, security, ergonomics and code compliance.

Prior, she was Southeast Director of Facilities for Andersen and Andersen Consulting (later Accenture). Responsibilities included strategic planning, annual space forecasting, capital budgeting and oversight, real estate negotiation and management, project management and security.

Christine was International Chairperson of IFMA 1990–1991, on the Board for eight years and an author/speaker on facility management throughout the USA and internationally. Awards include 1993 Distinguished Member, 1995 IFMA Fellow and 2001 Distinguished Author. In 2016–17, she served as Trustee for the IFMA Foundation.

1.7

The Beginning of Facility Management (FM) in Germany

From My Very Personal Point of View

Günter Neumann

It all began with a little red brochure by Herman Miller with the title “Do your buildings steal money from you.” PARTNER Corporate Design in Stuttgart, where I was a shareholder, was planning a cooperation with the American office furniture manufacturer. While studying their brochures, I was particularly excited by this little booklet. PARTNER developed and realized interior design concepts for companies. This also included the design of working environments. The “facility management approach” shown in the Herman Miller brochure was new to us. The fascination for this management approach quickly developed into a vision and from that a strategy on how to integrate this approach into the PARTNER portfolio. I told each of our business contacts that in the future we wanted to deal with facility management as a business aspect of our corporate design services.

One day I got a call from Dr. Josef Mack from Vienna, who almost jumped through the phone because he had finally found a person who could not only spell Facility Management correctly, but who obviously knew what the term meant.

The consequence of this telephone call was a meeting at the Würzburg freeway service station, where Dr. Josef Mack introduced Walther Moslener to me. A lively discussion followed, at the end of which a strategy paper was drawn up with the following key points:

- Establishment of a facility management association to spread the FM idea in Germany.
- Scientific processing of FM in an FM institute, which should be located in Switzerland for status reasons.
- Establishment of an FM consultancy as an operating company through which the other FM activities were to be financed.

With this strategy paper I returned to PARTNER in Stuttgart. At this point, each partner had the option to implement an idea in his own business area (play money, we called it). I succeeded in convincing the other partners of the FM idea, thus creating a unique selling point for PARTNER. The first chance to present the Facility Management idea to a larger audience was the PARTNER in-house exhibition called “PARTNER-6-Days” in June 1988. In addition to architecture and design, facility management was addressed for the first time. The FM presentation consisted of an optimized working environment and an IT installation with rooms in various planning stages that could be visually entered via Virtual Reality (VR) glasses. This was supported by lectures on all aspects dealing with the planning, construction, use, transformation, and utilization of building structures and facilities.

These presentations were taken up by the visitors with great curiosity but also with a lot of skepticism. Those responsible for real estate in the companies were afraid that the implementation of FM would mean having to rebuild the entire company. For the management consultants, it was “old wine in new bottles.” This meant the

Facilities @ Management, Concept - Realization - Vision, A Global Perspective, First Edition. Edited by Edmond Rondeau and Michaela Hellerforth.

optimization of internal business management by means of "overhead value analysis," a bread-and-butter business for many management consultancies at that time.

In the brochure "PART" published for this event, there was an article on facility management for the first time (see Figure 1). Research revealed that in his drama Henry IV of 1597, Shakespeare has Lord Bardolph say the following sentences, which today could be described as facility management.

For us, FM was a holistic and integrative methodological approach with a focus on: People, Place, Process. In doing so, we tried to bring these elements together with our existing portfolio, e.g., in:

- site analysis
- operational space planning
- architecture and engineering
- space management
- furnishing planning
- room installation
- maintenance and repair of structural and technical facilities
- budget planning

With this integrative approach, we have – in retrospect – overwhelmed the real estate managers of many companies. In Germany at that time, real estate was generally managed in different departments but not as assets. Life cycle considerations, value chains or outsourcing were foreign terms that did not exist in practice.

During the PARTNER in-house exhibition, the idea of organizing an FM congress to prepare for the foundation of an FM association matured (see Figure 2). At the same time, talks were held with IFMA in Houston. Furthermore, participation in IFMA's annual conventions in Atlanta and Seattle took place in order

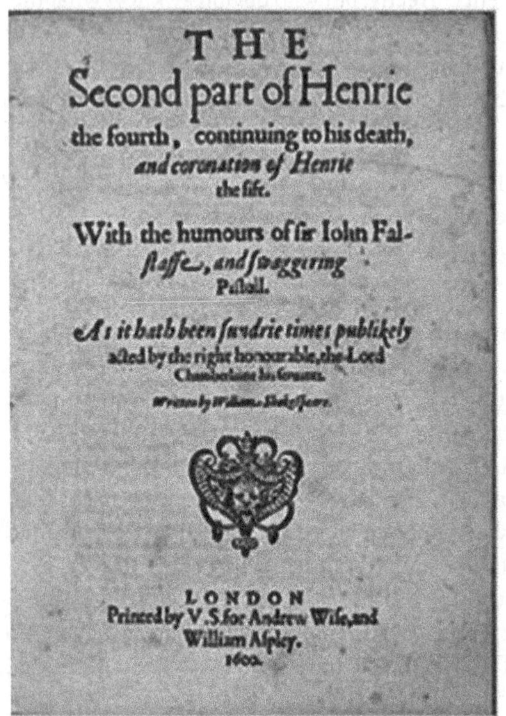

THE
Second part of Henrie
the fourth, continuing to his death,
and coronation of Henrie
the fift.
With the humours of sir Iohn Falstaffe, and swaggering
Pistoll.
As it hath been sundrie times publikely
acted by the right honourable, the Lord
Chamberlaine his seruants.
Written by William Shakespeare.
LONDON
Printed by V.S. for Andrew Wise, and
William Aspley.
1600.

We see th' appearing buds, which to prove fruit
Hope gives not so much warrant as despair
That frosts will bite them. **When we mean to build,**
We first survey the plot, then draw the model,
And when we see the figure of the house,
Then must we rate the cost of the erection,
Which if we find outweighs ability,
What do we then but draw anew the model
In fewer offices, or at least desist
To build at all? Much more in this great work,
Which is almost to pluck a kingdom down
And set another up, should we survey
The plot of situation and the model,
Consent upon a sure foundation, ...

William Shakespeare: King Henry IV, Part 2, Act1, Scene 3 (Lord Bardolph)

Figure 1 Facility management anno 1597. (*Source:* W. Shakespeare / Folger Shakespeare Library / CC BY SA 4.0).

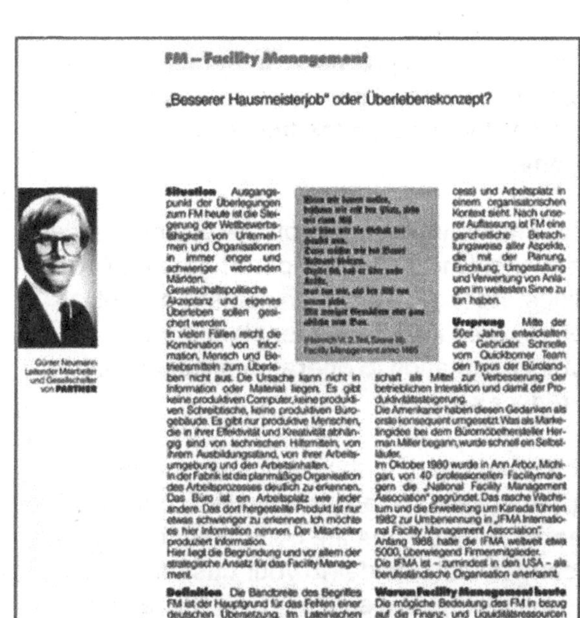

FM – Facility Management

„Besserer Hausmeisterjob" oder Überlebenskonzept?

Situation Ausgangspunkt der Überlegungen zum FM heute ist die Steigerung der Wettbewerbsfähigkeit von Unternehmen und Organisationen in immer enger und schwieriger werdenden Märkten.
Gesellschaftspolitische Akzeptanz und eigenes Überleben sollen gesichert werden.
In vielen Fällen reicht die Kombination von Information, Mensch und Betriebsmitteln zum Überleben nicht aus. Die Ursache kann nicht in Information oder Material liegen. Es gibt keine produktiven Computer, keine produktiven Schreibtische, keine produktiven Bürogebäude. Es gibt nur produktive Menschen, die in ihrer Effektivität und Kreativität abhängig sind von technischen Hilfsmitteln, von ihrem Ausbildungsstand, von ihrer Arbeitsumgebung und den Arbeitsinhalten.
In der Fabrik ist die planmäßige Organisation des Arbeitsprozesses deutlich zu erkennen. Das Büro ist ein Arbeitsplatz wie jeder andere. Das dort hergestellte Produkt ist nur etwas schwieriger zu erkennen. Ich möchte es hier Information nennen. Der Mitarbeiter produziert Information.
Hier liegt die Begründung und vor allem der strategische Ansatz für das Facility Management.

Definition Die Bandbreite des Begriffes FM ist der Hauptgrund für das Fehlen einer deutschen Übersetzung. Im Lateinischen (facilitas/-atis) bedeutet er Leichtigkeit, Umgänglichkeit, Freundlichkeit, Gewandtheit. Im allgemeinen Wirtschaftsenglisch werden darunter Betriebsanlagen, Einrichtungen verstanden.
Um Mißverständnisse zu vermeiden, sollten wir den Anglizismus auch in unserem Sprachgebrauch akzeptieren.
Die „International Facility Management Association IFMA" definiert FM als einen integrativen Management-Prozeß, der Mitarbeiter (people), betriebliche Entwicklung (process) und Arbeitsplatz in einem organisatorischen Kontext sieht. Nach unserer Auffassung ist FM eine ganzheitliche Betrachtungsweise aller Aspekte, die mit der Planung, Errichtung, Umgestaltung und Verwertung von Anlagen im weitesten Sinne zu tun haben.

Ursprung Mitte der 50er Jahre entwickelten die Gebrüder Schnelle vom Quickborner Team den Typus der Bürolandschaft als Mittel zur Verbesserung der betrieblichen Interaktion und damit der Produktivitätssteigerung.
Die Amerikaner haben diesen Gedanken als erste konsequent umgesetzt. Was als Marketingidee bei dem Büromöbelhersteller Herman Miller begann, wurde schnell ein Selbstläufer.
Im Oktober 1980 wurde in Ann Arbor, Michigan, von 40 professionellen Facilitymanagern die „National Facility Management Association" gegründet. Das rasche Wachstum und die Erweiterung um Kanada führten 1982 zur Umbenennung in „IFMA International Facility Management Association".
Anfang 1988 hatte die IFMA weltweit etwa 5000, überwiegend Firmenmitglieder.
Die IFMA ist – zumindest in den USA – als berufsständische Organisation anerkannt.

Warum Facility Management heute Die mögliche Bedeutung des FM in bezug auf die Finanz- und Liquiditätsressourcen eines Unternehmens wird oft unterschätzt. In den USA sind z.B.
- In der Jahresrechnung aller Unternehmen 25–50% des bilanzierten Anlagevermögens und 10–18% der Erfolgsrechnung grundstücks- und gebäudebezogen,
- 20% aller Unternehmen und Organisationen ohne Grundstücks- und Gebäudeinventar,
- 67% aller Unternehmen und Organisationen ohne Informationssystem, das ihnen aktuelle Daten zum Grundstücks- und

The starting point for FM considerations is to increase the competitiveness of companies. Socio-political acceptance and own survival should be secured. The aim is to burden the core business as little as possible with overhead costs. This is achieved by:

- Reducing facility-related costs
- Increasing the usability of plant and equipment
- Separating uneconomic assets in a timely manner
- Developing a safe, humane and functional work environment
- Establishing optimal workplaces to increase quality and productivity
- Determining and maintaining current data on buildings.
- Demonstrating the level of utilization
- Demonstrating in-house competence throughout the life cycle of land, buildings and facilities
- Developing and evaluating deployment alternatives.

Figure 2 Facility management from PARTNER's point of view – extract from PART 1, 1988. (*Source:* G. Neumann).

to gather experience for the establishment of an association and the implementation of events, as well as to build up an international network. The goal was to position the planned German FM association under IFMA's umbrella.

The Symposium

The mi-Verlag in Landsberg was won as co-organizer for the congress. On September 28–29, 1989, the first Facility Management Symposium in the German-speaking region took place in Frankfurt/Main. It was well attended with about 160 participants. The driving force for attending this congress was curiosity. Curiosity, what was hidden behind the term FM. The topics – analogous to PARTNER's service portfolio – were deliberately chosen with a focus on planning. We were familiar with these topics. In our projects, we repeatedly experienced a lack of willingness to link technical and business aspects in such a way that a clear added value for the company could be proven (1 + 1 = 3). With this principle, two single elements with a "lower value" such as a "dumb" CAD plan and an alphanumeric database become a powerful CAFM tool with a clear added value compared to the two input elements through an intelligent linkage. This kind of linkage has been expressed in various speeches and presentations.

To make the seriousness of our commitment clear, Donald A. Young and Eric C. Lund and their wives were invited to represent IFMA in Houston. Their presentations outlined the international aspects of FM and showed what opportunities German companies have through the implementing of FM structures. Examples were given of American companies with German subsidiaries where the American company uses FM but the German subsidiaries do not. A wasted potential (see Figure 3).

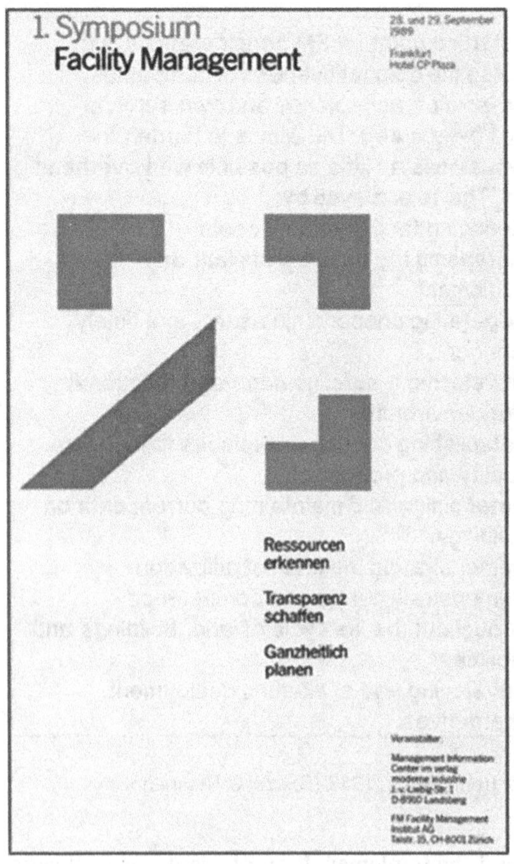

The aim of the symposium is to sensitize you to a topic which, in my opinion, is an important instrument for securing the future of your company.

On the one hand, because on the business side it sustainably increases the performance of your buildings, its facilities and equipment as well as the employees through better planning and constant adaptation to operational needs and thus reduces costs.

On the other hand, in the technical, creative area, the creation of an independent, coherent appearance visualizes the competence of your company to the outside (to your customers) and to the inside (to your employees). This aspect should not be underestimated, because in performing markets it is increasingly true that success = performance + image.

Take the chance to use FM as a strategic tool for a cost-efficient performance. Become a member of GEFMA.

Figure 3 Proceedings of the 1st FM Symposium – cover sheet and excerpt. (*Source:* FM Institute).

At the end of the symposium, there was a call for the foundation of a German FM association with the name gefma (German Facility Management Association). The gefma should and be based on the statutes of the IFMA and act as its German offshoot.

After the symposium, the conditions for the establishment of the individual companies and associations were quickly created. This included the drafting of contracts and statutes.

Foundation of FMC Facility Management Consulting GmbH

FMC Facility Management Consulting GmbH was the 1st FM consulting company in the German-speaking area. PARTNER GmbH and the Mack, Moslener, Gerhards group held equal shares in the company. Walther Moslener, who in the meantime had received his doctorate, became managing director. The first employee was the graduate architect, Brian Pendry. The first client was Aachen Münchner Insurance. FMC's service portfolio was similar to that of a facility planner. It was extremely difficult to compete with the large planning companies with their engineering services, which they were now selling as FM services, and with the corporate consultants who were trying to integrate FM aspects into their services. I sold my shares in FMC to the other shareholders in 2001 in order to concentrate fully on my activities in CREIS GmbH. This was followed by further changes of shareholders in FMC, which led to a service portfolio that included FM services accompanying planning as well as the analysis of existing FM structures and the optimization of existing service contracts. FMC is currently managed by Paul Stadlöder.

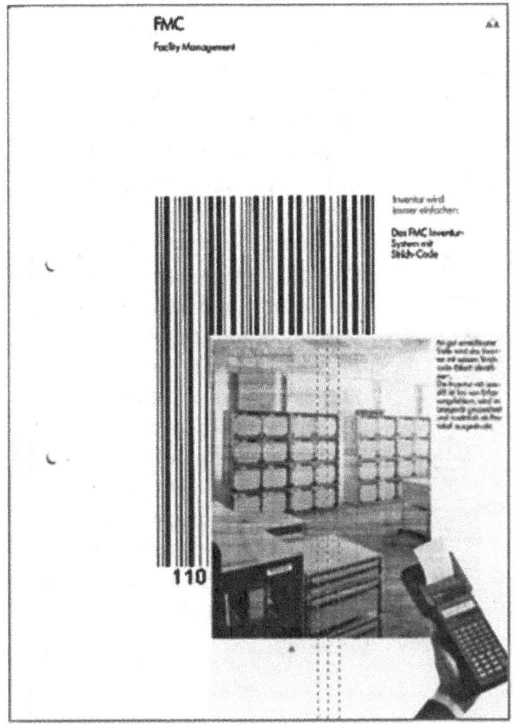

We are in a permanent state of radical change. The working world must recognize these dynamic factors. Faster adaptation of offices and their facilities to operational requirements, for example, requires new fixed asset management and more flexible logistics.

As a first step towards this, the previous invent-tory number is being replaced by a barcode label. It is EDP-capable, is read and enables the disposition of the inventory, in asset accounting and in room and occupancy planning. In a clearly visible place, the inventory is identified by its bar code.

The inventory with reading pen is free of recor-ding errors, is stored in the reader and then additionally printed out as a protocol. In the case of new acquisitions, it is the supplier's respon-sibility to place the barcode on the inventory as specified by the customer.

Figure 4 FMC inventory system with barcode. (*Source:* G. Neumann).

Foundation of the Facility Management Institute (FMI) in Zurich

With the founding of the FMI in Zurich under the scientific direction of Dr. Josef Mack, the ground for facility management in German-speaking countries was to be scientifically prepared. I was responsible for the management and Anton Wicki acted as Verwaltungsrat (in Switzerland for Supervisory Board). This was to lay the foundations for successful business models and products. The preparation and realization of the symposium was the first action. This was followed by a trade fair participation at the Hanover Industrial Fair and the publication of technical books. The follow-up to the activities showed that we were too early with our idea. The market was not yet ready for our holistic FM approach, and we were economically too weak to make and occupy the market (see Figure 4).

Foundation of the German Facility Management Association (gefma) in Stuttgart

The foundation of the gefma as the German offshoot of the International Facility Management Association (IFMA) took place on December 15, 1989 in Stuttgart. The founding members came from PARTNER and PARTNER contacts. They represented consulting companies, engineering firms and CAD manufacturers. The first board was formed by Dr. Wilfried Clauß of Drees & Sommer, Knud Hartung of Arge Immo, and Dr. Walther Moslener of FMC. In retrospect, this was already the first weak point, because there was an FM user neither among the founding members nor on the board. So, it was not surprising that the share of consulting and service companies in the membership grew disproportionately, because they expected a clear competitive advantage. Corporates as users tended to be the exception.

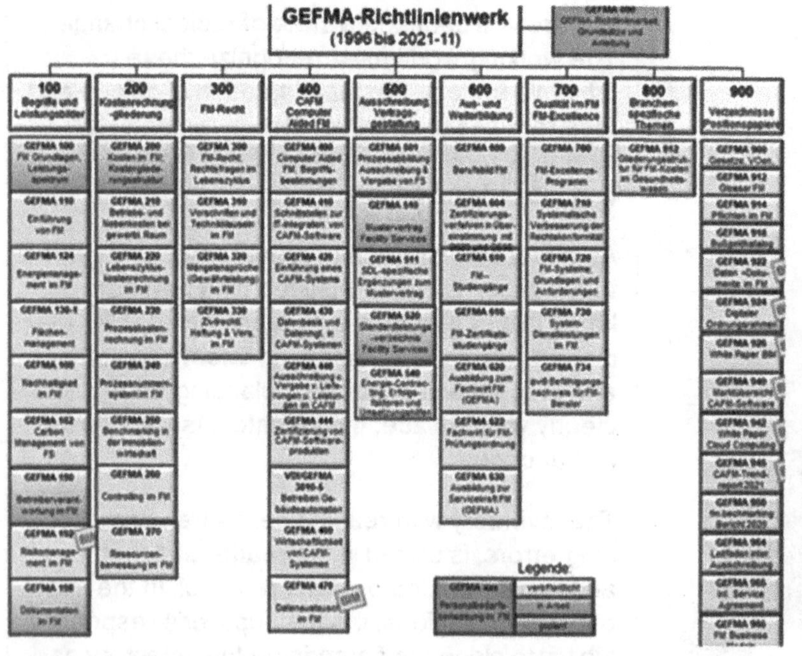

The Federal Association for Facility Management is involved in standardization work for facility management and provides working aids and technical support for all market participants. The guidelines developed are intended to provide legal certainty beyond the German market. Quality standards with certification have developed from some guidelines. For IT applications in FM (CAFM Computer Aided Facility Management), the German Facility Management Association provides working materials. A separate certification system certifies a minimum standard of software products.

Figure 5 Gefma guidelines – overview. (*Source:* gefma/Springer Nature).

The following boards of directors did not change anything either in order to gain more FM users as members and thus to follow the IFMA statutes. This led to a break with IFMA. gefma became a largely FM provider association. At this point, Walther Moslener and I left gefma and, with IFMA's blessing, founded IFMA Germany in Munich on December 12, 1996, which merged with Real FM in November 2006 as the professional association of real estate and facility managers. In my opinion, this was the first time that the two – also internationally – competing professionals (CoreNet Global and IFMA member) were organized in one association.

My leaving gefma did not affect its work on guidelines, which have a very good reputation in the market. This involved my work on guidelines that fitted in with my core CREIS business, such as:

- Operating and ancillary costs for commercial space
- Space management – basics
- Benchmarking in the real estate industry – basics, procedure, application

Many companies in German-speaking countries now follow these guidelines in defining, performing, billing, and evaluating FM services (see Figure 5).

At the same time, the FM market began to differentiate itself. Strategic FM was offered by specialized management consultancies. Otherwise, the German FM market was more operationally structured. In addition to holistic facility management, terms such as "commercial" and "technical property management" and "facility services" suddenly appeared. It took some time for these terms to differentiate themselves from one another.

Foundation of CREIS Real Estate Solutions

This differentiation was the motivation for me to found CREIS Real Estate Solutions in 1995 as a consulting and software company with the goal of developing and implementing standards for performance measurement of real estate and real estate-related services in asset, corporate real estate, property, and facility management.

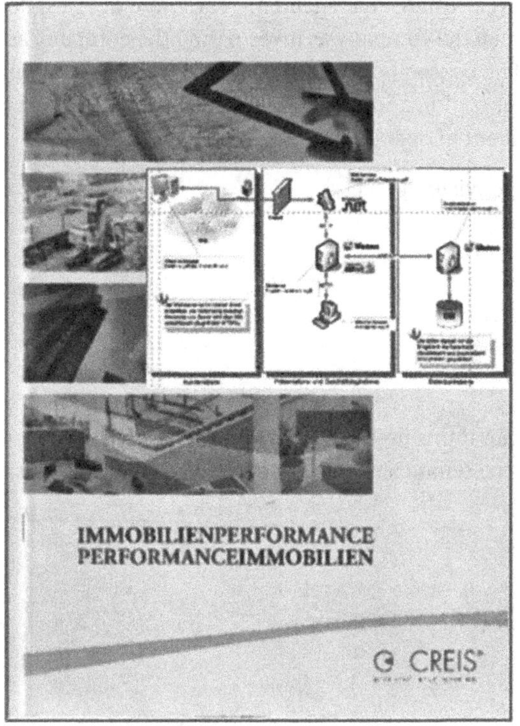

Our basic principles guarantee comprehensive performance measurement with robust market data. To effectively support your real estate business goals, our benchmarking is based on eight principles:

- **Anonymity and confidentiality**
- **Data validity and data stability**
- **Holism**
- **Viewpoint control**
- **Orientation to standards**
- **Scientificity**
- **Hierarchical structure**
- **Added value and handling**

The actual goal of benchmarking is to identify best practice. Key figures form the means of transport for this. For the extensive standard evaluations, the key figures were combined into corresponding groups in a consistent key figure system.

Figure 6 CREIS benchmarking system for evaluating FM performance. (*Source:* G. Neumann).

Based on these standards CREIS supports companies in improving their business performance. Hierarchically structured, value-oriented management indicators form the basis for effective controlling. The benchmarking method developed by CREIS has meanwhile become the standard for measuring the performance of real estate and services (see Figure 6). The following are evaluated:

- cost performance
- building economy
- consumption
- productivity
- sustainability

For corporates, investments in real estate generally represent secondary investments, which differ from market-oriented primary investments in the core business in that they only contribute indirectly to value creation. So far, yield considerations have hardly been taken into account. Nevertheless, real estate can be a sensible investment for companies for overriding strategic objectives if:

- know-how is to be secured through ownership of strategically important real estate (e.g., production and laboratory buildings),
- no real estate of the required type and quality is available on the market at the required time, or
- "resource potential" can be created with the investment in real estate and thus the market value of the equity can be increased.

However, it is essential for the market value-oriented view that companies do not see their real estate exclusively as an operating asset but complement its role as an asset with an investor-oriented view. However, these different perspectives must be manifested both in the organization of real estate management (corporate real estate management, facility management) and in the relationship between the company as investor and the company as user, taking into account corporate governance requirements. The consequences of inadequate implementation are often conflicts of interest that led to redundancies in the organization and thus to inefficient handling of corporate real estate. Performance here defines the degree of fulfillment of operational requirements in terms of:

- Usefulness (real estate as an operating resource in the service provision process)
- Flexibility (ensuring the flexibility of the core business)
- Cost performance (burdening the company with real estate costs)
- Space efficiency (space efficiency, space consumption)
- Sustainable corporate governance (Corporate Governance Code)
- Ecological sustainability (sustainable management)
- Social sustainability (employee satisfaction, motivation)

For real estate companies, real estate represents a pure asset. Even if this approach implies a one-sided focus on maximizing returns and increasing value, the property remains the tenant's means of operation. The more satisfied the tenant, the more sustainable the return for the investor.

Günter Neumann, MSc.
Managing Director,
Günter Neumann CREIS Consult
Bad Doberan, Germany

Work Experience

After activities in planning and realization of office buildings, the establishment of FM structures in the German-speaking countries followed from 1987, such as the foundation of the FM Institute in Zurich, the GEFMA, and the FMC in Stuttgart.

In 1995, the Corporate Real Estate Information System (CREIS) was founded with a focus on real estate and FM controlling and the establishment of the Real Estate Monitor (REMO) for performance measurement.

Main areas of experience:

- Development of a real estate strategy, definition and Balanced Scorecard (BSC), establishment of KPI systems.
- Establishment of value oriented CREM / FM structures (e.g., shared service centers, value-added partnerships / Supply Chain Management SCM)
- Establishment of real estate and service controlling (e.g., process cost accounting, cost-performance accounting, supplier audits)
- Implementation of benchmarking projects (strategy, organization, portfolio, services/SLA, costs, energy efficiency, productivity)

Education

- Interior design and carpenter's apprenticeship
- Master school in interior design and business administration,
- Certified Facility Manager (IFMA),
- Master studies at the EIPOS at the TU Dresden (MSc Real Estate).

Teaching:

- DIA at the University of Freiburg (2014–2018)
- HTW University of Applied Sciences, Berlin (2012–2021)

Publications (in German - excerpt):

- Space Management, Zeitner/ Marchionini/ Neumann, Springer Verlag, Berlin 2019
- Real Estate Performance Creates Performance Real Estate in Facility Management, VDE Publisher, Berlin 2012
- Guide to successful real estate benchmarking, Wolters Kluwer, Cologne 2009
- Real Estate Benchmarking in the Context of Increasing Returns in Real Estate Benchmarking, Reisbeck/ Schöne, Springer Publisher, Berlin 2009
- Supply Chain Management for the Real Estate Industry, Cross-company Management of Value Chains, VDM Publisher, 2009
- Real Estate and Service Controlling in Facility Management, Nävy, Springer Publisher, Berlin 2006

Contact:

Günter Neumann I gneumann@creis.net I www.creis.net

1.8

FM Down Under

Duncan Waddell

Evolution of Facility Management Practice

Through undertaking some exploration and research on this new area of business known as FM at an international level in the early 1990s, I became aware of an architectural practice in the UK known as DEGW that was thinking outside the "norm."

In 1982, Frank Duffy and John Worthington (two of the DEGW founders) established the journal *Facilities* to promote this, as yet unknown, FM profession in Britain, arguing that "facilities management was to architecture as software was to hardware in the world of the computer – an essential means of ensuring that clients" intentions, expressed in building briefs, should be reviewed, monitored, and updated throughout the entire lifetime of each building.

I thought that this was pretty radical thinking! It was an approach that was more than different contracting methodologies and responsibility structures. It wasn't just about FM outsourcing which was becoming a popular management pastime.

DEGW set up a consulting arm that delivered business management consulting and space management (design). The practice also consulted on building management, more widely known as facilities management (FM). ORBIT (a study delivered by DEGW) had claimed that as buildings were increasingly complex, proper FM was essential to ensure proper use of resources.

This captured my attention! There was no one that I could find talking about this notion at all here in Australia.

Another major influence in the growth of FM practice was through the USA, with two major office furniture manufacturers supporting the establishment of IFMA in the early 1980s with the focus on effective and efficient utilization of space using furniture (and fit outs) as a means of delivering better work environments. As the FM world grew, so did its impact across all built environments, capturing the sense of people, place, and process.

On reflection of this discovery, it's interesting to see how the concept and description of FM has evolved. We have all heard of the "lift pitch" or the "BBQ" discussion about "what do you do for a job"?

The early pitch I used to enable people to understand FM in simple terms was via a model developed by Tony Thompson (DEGW London) to speak of the management and delivery of the "soft and hard" management and operational services/products that an organization requires to support its core business purpose. This can be further categorized into a further five areas of management activities and responsibilities (see Table 1 below).

Facilities @ Management, Concept - Realization - Vision, A Global Perspective, First Edition. Edited by Edmond Rondeau and Michaela Hellerforth.

Table 1 This table is further detailed at source from papers by Thompson. It was further enhanced by Martin Pickard (FM Guru, UK) in his depiction of the FM functional responsibilities as shown via a MindMap in Feb 2014 entitled "What FMs Do."

FM explained in simple terms	*FM explained in 2023 terminology*
1 Corp real estate, building & construction	Built environment, building and const.
2 Strategic facilities planning	Strategic planning
3 General office and plant/asset services	General office & plant/asset services
4 Operations & maintenance	Operations & maintenance
5 Information management	Data, digitization, tech & analytics
Source: Thompson DEGW circa 1990s	*Source: Transcribed by Waddell 2010*

(*Source:* Adapted from https://www.biggerplate.com/uploads/pdf/HrT0FCcX_What-FM-s-Do-Feb-14.pdf)

With minor adaptations, the Thompson model can still be used to explain FM practice today with the initial concept of FM very much the same as envisaged by Thompson. The main difference is that the practice elements have become far more sophisticated and complex, with greater responsibility and accountability for the FMer.

Today we also embrace and practice (I hope) FM in terms such as sustainability, compliance, risk, and performance. These terms and other expectations of the FM function are overlayed with societal and organizational concerns. Drivers on issues and commitment to international government-endorsed programs such as the United Nations Sustainable Development Goals (UNSDGs) and the London Declaration with other climate initiatives also play a significant need for FM attention.

The Start of My FM Journey

In May 1986, a time indelibly stamped in my life as to when, like so many others, I "fell" into the field of Facilities Management in Australia. Not that I knew it at the time. I was completely oblivious to this emerging area of business practice!

Having finally completed my Bachelor of Business Studies in the 1970s, my early career had been involved in working with three industrial manufacturing organizations and I was looking for a change. I had applied for several roles in different sectors and accepted a role working with one of Australia's largest construction, development, and investment groups. Publicly listed and significantly profitable and successful, a new division had been created to focus on interior fit out works and building refurbishments.

Early leanings demonstrated that these two areas of activity were driven by different factors. Firstly, in office fit outs the need for effective workplace solutions was becoming more pressured in attempts to enable and/or drive higher productivity with significantly improved space utilization at the same time achieving those aims in the most cost-effective manner. On the other hand, building refurbishment was all about repositioning the asset in quality of building "grade," enabling an increase of rent, delivering a higher return on investment.

The early success of our division was testament to low vacancy rates, which prompted portfolio owners to look at reinvesting in existing building stock for better returns and with the lack of new stock lagging behind demand, reworking existing floor space to achieve better workplace solutions for the major corporates in sectors such as banking, insurance, legal, and accounting sectors and federal and state governments. The economy was strong, unemployment was low, until October 1987, known as Black Monday (or, in Australia, Black Tuesday), showed that things were a little too good to be true.

The outcome of Black Tuesday took a little longer to have an impact on the Australian market, albeit many lost significant values (decimation) in share portfolios immediately. It always takes some time for these events to have a direct impact in the construction sector as contracts have "terms" to completion, materials ordered and precommitments for space allocated and paid "up front" in various forms of deposit or bank guarantee. This impacted many of our clients as all large space users sought to downsize. In one case our building division was involved in construction of the office tower that the Melbourne Stock exchange was to be relocated. The heady days were over and there was significant market fallout in the economy and specifically the property sector.

During this early time there was a gentleman by the name of Malcom Campbell who, while working for an architectural practice, undertook an international research tour to see what innovation was occurring in the northern hemisphere in the property market. He returned with excitement about learning of the practice of Facilities Management (FM), especially from his time in the USA.

Campbell, being an entrepreneurial chap, set about establishing a professional Association to focus on connecting the potential market of FM and the opportunities associated with securing new work through strategic space planning for the architectural company practice. Given his role as Marketing Director for his company he sought to bring others "into the fold" and looked to Melbourne and Sydney being the two initial launch platforms for the Association. The practice was predominately focused on much improved thinking for space utilization (e.g.: blocking and stacking) via the use of the relatively new IT tool known as Computer Aided Facility Management (CAFM). The Australian FM association was formed and was predominately driven by suppliers to the field (furniture, carpet and other office fit out service suppliers). In Sydney the early participants were predominately Government Departments and in Melbourne included oil companies, insurance and banking organizations, notably large office space users

By early 1988, due to the stock market crash, every major corporate was looking to reduce office space, quit leases and improve utilization. The division I worked with was very busy reconfiguring space to reduce space demand and cut costs. The early role of FM in Australia had emerged through an economic crisis, with the coordination of this reconfiguration falling to a role not quite as yet known as the Facility Manager (FMer).

FM Arrives Formally in Australia

The Facility Management Association of Australia (FMA) acknowledges its establishment and continued activities since the year of 1988.

In 1988, the first one day FM Association Conference was held in Sydney and the keynote speaker was the current Chairman of IFMA, none other than Mr. Ed Rondeau.

It's interesting to note several facts about what the business environment was at the time of its foundation. The birth of the FMA came at a time and was driven by the problems that corporate Australia and Governments were grappling with, namely operating costs and better space management given the significant reduction in workforce numbers. The outcomes that were expected needed to be delivered rapidly. FM was in its infancy and was still an untested, unstructured practice. In most cases it was not understood what the benefits of a centralized coordinated approach to all office support activities were.

The second fact supporting the development of FM as a practice was driven by suppliers to the property and commercial office markets. In FM they saw an innovative way to market their products and services to the corporates and Governments.

Australia had a property market that was desperate for positive outcomes and with this new practice of FM being driven by the supply chain as the solution. It was seen as the best business "option" for the times. Unfortunately, the market often ended up with imported supply/contract solutions (many from overseas) that didn't necessarily suit the Australian marketplace. The voice of "FM Strategy" had not yet been developed by the organizations requiring the planning and delivery (now known as demand organizations).

In 1991, I left the division as the market had slowed considerably by 1990. The market would rebound some 20 months later, however the need for strategic space management and planning had continued throughout this period and continues today.

I undertook an international tour to learn more about this field of FM. I had decided that this management practice held a fascination for me as a business solution and that I was readily able to understand the concepts and relate them to what organizations were seeking FM to deliver in a more professional manner. I did a two-month tour, taking in Europe, UK (where I met DEGW), and the USA to see what these markets were doing in FM. My approach, and the research I was doing, was warmly greeted in all parts of the world I travelled through and everyone I met was welcoming and many from that trip became lifelong friends and colleagues.

In 1992 I established Australia's first FM consulting group. A struggle at first, in a management practice no one really understood, I eventually broke into the major corporates and achieved some very tangible, significant, and rewarding consulting appointments. I developed a great team working with me, who came from different backgrounds covering FM operations, architecture, space planning and management, and CAFM. Part of the secret was that successful FM requires a breadth of skills and is not directly associated with the practices of architecture, engineering or trade skills and our team covered many of those skillsets.

In 1993 I felt that we needed to educate the market as to the positive outcomes of FM management practice. I wasn't able to afford significant advertising funds, so I published Australia's first magazine dedicated to FM. I had no intention of being a publisher, however the vehicle enabled me to get a message to market. The magazine continues to be published today via another publisher.

During the 1990s Australia experienced a surge in FM "outsourcing" service contracting that matured during this decade, evolving from single trade or service delivery, through to the total integrated FM service provision which became progressively more adopted in the early 2000s.

This also became more sophisticated as contracts moved to take responsibility for transferring of employees, acceptance of financial risk with plant and equipment. The next level of development leads to the public/private partnerships (PPP's) and build, own, operate, and transfer (BOOT) models or other similar iterations. Predominately applied to public infrastructure (police stations, major rail stations, hospitals, sporting arenas, public housing, etc.) these styles of government contracting/financing methods recognized the integration of all FM functions with the overall obligations of the contracted entity/entities.

As FM management practice matured in Australia, it was important to continue my connections with the wider world of FM with annual trips to Europe, UK, and the USA. I had the pleasure of being invited to speak at many conferences and conducted training courses across Australia but also around the world.

These experiences enabled me to see the FM world with an eye as to what was different from the Australian market, what was emerging and being offered to the FM practitioner. It also enabled me to understand the different challenges and demands being made of the FMer of yesteryear versus today as well.

When I recall my visits to exhibitions in the USA across the years, it was very evident as to what the sector saw as being important lines of discussion with the FMer. An example is the transition from a heavy focus in the early 1990s on office desking, seating, floor coverings and fit out contracting slowly developing to the emergence of major integrated service providers in the 2010s and now with a solid focus in the 2020s on areas as sustainability, digital, data and technology solutions being offered today. There are still some common threads in areas, however I find that in retrospect the evidence of what was important in "the day" has significantly matured to a different focus in line with the professional expectations of the FMer of today. The floor area and mix of exhibitors at a FM exhibition is completely different today as to what was on offer in the 1980s

This "shift and lift" in knowledge required of the FMer has also evolved over the years being demonstrated by the topics being presented at conferences and training courses being conducted via FM Associations and Institutes as the needs of the market have evolved. An example of this is the area of Occupational Health and Safety. We all appreciate the need for this very sound focus; however, the topic is now much broader and deeper in content and has shifted significantly since the 1980s through to the 2020s. It is now very focused on the Duty of Care toward a wide variety of stakeholders within all organizations as well as the many areas of compliance practice, evidence and reporting required by statutory authorities and legislation, predominately as a FMer responsibility.

So, the views espoused by the research conducted by Duffy and Worthington are still on point for the practice of FM albeit much better understood as a knowledge-based profession that can claim its relevance and context in the complex world of the built environment.

Professional Management and Practice Standards in FM

In any profession today, it is recognized that the established "codes of practice" are clearly articulated and agreed to by all for those involved in that profession. This enables performance management and measurement to be identified, quality to be achieved with consistency and also deem what is considered best practice to be demonstrated or evidenced.

In 2012, the International Organization for Standardization (ISO) established a Technical Committee, ISO/TC 267, that is specifically dedicated to the development of standards and technical reports for the professional practice of Facility Management. Commencing under the guidance of the inaugural Chair, Stan Mitchell, some 52 nations from across the globe and their nominated experts in FM have developed the ISO 41000 series of standards and reports to enable and assist organizations in the practice of the management of facilities.

FM, as a professional practice, now has a sound base and methodology to underpin our field of endeavor, knowing that there is a consistent global understanding and agreement as to the expectations of the practice. These will also assists to act as a foundation for learning and development programs, career development, evaluation as to FM best practice and enable future professional directions to be explored with greater surety for all practitioners and industry on how to best support the structure, management and delivery of FM and associated services and products.

TC 267 looks to its current and future success and output being for the ongoing quality of the profession and encourages active participation in the review of existing and development of new standards.

FM People, Associations, and Institutes

One of the most enjoyable outcomes of my involvement in FM has been through the broad group of people I have had the pleasure to meet during my FM journey! The friendships that emerged over time have been full of sharing, learnings, laughs, and experiences.

I have held a number of Memberships with professional associations, the FMA being a significant part of my business life. As a founding Member and moving through to the honor of Chairing FMA Australia has been a wonderful experience to see FM being developed and recognized as a value driver in the built environment

I have also had the pleasure of representing Australia on the Board of Global FM, the federation of FM Associations from around the world. I was appointed to the independent role of Chair and now serve as Immediate Past under the Chair of John Carrillo. Global FM has provided a basis for both mature and emerging FM Associations to be connected in the common aim of growing the recognition of FM amongst individual members, as well as providing support and tools for all Members, such as the organization of World FM Day and the Global FM Market Sizing Study.

As Chair of ISO/TC 267, the invaluable experiences of current relationships and of those yet to come is a privilege and an honor to work with so many talented colleagues. The importance of the work the TC experts provide is invaluable to our profession and the profession should be ever grateful for their contribution past, current and in the future. It is the rock on which this profession will grow and be sustained globally.

There is no doubt that having had the opportunity of connecting and contributing through these and many other organizations has enhanced my professional and personal experiences. Thank you!

Duncan Waddell, B. Bus
Managing Director
FM Intelligence Pty Ltd
Melbourne, Australia

Duncan's background is in industrial manufacturing, property, refurbishment and repositioning of asset value and, over the past 38 years, ultimately in Facilities Management (FM). He developed a passion for FM due to the value drivers it can bring to organizations of all sizes and sectors and that it represents a coordination of common sense, business logic and innovation applied to the integration of people and the built environment. He started his own business in 1992 known as FM Intelligence Pty Ltd. It was the first strategy group in Australia to purely focus on FM. The business consults, operates, and provides learning and development globally across the world from its office in Australia.

He has been involved in promoting a greater awareness and supporting the professionalization of the FM as a strategic, tactical, and operational management discipline which currently includes the following active professional roles:

- Managing Director – FM Intelligence Pty Ltd
- Chair International Standards Organisation (ISO) Facility Management Committee ISO/TC 267
- Founding Member, Life Member and past Chair of the Facility Management Association of Australia
- Immediate Past Chair of Global FM (An international federation of Facility Management Associations)
- Member – Standards Australia – Facilities Management Committee MB-022
- Fellow of the Australian Institute of Company Directors
- Past Chair of the Corporate Real Estate Committee, Property Council of Australia
- Founding Publisher of the Australian facility management profession independent magazine. "FM Magazine"

He travels extensively worldwide speaking at seminars and conferences in addition to delivering consulting and training support services, highlighting the potential and demonstrating, through practical experience, the benefits of strategic FM practice across the built environment.

2

Contributors

Realization

The ***Realization*** describes the development of the concept in academia and the work environment, and the transformation of the professional association into a transnational activity, claimed in 1982 by the name change to International Facility Management Association (IFMA). The educational and professional activities, the founding of FM associations in other countries, their realization, and activities are characterized. Development of technology, scientific placement, academic classification, implementation of the profession in different areas and companies internationally are represented.

The following contributors provided their anthology and legacy on the realization and development of facility management:

2.1 Edmond Rondeau
2.2 Karin Albert
2.3 Yu Qingxin
2.4 Philip Lo
2.5 Steven Ee
2.6 Thomas Madritsch
2.7 Stormy Friday
2.8 Phyllis Meng
2.9 Barry Lynch
2.10 Kathy O. Roper
2.11 John Carillo
2.12 John Gileard
2.13 Jochen Abel
2.14 Nancy J. Sanquist
2.15 Joint Contributors:
Eunhwa Yang
Jun Ha Kim, and
Mathew Tucker
2.16 Doug Aldrich
2.17 Antie Junghans
2.18 Joint Contribution:
Eduardo Becerril and
Norma Pleitez

Facilities @ Management, Concept - Realization - Vision, A Global Perspective, First Edition. Edited by Edmond Rondeau and Michaela Hellerforth.

2.1

My Facility Management Journey

Lessons Learned

Edmond P. Rondeau

My work career began after I graduated from the Georgia Institute of Technology (Georgia Tech) in Atlanta, Georgia USA with a Bachelor of Architecture degree. After spending almost two years in the US Army Corps of Engineers as a Lieutenant with tours at Fort Benning, Georgia, and Cam Rahm Bay, South Vietnam, I returned to the US. I then worked for a number of architectural firms and when I passed my Georgia Architectural exams, became a registered Architect. In the late 1970s I went to work as a Staff Architect in the Office of the Campus Planner and University Architect at Auburn University, Auburn, Alabama USA.

Auburn University

At that time Auburn University had over 20,000 students with over 300 buildings on an over 1,000+ acre (404+ hectare) campus. As a Staff Architect I worked on a number of design and interior/exterior construction capital projects including new buildings, renovations and remodels of research labs, classrooms, dean offices, instructor offices, conference rooms, student housing, food service, IT, athletic facilities, veterinary operating theaters, library renovations, an air structure, football stadium addition, new Physical Plant facilities, a new School of Nursing, a new Electrical Engineering building, etc. This included working with university administration, finance, and accounting, purchasing, physical plant, deans, staff, professors, researchers, students, outside architects and engineers, contractors, vendors, City of Auburn police and fire department officials, and the State of Alabama Building Commission.

For capital projects, our Office was responsible for managing programming and design requirements for our campus customer with an Alabama based architectural and/or engineering firms selected by the Governor's Office (a political process which is not used now). There already was a Campus Review Committee with senior campus leadership including my boss and a Director of the Physical Plant who reviewed capital project budgets, schedules, and priority for the University President's review and approval. About the time I was hired a new Physical Plant Director was hired, he quickly saw that a number of Physical Plant maintenance and repair projects that were not on the Campus Planning Committee Agenda were being performed by his Operations & Maintenance (O&M) department such as painting, floor repair, wall repair, roof repair, mechanical repair, plumbing repair, electrical repair, paving repair, landscaping, etc. and that this work at times were demolished or removed by a capital project.

The Physical Plant Director met with my boss, the University Architect, to set up regular meetings to review with major staff of both departments the status of all large and most small campus projects and most O&M repair, replacement, and maintenance projects. These meetings were a success with sessions of positive communication,

Facilities @ Management, Concept - Realization - Vision, A Global Perspective, First Edition. Edited by Edmond Rondeau and Michaela Hellerforth.

savings to University expense funds and schedules, and the re-directing and planning of O&M staff maintenance and work order assignments in coordination with the University Architect Office projects.

Another agreement between the two departments was our major re-write and revision to the "Instructions to Architect and Engineers" as published by the Office of the University Architect and Campus Planner. The history of the design of campus capital projects at Auburn University by outside Architects/Engineers had been more design focused with little apparent focus on the overall campus design, maintenance and operations or later flexibility needs. With over 300 buildings, the Physical Plant Department had two full time roofing crews working on roof leaks on many different roof types that had been installed over the previous years. Also, the 300 buildings had over 40 different types of lamps/bulbs that needed replacement over time with some lamps/bulbs no longer available, had over 10 different types of door hardware manufacturers and locksets/key systems, and had over 12 different manufacturers of HVAC equipment and filters, some of which had no repair parts available.

Our combined task was to develop revised Instructions which provided Architects and Engineers with specific requirements and guidance in the development of new capital or remodel projects with thought to the overall campus design, the maintenance and operations of these facilities and the repair or replacement of building items. Essentially, we came up with a listing of requirements for the project specifications to help reduce initial and future costs and enable maintenance and repair of major and minor items from manufacturers that were available at the University location. These capital project specifications and bid package items included US manufacturers to be installed for:

- chiller plants and roof package units – Trane, Carrier or York
- cooling towers – Delta, Tower or Evapco
- electrical panels, switches, etc. – Square D, Westinghouse or GE
- elevators/escalators, passenger – Dover, Otis or American
- roof types and roofing material:
  - Flat –
  - Sloped
  - Special
- door hardware and locksets – Schlage, Sargent or Best
- door keying and lockset cores system – Best (only)
- Etc.

With these revised Instructions, we began capital projects with a specific design direction that helped the university, supported capital project design and construction, and supported reduced maintenance and operation costs, saved energy, and the maintenance and operations of these buildings and environments.

These new processes helped where while working on a new School of Nursing. Our budget would not support the renovation of the existing building HVAC system. In one of our project design meetings, a member of the team from the M&O department remembered that the adjacent Physics Department building had an existing cooling and heating system that was designed for an addition to the building that was not going to happen. He suggested that we could use the extra cooling/hot water capacity that could be directed underground via new underground piping to the Nursing School existing basement. This required that we had to re-work the controls of the cooling towers at the Physics building and provide appropriate new equipment in the Nursing School basement. This was preliminary engineered by the architect's engineer and the resulting pricing of this solution showed that this solution was within budget. The School Nursing later opened with a fully operational HVAC system for their cooling and heating requirement. This team solution resulted in a better teaching environment and a better use of university resources and facilities.

And all this was before we had heard of facility management.

<u>Lessons Learned</u>:

- University capital and maintenance funds were provided by the State of Alabama, and our new cooperative strategy helped to reduce planning conflicts which permitted the use of saved maintenance funds for other uses.

- The process provided closer cooperation between two University support departments which built trust and the sharing of information that led to better planning.
- This also led to inviting maintenance and operations personnel to University Architect capital planning meetings for their input into renovations and new building projects.

The Coca-Cola Company

An opportunity came for me to join the Coca-Cola Company in Atlanta, Georgia in 1980 as a Construction Manager in the Atlanta Office Complex Development (AOCD) Department. This department was responsible for the redevelopment of the existing 10+ acre corporate campus site near downtown Atlanta, and across from the Georgia Institute of Technology campus. In early 1980 the new 35-story corporate headquarters building, North Avenue Tower (NAT) was completed with only one unoccupied floor, the adjacent new 7-story Research Lab building was about 85% complete, and I spent my initial time working on the Research Lab building completing labs and office space. These two new buildings plus other existing buildings totaled over 1 million square feet (93,000 square meters). During this time I also received an MBA in Real Estate from Georgia State University in 1982.

One of my other responsibilities was to establish, coordinate, schedule, and manage Moves, Adds, and Changes (MAC) required on the Campus site. The NAT building (approximately 700,000 square feet) had 32 floors of new open plan Knoll systems wood panel furniture and the top 3 executive floors had traditional offices and furniture around a central atrium. Most of this MAC work was requested in the NAT and almost all the MAC work included the relocation/rearrangement of Knoll systems open plan furniture.

As I was the first person responsible to coordinate and manage the MAC process at the NAT building and the site, there was no defined process or history of working among the various separate campus support and service departments for MAC requirements including:

- AOCD Design
- Maintenance and Operations
- Fixtures, Furniture, and Equipment (FF&E – including Knoll furnishings)
- Telephone and Fax (including Bell Southern Bell and later AT & T)
- IT (IBM main frame)
- Word Processing (Wang)
- Mail, Printing, and FedEx, DHL, UPS
- Food Service and Refreshment Centers
- Records Management
- Landscaping
- Cleaning and Recycling
- Security and Shredding
- Safety
- Purchasing
- Finance and Accounting
- Human Resources
- Legal
- Travel
- Shipping and Receiving
- Parking
- Archives
- Executive Floor management
- Etc.

What I found for MAC work was that these departments and services had not been trained to coordinate these MAC requirements in an open plan work environment that the new corporate Tower required. Their previous services had been provided in a fixed office silo work environment where their services most often existed in silos with departments having individual vertical processes (see Figure 1).

Now with the new work concepts in the Tower, it became obvious to me that we needed to be able to coordinate all the MAC requirements to ensure a proper on-time process for our internal customers involved in each MAC. This led to a process of memos, phone calls, and face-to-face meetings (this was before e-mails) with various service departments to coordinate the various requirements of the approved MAC including timing, staging, signage, the move or construction requirements, special requirements including IT, and the post-MAC process and review. This led to an Integrated Vertical Services process for the coordination of all MACs (see Figure 2).

This MAC worked in the NAT and later in the Research Lab Building generally required using Knoll open plan panels and hardware from existing attic stock of Knoll items and the re-use of existing Knoll workstation items. Needless to say, we had no detailed inventory to test how new plan Knoll requirements compared with existing plan Knoll items. We asked Knoll if any of their customers had similar workstations, had issues with identifying attic stock against new planning, and did they have an inventory system of Knoll attic stock items?

From this question, we found out that one of the companies that we could contact was Texas Eastern Transmission Corporation in Houston, Texas. From this contact we also heard from Mr. William (Bill) Back about the National Facility Management Association (NFMA) that had been formed in 1980. I obtained an NFMA membership application and I and others at the Coca-Cola Company joined the NFMA in 1981. During that time technology/database software was not advanced enough to help us with our Knoll attic stock inventory.

I and others in our department also provided for the Columbia Pictures business unit some non-studio project management services in 1982 and 1983 in the Los Angles, California area (later sold to Sony). And we provided construction management services for the Wine Spectrum business unit in the Atlanta, GA area where we relocated this business unit's corporate headquarters twice in three years (also later sold to a privately owned wine organization). IT use at the corporate campus for office use included the development and installation of a central Wang word processing system and center and the beginning of department use of IBM Selectric typewriters with very small review screens. And in 1983 our department purchased a Bentley CAD System where internally MAC design and construction revisions to the NAT and the campus buildings could be made, tracked, and used for future planning.

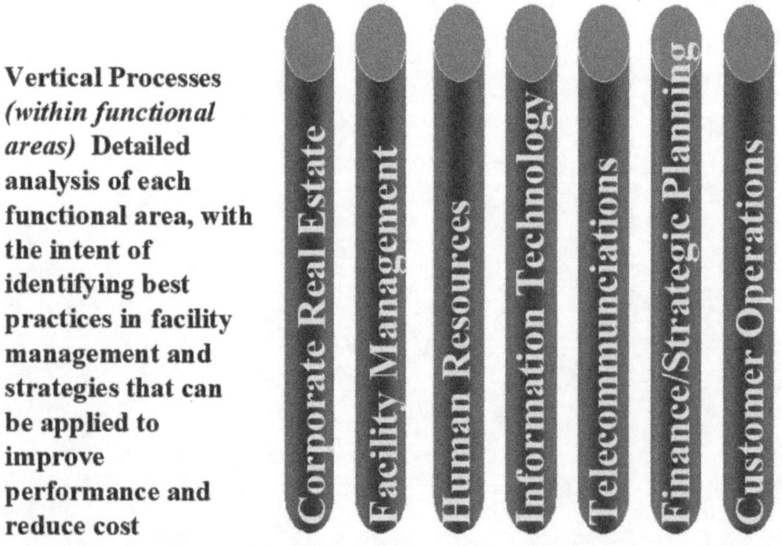

Figure 1 Functional service silos. (*Source:* E. Rondeau).

Horizontal Processes *(between functional areas)* Identification of opportunities for consolidation and/or integration of personnel, processes, materials, etc. between functional areas

Integrating Functions *(combining support functions)* Evaluation of management systems and other processes that apply to all functional areas

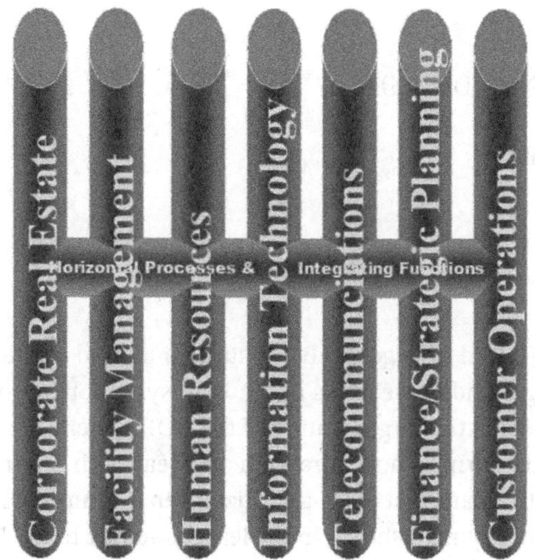

Figure 2 Integrated vertical FM services. (*Source:* E. Rondeau).

In 1983 we made a presentation to senior management on the benefits of organizing the related MAC and site capital project services into a Facility Management organization. But the time was not right, there was too much corporate history to overcome, and senior management was not ready for this change. But in time after I left the company these changes were adopted in part, and today the company has an FM department that uses best FM practices.

And we now knew about Facility Management.

<u>Lessons Learned:</u>

- Managing the new MAC program required networking and developing trust among the many support departments in the new corporate headquarters tower and on the site near downtown Atlanta.
- Developed and managed a yearly MAC budget to fund all MAC projects and to purchase FF&E (Furniture, Furnishings, and Equipment) for the Coca-Cola Headquarters and other campus buildings.
- Worked with many support departments in the re-design and re-installation of Knoll open plan office furniture and furnishings in the over 30 floors of the Headquarters Tower.
- Worked with the FMs in other business units to assist in capital projects including the Wine Spectrum, Columbia pictures, Research and Development, and the USA group.

Georgia Institute of Technology (GT)

Later in my career I went to work in Atlanta, Georgia, for the Real Estate Development Office (REDO) at GT (see www.realestate.gatech.edu) and I served as the first GT Real Estate General Manager (REGM) position in May 2005. This REDO department was established in May 2001, and I reported to the Executive Director who was also a Vice President of Georgia Advanced Technology Ventures, Inc. (GATV), an affiliated organization to Georgia Tech.

As the REGM I wore two hats – one for GT and one for GATV. The GT work included working on GT real estate–related requirements and working closely with the following GT departments and staffs:

- Legal Services
- Financial and Accounting Services
- Institute and Capital Plan Budget Management
- Capital Planning and Space Planning

- Facilities Maintenance
- Facilities Design and Construction
- Facilities Environmental Health and Safety (EH&S)
- GT Fire Marshall
- Office of Information Technology (OIT)
- Parking and Transportation
- Police
- Telephone
- Etc.

This included working with the Georgia Board of Regents (BOR) staff for all real estate requirements and leases. All owned property used by the 31 colleges and universities in the BOR system, by Georgia law, is owned by the BOR for the use of each member higher education organization of the BOR system.

The BOR approves leases with an initial term of not more than one year with a number of options were permitted. This could leave the Landlord with an amount of un-amortized Tenant Improvement (TI) expense and the GT Tenant would be liable for a lump sum payment shown in the lease schedule to the Landlord at move-out. GT not taking an option had not happened as far as we know in the past 10 years and we seldom had issues with Landlords who understood BOR lease requirements.

The BOR does not permit a University System of Georgia college or university to pay directly for TI for capital improvements in any leased space that would provide the Landlord with a windfall. The college or university can fund and pay for its Fixtures, Furniture, and Equipment (FF&E) items that can be removed for future GT use at another location. Should the Tenant Improvement (TI) funds offered by the Landlord not be sufficient for the GT Tenant's requirements, the college or university may "borrow" TI funds (with BOR approval) from a Landlord and pay the Landlord back as a "special monthly rent allocation" payment as would be stated in Exhibit B of the Rental Agreement. The total amount of TI borrowed, the % interest being charged and the length of term for the pay-back in total months must meet BOR reasonableness tests and must make good business sense. The GT Tenant must have the funds available to make these monthly TI payments for the agreed upon term.

The second hat that I wore as the REGM was as the Landlord for GATV real estate acquisitions, yearly budget development, expense review and management, design, construction, project management, property maintenance and operations, and property management services for properties owned and leased by GATV entities (see Exhibit C below for the GT and GATV real estate relationship). The GATV legal entities at the end of 2015 included:

- VLP 1, LLC
- VLP 2, LLC
- VLP 3, LLC (Technology Enterprise Park)
- VLP 4, LLC
- GT Real Estate Services, LLC

GATV was not funded by or managed by GT but was founded to support GT's real estate requirements. GATV has an independent Board of Directors that include a number of Atlanta based knowledgeable professionals in the real estate and economic development industry (see Figure 3). The GATV Board met 4 or more times per year which could include telephone conference calls and/or e-mail issues to vote on. The GATV Board approved an annual operating and improvement budget for the GATV properties that I as the REGM managed on a daily basis through the overall approval of my GT boss who was also the GATV Vice President.

GATV had contracted with Jones Lang LaSalle (JLL) since 2004 to provide outsourced property management and property maintenance services to the following GATV properties: VLP 1, LLC, VLP 3, LLC and VLP 4, LLC. JLL managed during my term at GT the day-to-day operations of these properties, addressed issues and tenant questions, and managed expenses, rent receipts, and other income based on an approved yearly budget for each property that began on July 1 of each year. Outsourced JLL personnel included two Building Engineers, a Property Administrator, and a Senior General Manager based off site who managed other JLL properties.

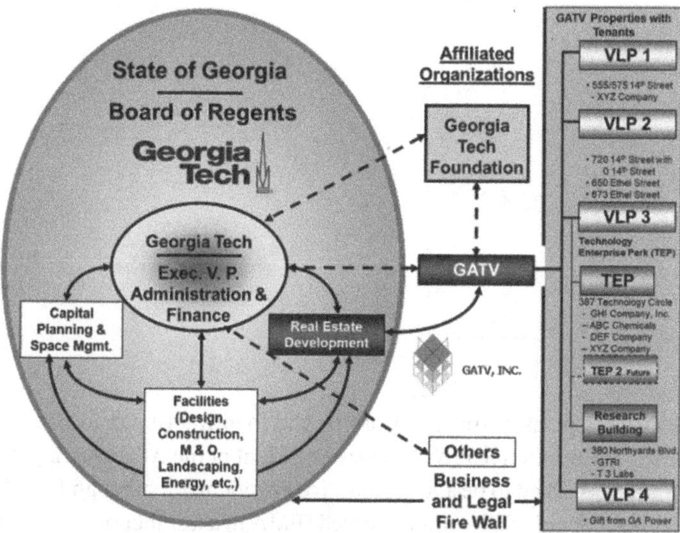

Figure 3 GT and GATV real estate relationship. (*Source:* E. Rondeau).

During my 10+ years at GT, I was responsible for working with GT customers for the acquisition of over 40 leases and for the acquisition of 6 purchased properties in and around the GT campus area. This included managing the yearly multi-million-dollar GATV budgets as a Property Manager for leases on over 20 acres of land and for over millions of dollars in design and construction budget. I worked with landlords, real estate brokers, real estate attorneys, real estate appraisers, property tax consultants, insurance agents, tenants, architects, engineers, specialty consultants, City of Atlanta staff, security vendors, landscape consultants, oil leases in Texas, GT and Atlanta police and fire department, etc.

I retired from GT at the end of 2015.

And we used Facility Management processes.

Lessons Learned:

- Develop your network of co-workers in your and other departments to help solve work and project issues before they become problems.
- Be transparent in your dealings with your customers and your project team members in all project decisions, especially in safety, contract and financial matters.
- Understand who in Finance and Accounting can help identify available project capital funds.
- Understand the process where real estate lease or property purchase projects are reviewed and approved by the Board of Regents of the University System of Georgia.
- Met and coordinated with GT and Atlanta police to establish panic buttons, security, camera, and safety reviews of all managed properties for student, academic, researcher, and visitor safety.

IFMA

My involvement with facility management began in 1981 when I was a Construction Manager with the Coca-Cola Company in Atlanta, GA, USA, and joined the National Facility Management Association (NFMA) in 1981. In 1982 I was one of the founders of the Atlanta, GA, chapter:

- Christine Williams (Neldon)
- Bob Euton

- Sam White
- Ted Stout
- Evelyn Blackstock
- Rob Stevens
- Gary Robinson
- Peter Ruys
- Robert (Bob) Browder
- Mike McGahagin
- Dewey Beardon
- Bill O'Connell
- Ed Rondeau
- Harold Troy

The Atlanta Chapter was chartered in 1982 (40+ years ago as of this printing) by the then International Facility Management Association (IFMA) President Charles White when Sam White was President of the Atlanta Chapter. I was elected the President of the Chapter in 1983 and was asked to be a Regional Vice President/Member and of the International Facility Management Association (IFMA) Board of Directors. I served IFMA in a number of volunteer leadership positions since 1983 nationally and I was elected as the 1987/1988 IFMA President (see Figure 4).

This leadership position led to meetings with other country groups forming FM associations including meeting with the Japan Facility Management Association (JFMA) in the USA and other associations in Europe in 1987. Also, my job at the time permitted me to attend a number of meetings at local chapters in the US, and with IFMA staff and Board approval, new chapters in the US and Canada were chartered.

As the Past President of IFMA in 1989, I worked with the IFMA Executive Director Melvin R. Schlitt to develop the basis for establishing the IFMA Foundation which was approved and funded in 1990 by the IFMA Board of Directors. The IFMA Foundation is the non-profit higher-education, research, and scholarship organization for IFMA and for FM. I was the first Chair of the IFMA Foundation Board of Directors in 1990 and began raising donations and gifts from companies and IFMA chapters to fund student scholarships for FM degree-related university programs.

I was later a member of the Facility Management Accreditation Commission (FMAC), an organization within the IFMA Foundation. The Commission set and developed academic FM standards for Associate, bachelor's, and master's degree FM programs in the US and internationally and approved degree programs (ADP) that are re-certified every 6 years. I served as a review team member to two FM universities in the US and to two candidate FM universities in Singapore in 2019.

My related FM academic text development experience has included being invited to be a co-author and/or a lead author for six books on real estate or facility management and am the author of the **Principles of Corporate Real Estate** manual which have been used as textbooks for numerous university FM programs in the USA. And, I have been invited to be a speaker or guest FM lecturer at many FM and RE conferences, programs and universities in the USA, Canada, Singapore, India, Hong Kong, China, Japan, Kuala Lumpur, Australia, Philippines, Austria, the Netherlands, Germany, Denmark, Spain, Italy, Brazil, Argentina, Mexico, and Cuba. And I have spoken and written for other associations including the AIA, the Building Owners and Managers Institute International (BOMI), the International Development Research Council (IDRC), and CoreNet Global (International Corporate Real Estate).

My teaching experience has led to the development of many presentations on FM for IFMA and other organizations. One of the teaching aides I developed is Figure 5 using the FM core responsibilities according to IFMA in 1984 and was modified for Figure 5 to reflect the current IFMA core FM responsibilities (see Figure 5).

Figure 4 IFMA Journal – Winter 1988. (*Source:* IFMA).

Lessons Learned:

- Managing and leading an IFMA Chapter, the IFMA association and the IFMA Foundation as a volunteer requires your understanding of the mission, vision, and goals of each component with a group of volunteers and staff to help provide insight and suggested direction.
- Many approved IFMA projects and programs for our members began as an idea that was nurtured with staff through a process of "nudging" with open discussion to build consensus.
- Networking with FM peers, related professionals and staff helped to build long-term relationships that helped IFMA and the IFMA Foundation grow and succeed.
- FM has grown into a global profession with members in many countries setting their own agenda and supporting the local members and programs.

Conclusion

My FM work and association experience has included working and teaching throughout the USA and in 28 countries. I have been fortunate to have been the co-author of 6 textbooks on Corporate Real Estate and Facility Management.

My experience has included all phases of FM and Corporate Real Estate for many large and some small organizations within the USA and internationally. This included

- Training and managing FM/RE staff and budgets,
- Developing FM and Real Estate strategies,

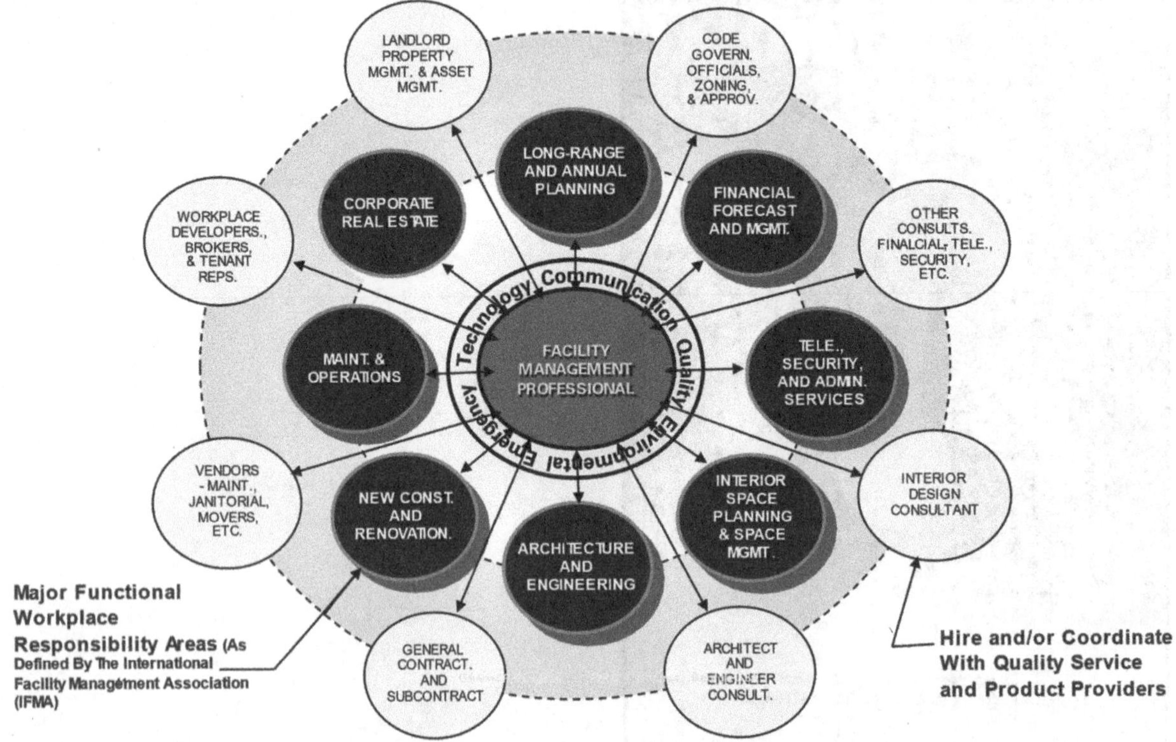

Figure 5 Facility management responsibilities. (*Source:* E. Rondeau).

- Leasing or purchasing real estate
- FM and RE databases and IT systems
- Working with senior management, customers finance and accounting, cash flow, internal audit, insurance, IT, FM maintenance and operations
- Hiring and managing architects, engineers, interior designers, consultants, vendors, and contractors.
- Hiring and working with attorneys,
- Manage FM Outsourcing vendor
- Providing project management services for moves, adds, and changes (MAC)
- FM services for furniture, moving, parking, food service, landscaping, janitorial, mail, telephone, Police and safety and security, Fire Department, energy conservation, environmental, sustainability, permits, and local and state officials.
- Meeting and working with FM professionals from around the world.

So, where is FM going and what is the future of FM? These questions will be reviewed and possible answers will be provided in Chapter 4, **<u>Summary, and FM Outlook.</u>**

See the next two pages for one article written by Ed and one article by Anne Fallucchi who quotes Ed Rondeau.

The Goal Is Not Survival; It Is to Develop and Excel

By Ed Rondeau, President, International Facility Management Assn.; Manager of Corporate Real Estate, Contel Corp., Atlanta, Ga.

There are companies spending millions of dollars on facilities, not to mention the advanced technical and computer aids required to assist their office staff. The assumption is that better facilities and greater accessibility to data, information, word processing and records will improve quality, save time, reduce cost and errors and help the company increase profits.

This is not always the case. Often, a company's commitment to productivity has not started with top management. The numerical side of office productivity has been addressed for many years, but many senior managers are only now beginning to assess productivity through analysis of their leadership roles.

In today's competitive corporate market, yesterday's inflexible, unimaginative managers are no longer viable. Managers must be more productive than their staff members. They must lead the way. To achieve this, it helps to review past daily, weekly or monthly reports as well as policies and procedures. Those which do not serve to enchance today's office technologies, interpersonal requirements or productivity should be revised or eliminated.

Senior management should ask if it is really attaining useful and meaningful information as a result of its efforts. By honestly assessing its capabilities, activities and objectives, management makes a positive commitment toward productivity. As a result, middle managers and staff are assured that statements from on high are more than mere window dressing.

INTERNATIONAL FACILITY MANAGEMENT ASSN. was founded in 1980. Its 1988 conference is Oct. 25–29 in Atlanta, Ga. Executive director is Melvin Schlitt at 11 Greenway Plaza, Suite 1410, Houston, Texas 77046. 713/623-4362.

"Because of today's variable economy, rife with takeovers, mergers, buyouts and consolidations, corporate managers should realize that they have no golden parachutes . . ."

Unfortunately, many companies employ senior management personnel who appear to be just names on memos. They tell the staff to work harder and produce more, but are seldom visible outside executive areas. But to meet today's challenging markets and increased competition, many senior managers are changing previous practices. They are moving from their armchair directors' roles to take part in the daily life of their corporations. They are getting a feel for the details and health of their company by keeping their ears open, asking questions and talking one-on-one with their co-workers.

Senior management, middle management and staff, in an on-going team effort, must encourage honest communication. This is essential to implement better physical work environments, improve and simplify workflow, provide responsibility with authority and involve everyone in the end product—regardless of the department. Communication is the only key to breaking down "turf" barriers. Through this process, senior management is visible and accessible and provides a vision that encourages and demands excellence, service and innovation as an on-going way of corporate life.

'Boundary Spanner'

Closer to my particular profession, a facility manager is a "boundary spanner" between all departments. This person usually deals with all levels of management and staff regarding long-range and daily requirements. Aside from the human resources group, the facility manager generally has an integrated view of activities in each department. He or she has well-rounded insight into a corporation's culture and an accurate sense of its efficiency and spirit.

Because of today's variable economy, rife with takeovers, mergers, buyouts and consolidations, corporate staff members should realize they have no guarantees or "golden parachutes." The company with the strongest leadership, most innovative programs, most realistic policies, team spirit and a hard-working staff and senior management "might" remain competitive and stay in business. But the objective is not to survive. It is to excel and develop an on-going program of productivity through leadership and excellence at all levels. Increased productivity becomes a natural part of the overall quality and service programs received and provided to customers by staff and management. TO

The Office magazine, Article by Edmond P, Rondeau, "The Goal Is Not Survival; It Is To Develop and Excel," page 96, January 1988.

November/December 1989

Editorial

The key word is 'management'

"Facility management in the United States has evolved and is continuing to evolve with a goal to provide management services which must meet corporate strategic, long-range and short-term business requirements," said Edmond Rondeau at the recent International Symposium on Facility Management.

The key word is "management." And perhaps "facility" should be "asset"—asset management, for it's bound to grow into a strategic business issue in the decade and century ahead.

Corporations must—and are starting to, as Rondeau stated—view their real estate and facilities as assets, especially since they represent, on the average, 25 percent of a corporation's total assets. That adds up to a sizable sum, and even a one percent savings in these assets, properly managed, can help sweeten the corporate bottom line.

Then, too, facilities management professionals must gear up to a business "management" mentality, so that they can effectively contribute to the productivity and profitability of their organizations. As corporations downsize and the mid-management balloon deflates, those who remain will have to be good at their jobs—more, they will have to excel.

Ed Rondeau, manager of real estate, Contel Corp., offered these sobering words at the symposium: "The future for facility managers and facility management. . .will not be easy. There will be tremendous opportunities for those who do their homework. . .and consistently perform. The magnitude of problems and issues will not be reduced but will increase as facility managers continue to move into an information-service economy. The challenge is to meet opportunities presented with professional and business know-how and manage them."

Anne Fallucchi
Editor-in-Chief

Facilities Design & Management magazine, Editorial, "The Key Word Is Management", Anne Fallucchi, Editor-In-Chief, page 63, November/December 1989.

Edmond P. Rondeau, AIA Emeritus, RCFM, IFMA Fellow
Ed Rondeau RE & FM Consultant and Educator
Atlanta, GA USA

Ed is a Real Estate & Facility Management Consultant and Educator. For over 10 years he was the Real Estate General Manager in the Real Estate Development Office at the Georgia Institute of Technology in Atlanta, Georgia USA. There he was responsible for the Real Estate acquisitions and Facility Management operations of assigned owned and leased properties for the university and affiliated organizations.

Previously he was the Director of Global Operations and Learning for the International Development Research Council (IDRC), an international corporate real estate membership association based in Atlanta, Georgia, served as the Director of Consulting Services for the Integrated Facility Management (IFM) business unit of Johnson Controls, Inc., and was the Manager of Corporate Real Estate for Contel Corporation. He also served as the Vice President of Property Management for the National Bank of Georgia, was the Vice President of Real Estate & Construction for Arby's, Inc., was a Construction Manager for the Coca-Cola Company, and was the Staff Architect at Auburn University.

Ed holds an MBA in Real Estate from Georgia State University and a Bachelor of Architecture from the Georgia Institute of Technology, was the 1987/1988 President of the International Facility Management Association (IFMA), served as the 1990 Chair of the IFMA Foundation, and was elected an IFMA Fellow in 1992. He is a retired registered Architect in Georgia and is a Retired Certified Facility Manager (RCFM).

He has spoken on corporate and university real estate and facility management issues throughout the US and Internationally.

2.2

The Institution Bauakademie in the Course of Time and Its Current Potential for Change

Karin Albert

If the term "Bauakademie" is used in Germany and far beyond the country's borders, the focus is immediately on the building of the same name and usually also on the seemingly never-ending discussion about its re-construction at "Schinkel Platz" in Berlin, Germany.

In order to make progress here, it is worth taking a brief look at the past and the purpose for which this building was built. The institution of the Bauakademie is quite wrongly lost sight of in the current debate. However, the question arises as to whether such an institution can be used to solve today's global problems, such as climate change, energy transition and solving the housing issue, all of which are linked to construction as a social task and could make a significant contribution. But first, let's look back.

The relocation of the Bauakademie to the current building is not a style of today, but it began during Schinkel's lifetime. Harald Bodenschatz attributes this to two aspects: on the one hand, to the separation of the building from the purpose of its use and, on the other hand, to the connection of the building with its spiritual creator.[1]

Both aspects still have an effect today. The discussion about the reconstruction of the building was and is conducted without a clear idea of its future use. The recommendation of the Budget Committee of the "German Bundestag" of 10 November 2016 on the provision of financial resources for the re-establishment of the building of the Bauakademie was also made without any orientation in this regard. Only in the dialogue forums on the reconstruction of the "Schinkelsche Bauakademie" (Status Forum, Ideas Forum and Scenario Forum) organized from February to May 2017 by the "Federal Foundation of Baukultur" and accompanied by the Federal Ministry of Construction did experts from the fields of architecture and urban planning, museum directors, representatives of reconstruction initiatives and Berlin politics discussed in a closed form, etc. about existing or possible new usage concepts. These forums also incorporated the results of a number of isolated initiatives up to that point.[2] Unfortunately, the results compiled extensively by the "Baukultur Foundation" were not subsequently used as a basis for further activities in this regard and were supported or elaborated on this basis. But with such a historic building at such a historically significant place as Schinkel Platz in Berlin, how can you make a decision on the shape and design of the building without having an idea of its content-related function?

At the time of its construction and completion in 1836, the situation was completely different. From the very beginning, the conception of the building was programmatically based on the unity of building and institution.

1 Bodenschatz, Harald: The red box. Zur Bedeutung, Wirkung, Zukunft von Schinkels Bauakademie, Berlin 1996, pp. 59/79ff.

2 Bauakademie-Journal, Förderverein Bauakademie Berlin, 83. Issue v. 01.09.2022, S. 3; https://www.foerderverein-bauakademie.de/files/foerderverein-bauakademie/pdf/Pa%20pdf%201000a%20Bauakademie-Journal%20.pdf.

Facilities @ Management, Concept - Realization - Vision, A Global Perspective, First Edition. Edited by Edmond Rondeau and Michaela Hellerforth.

The building was to accommodate two institutions: the "Allgemeine Bauschule" and the supreme control and examination authority for the state building industry in Prussia, the "Oberbaudeputation."

The Founding of the Royal Academy of Architecture and Its Absorption into the Technical University of Berlin

The Royal Academy of Architecture was founded in 1799 by Friedrich Wilhelm III. The development of the cities and the construction of all kinds of buildings were among the most important tasks of the emerging Prussian state. Well-trained building officials were needed for this. The Royal Academy of Architecture was intended to serve this purpose. It emerged from the Faculty of Civil Engineering of the Berlin Academy of Arts, which was primarily dedicated to architecture. The technical aspects of construction received little attention here. The training of the Bauakademie was to change. It should raise clever and skillful master builders and professionalize this profession. This made it one of the few training institutions that were clearly oriented toward professional practice. After their training, their graduates should have a particular understanding of public buildings, residential buildings, and commercial buildings. The aim was to close the gap between the "high magnificent architecture" and the "lower functional architecture." The trained master builders were mainly employed in the civil service. Public construction had developed substantially in previous years. With the establishment of the "Ober-Bau-Department" in 1770, the state building industry experienced for the first time a standardization and centralization and this was to be promoted and implemented by the well-trained building officials.[3]

Through a multiple reform process, the study requirements at the Bauakademie were gradually raised and the studies were reorganized according to the requirements based on the accelerated progress of the Industrial Revolution. Thus, the Board of Directors was replaced by an academic deputation under the permanent chairmanship of the President of the Ober-Bau-Department. In this way, the Bauakademie broke away from its close ties to the "Akademie der Künste." With the increasing demands on technical training, there had been increasing calls to divide subjects into different sectors.

Under the direction of Christian P.W. Beuth, who had already founded the trade school in 1821 and took over the directorate of the Bauakademie in 1830, the Bauakademie was renamed to "Allgemeine Bauschule." Together with the "Oberbaudeputation," to which Karl Friedrich Schinkel had belonged since 1810 and headed since 1830, it moved into the newly constructed building on "Kupfergraben" in 1836, almost 40 years after its foundation.

The revolutionary efforts of 1848 also had an impact on the "Bauschule" and its students. Their demands included, among other things, the abolition of compulsory teaching and the establishment of complete freedom of learning, as well as the re-transformation of the Bauschule into a Bauakademie. Although after the suppression of the revolution all liberal gains and civil rights were abolished and a constitutional monarchy was established, the demands of the building students, which the building officials also intermittently, were successful. The Bauschule was renamed the Royal Academy of Architecture and the training and examination regulations were revised. Although not all requirements were implemented, there were now technically differentiated degrees for master builders (land and beautiful construction, road, water and railway construction). A new course of study "Private Master Builder" was established. His training took place separately from the future building officials. This was in response to the expansion of private construction and a greater need for private design, planning and construction services.

Since the middle of the 19th century, the economic conditions in Germany had changed rapidly: development of technology and production processes and thus of labor productivity, increasing mass production and further differentiation of economic sectors. This age of an increasingly division of labor production as a prerequisite for

3 In 1804, the Ober-Bau-Department was transferred to the Technische Oberbaudeputation transformed.

industrialization and the continuous increase in productivity to its extreme, Taylorism, had begun. The increasing mechanization of the production and application of the natural and technical sciences in production also put the training at the Bauakademie under scrutiny. Above all, the separation of the construction subject into architecture and civil engineering was demanded. This demand was granted in 1875. At the same time, an organizational transformation began, which began in 1877 with the merger of the "Gewerbeakademie" and the Bauakademie and finally ended with the founding of the Royal Technical University of Berlin in 1879, into which the Bauakademie merged. The college had five areas: architecture, civil engineering, mechanical engineering (including shipbuilding), chemistry and metallurgy, general sciences. In this way, the training did justice to the increasing technical specialization and the division of labor production methods. The training of master builders was dissolved.

However, increasingly occurring interfaces are always an inevitable consequence of processes based on the division of labor. Technological development forced concentration on specific tasks and working in networks because the entire spectrum of knowledge and skills could no longer be controlled individually. An inevitable consequence to this day is the risk of loss of information, which is all the greater the less complex the tasks to be realized are considered. If one wants to avoid loss of information, the optimization of the overall process from planning to construction to operation and the classification of construction into overall social processes is crucial. In order to solve today's tasks, it is necessary to combine all the necessary technical expertise for the livable design of the built environment on a new level.

Karl Friedrich Schinkel – One of the Last Great Master Builders

Karl Friedrich Schinkel (1781–1841) was himself a student at the Bauakademie from 1799 to 1800. Under the leadership of Friedrich Gilly (1772–1800), with whom he soon became a deep friend, he was trained as a master builder. His teachers also included David Gilly (1748–1808), Carl Gotthard Langhans (1732–1808), and Heinrich Gentz (1766–1811). After the early death of Friedrich Gilly in 1800, Schinkel broke off his studies and took over the completion of the buildings begun by Gilly.

Schinkel was one of the last great master builders who, in the ingenuity of his own person, was able to combine the necessary specialist knowledge for planning, building, and operating buildings. He did not always fully implement this "master builder's principle" in his buildings. But in the construction of the Bauakademie he was involved in all phases, from the first drawing to the execution of the construction. This alone shows the importance he himself attached to this building.

The Bauakademie was the last large public building that Schinkel built. In 1830 Schinkel was appointed head of the superstructure deputation. One year later, Beuth took over the position of director of the School of Architecture. Both institutions were housed in different locations in Berlin. In this respect, the relocation of the customs facilities offered an opportunity to create a building on this site to accommodate both institutions. Since the building of the Bauakademie was built for the sole purpose of accommodating the general building school and the superstructure deputation, Schinkel de facto fulfilled a double function here – he was both builder and architect. In this respect, he designed the Bauakademie exclusively for its use.

Schinkel incorporated in the design of the Bauakademie his impressions, which he collected during his trip to the British Isles in 1826. Together with Beuth, who at that time took on the function of director of the Prussian Deputation for Trade and Technology, he was looking for machines suitable for introduction into Prussia. Here he was inspired by modern industrial architecture. The objectivity and practicality of these buildings, which nevertheless had a representative character, impressed him very much. He returned with a variety of insights: techniques for fireproof construction, the adoption of iron structures and brick vaults, the use of industrially manufactured and un-plastered brick and terracotta.

Under this impression, Schinkel broke away from the principles of the classical or Gothic style for the first time with the design of the Bauakademie and created a building that was clearly oriented toward the future. To this day, this building is regarded as an ingenious combination of tradition and modern. Also, in the construction it differed to previous methods. Schinkel first had the pillars and the massive false ceilings bricked up. In a later operation, the façade parts were used and installed – a first approach to skeleton construction. The Bauakademie is regarded as the first architecturally significant industrial building in Germany. It can also be described as Schinkel's leand gacy for modern construction.

In this respect, it is only too understandable that the Budget Committee of the German Bundestag in its meeting on 10 November 2016 expressly decided to make funds available for the re-establishment of the well-known form of the Bauakademie. This takes into account not only the historical significance of this building, but also the urban ensemble effect, which is restored here with the castle, cathedral, "Altes Museum" and "Friedrichwerderscher Kirche."

The German Bauakademie

At first it seemed as if after the founding of the GDR (German Democratic Republic) also known as East Germany, there should again be a unity between the institution and the building Bauakademie. Since 1950, under the direction of Richard Paulick, extensive plans had been made for the reconstruction of the Bauakademie, which had been destroyed by the effects of World War II. In this context, Paulick also planned an extensive conversion of the interior of the building in order to adapt the building to changing usage requirements.

The user of the building was to be the German Bauakademie, founded in December 1951, which emerged from the Institute for Urban Design and Building Construction at the Ministry of Construction and the Institute of Civil Engineering at the German Academy of Sciences.

Initially, it was mainly oriented toward architecture. The plenary had an advisory role. At the time of its foundation, the Bauakademie comprised five institutes and three master workshops. It employed 590 people. In the master workshops, which were managed by architects, projects for the reconstruction of war-torn cities were realized. In the 1960s, the focus of the Bauakademie's tasks increasingly shifted to the engineering sciences in order to enforce the industrialization of the construction industry in the GDR.

The tasks of this newly founded Bauakademie thus went far beyond the function of a building school, it represented the central research institution of the construction industry of the GDR. In the years of its existence, further institutes were successively assigned to it, so that at the time of the reunification of the two German states it included a total of 16 scientific institutes or academic and administrative facilities with about 4500 employees.

However, the German Bauakademie, called "Bauakademie der DDR" since 1972, never moved into the building of the same name, as the reconstruction of the Bauakademie was systematically deprived of financial resources starting in 1954. In 1961, it was decided to demolish the only fragmentarily rebuilt Bauakademie, initially with the perspective of reconstruction. For this reason, a large number of components were removed and secured, which are still available today for reconstruction.

Under the umbrella of the Bauakademie of the GDR, institutes for all scientific building areas were basically united. In them, both basic and applied research with a cross-sectional character was carried out. Their essential mission, the provision of all necessary technologies for the increasing industrialization of construction, was an indispensable prerequisite for meeting the huge challenges in industrial and residential construction. Of course, as a centra authority in a centralized state, the Bauakademie was also subject to the political doctrine of that state. But by concentrating building research under one roof and systematically bringing together the different aspects of construction, the construction process in its entirety could be regarded as a social task, with all the positive and negative consequences resulting from scarce resources and also under the conditions of an international embargo.

In the Scientific Council (right to award doctorates) and in the 12 subject-oriented sections (including specialist expertise from other scientific areas), the basic tasks of building research were defined and implemented in the institutes. In this way, scientific procedures could be developed that attracted both national and international attention. These include, for example: the drainage of damp masonry by active desalination, fluid technology for transporting heavy loads, the roof segment assembly of lightweight steel structures, etc. Using the fluid technology of the Bauakademie, for example, the Kaisersaal of the Grand Hotel Esplanade was moved to the Sony Center in 1996.

After the reunification of the two German states, the Unification Treaty of 31 August 1990, Art. 38, Science and Research, regulated in paragraph 1 that an appraisal of publicly funded institutions is carried out by the Science Council. This also applied to the Bauakademie of the GDR (Art. 38, paragraph 4). Transitional financing was valid until 31 December 1991. On this basis, the states of Berlin, Brandenburg, Saxony, Saxony-Anhalt, Thuringia, and the Federal Republic of Germany reached an agreement on the transfer of the institutes and other facilities of the Bauakademie of the GDR and set up a "transfer office of the Bauakademie."

Of course, the Bauakademie could not be integrated into the research landscape of the Federal Republic of Germany with all the institutes existing at that time. Therefore, immediately after the collapse of the GDR, a restructuring and reprofiling took place in the Bauakademie in a democratic process, combined with a self-evaluation. As a result, six institutions were dissolved, and two institutions were transferred to private legal forms. As a result, the number of employees has already been reduced by approximately 4300 to 1100. At the same time, an employment and qualification company was founded to give these employees a temporary professional basis in a "rescue company."

In the period up to July 1991, the assessment of the German Council of Science and Humanities covered a total of 10 institutes (Institute for Structural Engineering Berlin, Institute for Heating, Ventilation and Fundamentals of Construction Technology Berlin, Institute for Building Preservation and Renovation Housing Berlin, Institute for Urban Design and Architecture Berlin, State Building Research Institute Brandenburg, Institute for Civil Engineering Leipzig, State Research Institute Saxonia, Saxonian Building Institute Dresden, Institute for Industrial and Commercial Planning Halle, Institute for Building Materials Weimar), the building information, the further education center and the central facilities. The German Council of Science and Humanities indirectly confirmed the correctness of the path that had already been actively taken. In its statement of July 5, 1991, the German Council of Science and Humanities recommended that three institutes with transnational tasks, one state institute (Saxony-Anhalt) and four new materials research and testing institutes, and that groups from the academy institutes be transferred to these new institutes. For other individual groups, further funding in existing research institutions in the Federal Republic of Germany was proposed. However, the implementation of these recommendations proceeded very slowly, so that it was only possible to maintain the scientific potential of the Bauakademie at least partially.

For the training center, it was recommended that it be continued in the private sector. As early as 1990, representatives of this institution, together with representatives of the CAD/CAM laboratory of the Institute of Technology and Mechanization of the Bauakademie, founded a "GmbH" (Limited Company) with a focus on the further training of engineering specialists and a strong focus on mastering modern information technologies. As a minor shareholder, the transfer office of the Bauakademie also participated in this company foundation. This made it possible for the GmbH to continue the name "Bauakademie." However, this led to corporate law challenges at the time when the Bauakademie institution was dissolved, and its assets were allocated to the federal states as administrative assets. The legal successors of the former Bauakademie of the GDR were now the five new federal states and the Senate of Berlin. Thus, the newly founded GmbH was de facto no longer able to act. In a two-year process, the minor shareholder, the former Bauakademie, left the company Bauakademie, which since then has been operating exclusively as a private institution on the market. In the more than 30 years of its existence, it has developed into a group of companies and made an important contribution to development in the construction and real estate sector with technically differentiated consulting and training services.

But of course, neither this Bauakademie group of companies nor the other companies that have operated under the name Bauakademie in recent years are comparable to a Bauakademie in the sense of a Bauakademie or a research institution such as the Deutsche Bauakademie. Nor do they make that claim. Precisely for this reason, the question remains whether a Bauakademie in the sense of a scholarly society has potential for the present time and what contribution it could make to solving the upcoming national and international development tasks.

The Potential of the Bauakademie in Today's World

In order to be able to answer this question, it is worthwhile to think about whether building in Germany and in the world is still of existential importance and therefore necessary to deal practically, content wise and intellectually with the demands placed on construction within the framework of an institution Bauakademie.

Again and again, the discussion about the necessity of a building academy flares up. As early as 1992, representatives from various fields of architecture and urban planning, politics, business, journalism, and engineering sciences dealt with the question "Does Germany need a building academy?" at the prestigious building fair "Constructa" in Hanover. This question was also discussed at the aforementioned dialogue forums of the Federal Foundation of Baukultur in 2017. The "Bundesstiftung Bauakademie" also speaks on its homepage of the Bauakademie institution, describes its task, but leaves open how this institution should be designed in concrete terms.[4]

We live in a globalized world characterized by the continuous digitalization of production, climate change, and global migration. The ecological balance of the earth is threatened, the world's population is growing steadily, cities are exploding. All these challenges are associated with construction and require constructive solutions, the implementation of which requires the processing of the latest findings from a large number of scientific fields. The demands on construction are growing because, on the one hand, we deal responsibly with the built environment as the embodiment of comprehensive resources, and on the other hand we have to keep it livable. On the other hand, today we shape the working, living and living conditions of future generations.

This raises the question of who will coordinate these complex tasks, which require the interaction of different social forces and expertise, both nationally and internationally. Of course, there are many entrepreneurial and scientific institutions, associations and social institutions that deal responsibly with their respective topics and detailed questions. But a complex analysis of the global challenges facing society with a view to the contribution that construction can make to solving them does not seem to take place.

In 2008, the then Federal Chancellor, Angelika Merkel, appointed the German Academy of Sciences Leopoldina, founded in 1652, as the National Academy of Sciences of Germany. With its approximately 1600 members from more than 30 countries and almost all fields of science, it is a classic scholar society.

The aim of founding a National Academy was to create a legitimate institution which, independently of economic or political interests, would scientifically deal with important social issues of the future, communicate the results to politics and the public, and represent these topics nationally and internationally.

To this end, since 2009 the Leopoldina has had four classes (Mathematics, natural and technical sciences; Life sciences; Medicine; Humanities, Social and Behavioral Sciences) and a total of 28 subject sections, each assigned to a class. This association into classes promotes in particular the interdisciplinary discussion and the participation of the members in the work tasks of the National Academy.

In extensive discussions that the "Förderverein Bauakademie" was able to conduct with the Executive Committee and the Leopoldina Office in 2012 and 2015, it became clear that the construction sector is not represented in the sections, a fact that is not only incomprehensible, but also unjustifiable.[5]

Sustainable construction, which we strive for, requires minimization of energy and resource consumption and the lowest possible impact on the natural environment. The tasks to be expected in the coming years and decades

4 Pro Bauakademie, documentation of the Discussion at the Constructa '92 on February 8, 1992 in Hanover.

5 The Förderverein Bauakademie e.V. was founded in 1994, initially with the aim of accelerating and supporting the reconstruction of the Bauakademie building. According to the decision of the Budget Committee of the German Bundestag in 2016, he increasingly turned to questions of the establishment of an institution Bauakademie.

will place completely new demands on the construction industry of the industrial nations and their cooperative cooperation. To this end, the division of labor structures that prevail in construction today must be replaced by interdisciplinary process structures that are much more closely networked with each other. Here, at Bauakademie, a scholarly partnership of different disciplines, such as sociology, economics and computer science, and the energy and housing industry, could develop solutions which, in conjunction with adequate international research institutions, have a positive effect on securing the complex basic structural requirements of site-specific settlements such as water, energy and supply, infrastructure and employment, hygiene and health care. At the same time, however, the economic and entrepreneurial weight of Germany's construction and real estate industry can be strengthened.

In the course of intensive discussions with the Leopoldina's governing bodies, a possible establishment of a future Bauakademie was drawn up stimulated by the professional exchange, which could also be easily integrated into the structure of the National Academy (see Figure 1).

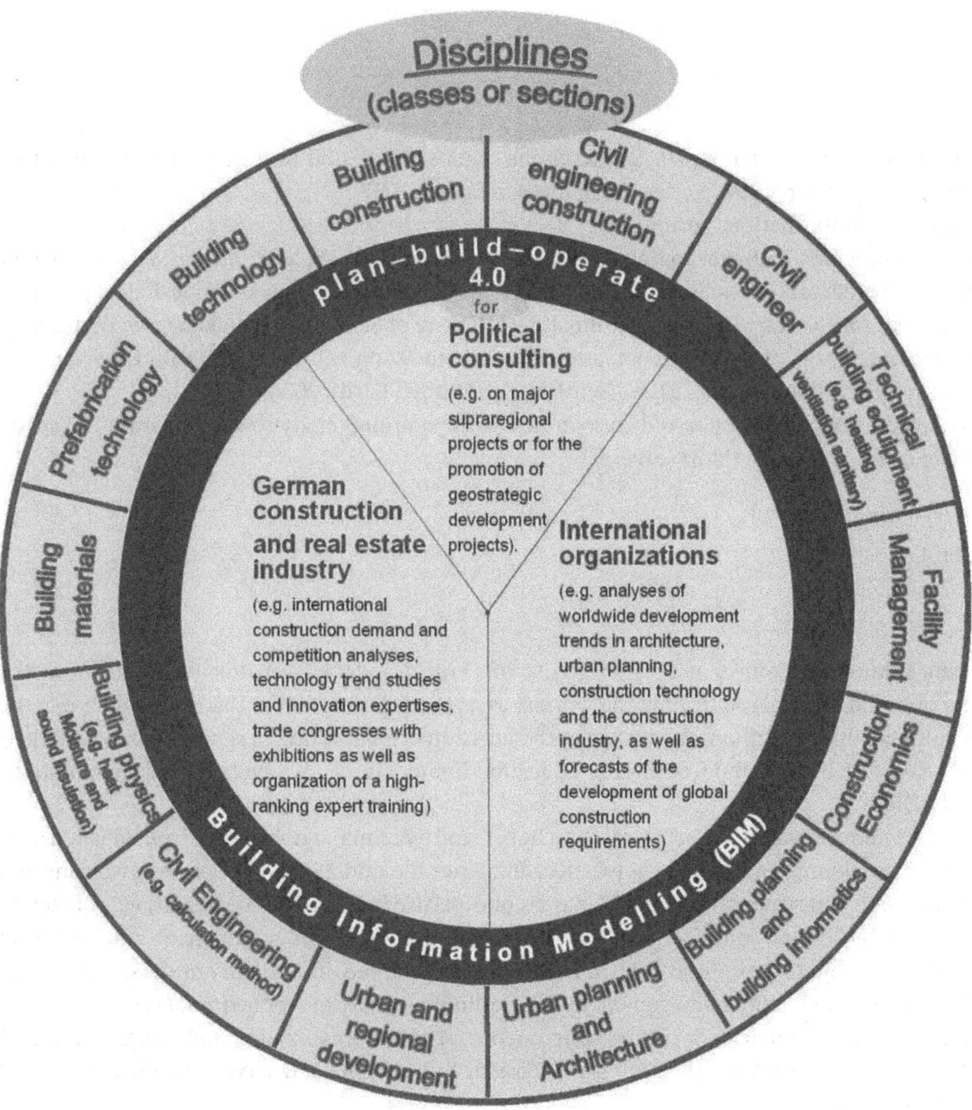

Figure 1 Building Information Modeling (BIM). (*Source:* K. Albert).

Although this structure from 2015 must be subjected to a further examination of the extent to which it still fully meets today's requirements, it forms a good basis for discussion, which already goes far beyond today's proposals on the tasks and structure of a future building academy.

The institution Bauakademie has always fulfilled a necessary function in its respective time and thus contributed to the stability, but also to the further development of the social building process. It would be desirable if the already existing social institutions would not only address the question, "Does Germany need an institution Bauakademie?" and answer it positively. Above all, it would be important to structure the Bauakademie in such a way that it specifically supports the interaction of the leading specialist expertise of the national and international construction industry as appointing academy members and thus promotes the constant further development of efficiency, productivity, and culture of construction from a national and international perspective. In this way, not only the Bauakademie building could be put to an important use, but also a large number of socially burning challenges could be solved. Undertaking this effort would be worthwhile – socially, ecologically, and economically.

Literature

Augustin, Frank (ed.), Mythos Bauakademie: Die Schinkelsche Bauakademie und ihre Bedeutung für die Mitte von Berlin; Förderverein Bauakademie e.V., Verlag für Bauwesen Berlin, 1997

Bauakademie der DDR, Bauinformation, Berlin 1983

Bauakademie-Journal, Förderverein Bauakademie Berlin, 83rd edition of 01.09.2022; https://www.foerderverein-bauakademie.de/files/foerderverein-bauakademie/pdf/Pa%20pdf%201000a%20Bauakademie-Journal%20.pdf

Bodenschatz, Harald, Der rote Kasten. On the Meaning, Effect, Future of Schinkel's Bauakademie, Berlin 1996

Pro Bauakademie, documentation of the discussion at the Constructa '92 on February 8, 1992 in Hanover

Sieber, Frieder; Fritsche, Hans, Bauen in der DDR, Beuth Verlag GmbH, Berlin 2006

Statement of the German Council of Science and Humanities on the non-university research institutions of the former Bauakademie of 05.07.1991, Düsseldorf (excerpt)

Dr. Karin Albert, Prof.
BAUAKADEMIE
Berlin, Germany

Prof. Dr. Karin Albert studied economics at the University of Leipzig and received her doctorate from the Humboldt University of Berlin. From 1986 to 1989 she worked as a research assistant at the Bauakademie of the GDR. After the dissolution of this institution, she took over the function of an authorized signatory in the market-based company BAUAKADEMIE EDV- and Continuing Education Center GmbH, which had been spun off from the Bauakademie der DDR.

In 1997 she was appointed managing director of this GmbH and was managing partner of Bauakademie Beratung, Bildung und Entwicklung GmbH for over two decades. Since the mid-1990s, she has been working as a consultant in the construction and real estate sector. Her areas of expertise include commercial aspects of facility management as well as educational consulting and personnel development for commercial and technical employees in the construction and real estate sectors. In 2015, she was awarded an honorary professorship in the field of business administration in facility management by the Berlin University of Applied Sciences.

Prof. Dr. Albert is in charge of complex consulting assignments for the reorganization and reorganization of entrepreneurial processes, e.g., B. for BASF SE, Stiftung Preußischer Kulturbesitz, DB Services GmbH, STRABAG PFS GmbH, and Bauerfeind AG.

In addition, under her responsibility, extensive projects were carried out in the field of educational consulting and personnel development for the long-term and goal-oriented development of the technical and methodological competence of specialists and executives in companies such as Deutsch Telekom AG, EnBW AG, DB Station & Service AG, BVG Berliner Verkehrsbetriebe and the public sector.

Since 1999, Prof. Dr. Albert has been involved in various committees of the German Association for Facility Management e.V. (GEFMA e.V.) and the Association for Real Estate and Facility Managers (RealFM e.V.). Among other things, she was instrumental in the development of the professional training of the facility manager and the functional and performance model. Since 2022 she has been an honorary member of the Association for Real Estate and Facility Managers (RealFM e.V.).

Since 1996 she has been teaching FM-oriented courses at the Berlin University of Applied Sciences.

2.3

Structured Growth of Facilities Management in China (1992–2022)

Yu Qingxin

Introduction

The year 1992 was an extremely important year for China. Mr. Deng Xiaoping, the chief architect of China's reform and opening up, visited Wuchang, Shenzhen, Zhuhai, and Shanghai, and made important speeches along the way. He proposed that China should firmly follow the market economy and insist on reform and opening up. The impact of this southern tour on China was enormous, and it triggered the golden 30 years of China's reform and opening-up.

The year 2022 is also an extremely important year but it is not a very good year. We call it the year of calamity. China is experiencing the third year of the epidemic, there are great challenges and changes in all aspects of the national economy.

In the 30 years from 1992 to 2022, China's facilities management and the Chinese economy have experienced very similar development context, which is structural growth. In 1992, we lagged behind the world in all areas of society, and there were still blanks in many fields. Over the 30 years, China has developed rapidly in many different areas, and emerged to lead and even surpass the world in many fields with the help of the global internet economy. The country enjoyed the best golden decade of development. This rings true with China's facilities management, which has gone through a very similar historical development process.

I – A Meaningful Conversation

Facility management first appeared in China around 2001, starting with a social tea gathering at the Tsinghua University Real Estate Alumni Association in Beijing. At that time, China's economy was at its peak, and the participants all talked about one topic: When will the real estate industry reach its peak? When will it start to decline or fall? After the downward spiral, where will the industry go from there?

This conversation sparked a long period of reflection, discussion, and practice among the participants. Afterwards, these young attendees became experts and eventually leaders in the real estate industry and in various fields or related segments.

I was one of the attendees. At that time, I was a property manager. After this conference, I started to search for answers with the same thinking and hope for the future. In the future, with the progress of urbanization and the development of the real estate industry, what kind of new face would property management take on? What kind of property management era would we enter? In the following years, I started to visit many developed cities in China, such as Shanghai, Guangzhou, and Shenzhen, and looked up a lot of information and books to try to find the answer. In the process of searching, I came across facility management.

Facilities @ Management, Concept - Realization - Vision, A Global Perspective, First Edition. Edited by Edmond Rondeau and Michaela Hellerforth.

Philip Lo

During an ordinary business gathering, I met two friends from Hong Kong, one of them was Philip Lo, IFMA Fellow. It was the first time that Philip came to mainland China to visit the state authorities of the real estate industry to recommend facility management. Unfortunately, the exchange did not go well due to language problems, or rather, due to differences in knowledge systems and in cognition. Philip, who was more or less frustrated, attended the gathering in the evening and it was at this gathering that we got acquainted.

After Philip started to introduce the origin, development, core business, and other related contents of facility management in North America and the West. He gave me the core elements of facility management: the three P's and one T in detail for the first time, I seemed to have found the answer. And I could say with certainty then that facility management is well suited for China's future growth in the real estate sector.

The reason is that during this period, China's industrial and commercial enterprises were booming. There were a lot of foreign and state-owned enterprises in China's manufacturing industry, as well as many private enterprises, which were growing comprehensively in all aspects of the economy showing unstoppable vitality. Facility management was just the right answer to this high-growth market.

In particular, China's high-technology was showing incredible power of growth. Facilities management would provide a place for the growth of two large industry clusters: traditional manufacturing and high-tech. This ushered in the future of property management and became an integral part of China's real estate management.

In this sense, Phillip was the first to introduce facility management to China. For facility management in China, Philip was a pivotal figure. For a long time to come, Philip had been playing the role of a "light bearer" and had made irreplaceable contributions to the industry and to the facility management profession in China.

Jumbo Lao Yu

This is where I need to introduce myself. In the introduction, I said that 1992 was an important year in which many people began to start their own entrepreneurial journey. In today's world, the well-known group of entrepreneurs of Chinese companies would be called the "*92 School*," meaning that they started their own business in 1992, and I was one of them.

In China, I founded the company known as Junhao and Jumbo Chains abroad. The people in the community or in the industry usually refer to me as Jumbo Lao Yu. *Yu* is my name and *Lao* in Chinese means "old" but when use it together with the name, *Lao Yu* means "*my good friend Yu*" as a more intimate manner to address an old friend.

In 1992, there was no such concept as "property management" in China. There were hardly any property management companies at that time, except in a few cities in the south. Even in the industry directory of the State Administration for Industry and Commerce, there was no such industry as property management. It wasn't until 1994 that the term appeared, and I was one of the first people in Beijing to start a property management company.

In fact, in the following 30 years, it has always maintained its identity as an independent third-party professional property management company. As a private enterprise, it has always been active in this market and we gradually entered the three main real estate fields of property management (PM), facility management (FM), and asset management (AM). The company has also become a well-known brand in the industry and is recognized by the market.

The Growth of Real Estate Management

Property management in China came from Hong Kong and originated in the UK. In order to study the origins of property management, I led a group of eleven people from the Beijing Property Management Association to the United Kingdom in 2011, where Octavia Hill, the founder of property management, once lived and worked, to study the origins of property management.

The development of real estate management from the UK to Hong Kong and then to China has a clear logical lineage. Afterwards, in their discussion about the emergence of asset management, experts in the field of real estate research at Tongji University in Shanghai and Tsinghua University in Beijing recognized that in the 1930s, asset management was born as the capitalist economy grew vigorously causing the entire urban form of production and manufacturing transactions to become more clustered. This was the first evolutionary journey of property management.

The development of property management, asset management, and facility management globally, and in China, is clearly evident. However, it is not accurate or sufficient to use any one term to express the management of buildings in cities with different classifications, different functions, and different ownership of property rights. Thus, we use a relatively neutral term, real estate management, to refer to a concept that includes property management, asset management, and facility management in China. This is the logic behind the emergence of real estate management in China.

II

Since the introduction of facility management in China, one of the most frequently asked questions is: What is the difference between property management, facility management and asset management?

In China, we usually define facility management and property management as the management of real estate held in private names. That is to say, the management to serve the owner's privately held residence is called property management. Asset management, on the other hand, is the process of increasing the value of assets, and the target of asset management is usually the investor or asset owner, who aims to make a profit. In the field of facility management, the service is usually provided to enterprises with industrial background, enterprises with production and operation activities, and even public facilities under the name of the government that do not aim at profit, but still have enterprises, production, and operation activities. We call this: facility management.

In terms of the standard definition of facility management and even the interpretation of related terms, China has not made any changes or innovations, but basically follows the standard translation of international facility management and interprets the concept as the standard term.

In 2013, China's Standardization Administration organized a team of experts to develop Chinese standards related to facility management. As of 2022, four national standards have been introduced, and these standards are equivalent to international standards. I myself have participated in both the development of the standards and later become a member of the evaluation committee for the whole standards. The basic concepts and definitions of facility management in China are fully in sync with the international community. However, because of the figurative nature of facility management in China, it is often used as an alternative term to equipment management, which is why facility management is often misjudged in the application process. This is one of the major irrevocable flaws of the Chinese translation process, making facility management known only to a few educated professionals at the cognitive level, but not to the general public.

In terms of education, we don't have large-scale training on how to promote facility management, and no large-scale education, training, or other related work at the level of the Chinese society. The early promotion was mainly done by me in Tsinghua University and Mr. Cao in Tongji University, who gave lectures on facility management at our universities respectively. I also started to work with my colleagues to start lectures on facility management at other universities in China.

Fortunately, in 2021, Philip Lo and Alex Lam, two well-known veterans in the field of facility management, and I jointly advanced the formal introduction of the facility management course to the Leiden Business Academy in China. We officially started the first training course in 2021, and continued to run it afterwards. This is the first systematic and complete introduction of a facility management course in China. Although this course leading to the professional designation of Facility Management Professional (FMP) is at the introductory level, it is of great significance to China.

I would like to thank the Leiden Business Academy for their support, as well as Dean Li Qun of the academy. Like many female leaders in the field of facilities management, Li's contribution to the knowledge and promotion of the field to the advancement of the industry has been irreplaceable.

In addition to training and educating professionals through formal lectures in university classrooms, more work is done through industry conferences to promote systematic knowledge of facility management.

In 2020, China's Standardization Administration organized one of the largest industry conferences on facility management since the concept entered China, with over a thousand attendees. More sessions were scheduled for presentations on aspects related to facility management.

For comparison, the first conference on facility management in China was probably in 2005 and was hosted by me. I invited experts from the US and people from the medical care and medical facilities industry in China to participate in the conference. The conference had around 50 participants and was called the International Symposium on Medical Facilities Management. It was the first international conference held in the field of facility management in China.

Judging from the number of participants before and after, the concept of facility management has been recognized by more people in the community than just professionals.

The role of the Royal Institution of Chartered Surveyors (RICS) in promoting professional conferences on facility management has been tremendous. In addition to the annual conference, there is a module on facility management at different thematic conferences. The RICS has also introduced a professional certification in facility management, which has lasted for several years. The academic standard, rigor and high academic level of its professional certification is beyond the power of other industry associations in China.

In addition to the RICS, Hong Kong Institute of Facility Management (HKIFM), Taiwan Facility Management Association (TFMA), Japan Facility Management Association (JFMA), Shanghai Property Management Association and other associations, all held relevant forums together with different organizations in China, which played a very good role in comprehensively promoting the sharing of knowledge and practices related to facility management.

The Shanghai Property Management Association, a local association in China, is also an active promoter in organizing annual forums for its members. Likewise, the rich local practical experience of Chinese facility management accumulated during this promotion has had a positive impact in the Asia Pacific region.

In fact, for seven to eight consecutive years, Jumbo Chains have sent lecturers each year to China's neighboring countries, such as India, Pakistan, Russia, Kazakhstan, Thailand, and Vietnam, to train professionals on the basics of facility management and the application of facility management in science and technology parks. This has had a positive impact on promoting relevant knowledge and achieved great outcome among the governments of these Asian countries.

I have also represented China to deliver presentations and lectures on China's facility management in the United States, South America, Japan, Hong Kong, and other places. I was installed a Fellow by the Hong Kong Institute of Facility Management and was appointed FM examiner by the RICS and I have also contributed to sharing knowledge in their training sessions.

IFMA China was established later, around 2015, and has played a slightly lesser role in promoting facility management than these associations. One of the main reasons is that at the beginning of its establishment, there were some mistakes in the design of the structure or the organization, which led to its failure to play its proper recognized role in the later development of facility management in China.

To this day, this organization is still paying the price for the mistakes made back then. Admittedly, this is a matter of great regret. But even so, IFMA China is still one of the major and indispensable driving forces of facility management. However, in 2022, we are pleased to see IFMA China embarks on a renew strategy to support FM in the country.

III

Facility management is often a source of pride and admiration to its peers because of its complete and huge product system. IFMA provides a highly abstract categorization of the three P's and one T, which refers to People, Process (business), Place, and Technology respectively. In China, to address how to combine facility management with Chinese local practice, and how to transform and make plans for the product system, we analyze and reorganize from the perspective of the three Ps.

The first P is for People as well as business. For these businesses, we disassemble them strictly according to the production chain of enterprises or the industrial chain of industries. A service chain was formed to offer support for the industrial chain and between the links of the industrial chain. Furthermore, the support includes administrative support, industrial services, corporate services, etc., respectively. These services are then sorted out to form a complete service system again.

The second P is for Process, or the whole business life cycle of real estate held by enterprises. Based on the five core demands of site selection, project operation and maintenance, spatial services, performance, and evaluation, we split, categorize, and organize the related business of FM again. The result formulates the knowledge map (brain map) of Jumbo Chains. If it is categorized and expressed in the way of the periodic table of chemical elements, a product classification map is created. If it is divided according to roles, a facility management role structure map of the product is created. If you follow the dimensions of FM company growth, a quadrant diagram is derived. This becomes the strategic FM map of my company.

The third P is Place. Facility management mainly targets the real estate properties owned by industrial and commercial enterprises, which are broadly classified by function into five categories. The first is the decision-making headquarters and office space, which are its main representation. The second is the research and development center. The third is manufacturing sites, which is more often reflected by factories. The fourth is logistics and warehousing. The fifth category is the sales terminal. According to such a classification of *Place*, the complex business of FM is further analyzed by breaking down the various business components so as to be understood, organized, and refined, in order that we can better serve the occupants.

Such analysis helps the specialization, refinement, and standardization of each business unit (BU), which then forms a complete Standard Operating Procedure (SOP). The different systems, subsystems, and complete service systems composed of SOPs, furthermore, help to transform these knowledge-based product services into information technology products through digitalization. It is a relatively perfect combination between the practice of facility management in China and the Internet within other fields of technology.

Such a service system not only helps facility management enterprises to grow in a more specialized and sophisticated way, but also helps customers to participate problem resolutions together, providing a complete database of sample solutions, which in turn helps to meet the FM needs of customers and achieve their strategic goals. In this way higher value is created for customers.

IV

Facility management is a complete service product system based on the whole life cycle. It is presented professionally by SOP, which not only brings long term growth and profit for FM enterprises, but also creates higher value for customers and society. The presentation of these values is based on the growth objectives of the industry or enterprise, and the many needs formed in the process of development to present. In China, we divide the behavior of industries to realize their own construction and achieve their strategic goals into two groups of systems. One is the supply side and the other is the consumer side. Facility management plays an irreplaceable role in supporting the continuous change and innovation of both the supply and consumer side of enterprises to create value together with customers.

On the supply side, it is applied to many aspects from land, real estate, and construction, to labor supply, and employee services, etc. It also has a large number of applications in energy management, including the integrated application of five energy-saving means, such as energy-saving from design, structural energy-saving, energy-saving from IT technology, energy-saving from products, and energy-saving from behavior, which greatly reduces the cost of production and manufacturing.

In the non-core aspect of spare parts supply, warehousing and logistics, facility management improves business processes, increases efficiency, and reduces the logistics costs of enterprises. In terms of finance and taxation, through financial balance and other means in the effective FM strategy, the budget and implementation can be controlled scientifically and rationally in a sustainable manner. Its value in the supply side is presented in the practice of Jumbo Chains, which has a large number of cases that are commended by customers. In terms of value creation, different combinations of solutions together constitute the support system for value creation (see Figures 1, 2, and 3).

China's economy has been growing at a high rate, but facility management in China is relatively limited on the consumer side. Although there are many noteworthy cases, it does not constitute such a support system which

Periodic table of Products for Industrial Facility

PM						Facility Service				Support Service		Industrial Support		
Public Space	**Security**	**Cleaning**	**Gardening & Greening**	**Parking**	**Maintenance**	**Project Management**	**Energy Management**	**EHS Management**	**Safety Management**	**Space Service**	**Administrative Support**	**Production support**	**Investment & Site selection services**	**Industrial operation**
Road Maintenance	Sentry	Normal Cleaning	Greening	Parking lot Operation	Operation & Maintenance	Decoration Management	Energy Assessment	Water Treatment	Industrial Safety Management	Moving Service	Conference Service	Professional equipment operation	Marketing & Promotion	Policy advisory service
On-site Inspectation	Temporary	Specific Cleaning	Pruning	Charging	Equipment Inspection	Acceptance Inspection	Energy Reconstruction	Sewage Treatment	Public health risk Management	Daily Cleaning	Recruitment Service	Professional equipment maintenance	Investment promotion service of the park	Industrial Investment promotion service
Municipal Facilities Maintenance	Central Monitoring	Recycling	Green plant Leasing	Car washing Service	Centralized monitoring & emergency command center	Transformation & Renovation	Energy Saving	Environmental Monitoring	Environmental risk Management	Equipment Maintenance & repairing	Fitness Center Service	Financial leasing service for equipment	Enterprise site selection consulting service	Industry Incubation service
Equipment Operating & Maintenance	Patrol	Temporary	Daily Maintenance		Equipment Maintenence		Energy Metering		Disaster risk Management	Cleaning service of Production line	Mailing Service		Rental & sales agent service	Industry fund
	Emergency Security	Thorough Cleaning	Maintenance of Horticultural facilities		Equipment Repairing		Transfer service for Electric meter		Social security risk Management	Communications Service	Catering Service			Pilot test platform
	Parking	Cleaning after Decoration					Collection & Payment Service			Plant Maintenance	Procurement Service			
		Disinfection & Pest Control								Initial Cleaning	Commuter Service			
		Waste Treatment								Apartment Management				
										Visual Identity Management				

Figure 1 Periodic table of products for industrial facility. (*Source:* Y. Qingxin).

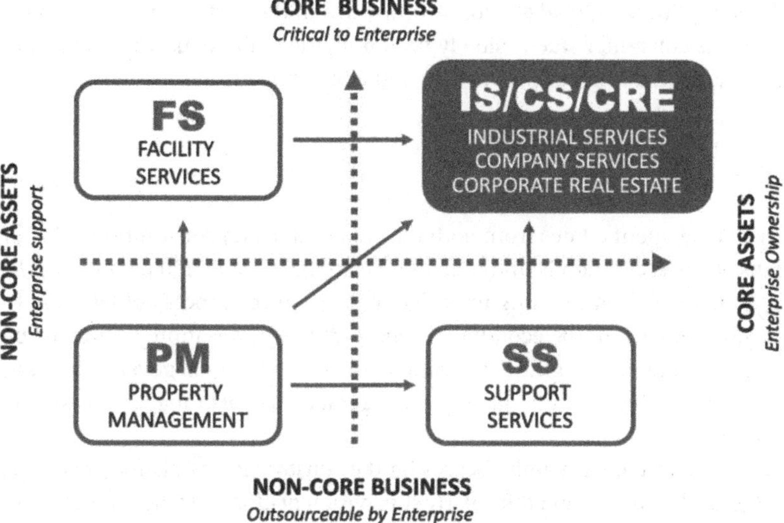

Figure 2 FM role structure map for enterprise services. (*Source:* Y. Qingxin).

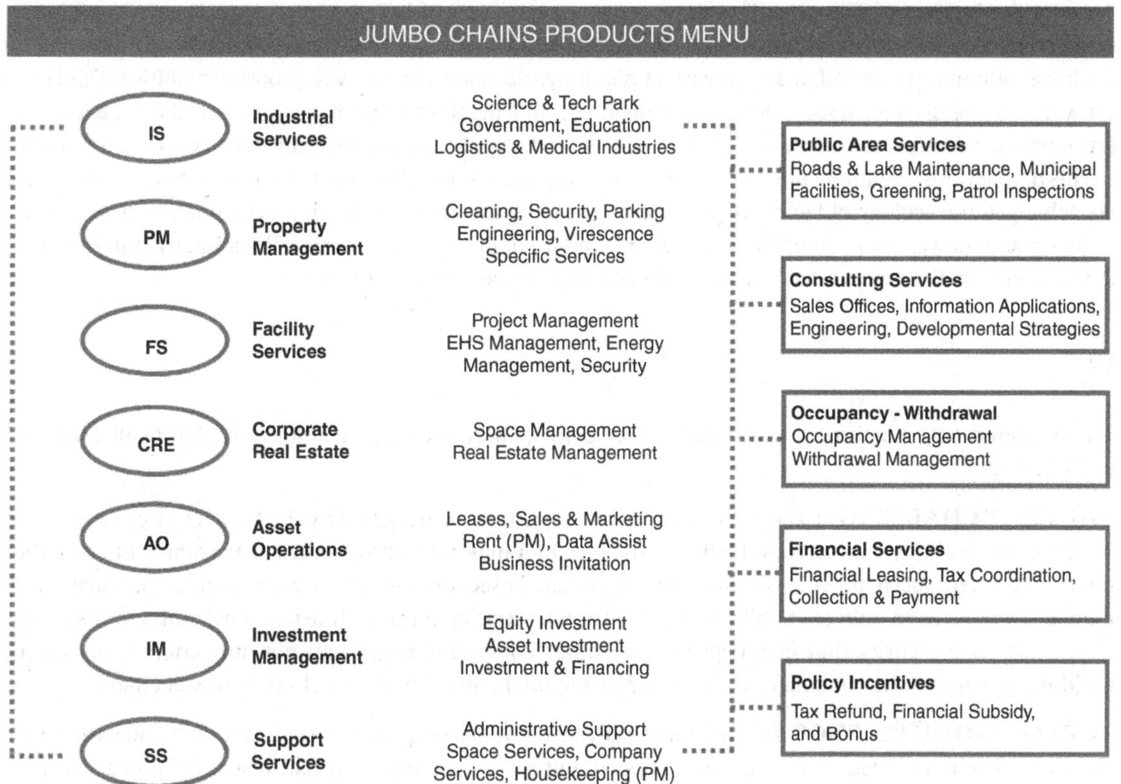

Figure 3 Jumbo chains product menu. (*Source:* Y. Qingxin).

offers many solutions like on the supply side. With the global economic downturn in the past two years, China is no exception. Exploration and practice on the consumer side is slowly becoming the main focus after 2019, and we look forward to better results. Jumbo Chains is doing our part to contribute in this area.

V

The scenario of value creation in facility management comes from both the supply side and consumption side of the industry, but the most effective means of value creation is undoubtedly the application of high technology. In the 30 years of reform and opening-up, China has been learning to bridge the gap. In the process of learning to catch up, it has made a lot of important achievements in the economy. However, the world is more impressed by China's achievements in field of science and technology, digital life, digital management, digital governance, and other aspects of society. Benefiting from these achievements, facility management has also made remarkable development in its technological applications.

In terms of technology application, facility management mainly focuses on the construction of platforms such as operation management platform, decision control platform, and third-party supervision platform. At the same time, in the construction of different scenario-specific management pivot software system, facility management fully implements the three aspects of efficiency, effectiveness, and efficacy of the enterprise value presentation. In the use of mobile technology on the work side and consumer side, the application of APPs and mini-programs makes users and executors have a better working experience with the input and output of data flow directly linked to the cloud technology.

As for the Internet of Things (IoT) technology, a large amount of IoT and interconnection information directly support applications such as system integration, algorithm, and Artificial Intelligence (AI) in the platform end through the huge computing power offered by private, public or hybrid clouds. In this way, proprietary software and integrated software systems of directed scenarios can be structured, allowing third-party management by the FM company, property owners, industry associations and government, to let the applications to be availed in various Chinese cities.

This digital transformation and upgrade have not only created excellent value for the customers and society, but also changed the ecology of facility management. It can be asserted that these practical applications provide the industry with newer, more complete and higher standard data support in terms of benchmark, but in many areas, it also leads the direction of the development of facility management in the world.

VI

As we entered 2022, the third year of the global pandemic, I perceive four challenges faced by facility management in China today.

THE FIRST CHALLENGE: The public health security concerns triggered by the COVID-19 epidemic have verified the vulnerability of society and cities in the face of a major catastrophe. It has become a huge challenge to build a resilient social prevention and control system based on the safety management practices of facility management so as to address public health security, as well as natural disasters, environmental security, and large-scale civic security that may occur at any time. These challenges will not only come from a particular building, or a particular customer, but be posed by regional and even city-level safety management.

THE SECOND CHALLENGE: As the Chinese government has proposed to improve urban and community governance capacity of cities, to provide richer social and public services to urban residents in order to achieve a gradual increase in people's satisfaction, urban management and community management has thus become a huge challenge in China's urbanization process. Facility management companies can meet the challenge and offer solutions. Jumbo Chains has such practices in facility management in both old and emerging cities in China, with a complete total facility management (TFM) practice. The antiquated systemic problems encountered are in urgent need of being dismantled one by one and given perfect answers (see Figure 4).

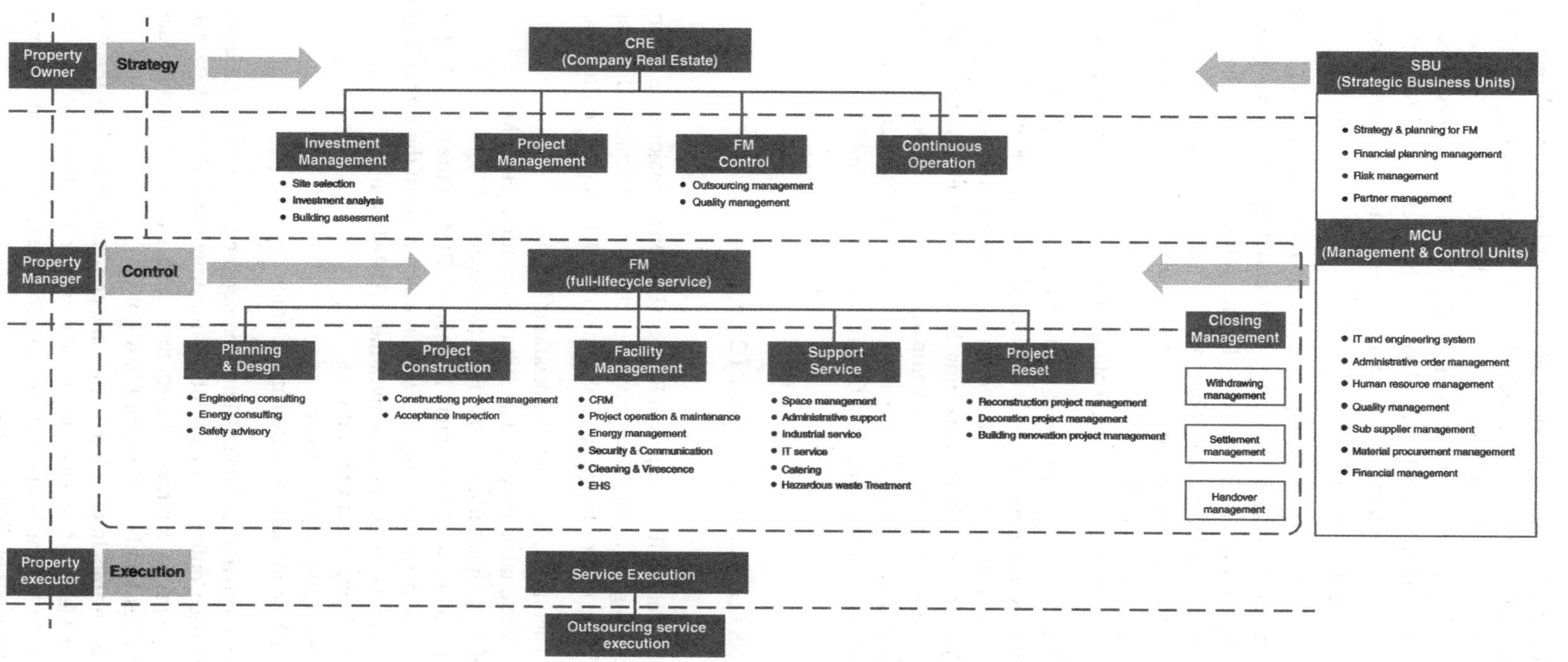

Figure 4 Structural organization & relationship. (*Source:* Y. Qingxin).

THE THIRD CHALLENGE: In cities, communities and new urban areas, usually within a scale of 100 square kilometers, there is a complete range of real estate and a diversity of customer types and needs. Under such circumstances, how can we integrate FM with AM and PM, and at the same time, effectively organize and collaborate with the country's governance structure. As an example, for the third-party organizations such as NGOs and commercial and industrial operators who are the key players in the operation of new urban areas? This is not so much a process of integration and innovation of facility management, but a social practice of facility management in such a large-scale and complex urban scenario.

FOURTH CHALLENGE: In the face of such a complex application scenario on a large urban scale, an ecological body is formed by the comprehensive application of science and technology, the new technologies collected in digital life, digital management and digital governance, as well as the integration and connection between different technologies. Not only should we support the orderly implementation of various management services, but more importantly, we should establish an open system, so that this system becomes a self-consistent, coexisting, progressive, and continuously optimized ecosystem. Again, this is a challenge for facility management and IT practitioners. There is no doubt that this challenge is an exciting one, because the power of this challenge lies in the future of facility management.

LAST BUT NOT LEAST: During the promotion and application of facility management in China from 1992 to 2022, Alex Lam, IFMA Fellow, is considered to be an influential and extremely important figure in China. Alex has contributed a lot to the introduction of FM management knowledge and technology, as well as to FM education, technology, and the building of associations in China. He has been both a mentor and a friend to me, and has helped me tremendously in the development of my company JUMBO CHAINS. I would also like to thank the many professionals in the facility management industry from Europe and the United States for their tremendous and outstanding contributions to the development of facility management in China.

Yu Qingxin
President of JUMBO CHAINS Real Estate Management (Group) Co., Ltd.
Beijing, China

Mr. Yu Qingxin graduated from Beijing Technology and Business University, majoring in Enterprise Management, and furthered his study at the MBA program at Tsinghua University School of Economics and Management. In 1992, Mr. Yu established JUMBO CHAINS Real Estate Management (Group) Co., Ltd. in Beijing, in which he serves as the president.

- Member of International Facility Management Association (IFMA)
- Fellow of Hong Kong International Facility Management Association (HKIFMA)
- Member and Examiner of Royal Institution of Chartered Surveyors (RICS)
- Vice President of Certified Commercial Investment Member (CCIM) of the China Region
- Vice President of Building Owners and Managers Association (BOMA)of the China Region BOMA
- Investment Expert of International Commercial Investment Association
- Expert of National Facility Management Committee
- Senior Property Manager with 30-year experience
- Member of China Property Development and Research Center

As early as 2006, Mr. Yu introduced the concept of facility management (FM) to China, and successfully organized the First International Exchange Conference on Facility Management in China in 2007. Mr. Yu has been teaching FM in science and technology parks, commercial asset management, and real estate management at Tsinghua University and served as a professional examiner for FM major.

Mr. Yu has been invited to various international academic conferences such as the World Workplace Annual Conference of International Facility Management Annual in the US and Asia, the South America Facility Management Forum in Sao Paulo, Brazil, and the Annual Conference of Facility Management in Hong Kong, China, promoting the development of FM in China.

2.4

40 Years of Facility Management

Philip Lo

How It All Began for Me

Facility Management is my second life. My first life was that of an architect. After my graduation from university in the 1970s, I practiced as a professional architect, and even established my small architectural practice in the City of Edmonton in Canada.

As an architect, I have always been fascinated by technology. After my graduation from architectural school, I was fortunate to join an architectural firm in Montreal, Canada, Bobrow & Fieldman, who was a pioneer in the use of CADD Technology. Back in those days (1974), it was a huge investment for a small architectural firm, where they not only had to invest in very expensive hardware and software, but even had to purpose build a specially air conditioned room in the basement of their office, in order to accommodate this new technology. That was my first encounter with CADD, which had a significant impact on my career in Facility Management in later years.

After eight years of practice as a professional architect with my own practice in Edmonton, I decided to move to Vancouver, as it was a much friendlier environment to live in for the family, but definitely not an easy one for a small architectural practice to survive. I would either had to join the ranks of small architectural firms (usually one man outfits, carrying out small residential renovations projects which I did not find challenging), or be part of a large architectural practice (which I would find extremely hard to adjust to after years of practicing as the sole principal of a small firm). So, I had to find a way to survive and thrive.

As an architectural firm, I was often involved in corporate renovations and interior projects, and one of the first questions I always had to ask my clients was "do you have any existing drawings of your space ?" The normal answer was usually, "yes, but they are out of date and incorrect" or, "I do not know where they are and it would take me forever to find them" Whenever there is a need, there must be a solution somewhere !!! I saw a problem there and one where CADD might just be the answer? By 1986, CADD was gradually becoming more affordable with the growing popularity of AutoCAD by Autodesk among architects, so I invested in my first CADD System, which comprised of a large desktop computer, AutoCAD, and associated software, as well as a large expensive and clunky ink "plotter." All these equipment took up most of the space in my tiny 250 sq. ft. office in the Art Deco Marine Building in downtown Vancouver which I loved. Despite this humble beginning, I was in business!!!

The idea was to convince building owners, for me to measure up each floor of their office building, and then using CAD to draw up the various floor plans of the building, plot them out, and then print them in a small conveniently scaled format, to form a neat handbook. The key selling point was that after the initial documentation

Facilities @ Management, Concept - Realization - Vision, A Global Perspective, First Edition. Edited by Edmond Rondeau and Michaela Hellerforth.

work was done, any future updates and changes could be done easily and neatly through CAD, keeping track of all the changes throughout the LIFE of the building. That was my first encounter with LIFE.

Cycle Management of a Building

Building owners liked this concept, and soon, they were asking me to put in additional information on each floor plan in the booklet, such as names of the tenants, their square footages, their monthly rental rates, as well as their lease expiry dates, etc. and the business gradually flourished. I named the company PLANEX, as it contained first letters from the initials on my name, and then adding the letters AN, would well reflect the nature of the business which was documenting floor "PLAN"s of buildings with CADD. The suffix EX that was added to the name could stand for EXcellence, EXact, and Expansion, etc. which are all good attributes for a young growing business. So, my first FM business, Planex, was born.

A fateful day happened in the summer of 1988, when the local Autodesk dealer introduced me to Bruce Forbes, CEO of ARCHIBUS, who at that time was promoting his new software personally, introducing it to architects who were using AutoCAD during those early days of CAD Technology.

The ability to INTEGRATE CAD drawings, with DATABASE information blew me away. This fitted in exactly with what I was doing, which was documenting floor plans with CAD, but I needed to type in all the other information manually on each floor plan. The ability to click onto a drawing, and automatically retrieving all the information about the space, such as square footage, leasing information, and tenancy information was exactly what I needed, and formed the basic concept for FM Space Management.

Between the years 1989 and 1992, my firm in Vancouver, Planex FM, implemented several basic Space Management projects using ARCHIBUS, for the local universities, hospitals and the Workers Compensation Board, etc. and was establishing a fairly successful small practice. I started hiring computer programmers and CAD technicians, and not architects, for my business. On a fateful trip back to Hong Kong in 1991, one evening, I took the Peak Tram, which was a popular tourist attraction, up to the Peak, the highest lookout point on Hong Kong Island. There, overlooking the entire North shore of Hong Kong Island, with all the bright lights of the city of a few million inhabitants and hundreds and thousands of buildings below, it dawned on me that my calling was to promote FM and FM Technology back to Hong Kong, the place where was born. I made the bold decision to start a FM practice in Hong Kong, thousands of miles away from home, not knowing a single client, and without ever having worked in this "foreign" land before. This was the start of my "second" life, a life in FM.

My Introduction to IFMA

Bruce Forbes introduced me to IFMA, and I attended my first IFMA conference in Seattle in 1989, which was just a short drive from Vancouver. It opened up my eyes to the World of FM, which in those days, was more of a showcase for Office Design and Office Furniture. I still had no idea what FM was all about, but was extremely attracted to the use of software in the management of office space. With limited knowledge in FM and Technology, I started doing projects in Vancouver using the Archibus software. Due to the lack of support and with limited communications (pre-internet days), I basically stumbled my ways through the maze of FM technology. Eric Teicholz's book on Computer Aided Facility Management also had a great influence on me, and taught me a lot about the application of technology for FM, particularly from an architect's perspective. I owe a lot to Eric's insights and knowledge sharing and it is only fitting that I nominated Eric to become an IFMA Fellow, an honor which he well deserves.

In February 1992, I went to Hong Kong and established a company called Planex Hong Kong Limited, and started promoting the concepts of Facility Management, as well as the use of the ARCHIBUS software to perform tasks such as Space Management, Lease Management, Telecommunications Management, and Maintenance Management to the local community. It was not easy, as nobody had ever heard of the term Facility Management, and many even thought it was about the maintenance of underground utilities and sewer systems.

In the fall of 1992, Dennis Longworth, CEO of IFMA at that time, together with Don Young, Director of IFMA Marketing, was on a trip to Japan, attending the Japan FM conference by JFMA (Japan Facility Management Association), which was already established at that time. He suggested that if I could gather a group of FM practitioners in Hong Kong, then he would be glad to extend his trip and pay a visit to Hong Kong, and to give a talk introducing the concept of FM, and the IFMA organization to this group. By that time, I have already spent about 6 months in Hong Kong, introducing the concepts of FM to the local community, and started to get to know a few people managing different types of facilities. I managed to send out an invitation to about 15 people to this meeting, ranging from architects, interior designer, engineers, project managers and academics, who were involved with facilities such as universities, hospitals, and airports. After an introduction of IFMA to the group, Dennis asked how many of us would like to join and become members of IFMA. Ten of us raised our hands, which just made the quorum necessary to officially form the FIRST chapter of IFMA outside of North America, and IFMA Hong Kong Chapter was born September 14, 1992, making the organization truly INTERNATIONAL!

The Growth of IFMA in Hong Kong

After the establishment of this formative IFMA Chapter, which none of our initial 10 founding members quite understood what the concept of the 3Ps (People, Place, Process) was really all about, we were really on our own as how to grow a new unknown profession, with total freedom and direction to grow the chapter in whatever way we see fit. Since I initiated the first meeting and invited everyone to attend, I was unanimously voted to become the Chapter's first president. One of our first objectives was to learn more about what FM was really all about, and to grow our membership. To achieve these objectives, we organized monthly luncheon meetings at the Hong Kong Club, where one member would give a talk about their facility each month, or on how they manage their facilities, be it an airport, university or hospital or commercial operations, etc. In order to grow our membership, each member was asked to bring along a guest, who could be a friend or a colleague, as long as they are in a similar profession, and then we will try to get them interested to join and become a member of IFMA. I used the presentation which IFMA left behind for me (prior to the days of PowerPoint) to show the many potential prospects who subsequently joined IFMA as members.

We also organized our FIRST FM conference in Hong Kong at the Regal Kowloon Hotel on April 30th, 1993, and invited speakers from the US, the UK, Australia as well as local speakers to speak on the topic of "Facility Management – International Practice and the Prospects for Hong Kong." This was shortly followed by another conference in September at Ocean Park. During the following years, we continued to organize annual conferences, inviting many local and international speakers, to continue to education us about the practice of FM and to spread the gospel of FM. Many of these speakers from IFMA remained lifelong friends, and I am so indebted to every one of them, in sharing their knowledge and enthusiasm about FM to our group of pioneers in Hong Kong. As a result, our membership grew steadily to over 200 members in the subsequent few years.

The Hong Kong facility management conferences were very successful, and we continued to organize many FM conferences and workshops and later changed the conference title to the Annual Asia Facility Management Conference, though always held in Hong Kong, as there were no other chapters in Asia.

Despite all our hard work and effort in promoting FM, the 1990s were indeed a very difficult period for Facility Management to be recognized and accepted into the mainstream practice in Hong Kong. During those years, Hong Kong was enjoying tremendous economic growth. With ever rising property values, most property and facility owners were more concerned about how much and how quickly they can sell or lease their assets for. No one paid much attention to quality maintenance and management, and much less to FM concepts such as life cycle management, assets value enhancement and other long-term benefits of FM. Facility Management, or Maintenance Management, as it was often perceived as in those days, was simply a necessary evil, which they were trying as hard as they can to minimize such expenditure, just to keep the facilities operating at a minimal cost without breaking down. The concept of FM was indeed a very hard sell.

However, in 1998, there came a sudden and significant change. The Asian financial crisis which started in Thailand decimated the Hong Kong property market. Property owners were suddenly concerned about how they can make their properties more attractive and be able to offer better quality and value to their potential customers. Also, managers became more concerned about how to operate their facilities efficiently and effectively. With influence from the UK, the concepts of Out-tasking and Outsourcing started to emerge, as a means of reducing overhead and letting other professionals worry about the problem of running a facility. All these concerns relate back to Enhancing Value of the Properties and Operating Facilities more Efficiently and Effectively, thus turning to the ready acceptance of Facilities Management, as an alternative approach to traditional Property Management practices.

Promoting FM Technology in the 1990s

I look back with a smile on how I promoted my FM business in the early days. Carrying my heavy blue and white screen laptop running on DOS, which was an expensive novelty in those days, I went around promoting Facility Management and the concept of using Technology to manager facilities to various prospective clients. Most of the clients I met were with major corporations, universities, private developers, and government departments. It was not an easy sell, but most of them were intrigued by the amazing ability to link graphical floor plans with database information, and that they can access much needed facility data "visually," and generating useful reports graphically. so a few of them were brave enough to allow me to try to do "trial" projects for them.

The early projects were met with both success and failures. Success because we implemented the projects successfully, but failures mostly because we were slightly ahead of our times. One of the more memorable projects we implemented in the early days, was one for the Hong Kong Housing Authority, one of Hong Kong's largest government departments, responsible for providing housing to over three million people in Hong Kong. In each of their housing projects, there is usually a shopping center associated with the development, for the convenience of residents living in these massive housing units. We elected to use the Lease Management Module of Archibus, to demonstrate to them the ease of accessing Leasing information graphically, which would be an extremely useful tool for prospective tenants, trying to identify suitable store locations available in these shopping centers.

With a limited budget, we agreed to do a trial system for 16 shopping centers. We had to draw up outline layouts and floor plans of these 16 shopping centers, many requiring on site measurements for verification, and then identifying each rental unit through polylines. Then using software, we linked appropriate leasing information and database to each of the units. The most challenging issue was that they needed the system to use and show Chinese characters, as much of the database information were in Chinese. With much difficulty, through our skillful technicians, we developed a Chinese version of the system. The system was a huge success!!! We were even asked to demonstrate the system at the Housing Department's annual showcase exhibition, where we had a desktop system setup together with a big printer which instantly printed out graphical reports and plans based on interested retail tenants' enquiries.

The following step was to roll out the system to the over 100 shopping centers for the Housing Authority, which would have been a huge massive project for a young company. We were asked to discuss the project with the Authority's IT Department (known in those days as the Computer Department). However, even after initial discussions, they dismissed our proposal, they found that the system we were proposing was totally incompatible with their existing systems. They were using a mainframe CAD software called CADAM as their main CAD drafting system (we were using AutoCAD) and the computer system they were using was an IBM Mainframe AS400 system (while we were using the simple DOS operating system). So, this huge potential project met a premature death. It is interesting to note that a few years afterward, the entire department switched over to using AutoCAD and Desktop computers replaced the old main frames.

Unfortunately, we were just a little bit ahead of the times.

Another similar project demonstrating the impact of Technology on Facility Management is an early project we did for the MTR (Mass Transit Railway). On top of several of their subway station locations, the MTR also developed housing units together with shopping centers. We used a similar approach to demonstrate the effectiveness of a "visual" lease management system integrating floor plans with their leasing information. We completed a project for 4 Shopping Centers, and it was successfully implemented. The system was installed and used on the ground floor administrative office in one of their projects. One day, the Director of the MTR dropped by and was so impressed by the system that he asked if we could link up the system to his computer (which is on the 20th floor of the office building next door), so that he can access the information directly from his desktop computer? During those pre-internet days, the only possible way was through physically linking the computers through LAN (Local Area Network). We checked with the local cable service provider and the cost of running such a line from one building to the 20th floor of a building next door, was three times the cost of our entire project, so the idea was dropped and unfortunately, without the director's support, the project also died after a couple of years of use. With the emergence of the internet a few years later, it would have solved all these problems. Again, we were a just little bit ahead of our times.

During the late 1990s, technology played an ever increasingly important role in Facility Management. The ability to distribute and communicate Facility database and information easily through the Internet anywhere wirelessly was critical to the rapid adaptation and acceptance of Facility Management. Therefore the word Technology, was appropriately added to the 3Ps (People, Place & Process) in IFMA's new definition of FM in the late 1990s.

Facility Management – East and West

Facility Hong Kong is the perfect confluence of East and West. With a mainly Chinese population, embracing Chinese culture and values, but at the same time, being a British colony for 99 years, Hong Kong embraces the Western culture, but with a mix of Chinese tradition and heritage. Therefore it was very interesting to see how Facility Management evolved in this very special city. Through IFMA, I introduced the term "Facility Management" to Hong Kong in the early 1990s. During that time in Hong Kong, the Management of Buildings was mainly referred to as Property Management, and it was mainly applied to the maintenance and operations of Residential Properties. In the case of larger institutional facilities, such as universities, they often called it Estates Management, which is a very British terminology. The Institute of Housing was a very well respected organization in Hong Kong, with its roots from the UK, and it promoted the quality management of properties, especially applied to the thousands of public and private housing (apartment) units in Hong Kong. So Property Management was very dominant in Hong Kong in those days.

In North America, during the 1980s when IFMA was first started, Facility Management was mainly focused on the management of Corporations facilities and office interiors. I remembered that the first time I went to the Seattle IFMA conference in 1989, most of the major exhibitors and sponsors were Furniture manufacturers such as Steelcase, Sunar Hauserman, Hayworth, and Herman Miller, and many of the exhibits were focused on Office Space Management and Interiors. Even in the booth showcasing ARCHIBUS, it was mainly focusing on Space Management, (may be because Bruce Forbes started ARCHIBUS from the architectural practice of Jung Brennan focusing on Space Management for corporate clients.). However, this was the aspect of FM, and the integration of CAD with database, which impressed me and led me to the field of FM.

To put it in an oversimplified way, initially FM in North America was more focused more on Space and Interior Design and Management of Public and Corporate Facilities, and FM in Hong Kong, due to its British influence, was mainly focused more on Operations and Maintenance Management of Residential, Commercial and Retail Properties. This became more apparent in 2000 with the formation of HKIFM (the Hong Kong Institute of Facility Management). The initial makeup of the local IFMA membership was mostly architects, interior designers, and managers of corporate and institutional facilities such as hospitals and universities. Whereas the makeup of the initial membership of the HKIFM was mostly Engineers and Building Surveyors (which is a professional much more prevalent in the UK), and many property management companies (of residential, commercial and retail properties) joined the membership. But the objectives of both organizations are basically the same, which is to educate, improve, and promote Quality Management of Buildings and Facilities. With the growth of the IFMA Hong Kong Chapter and the HKIFM in recent years, the term Facility Management is becoming more of an amalgamated terminology embracing all aspects of the Workplace, and being readily adopted and accepted by the community, in whatever way it applies to their facilities. We see many public and institutional organizations, changing their departmental name, from Estates Management or Property Management to Facilities Management, and many property management companies and corporate departments, changing their names to reflect the Facilities Management nature of their work.

The HKIFM also played a major role and should be credited with the promotion of FM in Hong Kong as well as in China. As mentioned, it was initially formed by a group of local Surveyors, mostly Building Surveyors, whose practice was more focused on the operations and maintenance of buildings. In a similar way that the Surveying profession was introduced to Hong Kong through the RICS (Royal Institute of Chartered Surveyors) and after many years, Hong Kong established its own HKIS (Hong Kong Institute of Surveyors) in 1984, adapting the profession to more local practices. Similarly, the profession of FM was introduced to Hong Kong through IFMA, and after a few years, the HKIFM was established as the local FM Association promoting the practice of FM in a much more localized manner, adapting the practice to suit more local needs. They also managed to solicit the support of government departments, which greatly helped in the promotion and adoption of the profession by the Hong Kong community.

As part of its effort to promote FM in China, through the assistance of government grants PSDAS, the HKIFM also organized FM conferences in different cities in China every year; this greatly helped in promoting the concepts of FM to China

The Growth of FM in China

Over the last 40 years, the transformation of China has been phenomenal. There has been mass migration of people living in rural areas to urban areas in China, and with that, the construction of many new buildings and facilities. However, even in the early 1980s, when I first stepped foot into China, there was hardly anything even remotely called Property Management, much less Facility Management. At that time, buildings were very poorly maintained, and everyone was focusing on building new buildings, and no attention was paid to maintaining the existing buildings. People were focusing on rapid growth of the residential and industrial sectors and providing new housing and facilities for the new industries. Providing long term quality management for buildings and facilities would be the last thing on their mind.

Most interesting and perhaps most telling of the perception of Property Management in China in those days, was the term used for Property Management, especially at some of the major universities in China. The Department responsible for the Operations and Maintenance of the university's facilities was known literally as "The Behind (the Scene) Hard Working Service Department" (后勤服务部) which refers the low level team busy providing maintenance as a back of house service. It was considered as labor intensive work for low level worker responsible for the operations and maintenance of the buildings. The change in perception of FM which is to raise the work from the Boiler-room to the Boardroom was extremely difficult. Concepts of proper Property Management was initially introduced from Hong Kong through organizations such as the HKIH (Hong Kong Institute of Housing) which conducted courses and introduced concepts of proper quality management for buildings to China, but was mainly focusing on housing managements activities such as Maintenance, Security, Cleaning, and Landscaping Services.

With the support of IFMA, we organized the First Facility Management Conference in China in 1997 in Shanghai. At that time, FM was literally unknown in China and was a foreign concept barely understood by a few of the multi-national firms operating in Shanghai.

After much thought and deliberation after the conference, I thought that one of the best ways to introduce FM to China, was through the then growing Property Management industry, so I started my mission of introducing FM to China, through talking to the various Property Management Associations in several of the major cities in China.

From 2001 to 2003, together with Oscar Chan (IFMA Fellow who unfortunately passed way a few years ago), we started talking about FM to key people in the various Property Management Associations in Beijing, Shanghai, Chongqing, Chengdu, Shenzhen, and Guangzhou. However, it was a hard concept to for them to understand, especially in China in those days, and most of the people we spoke with dismissed us as promoting foreign concepts which were not appropriate for use in China. However, there is an individual whom I must mention here, Mr. Yu Qingxin, who at the time, was the young President of the Beijing City Property Management Association. He owned and ran a small property management company in Beijing at that time and he took an immediate interest in the concepts of FM. We communicated continually after our initial meeting in 2002, eventually leading up to the first of our many World Workplace visits together, which took place in Dallas, Texas, in 2003. It opened his eyes to the world of FM. Over the last 20 years, he actively promoted FM and transformed his own company, from a traditional Property Management company to an innovative Asset Management company, embracing and practicing all aspects of FM. He also worked with various government departments actively promoting concepts of FM, leading to the participation in setting up ISO 41000 standards for Facility Management in China as well.

Multinational management firms such as JLL, CBRE, and Savills also helped promote the concepts of FM in China. They applied good FM practices to many of their corporate and global customers in China, thus enabling many of these organizations to see the value and benefits of FM in the Chinese community.

After many years, the Shanghai Chapter of IFMA was eventually established in 2010 and the first World Workplace China conference was held in Shanghai in 2013. In subsequent years, other chapters were established in Beijing, Shenzhen, Guangzhou, and recently in Suzhou. However, membership numbers are still very limited at the time of writing, however, the potential of FM in a large country such as China is unlimited, and with proper management of the new IFMA China, I look forward to seeing significant IFMA membership growth in China, well exceeding the numbers in many of the developed countries.

IFMA China is also actively promoting educational course in China, starting off with IFMA's FMP courses, and introducing credentials such as CFM, FMP, SFP to the Chinese community.

The Future of FM in Hong Kong and China

For many years, Hong Kong has long been China's window to the world. The rapid modernization of China was to a great extent through leveraging Hong Kong's strategic position as China's gateway to the West. With the sovereignty of the former colony reverting back to China in 1997, Hong Kong is becoming much more closely integrated into the overall Chinese economy, and being part of the Greater Bay Region (Hong Kong, Macau, Zuhai, Shenzhen, Guangzhou) economy as opposed to its former status as a major single financial hub. As with many practices and professions which were originated from the West, Facility Management was initially introduced to China through Hong Kong, but has now transformed to becoming a profession very much on its own, adapting rapidly to the Chinese culture and practices.

Since joining the WTO, China has been actively participating in many of the major global issues, such as Global Warming, Environmental Health and Safety, Sustainability, ESG, and Smart Cities, which are all issues closely related to the practice of FM. The COVID pandemic in particular, over the last two years, has significantly impacted and changed the Working Environment forever. FM demonstrated its value by showing the world how the Workplace can quickly adapt to such changes. China is affected equally by all these issues, and I am sure the value of FM is being quickly recognized, a profession which embraces all these issues.

Through technology, the world is growing much more connected and integrated. FM is also rapidly changing and evolving, from the initial years when I was involved with IFMA, where FM was mainly about office interiors and building operations and maintenance. Now FM is a profession which is involved in all aspects of our daily lives. It is constantly evolving and embraces many aspects including Workplace Management, Environmental Management, Energy, Management, etc.

Somehow, I regret retiring at these most exciting times for FM, but I am also happy and honored to have been given an opportunity to participate in its initial growth through the early years. especially in Hong Kong and China. For FM, it is only a beginning in Hong Kong and China, it will be most interesting and exciting to see how this profession will evolve and mature in the coming years.

Years ago, even before I went back to Hong Kong, I went to see a fortune teller, who told me that my future was to be one which involves connecting the East with the West. In retrospect, I think he was very correct, as I see that over the past 30 years, in my small little way, I have connected the East with the West through FM. It has been a most exciting journey and I cannot wait to see how the next chapters unfold.

Philip Lo, IFMA Fellow, Hon Fellow HKIFM,
Vancouver, BC Canada (formerly Hong Kong)

Philip Lo was born in Hong Kong and received his degree in Architecture from McGill University in Montreal, Canada. He subsequently practiced and set up his own architectural practice in Canada for over 10 years. His practice involved a lot of corporate design and interior projects which led him to Facility Management. However, it was technology which interested him to move to Hong Kong to start his practice in FM technology.

He established the first chapter of IFMA outside of North America in Hong Kong in 1992 making the organization truly international. His active involvement in the promotion of IFMA and FM in Hong Kong and China has earned him the honor of being awarded as a Fellow of IFMA, as well as an Honorary Fellow of the Hong Kong Institute of Facility Management.

Philip was active in the promotion of the profession through teaching courses on FM Technology at several universities in Hong Kong, speaking at various FM conferences internationally, while carrying out an active practice implementing different FM technology systems for various corporate and institutional clients.

Philip recently retired but plans to stay active in the field of Facility Management, particularly in Asia.

2.5

Missional Journey Toward Raising the Recognition of Facilities Management as a Business Advantage to Organizations in Singapore

Steven Ee

Through this writing, I hope to share with you my missional journey on how I contributed toward raising the recognition of facilities management in Singapore and to serve as a guide to advance the reader's career in facilities management. My writing is organized as follows:

1) Before Adopting the Facilities Management Approach
2) Research into Facilities Management
3) My Journey during the change of FM arena in Singapore
4) Sharing of Learning on Facilities Management
5) Moving Forward

1) Before Adopting the Facilities Management Approach

In the 1990s, Facilities management (FM) was still a very new concept and unheard of in Singapore. For those of you who have not heard of my country, Singapore, it is a sunny island in South-East Asia. We are a thriving global financial hub, and we take pride in our stability and security.

As mentioned, the practice of FM was hardly known in the 1990s, in Singapore. That was the time when I completed my studies in Building Services Engineering and entered the working world. My entry into the practice is quite accidental. My first job was as a Sales Engineer in a water treatment chemical company, responsible for cooling water systems, chilled water systems, boiler systems, and waste treatment systems. Being an active and inquisitive guy, I was bored with the predictable and uninspiring water treatment chemical job.

I was young and full of drive, looking for a challenging career when I joined a semiconductor-wafer fabrication organization. I was involved in the construction of the operations, and the installation of the facilities till the final maintenance of the first-of-the-kind class 1 "clean room" in Singapore. I was obsessed with the heating, ventilating and air-conditioning systems. It has become my strength and area of expertise.

Being first-of-the-kind, I was introduced to new technologies such as Supervisory Control and Data Acquisition (SCADA) and the Integrated Workplace Management system. I was impressed with the efficiency and effectiveness in automating building operations and maintenance.

However, I noticed that the practice of FM is wide and varied, different from organization to organization. Most of the time, it is limited to building maintenance, repair, cleaning, and handling the bolts and nuts of running facilities.

The FMs are often not involved in the initial project. We just inherited the "output" of the completed project. We were not consulted on the specification, or involved in the design of the facilities. We were only tasked with the operations and maintenance. Yet we were responsible for meeting the stakeholders' satisfaction with the facilities.

Facilities @ Management, Concept - Realization - Vision, A Global Perspective, First Edition. Edited by Edmond Rondeau and Michaela Hellerforth.

Sometimes the maintenance of the facilities was overlooked in the design. In one of the organizations that I have worked in, the chiller plant room was built on the roof, aesthetic, beautiful, but it was impossible to carry out maintenance as there was no access to it. In the end, FM has to bear the unbudgeted "cost" to tear down one side of the brick wall and erect a collapsible door.

FM is often depicted as a technical function and viewed as a cost center. Facilities breakdown is viewed as FM's failure in maintenance thus causing inconveniences and bleeds the organization's profit. FMs are not given the chance to justify ourselves; sometimes it could be the facility's capacity is not sufficient to meet the needs or the installation was not carried out properly. I felt that the work in managing facilities was undervalued, and I believe that there should be an effective approach to this practice.

2) Research into Facilities Management

There were no tertiary programs on facilities management, offered as a distinct discipline, in the 1990s in Singapore. Thus when I did my Master of Science in Project Management, I took the opportunity during my thesis to focus on FM. The title of my thesis was "The Importance of Incorporating Facilities Management Throughout Project Life Cycle" intending to research the factors that attribute to the advanced practice of FM in the United States of America, the United Kingdom, and Japan, also the benefits and advantages of adopting the practice in Singapore.

Many questions guided my research on FM, on the concept, the practice, the challenges, the frustrations, and how the practice of FM can be introduced and advanced in Singapore. The five key questions that directed my research focus were:

- How to articulate the concept, benefits, and advantages of the practice of facilities management?
- How may FM enable a safe, secure, comfortable, productive workplace that complies with the legal requirements?
- How may FM add value in a way that matters to their organizations?
- What knowledge and skills are needed to enable FM to be a strategic resource to its organizations?
- How may FM be a distinct discipline and profession in Singapore?

It was during my research that I discovered the International Facility Management Association (IFMA) from the United States of America, and their involvement, as well as contributions to the advancement of the practice of FM.

I enjoyed the writings of FM experts such as Prof. Ilfryn Price, Dr. Richard Sievert, David Cotts, Edmond Rondeau, Martin Pickard, and David Casavant. I was inspired by them (I called them my mentors) to press on with my thesis' objectives even though my advising professors were against it.

Thanks to all my "mentors," I passed my thesis with Distinction. Encouraged, I wrote an article based on my thesis and submitted it to IFMA's *FMJ Magazine*. To my surprise, it was published. Gradually, I was invited to speak and present at conferences. I remember my idea of FM being involved in the project life cycle stirred interest and arguments from other professionals. They are astonished by this idea as most view FM practitioners with no specialty.

Through these lively discussions, I realized that it is important for FM to be established as a distinct discipline and gain a unique identity to be recognized as a professional. Only then can we fulfil our mission as the steward of the organization, overseeing the facilities' operations and ensuring that the facilities support the business operations.

With this strengthened conviction, I stepped out to introduce this relatively new practice to my country.

3) My Journey During the change of FM arena in Singapore

In 2005, Singapore's Building and Construction Authority (BCA) launched the Green Mark certification scheme as a national sustainability strategy to accelerate Singapore's green building agenda, requiring 80% of all Singapore buildings to reduce energy use and emission production and to achieve a Green Mark Certified rating by 2030.

Riding on the FM wave in Singapore, I founded FMS Associates Asia Pte Ltd. (FMS), a facilities management training and consultancy company, with a mission to equip FM practitioners with essential knowledge and skills that will enhance the individual and organization's performance. The key focus is to elevate the practice of FM to be recognized as a value-added function and business advantage to the organization.

My partners and I did not have any experience in running a business, it was our weakness. However, we were so driven by our passion that we decided not to let our weakness hinder us and we pressed on. That spirit has translated into our motivation's tagline for our learners and clients: "To start, you don't have to be good. To be good, you must start."

The following year, FMS gained the appointment as IFMA's approved provider for its Facility Management Professional (FMP) credential program. I conducted the first class of FMP in July. I still remember explaining this new concept of FM to the learners. Differentiating FM from property management and introducing them to the business of FM.

To further push for the recognition of FM, servicing as the Vice President (Education) of the IFMA-Singapore Chapter, I facilitated the signing of a Memorandum of Understanding (MOU) between the IFMA Foundation's Accredited Degree Program and BCA for closer cooperation, as well as the signing of an MOU between IFMA and BCA, which not only positioned IFMA and its Singapore Chapter as the region's leading FM organization but also led to the first IFMA-BCA co-hosted a conference in Singapore.

I was so proud when those learners that I coached went up the stage to collect their Certified Facility Manager (CFM) credentials from IFMA's president during the conference. It was so satisfying to celebrate their achievements, recalling the hard work they put in on those Saturday morning gatherings to discuss their project works for the CFM credentials.

In 2011, FMS partnered with Singapore's Employment and Employability Institute to develop a curriculum – Competent Facilities Practitioner program, to help those who are interested in transiting into this FM profession.

In 2014, Singapore's Prime Minister outlined plans to turn Singapore into the world's first Smart Nation and launched the Smart Nation Programme Office to drive the national effort to transform Singapore into a Smart Nation. These Smart Nation Initiatives were launched to harness Infocomm technologies, networks, and big data to create tech-enabled solutions to keep pace with the Fourth Industrial Revolution and implemented them on a national scale.

I published my first book in 2015, titled *Value-Based Facilities Management – How Facilities Practitioners Can Deliver a Competitive Advantage to Organisations*. It was intended to provide a roadmap to guide and encourage FM practitioners to contribute value to their organization. I believed top management will only give recognition and respect to the FM practitioners if they can contribute value to the organization. To me, respect is earned and not demanded. Thus being the steward of the organization's largest investment – the building and its facilities, have we done our part and given our organization, the best return on investment?

The book guides FM practitioners to focus on the three values – Identify, Sustain, and Contribute values – which are illustrated by the I-S-C FM Value Model. It directs practitioners to "Think Strategically, Translate Tactically, and Target Operationally."

I am honored to receive praises globally from renowned FM experts and authors who have attested to the concept, models, and ideas shared in the book.

> *This easily understandable book should be taught in all FM certification and college courses, and be found on the bookshelves of all of us who manage these valuable public and private assets.*
>
> – David Cotts, PE, CFM, IFMA Fellow

As prophesized by Mr. David Cotts, my book was eventually adopted as a recommended textbook for FM students in Singapore.

In 2016, FMS was awarded the "FM Training Institution, Enterprise Excellence Award" by the IFMA-Singapore Chapter, in recognizing the company's contribution toward FM learning and development in Singapore.

I was thrilled when the new international standard for Facility Management was published in 2018. Stanley Mitchell, Chair of ISO/TC 267, the technical committee developing the standard, said: "ISO 41001 will help to clarify the 'what' as well as the 'why' facilities management is a strategically important discipline to all organizations in the management, operation and maintenance of the workplace, its assets and operational efficiencies."

Working with an associate whose strength was in the management system, we led an education institution to be the first in South-East Asia to obtain the ISO41001:2018 certification. We celebrated this big step of advancement in FM by organizing an FM Leaders' Summit, to gather like-minded leaders and aspiring leaders in the field of FM to learn and exchange views on the advancement of this practice. The response to the FM Leaders' Summit was overwhelming. We had FM practitioners flying in from Australia, Brunei, East-Timor, India, Indonesia, Malaysia, Nigeria, Thailand, Ukraine, and the United Arab Emirates to support us. It was such an exhilarating event.

COVID-19 was a curve ball that hit us all unexpectedly. FM scrambled to stop the spread of infections in the workplace and do our best to provide our colleagues with a safe and healthy workplace. During the lockdown, FM was classified as one of the essential services that was allowed to continue. We were deployed to hold the fort. First to go in, last to go out.

4) Sharing of Learning on Facilities Management

As an FM trainer and speaker for over 15 years, I would like to share some of my insights into the success of an FM practitioner.

- **Facilities Project Management**

Through my years of working as an FM practitioner, I realized that project management is an important skill for FM. I remember in one of my appointments, on the first day of work, I was tasked to convert a canteen into an IT office. The work involved was not only the physical conversion of the space but also taking care and getting the buy-in from the stakeholders. Imagine how much my colleague disliked me as my first job when I joined the company was to take away their canteen, their entitlement.

FM jobs are never just the physical facilities or place. We are the ones to provide our colleagues with a safe, secure, and healthy working environment. We have to meet our stakeholders' needs and expectations. Achieve pleasant users' experience.

Here I would like to share the success formula for facilities projects.

TCQ + PPP + SS

Where:

T	: Time (dateline, the completion date)
C	: Cost (budget, resources allocated)
Q	: Quality (meeting the specification and scope)
PPP	: Considering and validating the People (stakeholders), Place (space, accommodation), and Process (the intended workflow)
SS	: Stakeholders' Satisfaction (enabling pleasant experience)

- **Effective and Efficient Facilities Management Operations**

Figure 1 is a model I used to illustrate the application to achieve effective and efficient facilities management operations.

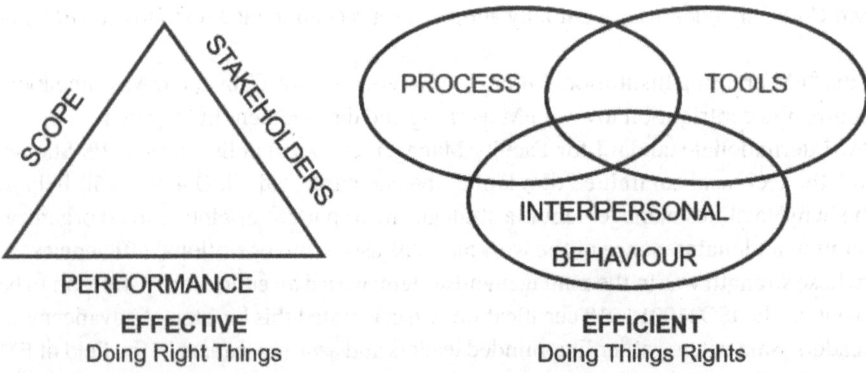

Figure 1 Effective and efficient facilities management operations. (*Source:* Steven Ee).

In a nutshell, effectiveness means *doing the right things, for the right reasons*, whereas efficiency means making work better. The key point was to address the importance to ensure what FM is doing is effective than doing it more efficiently in the first place, without considering doing the right things, may end up doing the wrong things better.

- **P.R.I.S.M Leadership for FM**

The acronym P.R.I.S.M. was intended for easy remembering and applying the essential skills to be a successful FM leader.

Process	A leader needs to identify and establish FM processes that support the achievement of the organization's strategic objectives. The process-based approach enables the responsible members to be clear of the activities and resources needed to transform into the required performance outcome.
Relationship	A leader needs to be skilled in interpersonal relationships, building relationships with the stakeholders, facilities users, contractors, vendors, and team members.
Integration	A leader needs to identify where processes and resources can be integrated to achieve optimum resource utilization and productivity.
Strategic	A leader needs to consider the big picture and long-term objectives to translate the strategy into tactical action and operations targets.
Missional	A leader needs to be mindful of the mission of FM, which is to manage the facilities and support services to facilitate the business operations' success.

5) Moving Forward

The evolving trends of cost containment, process management, workplace arrangement, legal compliances, users' needs and expectations, advancement in technology, and performance improvement, growing environmental pressures, security, and other activities in facilities have a big impact on widening the scope of facilities management. I believe that FM leaders must be the drivers to influence the advancement of the practice of facilities management. To do that, FM leaders must be anticipatory, agile, and adaptive in keeping pace with organizational changes in facilitating business operations' success.

I believe that Facilities Management has the potential to be a powerful position within an organization because FM has the knowledge of facilities performance that facilitate business operations success, and I believe that the level of recognition and respect for facilities management is determined by the ability of its practitioners to demonstrate value contributions to their organizations.

FM must have its own distinctive knowledge base to evolve a unique identity among professionals if it is going to be sustained and evolved. I believe that in essence, FM should only manage things which FM is good at and better than other disciplines. FM needs to evolve from centering on the facilities process to embracing the management process. If we do not establish our unique identity in Facilities Management, others will.

If you are new or transiting into the field of facilities management, I hope this writing has inspired your passion and contributed to your knowledge and skills in facilities management. If you are a seasoned practitioner, I hope you are inspired to join me to raise the recognition of facilities management to be a value-add function and business advantage to the organization.

Quoting my friend, mentor, and partner, Dr. Gan Cheong Eng, "Facilities Management is a goldmine profession."

To your success in FM!

Steven Ee, CFM, IFMA Fellow
Managing Director
FMS Associates Asia Pte Ltd Singapore

Steven Ee's mission is to raise the recognition of facilities management has taken him around the world. His workshops and executive coaching have challenged facilities practitioners to raise the bar and earn respect as strategic resources for their organization.

He is an IFMA-Fellow and a Certified Facility Manager awarded by the International Facility Management Association (IFMA), and IFMA Approved Instructor and IFMA Subject Matter Expert for its Facility Management Professional program.

He is the co-developer of the Competent Facilities Practitioner program and has developed the Workforce Skills Qualification program for facilities management in Singapore.

His sincere, multidisciplinarian approach has translated complex ideas into practical results and has inspired other practitioners to pursue excellence in the profession.

He is s regularly invited to speak at seminars and conferences attended by over thousands of facilities practitioners in Singapore, Malaysia, Brunei, Thailand, Indonesia, the Philippines, and China.

He can be contacted via steven@fms-1.com.

2.6

The Austrian Success Model of Excellent Facility Management Education and Proactive Work by Professional Associations

Thomas Madritsch

Introduction

The Austrian Facility Management Industry and the Role of Professional Associations

The Austrian facility management industry provides jobs for around 200,000 people and contributes an added value of almost EUR 20 billion to the country's gross domestic product. Together, the industry's businesses ensure the well-being of more than 2.5 million clients, customers, and users. The two professional associations Facility Management Austria (FMA) and International Facility Management Association Austria (IFMA) aim to bolster the spread and adoption of professional facility management services in the country. A significant contribution to achieving this goal is made by the 16 training and higher education institutions in Austria that offer a comprehensive range of programs in the field, from practical on-the-job facility management training courses to full-fledged academic degree programs. Jointly with the training and higher education institutions, the professional associations aim to support their members in the pivotal task of educating a future cadre of facility management professionals. In addition, the associations see themselves as an essential forum for communication between facility management professionals, clients, and customers domestically and internationally.

The Industry Association Facility Management Austria

The nonprofit organization Facility Management Austria (FMA), founded in 1995, sees itself as the Austrian network for businesses active in the field of facility management. The membership includes well-known nationally and internationally active companies, organizations, associations, and educational institutions, as well as individuals pursuing facility management education or training. The FMA has been enjoying steady annual growth and, as of 2022, counts 300 businesses among its membership.

International Facility Management Association Austria

The International Facility Management Association Austria (IFMA Austria) is the Austrian branch of the globally active International Facility Management Association (IFMA) and offers individual memberships. It was established in 1998 as the Austrian branch of IFMA International. IFMA was founded in 1980 and is headquartered in Houston, USA; the international network is active in over 100 countries and has more than 20,000 members organized into 127 chapters and 16 councils. One of the highlights of the association's work is the annual IFMA World Workplace conference. IFMA Austria currently counts 315 individual members in its ranks.

Facilities @ Management, Concept - Realization - Vision, A Global Perspective, First Edition. Edited by Edmond Rondeau and Michaela Hellerforth.

Cooperation Between the Associations

For decades, the excellent networking between the two respective executives has been the cornerstone of cooperation between the two associations. For example, the chair of the FMA is also the deputy chair of the IFMA, and the chair of the IFMA is, vice-versa, deputy chair of the FMA. This reciprocal structure has been tried and tested over the past 20 years and has proven its worth time and again. In addition, collaboration and cooperation between the bodies are further enhanced by the reliable assistance the central administrative office provides, which supports both associations in a coordinating capacity.

Unique Range of Networking Opportunities

Together, FMA and IFMA Austria afford their members national and international networking opportunities and professionally support the interests of all those committed to the concept of facility management. In topic-specific task forces and market-oriented communities of interest, the members research, develop, and promote today's issues in facility management. The network acts as an organizer of events and excursions, as well as a partner in the Austrian facility management standardization drive and cooperates very closely with educational and training institutions.

Communities of Interest

Communities of Interest were launched in 2014 as key platforms for in-depth work. They reflect the current state of the market and offer members the opportunity to collaborate on more specialized interests under the umbrella of the associations (see Figure 1). The following communities of interest are currently active:

- Facility Management Service Providers (Comprehensive facility management service providers, facility service providers, facility management consultancy and planning)
- Education & Training / Students
- In-House Facility Management

Task Forces and Facility Management Round Tables

Concurrent association internal task forces offer a venue for discussing overarching issues and developments in facility management. These provide a platform for Members to get involved, exchange information, compare experiences, and participate in discussions. Current topics of interest are introduced through keynote speeches and then debated with the participants. The first two initial focused task forces covered the issues focused on the themes "Quality versus Price" and "Price Dumping in the Building Services Industry." Additionally, facility round

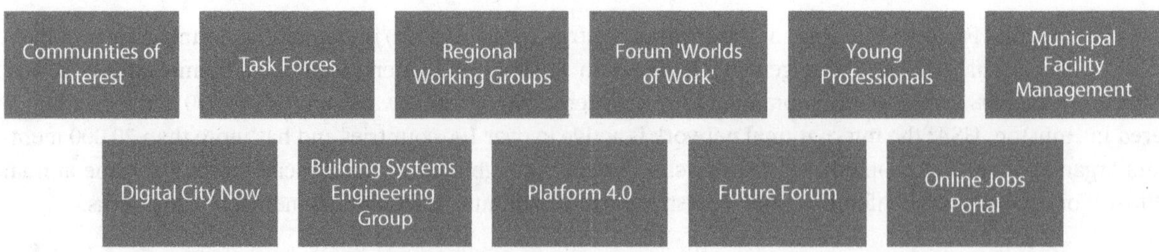

Figure 1 Overview of networking opportunities. (*Source:* T. Madritsch).

tables have addressed topics such as "Indoor Mapping, Indoor Viewing, Indoor Navigation" and "Documentation at Property Handover," among others.

Current association task forces include the following:

- **PowerPack Properties**[1]
  The Building of the Future
- **Facility Management Seal of Quality**[2]
  Planning, Building, and Operating Properties following Facility Management Values

Customer Proximity through Regional Working Groups

An indispensable part of the organizational structure of the Austrian network for facility management is its regional working groups (RGWs), which establish a country-wide presence and ensure that members, potential members, and the facility management industry within a region have a dedicated support network (see Figure 2). Among other things, the association's regional working groups serve as an instrument for promoting and spreading the concept of professional facility management and for strengthening the market presence of the associations in Austria as a whole. Regional leaders and the program managers of the event series Facility Management Executive Retreat (*FM-Executive Treffen*) and Current Topics in Facility Management (*FM-Aktuell*) are in close contact, jointly drawing up the annual schedules of events of individual regions, which are further coordinated with the central office. This ensures that the events do not overlap in dates or topics. The result is a calendar of events and activities that is highly attractive throughout the year. Moreover, individual members are assigned to a regional group based on their location and encouraged to address any concerns they may have to the responsible regional group leader. Current issues and trends are considered within the regional working groups, particularly those of local importance.

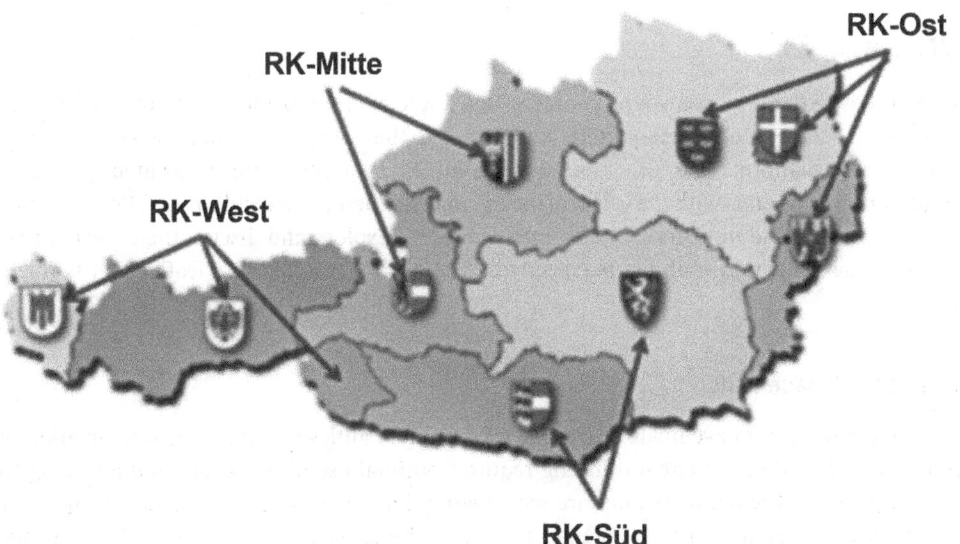

Figure 2 Austrian regional working groups. (*Source:* Adapted from Austrian Network for Facility Management).

1 www.fma.or.at/fileadmin/uploads/FMA/dokumente/fachliteratur/PowerPack/Folder_PowerPack_06_web.pdf.

2 www.fma.or.at/fileadmin/uploads/FMA/dokumente/fachliteratur/Qualitaetssiegel/Folder_Qualita__tssiegel_07_web_final.pdf.

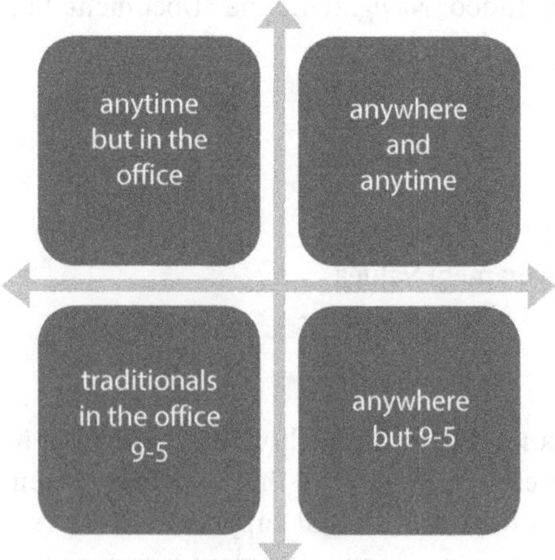

Figure 3 Chart: Working time models by locality and time. (*Source*: Adapted from IFMA Workplace Evolutionaries).

Forum "Worlds of Work"

The forum "Worlds of Work" was founded in the spring of 2020 to raise issues and develop solutions relating to the workplace and new working environments. The new platform recorded a highly successful start when, at the initial virtual kickoff, about 60 experts were able to exchange ideas and discuss solutions to current problems. To ensure that international approaches and developments are considered and integrated, the forum is in constant exchange with the IFMA Workplace Evolutionaries, a global community with over 1000 members concerned with similar issues (see Figure 3).

Ultimately, the local network of experts on the topics of the workplace and new worlds of work continues to expand and serves to facilitate exchange and increased public visibility of these highly relevant considerations. Subsequently, three working groups were established to deal with the issues of managing cultural change, designing work environments, and leveraging technologies. A highlight of the forum's work so far was the publication of the paper *Aufbruch zu neuen Arbeitswelten*[3] ("On to New Worlds of Work"). Currently, the following working groups are active within the context of the forum:

- Agile Utilization of Spaces
- Optimizing Hybrid Environments
- Sustainability and the Circular Economy
- Contemporary System Landscapes in Facility Management
- Centralization vs. Decentralization

Young Professionals Initiative

The Young Professionals Initiative (YPI) targets young people who want to commit to and benefit from participation in a strong network within facility management. Both newcomers to the industry and more seasoned professionals are welcome to join the platform. The only requirements are that members are committed, young or young-at-heart, and have some touchpoints with the world of facility management. As a result, Young Professionals from a wide variety of backgrounds and industries have come together to explore and discuss the topic of crisis management in facility management from multiple perspectives. Their work resulted in a leaflet that is freely available for download.[4]

Platform "Municipal Facility Management"

Increasing demands on the public sector have made it necessary for municipalities to pursue the optimization of their property operations actively. Future-oriented thinking requires optimal use of properties and an in-depth focus on life cycle costs. More and more municipalities are recognizing that real estate is a strategic resource for their core business area and that efficient real estate management contributes significantly to their bottom line.

3 https://www.fma.or.at/fileadmin/uploads/FMA/dokumente/forum_arbeitswelten/Forum_Arbeitswelten_Aufbruch_zu_Arbeitswelten_10_final_web.pdf.

4 https://www.fma.or.at/fileadmin/uploads/FMA/dokumente/bildung/Young_Professionals/YPI_Flyer_Krisenmanagement_Brand_05_final_09721_web.pdf.

Competent implementation of such a strategy requires professional facility management services. Thus, facility management is increasingly important for municipalities, and an entrepreneurial spirit is more in demand than ever. Consequently, the interest community also chairs the Facility Management Working Group of the Austrian Association of Cities and Towns.

Platform "Digital City Now"

According to estimates, further digitization in the Austrian construction and real estate industry could increase productivity by approx. 20–30%. To achieve this goal, the industry must engage in a collective effort. Above all, the needs of SMEs must be addressed, as they frequently operate in a single section of the process chain due to the strong fragmentation of the industry. As a result, they rarely have the necessary investment strength to participate in innovation processes. The platform *Digital Findet Stadt* ("Digital City Now") has set itself the goal of changing this state of affairs.

Digital City Now is Austria's largest platform for digital innovations in the construction and real estate industries and is responsible for an array of innovative solutions. With a network of over 300 companies, advocacy groups, and research institutes, the platform helps promising innovation projects reach market maturity and strengthens the innovative power of Austrian SMEs. Distinguished through its unique network structure and its broad, strategic orientation, the Austrian platform is a pioneer in the German-speaking region.

Acting as an interface between the worlds of research and business, Digital City Now is shaping the digital transformation with a network of over 300 companies, advocacy groups, and research institutes, thus contributing to an increase in resource, energy, and cost-efficiency in the construction and real estate industries.[5]

Building Systems Engineering Group

The Building Systems Engineering Group aims to create better buildings through efficient use of technology, optimized processes, and better communication. In order to achieve this goal, the group consults and coordinates with representatives from the fields of investment, architecture, planning, research, as well as facility management, and specialist construction businesses. Together, these stakeholders define pertinent issues and develop appropriate responses. The Building Systems Engineering Group provides a neutral platform for the development of solutions for better buildings, improved communication, and smoother implementation of projects. Here, too, the results are made available to members via presentations and panel discussions such as on "The Path Towards Better Buildings." In addition, the working group maintains a blog on its website where members can submit contributions.[6]

Platform 40

The Platform for the Future – Design. Construct. Operate was formed at a symposium on May 9, 2016. It is open to all those involved in designing, constructing, and operating properties and should be understood as a point of contact, a platform for visionary ideas, and a tool for outward representation. Its watchwords are Jobs, Industry, and Export. The platform wants to contribute its part to ensure the industry remains competitive and provides high-quality jobs for future generations through innovative value creation both domestically and for the export market. A primary focus is placed on digitizing the value chain of designing, constructing, and operating properties in order to create data-based control loops that help optimize the life cycle in all its facets.

---

5 www.digitalfindetstadt.at.

6 www.fma.or.at/netzwerk/tga-gruppe/blog-der-tga-gruppe.

The Future Forum of IFMA Austria

The Future Forum of IFMA Austria was founded in the fall of 2016 to develop solutions that add value for all facility management industry stakeholders. Both seasoned professionals and newcomers to the field are invited to get involved in the activities of the association. The first round of the Future Forum resulted in the introduction of a Facility Management Seal of Quality and a white paper series on relevant issues such as *Die Zukunft des Facility Managements – Vision FM 2030* ("The Future of Facility Management – Vision FM 2030"[7]) or on the user-perception of the qualities of buildings.[8]

Online Jobs Portal

In times marked by a competitive job market and low unemployment rates, finding highly qualified employees is a major challenge for many companies. To alleviate this problem, the associations have developed an Online Jobs Portal on which FMA and IFMA members can advertise their vacancies free of charge. The platform is available to unaffiliated businesses for a cost-based fee.

International Collaboration

To increase networking, the Austrian members of IFMA Austria and Facility Management Austria (FMA) also work very closely with international facility management associations. They seek to foster a broad exchange of knowledge and take advantage of synergy potential. Some examples of this exemplary cooperation are described below.

Founding of FM3

As part of EXPO REAL 2014, Facility Management Austria (FMA), the German professional facilitate management association RealFM e.V., and the Swiss Association for Real Estate and Facility Managers (SVIT FM) sealed their transnational cooperation and founded FM3. The aim of their cooperation is to anchor the understanding of facility management even better in politics, business, and teaching throughout Europe and to further intensify the joint development of competencies, methods, and tools for facility managers and real estate experts. Moreover, the bundling of knowledge is expected to increase the efficiency of the associations and increase the added value for their members. Through FM3, the three professional associations intend to make even better use of common resources and dovetail key activities that have already been successfully launched.

Facility Manager Job Profile Brochure

Another result of international collaborative efforts between the Facility Management (FM) industry associations in Germany, Austria, and Switzerland is the development of a "Facility Manager" job profile brochure. This job profile guide reflects the international concept of facility management and presents a discipline that plans, controls, and organizes the implementation of support processes for the most diverse core processes in industrial and service companies as well as in public organizations and institutions. The associated brochure provides information about the diverse tasks, areas of application, skill requirements, qualification opportunities, networks, and levels of responsibility in facility management.

---

7 www.ifma.at/fileadmin/uploads/FMA/dokumente/fachliteratur/White_Papers/ifma_Zukunft_FM_10_web.pdf.

8 www.ifma.at/fileadmin/uploads/FMA/dokumente/fachliteratur/White_Papers/NutzungBuerogeb_10_web.pdf.

Active Collaboration with EuroFM

Thanks to their enduring membership in EuroFM, the European network for facility management, the FMA and IFMA Austria are involved in a lively European exchange of opinions and experiences. Members such as Albert Pilger, Thomas Madritsch, and Alexander Redlein have actively contributed to the further development and spread of facility management in Europe as board members of EuroFM. A.o., longstanding board member of the FMA and IFMA Austria, Albert M.M. Pilger, served as chairman of the EuroFM board from 2007 to 2010, Alexander Redlein was elected Treasurer of the EuroFM board in 2012, and Thomas Madritsch led the Education Network Group (ENG) from 2005 to 2010.

The open network of experts, scientists, professors, practitioners, and researchers results in an excellent resource for science, education, and practice. The FMA and IFMA Austria members also have access to the EuroFM members' network. At the core of EuroFM are four groups that initiate projects and activities (see Figure 4).

EuroFM members engage in an open exchange of information and experience through meetings, seminars, and workshops at the European Facility Management Conference and collaborative research projects. Also worth mentioning is the electronic facility management magazine *European FM Insight*, whose first editor was Albert MM Pilger, which has achieved a worldwide readership of 65,000.[9]

Collaboration with IFMA International

IFMA Austria is part of the worldwide, individual membership-based facility management network of the "International Facility Management Association" (IFMA) based in Houston, Texas. At the annual highlight, the IFMA World Workplace, the representatives of the facility management associations gather to exchange information about the latest developments and findings in the facility management industry. Particularly noteworthy are the contribution to the further development of the association by Albert Pilger, a founding member of the FMA and IFMA Austria, in serving on the IMFA Board of Directors, and the international contribution by Andreas Rohregger, a graduate of the Facility Management course at the University of Applied Sciences Kufstein Tirol, through his tenure on the Board of Trustees of the IFMA Foundation. The special contributions are also reflected in the fact that Albert Pilger and Thomas Madritsch were honored as IFMA Fellows and the latter also as Educator of the Year 2011 for their achievements in establishing and further developing FM.

PNG – Practice Network Group
ENG – Education Network Group
RNG – Research Network Group
CAG – Corporate Associates

Figure 4 Four EuroFM are groups that initiate projects and activities. (*Source:* EURO FM).

9 www.eurofm.org/news/european-fm-insight.

Collaboration with IFMA Europe

IFMA Europe was founded in 2016 to provide the best possible support for the concerns of its European members and to promote facility management in this region. IFMA Europe offers support in knowledge transfer, education, events, and networking. The Austrian members regularly participate in the annual WWP Europe conference as speakers and listeners.

Facility Management Education in Austria

High Density of Training Institutions

With the progressive development of the facility management market, the technical requirements for the profession are also increasing. Through the associations' active promotion of higher qualifications, the demand for the latter has seen a steady increase so that, currently, 16 training and higher education institutions in Austria offer various training and further education opportunities in facility management. As a result, Austria has one of the highest densities of facility management training centers. All major Austrian facility management education and training institutions are members of the FMA. In the sense of practice-oriented training and further education and the optimal solution to future problems, the association members provide important impulses for the design of corresponding training content (see Table 1).

The Three Academic Training Institutions

One can proudly say that with three academic training institutions in Austria, facility management has emerged as a science of its own (see Figure 5). The University of Applied Sciences Kufstein Tirol was the first to offer

Table 1 16 Austrian higher education institutions with educational offering in facility management Adapted from.[10]

ARS Academy	www.ars.at
ATGA	www.atga.at
Austrian Standards Plus Ltd	www.austrian-standards.at
Vocational Training Institute Vienna	www.bfi.wien
Business Circle Management	www.businesscircle.at
Die Weiterbilder	www.dieweiterbilder.at
Danube University Krems	www.donau-uni.ac.at
Pilger Facility Management Academy	www.pfm.at
University of Applied Sciences Kufstein Tirol	www.fh-kufstein.ac.at
University of Applied Sciences Vienna of the Vienna Chamber of Commerce	www.fh-wien.ac.at
Ghezzo Ltd	www.ghezzo.at
HTBLVA Graz – Ortweinschule	www.ortweinschule.at
University of Technology Vienna	www.tuwien.at
WEKA-Academy I WEKA-Publishing House	www.weka.at
WIFI Upper Austria	www.wifi.at
WIFI Styria	www.wifi.at

10 www.fma.or.at/bildung-karriere/ausbildungsinstitutionen.

Danube University Krems
Master of Science in Facility and Property Management

University of Applied Sciences Kufstein Tirol
Bachelor of Arts in Business – Facility & Real Estate Management
Master of Arts in Business – Facility & Real Estate Management
General Management MBA in Facility Management

University of Technology Vienna
MBA in Facility Management

Figure 5 Three academic training institutions in Austria. (*Source:* T. Madritsch).

academic facility management training in the German-speaking region with bachelor's and master's programs starting in 1997. Due to their high-quality standards, the degree programs were also the first to achieve certification as IFMA Recognized Programs in Europe. In total, approximately 350 facility management students are enrolled in academic programs every year.

FMA and IFMA Austria's Educational Program Databank

Due to the sound cooperation between the corporate sector, associations, and training institutions, a number of high-quality training and further education offers have been developed in recent years. In fact, the educational offering has become so large that a database was created to provide a comparative overview that enables prospective students to make an informed choice and select the program best tailored to meet their needs and interests. This database distinguishes between the following three higher education and continued education categories (see Figure 6):[11]

The Educational Achievement Prize of the FMA and IFMA Austria

The Facility Management Educational Achievement Prize was created in 2002 to highlight the level of excellence of facility management training in Austria and to augment the exchange between the industry, training centers, and professionals at the start of their careers. Since the first call for submissions for consideration for the FM Educational Achievement Prize in 2002, 357 theses have already been submitted. The competition provides an annual insight into current research and training activities in the field of facility management. In addition, students and researchers are provided with the opportunity to discuss their findings with experts from the facility management industry (see Figure 7). Students and recent graduates can submit theses in three categories:

- Category A: Master's theses, diploma theses, and dissertations

11 https://www.ifma.at/bildung-karriere/ausbildungsdatenbank.

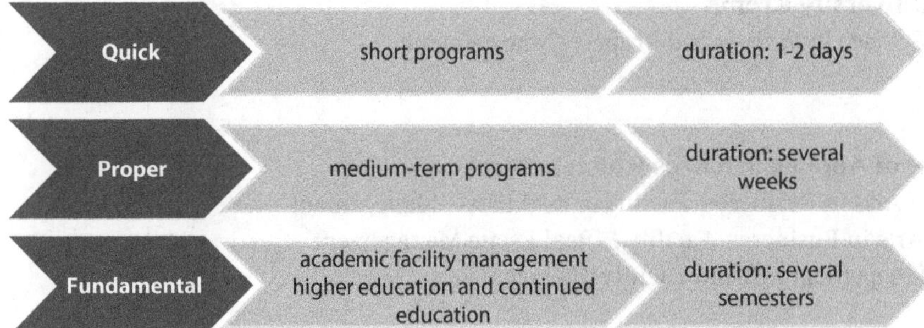

Figure 6 Higher education and continued education categories. (*Source:* Adopted from FMA and IFMA Austria).

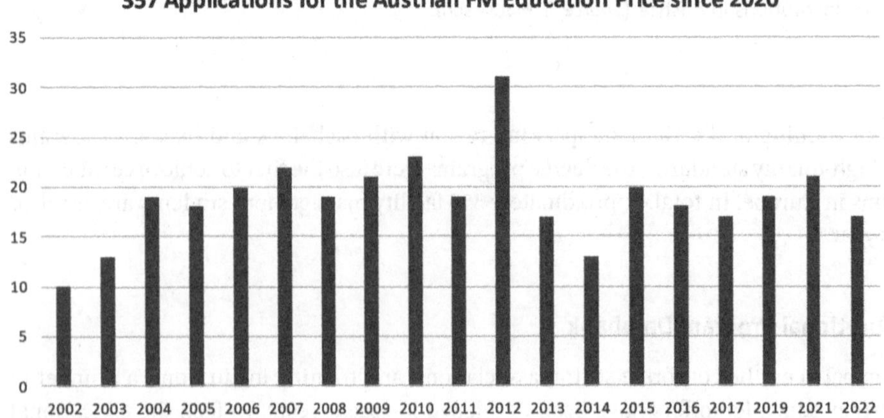

Figure 7 Chart: Submissions per institute of higher and continued education. (*Source:* Adopted from FMA and IFMA Austria).

- Category B: Bachelor theses
- Category C: Project reports, final exams, module reports, and internship reports

Facility Management Certification Programs

Figure 8 IFMA Certified Facility Manager logo. (*Source:* IFMA).

IFMA offers the following certification programs for facility management professionals:

Certified Facility Manager (CFM)

IFMA's Certified Facility Manager (CFM) certification program was developed to meet the need of facility managers for a professional title that not only brings additional credibility to facility managers but also promotes their professional recognition. An assessment tool is used to evaluate specialist competence from facility management in 11 areas to confirm the expertise acquired in practice. Holders of this professional title must undergo recertification every three years (see Figure 8).

Facility Management Professional (FMP)

IFMA's Facility Management Professional (FMP) certificate program provides initial professional qualifications for career starters in the facility management sector. It certifies the specialist skillset required for accelerated entry into the profession (see Figure 9).

Figure 9 IFMA Facility Management Professional logo. (*Source:* IFMA).

Sustainability Facility Professional (SFP)

The Sustainability Facility Professional (SFP) certificate program is designed for facility managers who seek to acquire in-depth knowledge of sustainability aspects in FM. Emphasis is placed on the development and adaptation of sustainability strategies in FM, their implementation in the facility management organization, and the sustainable operation of real estate (see Figure 10).

Figure 10 IFMA Sustainability Facility Professional logo. (*Source:* IFMA).

Certified Object Manager (certOM)

In a joint project with RealFM e.V. Germany and SVIT FM Switzerland, IFMA developed the Certified Object Manager (certOM) certification process for experienced professionals with several years of applied experience in property management (service controllers, property managers, facility service managers) to certify skills acquired explicitly and implicitly. The certification process involves a two-stage procedure, which first assesses existing competencies and, if applicable, identifies deficits that must be addressed before certification. The certification program offers candidates the opportunity to take appropriate measures to acquire these requisite skills on a case-by-case basis prior to certification. The certification is available at two levels, basic and advanced (see Figure 11).

Figure 11 Certified Object Manager logo. (*Source:* Joint project with RealFM e.V. Germany and SVIT FM Switzerland, and IFMA).

Successful Models and Success Stories

Development of Facility Management Standards

As a result of the networking and activities by the various facility management associations in Austria, it was recognized early on that a significant contribution could be made to advance the professionalization of facility management standards. Thus, in cooperation with the standardization institute, ÖNORM A 7000, the first Austrian facility management standard, was published in 2000.

A 7000	Facility Management Basic Concepts	December 01, 2000
A 7001	FM Agreements in the Use Phase	January 01, 2003
A 7002	Requirements for Facility Managers	December 01, 2001
A 7010-1 bis A 70010-4 Property Management		2005–2014

This partnership and due to the pioneering work and the experience gained, Austrian members of the association were able to work on and help shape the European facility management standard EN 15221-1 to EN 15221-7.

This EN 15221 series of standards has made a significant contribution to establishing a uniform understanding of the facility management model throughout Europe. Ultimately, this series of standards was not only taught and applied in Europe, but also gained worldwide significance within a short period of time. Consequently, this also led to the worldwide series of standards ISO 41011 and following, which finally defined facility management as a management discipline.

In January 2008 EuroFM signed at the University of Applied Sciences Kufstein Tirol a Partnership Agreement with CEN, the European Committee for Standardization.

Congresses and Specialist Conferences

The annual specialist meetings and conferences are an essential element of networking and knowledge transfer and have been organized annually by professional associations and training centers for the last 25 years. The three target groups, students, academics, and professionals are served here in different formats (see Figure 12). Of course, each event category is open to all target groups and enables the transfer of knowledge and exchange of information across the target audiences.

Since 2011, the University of Applied Sciences Kufstein Tirol has organized an annual international WinterSchool. Up to 250 students from European and international partner universities collaborate in this event and develop sustainable real estate projects in competition in small groups. In the past years, the event has led to the development of projects on health care facilities, serviced housing, innovations to reduce real estate vacancies, or social projects such as a children's hospice. This experience has enabled participants to learn to apply facility management in planning and gain different perspectives of facility management in an international context. Peer learning by students in teams and coaching by experts from the university and practice thus constitute the core of the learning and teaching concept of this compact week of facility management immersion and exchange. The WinterCongress started in 1998 as FM Talks and includes lecture series and discussions on current facility management and real estate management trends such as digitalization, service excellence, and sustainability. International experts and professionals from the industry present their innovations, ideas, and research results and engage students and community participants in in-depth conversations about these topics and other areas of development. Networking is an integral component of this event, given its interdisciplinary platform.

The International Facility Management Congress (IFM) at the University of Technology Vienna (TU Wien) has become a key event in the domestic and international facility management industry as a scientific congress for facility management and is increasing in popularity year after year. Top managers from the corporate sector and highly qualified researchers from Europe, the USA, and Asia meet for two days at the University of Technology Vienna to exchange ideas and present creative new facility management approaches.

The Austrian Facility Management Day congress is organized by the FMA and IFMA-Austria and has developed into the annual industry meeting point for facility management providers and customers in Austria. At this one-day event, more than 250 participants can expand their industry network and find out about the latest developments and knowledge of the facility management industry. The event provides in-person and online attendance options.

Research Projects and Results

With their research results, the three academic training centers also significantly contribute to the industry's further professionalization. The Institute for Energy, Facility & Real Estate Management at the University of

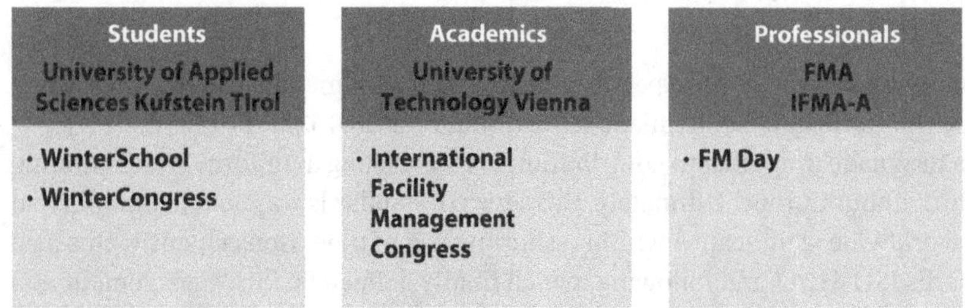

Figure 12 Annual specialist meetings and conferences. (*Source:* T. Madritsch).

Applied Sciences Kufstein Tirol focuses on the one hand, on the collection, analysis, and interpretation of real estate-related data and, on the other hand, on the analysis and determination of the potential of renewable energies and the business areas of the energy and sustainability industry. The acquired know-how is available to corporate partners for practical use. Some areas of competence are listed here as examples:

- **FM & REM processes**: Analysis of the maturity of facility management and real estate management processes
- **Life cycle-oriented real estate development**: Calculation and optimization of life cycle costs in the context of architectural competitions and real estate developments
- **Real estate prices**: Analysis and interpretation of real estate prices and development of real estate market ratings
- **Behavioral real estate**: Collection and analysis of user decisions and behavior in the context of real estate
- **Residential construction**: Collection, analysis, and optimization of life cycle costs in the area of residential building (construction)
- **Smart Meter research & trends**: Analysis of the use of Smart Meters
- **Building-integrated photovoltaics**: Measurement methods and analysis building-integrated photovoltaic systems
- **Energy & communities**: Survey, analysis, and calculation of the future energy needs of communities and development of corresponding community and regional master plans
- **User satisfaction**: survey, analysis, and optimization of user satisfaction in buildings
- **People – building – technology**: collection, analysis, and optimization of the interaction of people, technology, and building (construction)
- **Healthcare real estate**: Analysis and optimization of operating costs for special purpose real estate

Outlook for the Future and Conclusion

The Austrian Network for Facility Management for Climate Protection

Sustainable buildings and operations are decisive factors for achieving Europe's ambitious climate goals. In line with the Austrian government program 2020, FMA and IFMA Austria are committed to creating a climate-neutral building stock and operation by 2040. The associations see the corresponding social and economic challenges as an excellent opportunity for the facility management industry and, therefore, seek to do their part and make valuable contributions to support the national government's agenda. Much attention is paid to the CO2 countdown initiative of the Austrian associations.

Sustainability Week

The topic of sustainable buildings and sustainable building operation is also discussed on the various association platforms through member contributions and input and has resulted in the development and presentation of innovative solutions. Particularly worth mentioning is the Sustainability Week repeatedly hosted by the Facility and Real Estate Management program at the University of Applied Sciences Kufstein Tirol, where students and business professionals present and discuss a wide variety of sustainability topics over the course of a week. The increasing social demand and the changing requirements affecting companies and politics have led to a new format of this event since 2020: the Sustainability Week now serves as an important platform for a holistic approach to sustainability. The event involves a week of innovations, keynote speeches, and applied workshops on topics ranging from sustainable construction and renewable energy production to work-life balance and self-care. Students from different departments organize and host this week for students, interested schoolchildren, and citizens from the region. With the addition of live-streamed lectures, the event has expanded its reach beyond the region and now includes approximately 1,300 participants seeking education, inspiration, and the opportunity to discuss and collaborate in sustainable development. An additional feature of the event is the podcast series on sustainability developed by the university's Facility and Real Estate Management program.

Conclusion

Due to the excellent cooperation between training institutions, professional associations, and the industry sector, a unique variety of platforms and exchange opportunities have been created in Austria to further develop facility management. The active association life gives newcomers to the industry and students the opportunity to integrate and work together from an early stage and, at the same time, serves to network seasoned professionals, educators, and corporate partners. Moreover, the fluid exchange of information enables the associations to identify the evolving needs of consumer groups and future trends in real-time and facilitates a rich exchange between members. This successful long-term effort by the facility management associations is best represented in the fact that facility management has become a highly respected occupational sector in Austria that is consistently in high demand.

Prof. (FH) Dr. Thomas Madritsch, IFMA Fellow, MRICS
Managing Director, (CEO)
University of Applied Sciences
Kufstein, Tirol, Austria

Thomas Madritsch began directing the University of Applied Sciences Kufstein Tirol in 2011, following more than ten years as Director of Studies and fifteen years of professional experience in facility management. After many years of practice, Thomas wanted to contribute to the development of Facility Management as a distinct management discipline in Austria and Europe and establish it as a recognized profession with an academic education.

During his leadership, the University has increased its standing to become a highly ranked academic institution with 2200 students from 50 nations and 210 international partner universities all over the world. Thomas holds many key positions on national and international committees and advisory boards, as well as with professional associations. He has published a wide range of papers and journal articles based on his combined practical and academic experience in real estate and facility management. Thomas has lectured and given presentations on an array of issues in facility management at universities and conferences throughout Europe, Asia, and the USA. The quality and scope of his work in this field has been recognized in awards from respected organizations both at home and abroad.

Beyond his professional and academic interests, Thomas enjoys cross-country skiing, cycling, traveling, and exploring other cultures. Yet a key aspect of his life is also his family – his wife and two children.

2.7

A Perspective on Facilities Management (FM)

Stormy Friday

Welcome to Facilities Management

The year was 1982 and yours truly considered herself to be a hotshot management consultant whose most recent accomplishment was the reorganization of the 200-person US Environmental Protection Agency (EPA) Human Resource (HR) Department. Then came the call. The individual who engaged me to work on the HR project suggested to the new EPA Administrator that I might be a candidate to help with other reorganizations.

The meeting with the Administrator was memorable. More so than any meeting I have had after that day. She boldly told me she had a difficult organization that did some "light dusting" and other "housekeeping chores" and asked if I would be interested in helping her "fix" the problems within the organization. She called the organization Facilities and Support Services and asked if I would become the Director in an administratively appointed position that did not require confirmation by the Senate but approval by the White House. I had fond memories of collaborating with my dad on house maintenance projects but knew nothing about light dusting and other housekeeping duties. Could light dusting really be the organization's main purpose and function? Would I be able to tackle the problems within this organization?

Against my better judgment at the time, my consulting bravado jumped into high gear and without much hesitation I said yes, I would take on the assignment. As it turned out the organization did more than just "light dusting" and little did I know how problematic the organization really was. My introduction to FM was off and running.

The organization had been a resting place (a kinder, more gentle description than what it really was) where previous EPA administrators "parked" political individuals they weren't certain where to place. These individuals weren't making a huge contribution to the success of the organization and needed to be managed. I inherited over two hundred (200) staff in addition to more than two hundred (200) contractors and had responsibility for four (4) million square feet of headquarters space, ten (10) regional offices, twenty-six (26) laboratories, and several overseas properties.

My first day on the job I arrived in my basement office early and was eager to get started. I had not been at my desk for more than 20 minutes when someone yelled out get down and then shots rang out. As it turned out two security guards had an argument over a woman and sadly, one killed the other. I learned shortly thereafter that security was part of my bailiwick. It didn't take long for me to disarm the security guards. My introduction to FM was complete.

My experience as an FM practitioner was the best of times in my FM career and the worst of times. Our FM organization did many wonderful things from having the first customer service program for a Federal FM department to being recognized by the US Senate Appropriations Committee for outstanding budget management, which resulted in significant budget increases. We implemented the first master plan for a Federal FM organization and

Facilities @ Management, Concept - Realization - Vision, A Global Perspective, First Edition. Edited by Edmond Rondeau and Michaela Hellerforth.

became active in the Federal Facilities Council. A staff that once was relegated to being in the basement for many reasons, became well known throughout the agency for its superior service and customer orientation.

The worst of times occurred during a change in administration at the White House which resulted in a return of former senior officials at EPA. Much to their dismay, I was asked to stay on in my position, but within a year was told I was being removed from my job because "I was doing too good a job and it couldn't be tolerated." Talk about feeling undervalued and deflated. I learned the problem stemmed from the outstanding reputation my organization had earned around Washington. Staff on Capitol Hill called me instead of the Assistant Administrator for Administration and my name appeared in the Washington Post for improvements and changes our organization was making.

Shocked and dismayed they moved me to head up Information Technology (IT), which I knew nothing about. I called the office the "Aloha Suite" and vowed I would be gone within a few months. That is exactly what I did.

Changes in Facilities Management

After my departure from EPA, it became my mission to help FM leaders build capacity within their organizations to manage both the technical and management side of FM. Based on the experience with my own organization and those of other Federal FM organizations I saw a plethora of technical knowledge but limited or no management skills. FM staff were well-equipped to oversee technical problems, but were ill-prepared to manage the "management" side of FM.

Fast-forward to 1989 when a client of the high-tech design build firm where I was a marketing representative told me that if I had my own FM consulting firm, he would hire me to develop a strategic plan for his organization. Forty-eight hours later The Friday Group was born and has been aiding FM organizations ever since.

Since that fateful day in 1989 I have been fortunate to fulfill my goal. The Friday Group has helped countless FM organizations across the globe with organization reengineering, strategic planning, strategic sourcing and customer service and marketing. Throughout the firm's history we have created academic, training and certification programs and authored many books for the industry. Personally, I have been fortunate to speak, train and consult with FM organizations in over thirty (30) countries and allowed my passion to develop FM organization management strength to guide the work of the firm. Serving on the IFMA Board of Directors, the BOMI Board of Trustees and the ProFM Commission, I have seen the steady growth and development of the FM profession. The Friday Group has supplied the lens through which I could see and take part in the growth and advancement of the profession. As a keen observer of the changes over the past 30 years, it is important to share these observations with others who are forging the pathway to FM's future.

Customer Service and Marketing

First and foremost, the FM profession now fully embraces the importance of customer service and marketing as a core capability of the organization. It has been a long struggle to get FM leaders to understand and value the importance of customer service. Originally FM organizations were "dictatorial" in their approach to service delivery. They created rules and regulations that were enforced within departments in companies and other institutions without much dialogue with service recipients. The term "customer" did not resonate within the FM vocabulary. How could employees be considered customers and why would they need to provide input to structure FM service delivery?

When I reminisce about the old days in FM, I chuckle about the year 1993. I submitted a presentation topic to the IFMA program committee, "Do You Tell Your Customers How Much You Contribute to Their Work Environment?" I was told the committee was not in favor of the topic but agreed to let me speak. They informed me that I should not expect many attendees at the session. My room at the conference was sizeable and it began to fill up slowly. Suddenly there was a tremendous influx of participants and then it became standing room only in the room. 200 plus FM practitioners were curious about a topic that was foreign to them. The FM world had an awakening about customer service.

Marketing is the companion to customer service and the concept of marketing also was slow to catch on in FM. FM executives missed opportunities to learn about corporate executive "hot buttons" and the best way to make the business case for their ideas and plans. Theirs was a world of tedious charts and graphs and FM organization weren't stellar at making hard-hitting presentations. They also had a reluctance to seek advanced endorsement among supportive executives for their proposals and plans.

The good news is FM leaders have come a long way and are now skilled in the art of marketing. It has taken a while, but they fully understand and manage the process to develop a strong business case, obtain support for their recommendations and make powerful presentations that result in approval for their goals and plans.

Strategic Planning

The road less traveled has always been my mantra on FM strategic planning. For years FM leaders said they considered strategic planning to be a luxury they could ill afford and did not spend time on it with their staff. Often, they would equate strategic planning with capital and master planning and say they had it covered, not understanding that strategic planning incorporates plans for FM organizations in areas that go beyond the bricks and mortar responsibilities.

Over the years FM leaders have accepted the value of strategic planning. They realize a strategic plan cements the relationship between corporate strategic and business plans and theirs by supplying a pathway for the FM organization to follow in supporting corporate goals. FM organizations now incorporate strategic planning into their operational practices and understand the significance of establishing organizational goals as well as those associated with technical service delivery.

Organization Effectiveness

Without extensive training or prior experience with traditional management concepts FM leaders often allowed their organizations to develop and grow without benefit of a rationale for how it should function or be structured. They did not solidify mission requirements as the guide to how the organization should be configured and run. HR organizations often were at a loss as to how they could aid FM leaders because their depth of understanding about FM was limited. FM leaders did not spend time educating HR organizations about their role, responsibilities, and requirements so the relationship faltered. As a result, FM leaders had difficulty selling their organization scheme and staffing requirements because they failed to supply hard documentation on the correlation between service requirements, organization performance, and staff complement.

Over the last 30 years there has been a dramatic change in the way FM leaders design their organizations. FM leaders have developed structural/functional analysis skills and now are able to craft FM organizations that meet current and future business mission requirements. Their organizations are designed to be stable yet sufficiently flexible to adapt to revised corporate mission and direction changes that dictate modifications to the service delivery structure.

Legacy Planning

In the early years, FM leaders spent little time creating a legacy for their organizations. They did not have a vision for developing staff knowledge, skills, and abilities (KSA) to meet future FM requirements to support the corporate goals. FM leaders did not focus on creating career tracks for technical and managerial staff and were not fully invested in the need for staff who were trained or had FM credentials. They did not create opportunities for potential leaders to develop and display talent or set up criteria for upward mobility within their organizations.

The FM profession now is light years ahead of where it was 30 years ago. Today, most FM organizations applaud and frequently require FM certification for current staff and new hires as well as conduct robust staff development and training. They perform skills analysis to figure out gaps in staff capabilities and create training and development plans to fill these gaps.

Managing with Metrics

In the not-too-distant past, The Friday Group was asked by the Chief Financial Officer (CFO) of a major hospital to validate his recommendation to downsize the FM organization. He believed the organization was overstaffed and based his conclusion about downsizing on an analysis conducted by an external consulting firm that used hospital-based key performance indicators (KPI) to decide performance and staffing requirements. This CFO did not understand that the hospital-related KPI were not relevant to the FM profession. As a result of our analysis that used FM-related performance indicators, we were able to prove that the organization was understaffed and unable to deliver the level of service hospital administrators were seeking.

FM organizations struggled for years trying to justify their case with respect to staffing and funding for equipment, projects, and basic operations without hard metrics to substantiate their need. It took a while for FM leaders to appreciate the relevance of key indicators their corporate executives used to measure FM organization success and to apply FM industry benchmarks as performance guides. When FM organizations started using KPI against which they could prove superior performance, the stature of FM organizations rose dramatically.

Seat at the Table

It took a long time for FM leaders to be recognized by senior executives as having the ability to play a significant role in business decision making. It used to be standard knowledge that FM organizations were “the last to know” when a corporate decision affected their business operations. When asked to participate however, FM organizations could document how their contribution played a vital role in guiding corporate outcomes. FM leaders could play a major role is helping corporate executives analyze mergers and acquisitions from a bricks and mortar perspective as well as demonstrate the impact new personnel would have on the need for additional FM services. Their expertise was beneficial to real estate decision making in determining buy, build, or lease opportunities.

Gradually, through demonstrated performance, sheer perseverance, and internal customer support FM organizations became known and respected for their analytic capabilities.

Stormy Friday, MPA, Hon. FMA, IFMA Fellow
President, The Friday Group
Arnold, Maryland USA

Stormy Friday is the founder and President of The Friday Group, a firm specializing in strategies and solutions for the facility management profession since 1989. Ms. Friday began the firm after years of diversified management and facilities experience, and as a consultant. An internationally recognized speaker in the field of facility management, she has visited over 30 countries to consult, train, and speak on FM trends; organization development; productivity and motivation; marketing and customer service; strategic planning, and outsourcing alternatives.

Ms. Friday co-authored a book entitled Quality Facility Management: A Marketing and Customer Service Approach (Wiley) and authored Organization Development for Facility Managers: Tracing the DNA of FM Organizations (AMACOM; Dog Ear Press). Formerly on the IFMA Board of Directors, she was named IFMA

Fellow in 1999 and is the 2002 recipient of the IFMA Distinguished Author – Instructional Materials Award. She has been a Contributing Editor for Building Operating Management magazine and currently writes articles for FacilitiesNet. She often appears on podcasts for the FM industry.

Ms. Friday served on the BOMI Board of Trustees as Chair of the Academia Task Force for seven years. She served as a facilitator for the joint BOMI/Trade Press Master Facility Management (MFE) web cast program and from 2006 to 2007 wrote a weekly blog for FacilitiesNet. She also served as the facilitator of the joint BOMI/Trade Press Facilities Professional Leadership Series, a web-based educational program.

Ms. Friday was one of the founding contributors in the development of the Pro FM credential and currently serves as Chair of the ProFM Institute Commission.

2.8

My 40-Year Odyssey in Facilities Management

Phyllis Meng

My facility career spans over 40 years, and there have been many changes and perceptions of facilities during that time. I started out to be a Certified Public Accountant (CPA) but ended up "falling" into facilities which is a whole lot more interesting than sitting behind the desk working with numbers. When I fell into facilities, it was perceived as strap on the tool-belt. In some cases, facilities management is still perceived as maintenance rather than management.

When I graduated from high school in 1961, I wanted to be a CPA. At that time, females were not CPAs, but I am hard-headed so I went in that direction. Life took many twists and turns, I never made it to CPA but did make it to Certified Facility Manager (CFM) in 1993. I was one of the first to obtain my CFM when it was offered. The questionnaire was grueling because it asked about my education and experience in various facets of facilities management.

I was hired at the County of Ventura, California, where I recorded transactions in full sets of manual bookkeeping books. The department I worked in obtained and utilized CETA youth employment grants from the Department of Justice (DOJ). The county reported their financial results on a cash basis. But to request reimbursement funding form DOJ the reports had to be in the accrual reporting format. After years, I applied for a position of Inventory Control Clerk for the County. This position was to locate and physically tag all fixed assets of the County and ensure that the fixed assets reported for each department was correct. I sent out a yearly audit for the departments to verify that they in fact had the fixed assets assigned to them.

The County had just completed their Government Center complex that had a 425,000-square-foot headquarters building with nothing but Herman Miller systems furniture and no private offices. This was early in the 1980s. The other 400,000-square-foot building and courtrooms had conventional furniture which was over the fixed asset threshold, so all desks and computer had to be tagged and logged.

But how do you tag and keep track of the various components of systems furniture. I scheduled a meeting with the Department head and asked the question as to who was going to keep track of the various parts and pieces of the systems furniture which was quite a new concept. He indicated that was a good question. He had been wanting to make changes because the building is just now being occupied and the people overseeing the project were going away. I was reassigned to a different department and assigned the task of keeping track of all of the furniture in the complex, as well as systems. The installation company had taken an inventory of the various components of systems furniture and set up a database to keep track of the pieces. I had to learn the computerized system and then keep track of components removed and then installed into a specific workspace.

It did not take long for a department to want changes to their layout. The original design and installation contractor was gone and there was no one. I obtained a set of templates for the Herman Miller systems furniture and ¼" graph paper to design the layouts. This was before AutoCAD and all drafting was done by hand. As word spread that

Facilities @ Management, Concept - Realization - Vision, A Global Perspective, First Edition. Edited by Edmond Rondeau and Michaela Hellerforth.

changes could be made to the cubicles, I got busier designing workspaces with templates and graph paper. If the 11 × 14 graph paper was too small, I just scotch taped several pieces together. I would specify the components to be removed and the ones needed for assembly of the new cubicle. As a result of the demand for reconfiguration of cubicle components I hired three people just to do the moves, adds and changes to the systems furniture cubicles.

This was the Administration building for the County of Ventura. The furniture had to accommodate the services of a county administration office. The stand-up customer counters were Herman Miller, the 80-inch-high panels for department heads were Herman Miller panels, and there were other unique uses of the furniture. Since this was the early 1980s and no one had heard of systems furniture, Herman Miller took full advantage of asking my supervisor and myself to conduct tours of our unique facility. We started to become tour guides for perspective Herman Miller customers. We were spending more of our time being tour guides than doing our job. We finally had to indicate that only companies with a presence in Ventura County will be provided a tour of the facility. At the same time, Herman Miller offered me a trip to Zeeland, Michigan to view the manufacturing factory in Zeeland and the chair factory in Holland. During the same trip as a guest of Herman Miller, I attended the Facility Management Institute (FMI) in Ann Arbor, Michigan, in the early 1980s.

Prior to that time, I was unaware of what was called facilities management and the management of spaces as well as buildings. It was a several-day course where we visited an acoustics laboratory and discussed the various aspects of facilities management. The instructors were fantastic and really learned a lot during that time. I cherish and use the huge manual that we received. It was at FMI that I was introduced to the new organization of IFMA. There were some great visionaries at the FMI that were shaping this new profession. I met the representative for the LA area and worked in trying to start a chapter. This is where I met Art Hahn who was the first President of the LA chapter. Unfortunately, I was in Ventura and I knew that the County would not support me in helping to start a chapter in the LA area. So my dreams of joining IFMA would need to be put on hold.

I went from the County of Ventura to a dealership that had just obtained the contract for the systems furniture design for the County of Riverside new multi-building complex and all UC campuses. Unfortunately, the systems furniture was Westinghouse not Herman Miller. Boy what a difference in systems and how they are designed, built, and installed.

In October 1987 (right after the Whittier earthquake), I was hired at the Los Angeles County Transportation Commission. They were growing fast, I was number 99 and in a few years they expected several hundred. I was hired to do what we now call moves, adds, and changes. The standards were Herman Miller tables with tub drawers and a pencil tray with a free standing file cabinet. We were hiring and stacking people into an older building in downtown Los Angeles.

At this time, I joined IFMA and became active in the Los Angeles chapter. As my job grew in 1988–1989 I was part of a team to oversee the tenant improvement of leased space that continued to grow. I represented the tenant in this project and at that time, I learned the lesson to never trust an architect or engineer. The architect was to build this fantastic facility or space that is very difficult to complete, to utilize and maintain. The work had to be expedited because our lease was ending in our existing building. I oversaw the TI work, the relocation of staff from the existing to the newly built facility. This project took a lot of selling to the staff members. They were in offices with usually more than one person in the office. Now they were going into cubicles with 65-inch-high panels. No one had a private office and that caused concerns for some staff members due to confidentiality issues. It took a lot of "selling" for the staff to accept the new cubicles.

Right after that project was complete, we took over two more floors of the building and completed tenant improvement work there as well. When we relocated out of the building in 1995, we had leased, subleased, or rented almost the entire building. I had to keep track of at least 15 different leases with 15 different escalation charges and costs per square foot.

As I was growing at my job, I also wanted as much information and help as possible. I became active in the Los Angeles IFMA chapter. It was great to go to meetings with my "shopping" list of things to ask some of the attendees. They were a great support group.

I started going to IFMA conferences in 1994 (St. Louis). That is where I learned about the public sector council of IFMA. At that time, the members were from GSA and the council was very successful. They started what was called "list serve" where anyone could send an email and ask a question. The other members would respond with information to assist them. One time I asked about a scope of work for cafeteria services and within an hour, I had a stack of scopes of work and was visited by the contractor for the State of California. Talk about support!!

I became active in the Public Sector Council serving as Newsletter editor and won the 1996 IFMA Council Achievement Award for the newsletter that I produced. I went through the various offices of the Council and was Council President in 2005 when the Council won the IFMA Council of the Year Award.

I craved educational classes to learn more about facilities management, what it is, and all the skills needed for the profession. In 1991 at the IFMA conference in San Diego, I took the four- to five-day Project Management class rather than attending the conference itself. I was able to justify the educational course but not the conference itself to my superiors. In 1994, I took the Operations and Maintenance course in St. Louis. Again the class was at the same time as the conference. The course was not as polished as the previous course I took. I still have the manual and hand outs which are very poor and almost unreadable. Nothing like what IFMA provides today. I felt it important that the facility professionals learn information from a structured setting. When I was growing in my facilities career, it was OJT (on-the-job training). So becoming an instructor and to help others learn about the profession was very important to me. In 1996, I took the Train the Trainer course that IFMA provided in Houston. Glen Jay was the instructor. He was very good at helping us to become great instructors for IFMA. Up until this time, I had never facilitated an educational course or provided a presentation for IFMA or anyone else. My first class was with Rich Faneilli in Orlando, Florida. That first experience was excruciating because I really boomed. Rich really helped to mentor me and to make me a better instructor. But it took many years to gain the confidence to be a good instructor.

As I was growing in my facilities career, it was sometimes difficult because being a female most males at that time did not think that females knew what they were talking about. Also, facilities management is still to this day considered strapping on the tool belt. A lot of times, I was the only female at project meetings. When I was doing the tenant improvement for the first two floors of the new building in 1988–1989, I was the project manager, and the construction people did not want to listen to me. But since I was the tenant representative, they had to listen to me and follow my instructions.

From the time I joined IFMA, the members and the association provided me with guidance, consultation and assistance to grow my career. As I grew from Program Chair to Los Angeles Chapter President, I continued to network and meet other members who provided much needed advice and recommendations. I have attended every World Workplace since 1994 and have provided at least one presentation at every World Workplace since 2001. By going to WWP I met with vendors who could provide the goods and services I needed and could help me with specifications and scopes of work. I met practicing professionals that I still call on today. There are members that I met throughout the years who I still keep in touch with.

I was IFMA Foundation chair for 1996 and 1997. When I was appointed Foundation Chair there was no Board of Trustees, there was by-laws that needed review and updating but I had funding for research grants. Most members did not know about the Foundation and the good work that it was doing. I started a publicity campaign to inform the members of what the Foundation was doing. I went to various regions and made presentations to their members about the benefits of the Foundation and what it was doing to help the profession. I also started the Foundation Ambassadors which were used for fund raising. There are still IFMA Foundation Ambassadors, but their current task is different than what I had envisioned. I was able to triple the amount of grants and funding for the IFMA Foundation due to my efforts.

In 1998 I was awarded the IFMA Distinguished Member Award for my leadership in the chapter, council, IFMA Foundation and on the IFMA Board.

In 2002, I received my IFMA Fellow designation based mostly on what I had done to increase the awareness and contributions to the IFMA Foundation. Also, I had coordinated the first mega regional conference, President of

LA chapter, as well as a member of the IFMA Board of Directors along with facilitating the IFMA classes. When I was awarded my IFMA Fellow it was the largest Fellow class with eleven members recognized as Fellows. It was interesting because of eleven Fellows there were only two females me and Peggy McCarthy. Females are starting to enter the field of facilities but it is a slow process.

I am very passionate about facilities education because as I was growing in my facilities career there were no classes to attend to find out about the mysterious world of FM only on the job training (OJT) which can be painful. After I attended the IFMA Train the Trainer course and got my feet wet. I started facilitating more courses. I started with the FMP courses. These courses and my students made me realize how my organization should operate. I had never heard of strategic or tactical planning before and what it entailed. I was able to learn as a trainer what was successful in an organization and what was not. I began to look at my own organization and see some of the issues/problems and how I would recommend making it better.

I continued to grow in learning more about facilities. I had started out with templates designing offices, relocations both large and small, tenant representative on tenant improvement projects then into writing scopes of work for contracts. I relied on our maintenance staffing to help me to understand the equipment, how it works, and what is needed. They were very supportive of me because I asked them questions and took their recommendations. Later on I supported them against a subject matter expert who had never worked in facilities but was from construction. He believed that according to the experts as certain set of events/condition was the way to go. According to the hands on maintenance people this is not the case. The maintenance people were correct and the subject matter expert (no matter who indicated otherwise) was incorrect.

I continued to expand my educational efforts and could now instruct the CFM Exam review and the met the requirements to facilitate the SFP course. Along the way, I became a BOMI (Building Owners Management Institute) instructor and instruct the Budget and Accounting course. These additional courses helped me to continue to grow in my understanding of the requirements of facilities, facility managers and building occupants. In 2009, I was known as the Finance guru because most of my presentations addressed the financial aspect of facilities. So I was asked to develop an online class for the UCI (University of California, Irvine) extension Facilities Certificate program. I have been instructing this online course since 2009.

When my boss (Deputy Executive Officer, General Services) passed away of cancer in 2009. His position was opened for recruitment. I thought his second in command would apply for the position. By that time, I had over 20 years' experience in facilities and knew where I wanted to take the department. So, I applied for the position. I was surprised to be invited for an oral interview because this was many steps over my supervisor position. I thought I had blown the oral interview because I was not comfortable with my answers. But that was not the case. I was invited to interview with the Executive Officer who would be my boss. I was hesitant but I showed where I wanted to take the department (strategic planning) and how I planned to restructure the reporting and jobs of the facilities department. This position was responsible for the facilities maintenance of the 625,000-square-foot high-rise headquarters along with all of the building projects plus the mailroom, copy center, records management, the library, rideshare, and travel. I also had visions of structuring the other departments and their reporting structure. To my amazement (while I was teaching a CFM Exam review course), I was offered the job which again was several levels over my current position. The only issue is that the salary was lower than my previous boss whose position I was assuming, and it was still lower when they advertised the position after I retired. Was that because I was female? I will never know.

After becoming the Deputy Executive Officer, my first priority was to put in place a strategic and tactical facilities plan and reorganize the facilities function. There were supervisors reporting the Deputy Executive Officer rather than Managers or Directors. I had more important things to do than to deal with a problem custodian. My other priority was to obtain a subject matter expert to confirm that we needed to replace the cooling towers as soon as possible because due to deferred maintenance and other factors they were not going to last through another summer.

Very quickly, I was able to obtain approval for the capital project to replace the cooling towers plus other major pieces of equipment that would fail shortly due again to deferred and reactive maintenance. With the approval of the capital project, I also obtained approval to hire additional staff so that more preventive maintenance work orders could be completed rather than go into practicing deferred maintenance again. To complete the various building equipment replacement projects, I hired an engineering firm to complete the various projects. I also hired an in-house Project Manager to oversee the various projects that they are continuing to complete.

When I retired in 2015, the department was much different than when I took over in 2009. There were clear reporting lines up the hierarchy chart. There were better union relationships with the various unions. I initiated several sustainable projects. The biggest one was daylight cleaning and reducing custodial services. This allowed staff to be reduced during a 20% budget cut requirement. There are now better customer relations between the department and the building occupants.

Now that I am retired, what am I going to do now? This was about the time that the IFMA Foundation was developing the Global Workforce Initiative (GWI). I worked with the Foundation trustees along with Chaffey College to develop the curriculum for the Chaffey College Associates degree in Facilities Management. The intent was to use the IFMA Essentials of FM class and have an instructor go through the slides explaining to the students what each meant because nobody knew about the facilities management field. There was an instructor who had been working over a year with the team to develop and approve the curriculum. Two weeks prior to the start of the course, I was contacted and asked if I would like to teach this Community College course because the business instructor had been promoted to Dean. I have never taught at a Community College and the onboarding in two weeks was remarkable. So in the fall of 2019, I started teaching the IFMA Essentials of FM and a Project Management course for Chaffey College about 65 miles (with traffic) to Chaffey College. I also developed exercises each week for the student's homework as well as inviting subject matter experts to come and talk to the students about facilities. Then a similar course format was started at West LA College. Since I was already doing the courses at Chaffey and IFMA requires an approved instructor for the Essentials course, I was asked to facilitate these courses as well. I again had to go through a different hiring process.

Since there was no active local chapter in the area, I started a Student Chapter that became a success. We were able to hold a Vendor fair for the Chaffey College students. From the vendor show we were able to send several students to WWP in Phoenix in 2019. The chapter received Student Chapter of the Year in 2019 and I received the Distinguished Educator Award in 2019.

The Chaffey program is an award-winning program that continues to be successful. There are students who have served internships and hired by ABM, JLL, and other smaller organizations. I see the GWI continuing to expand to provide education and training to the community college students and to get them excited about this challenging career.

The Chaffey course content that I developed has been utilized several times by the IFMA Foundation in starting new programs at San Mateo, Hot Bread Kitchen and various other community colleges.

As a result of the Chaffey College successes I was awarded the IFMA Foundation Award of Excellence.

Several years ago, UCSD (University of California, San Diego) was looking to offer their facilities certificate classes both in the classroom as well as online. There was one instructor who was doing quite a few and I had contacted them in the past. They asked me to develop an online course for Operations and Maintenance. Since I was already teaching online at UCI, it was fairly easy to put the class online. The difficult part was the development of the course material and homework assignments.

I am glad that I am retired and can look at facilities management from a distance. It has been rough on the current facilities professionals over the last two years. They had to find plexiglass and design work areas away from others. They had to change the cleaning protocol for the building. They had to deal with empty buildings. They had to adjust the maintenance tasks to make the indoor air cleaner at a higher volume.

I saw a trend prior to retiring that with the internet and most everything electronic, that employees were working at home. There were a few companies that adopted that format before they were forced to by the pandemic. I see a changing role for the facilities professional because of climate change issues and vacant space with employees working from home. The company will save money on lease costs or rent if own the building because they will not need to have space for all the employees who are now working from home along with the cost of electricity and water.

The facilities manager will need to be the expert and implement sustainable projects while still maintaining the extra cleaning protocols established during the pandemic. These require different skill sets than previously required in the old way of thinking. The pandemic has helped the facilities professional because now everyone knows that we are there just not some invisible creatures, that what we do is important to the health of everyone, and we should be leading the way on sustainable projects to reduce costs and be a good steward of the environment.

Phyllis Meng, CFM, SFP, IFMA Fellow
Newport Beach, CA USA

Phyllis has over 40 years of experience in facilities management starting at the County of Ventura and ending up retiring as the Deputy Executive Officer, General Services for LA Metro where she oversaw over 100 employees and was responsible for the 28-story high-rise headquarters, mailroom, copy center, records management, travel, and rideshare.

During her facilities journey she received her Certified Facility Manager (CFM) certification in 1993 and Sustainable Facility Professional (SFP) in 2012. Part of her growth in the facilities field was her association with IFMA, its chapters (LA Chapter President 1995), councils (Public Sector Council President 2003–2005), IFMA Foundation Chair (1996–1997), and IFMA Board of Directors (1998–1999).

Phyllis is passionate about educating up-and-coming facilities professionals and became an IFMA certified instructor in 1996. She facilitates all IFMA course offerings as well as being part of the team that recently updated the FMP and SFP courses. She attended every WWP conference from 1994 to 2022 and spoke at every WWP conference from 2001 to 2021.

In Toronto in 2002, Phyllis was honored to be designated as an IFMA Fellow because of her contributions to IFMA and the facilities field.

When Phyllis retired it allowed her to increase her educational offerings. She has been able to offer teaching the IFMA courses for the local Southern California chapters. She was tapped by the IFMA Foundation to instruct the first Global Workforce Initiative model of offering a Facilities Management degree at Chaffey College, a community college in Rancho Cucamonga. This led to duplicating the model at West Los Angeles College in Culver City.

Since 2009, Phyllis has been an instructor at UCI (University of California, Irvine) in the Division of Continuing Education teaching Financial Analysis to Facilities Professionals, a course offered in their award winning Facilities Certificate program. Phyllis was then asked by UCSD (University of California, San Diego) Facilities Certificate program to develop an online Operations and Maintenance Course for them.

During her career Phyllis has won awards from IFMA. Her awards include: 1995 – Distinguished Author; 1996 – Council Achievement for Public Sector newsletter; 1998 – Distinguished Member; 2005 Council of the Year; 2019 – Education of the Year. From UCI she received the UCI Extension Advisory Board Award.

2.9

How to Make Good Facility Decisions

Barry Lynch

When retired athletes are asked what they remember most about their careers, they say, "the people." I'm the same way when I think about my FM career – I recall the people, not the budget presentations, close calls in making deadlines or problems solved. What I've learned from various mentors and situations along my career path is what I would like to share; in this brief; a decision-making process that can be applied in multiple situations, and I would like to offer insights in applying that process to address new situations.

Jim Cogdill was my first mentor and head of Facilities at R. J. Reynolds Tobacco Company (RJR) in the 1980s (for the setting – picture *Mad Men*[1] – guys in dark suits with narrow ties, everyone smoking, secretaries bringing in coffee on trays). He hired me away from Gensler Architects and had a relaxed, slow, southern drawl that lulled "them corporate Yankee boys" into thinking he wasn't smart, but he was sharp as a tack. His style was unparalleled. For example, after we finished rehearsing an executive presentation he would take a drag from his cigarette, think deeply for a while and proclaim, "well if it don't work out, I can always go back to Tennessee and raise hogs." I eventually found out this was a complement, but it took a little while to decipher this and other sayings like "bless his heart" (did not mean what you would think). Jim was a visionary, a former plant manager who worked his way up from junior industrial engineer on the production floor to corporate facilities director.

Executives had recently challenged director-level management to come up with breakthrough improvements within their span of control and Jim invented the term "Facilities Express." We were in a kickoff meeting where he was challenging his direct reports to figure out the details of how to deliver projects faster and cheaper while reducing operating cost when his secretary came in, whispered in his ear, and escorted him out of the room. We were perplexed but continued our discussions. Jim returned, after what seemed like an eternity, and was visibly shaken. "KKR (Kohlberg Kravis Roberts & Company) has made an offer for a Leveraged Buyout of the Company."[2] Everyone looked blankly at each other. Most people didn't know what it meant. Jim said "I've talked to my boys (not sons) and they say it's all over for facilities, which are considered no more than just another cost to be cut by them Leveraged Buyout Boys." He explained that there would soon be a bidding war and that the winner would purchase the company's stock with cash received from bond sales. The bonds would be paid back from increased operating profits resulting from cost-cutting. "I've got two kids in high school and I'm looking at college bills. I'm not about to give up." He went around the room and questioned each person, asking "are you in, or do you want

1 *Mad Men* is a US television period drama about Don Draper, an advertising executive heading a Madison Avenue, New York, firm. The series is set in the 1960s and is known for "showing it like it really was." The series is available on multiple streaming platforms.

2 The Leveraged buyout would be Chronicled in book *Barbarians at the Gate* by Byran Burrough and John Helyar, published by Harper Collins, 1989, and was made into a 1993 made-for-TV movie by HBO. RJR was eventually purchased by the Leveraged Buyout firm Kohlberg Kravis Roberts & Company (KKR) and was broken up and sold off in parts after the bonds were repaid (11 years).

Facilities @ Management, Concept - Realization - Vision, A Global Perspective, First Edition. Edited by Edmond Rondeau and Michaela Hellerforth.

a retirement package?" The process had the feel of a North Carolina Bible study session, and everyone knew how to testify. I stepped up and stated that I had been with the firm less than a year and that, (at that time in the 1980s) no company would hire someone at my level who was job jumping or let go with minimal time spent at a position. I affirmed that I was in. I had to make things work.

At our team's first rehearsal for an executive committee presentation on what he called "the stacking plan" (actually a strategic facilities plan) Jim grew agitated. Although he never swore, you could tell from his strong language he was upset. "Hoss[3] you know better than this! What in the world are you thinking? How many presentations have you done, and we're here at the last minute with this? You're going to get us all fired! These cats don't care about past performance. They want to know what you're going to do in the future!" I explained that without trend analysis, executives wouldn't have a context for the numbers we used in deriving costs. "Context..." he said as he took a drag on his cigarette and looked at the carefully crafted stacking plans that were arranged on the walls of our War Room showing how facilities changed on almost a weekly basis. He stood up, walked around the room and examined our project cost graphs and other content that provided context. Dead silence. Standing. Thinking. Smoking. After a while, you could see him relax. "Do it again – do your presentation again."

After the second time he said, "you and me are going to have lunch with Jerry (Jerry Gunzenhauser, our CFO) and you're going to explain your thinking." I knew that timing was crucial and I wanted to soften Jerry up first. At lunch, I started off by explaining that I had taken to heart his management briefings where he emphasized the new focus was After Tax Cash Flow, not Net Income. I had figured out if we changed the timing on our stacking plan, we could save close to a million dollars on the One Triad lease, even though we would have to pay an $800,000 penalty (more than offset by depreciation write-off). Jerry was elated. Since he was in a good mood, I then explained that in order for executives to understand the thinking behind our stacking plan, we needed to show trend analysis. He said "show me" and we walked over to the War Room. After a major league grilling, he smiled widely and said, "you're good." Then he got serious and said, "I'm going back to my office and I'm going to decrease the lease budget. Once I do that, the money is gone forever. Are you dead sure?" I gulped, said a prayer, and squeaked "yes." Welcome to the big time, Barry Lynch.

The War Room – A large conference room (see Figure 1) that was used to communicate the complex chain of projects involved in vacating $29 million in leases. The schedule wall in the background used colors to show the renovation status in a 17-floor building where groups were relocated internally, freeing up space that could be renovated for groups moving in from lease space. Each row refers to a floor and the columns show the status each

Figure 1 The War Room.

3 Jim called all his direct reports "Hoss" if they were guys and "Honey" if they happened to be gals.

week. Company executives received the first preview, followed by corporate leaders, divisional leaders, department managers and finally, line personnel like the cleaning crew, day porters, mail crew, IT, and others who had to adjust schedules on a weekly basis. Not shown – the other three walls with additional information.

"Our thinking" was the decision-making process I started formulating when I was in grad school when I got together with a bunch of other students. We had been whining over beers that we weren't being given a well-rounded education, and we responded by developing a self-study class about decision-making. We bemoaned the fact that in the business school we were learning equations, debits, credits and organizational behavior, but didn't have the chops to present to an executive team (this was way before Power Point). In my other life as an architecture student, we were learning how to solve problems and design, but not how to "package the deal." You might have heard about the agony architecture students go through in "design crits" – a process where design students put their drawings on a wall, mumble some explanation and get absolutely excoriated by the professors. My fellow decision-making students and I weren't dumb and we knew that the greatest architects had the best presentations, which were really sales pitches.

We wrote letters to famous architects[4] asking for advice and every week we did research and shared our findings. This is where we learned about "Burton's Rule of No," which states, "always say no when presented with a request. If it's a good idea, the presenter will be back with a better pitch and better reasoning. If it's not a good idea or the presenter is not passionate about it, they won't be back." You would think that Michael Scott from *The Office*[5] came up with the theory, but "Burton's Rule of No" was actually found in a prestigious management journal. FM's should understand this law when getting rejected at presentations. I've seen it used at high level corporate budget presentations with great success and I've also seen it applied at the team level where it absolutely killed the team spirit.

I resolved to formulate my own process to assist in making good decisions and spent several years watching more experienced professional team members inexplicably succeed and fail in different situations. One of things I noticed was, again, context matters. Senior professionals would give the standard pitch and analysis to business executives as to non-profit leaders and then wondered why they didn't always succeed. If you are a fan of the IFMA Member Exchange,[6] an on-line forum where IFMA members pose questions and look for insightful answers, you will notice that in quite a few posts the questions don't receive immediate answers but garner more questions because the people with answers are looking for context.

As I entered the turbulent waters of the Leveraged Buyout I was concerned because management was laser-focused on budget line-item cost reduction. The proposals I was floating up the management chain often were nuanced – like paying an $800,000 lease termination penalty, getting a $1.6 million hit on earnings, but still coming out a million dollars ahead on After Tax Cash Flow or implementing changes in project delivery or facility operations that did not necessarily produce line-item cost reductions, but produced future cost avoidance.[7]

The decision-making process I developed evolved into four steps:

- Where we've been
- Where we're headed
- How to get there
- Consensus Building / decision-making

4 For the younger crowd – In the not too distant past, people wrote letters instead of sending e-mails.

5 *The Office* is an American TV series (2005–2013) that is currently available on the Peacock network. The series is based on the British TV series of the same name and mocks inept management at Dunder Mifflin, a fictitious paper company that is in a slow and steady decline as the world moves to a paperless office environment.

6 "The IFMA Experience Exchange" is a members-only discussion forum of the International Facility Management Association, the world's most widely recognized association for facility management professionals, supporting 24,000 members in 105 countries.

7 Cost Avoidance – not all companies recognize future cost avoidance. The simplest example is implementing an energy saving plan that will decrease future energy usage (and hopefully cost).

It's actually very similar to the process outlined in IFMA's Strategic Facility Planning White Paper,[8] but can be applied to a wide range of situations in both your professional and personal life. I'd like to give two examples or case studies of applying the process to business problems that Facility Managers may be facing:

- **How to set yourself up for success when the consultants come looking for a 10% cost reduction** – selected this scenario because I gathered expertise in this area under dire circumstances (if I did not produce, my job and department would be gone).
- **New Facility Management vehicle purchase. Gasoline (petrol) or electric?** I am interested in this because I have never seen or read an apples-to-apples comparison. I wanted to find out what the deal is because it is an area of my interest, which I'm sure other Facility Managers share.

Cost Reduction When the Consultants Come Looking for a 10% Reduction in Your Budget

The financial winds of change have been picking up speed. Change feels imminent – like sailors preparing for a coming storm on a perfect, sun-drenched, flat sea. They know the storm is near, but everything looks ok – for the moment. Cost Reduction efforts may be on the agenda for some Facility Managers. Everything may seem now, but...

Where we've been – While working on "Facilities Express" I called around to friends and former classmates for advice. Those that had gone through the experience of a major cost-cutting exercise told the same story. Outside consultants and executives from other divisions come around looking for cost reductions. You need to tell them a story, not just review your budget report. Facility costs are often viewed as an easy mark and a 10% overall budget reduction is viewed as an attainable goal. However, if you've been doing a good job managing or reducing cost, it's difficult to get the 10% reduction. Because we had a few years of cost reduction success at RJR, it was vital to show what had been achieved, and what future, reasonable goals might be. My contacts advised that every component of every line-item cost had to be scrutinized because there were often hidden gems that were not apparent when viewing aggregate costs. In one situation, we used the Building Owners and Managers Association (BOMA) Experience Exchange Report[9] to show that we were running our facilities efficiently, having the same overall operating costs as leased buildings which have to be efficient in order to survive.

After we showed how efficient and effective, we were for overall cost, we had to repeat for all the components in the BOMA cost structure. At the last minute we were running about $100,000 short of our overall goal and had to come up with a plan by close of business that day. We broke out the BOMA Experience Exchange and looked at their deep dive data. We noticed that our cleaning materials cost was far above average. We then met with the cleaning coordinator who asked the contractor to show us where they stored materials. We found stores of paper towels and toilet paper stashed throughout the complex (1 million sq. ft.) as the contractor had been hoarding a set order quantity that was delivered on a regular basis, regardless of use. We inventoried every stash and came up with 80% of the required reduction with just this one line item. While I realize that cost benchmarking is not very useful in today's environment, the small things are important and when you manage them well you prove you are a good manager. The graph below paints the picture that total cleaning costs were lowered by decreasing the total number of facilities as well as the number of cleaning staff (see Figure 2).

8 "Strategic Facility Planning: A White Paper," published by the International Facility Management Association and available in their website at IFMA.org.

9 BOMA EER BOMA Experience Exchange Report slices and dices the BOMA cost structure in numerous ways. This report was used because the timeframe pre-dated IFMA Benchmark reports. Reports show median and average costs by city, building size, location within a city (downtown, suburban), and so on.

	2014	2015	2016	2017	2018	2019	2020	2021
BUILDING A	1,127,348	803,016	803,000	803,016	588,390	615,000	615,000	615,000
BUILDING B	296,704	364,781	311,300	250,553	201,000	130,000	130,000	130,000
BUILDING C	226,625	79,994	106,500	104,515	89,348	92,000	92,000	92,000
BUILDING D	69,508	24,630	11,700	9,734	14,598	16,000	16,000	16,000
BUILDING E	58,515	58,515						
BUILDING F	34,340	34,340						
TOTAL	1,813,040	1,365,276	1,232,500	1,167,818	893,336	853,000	853,000	853,000

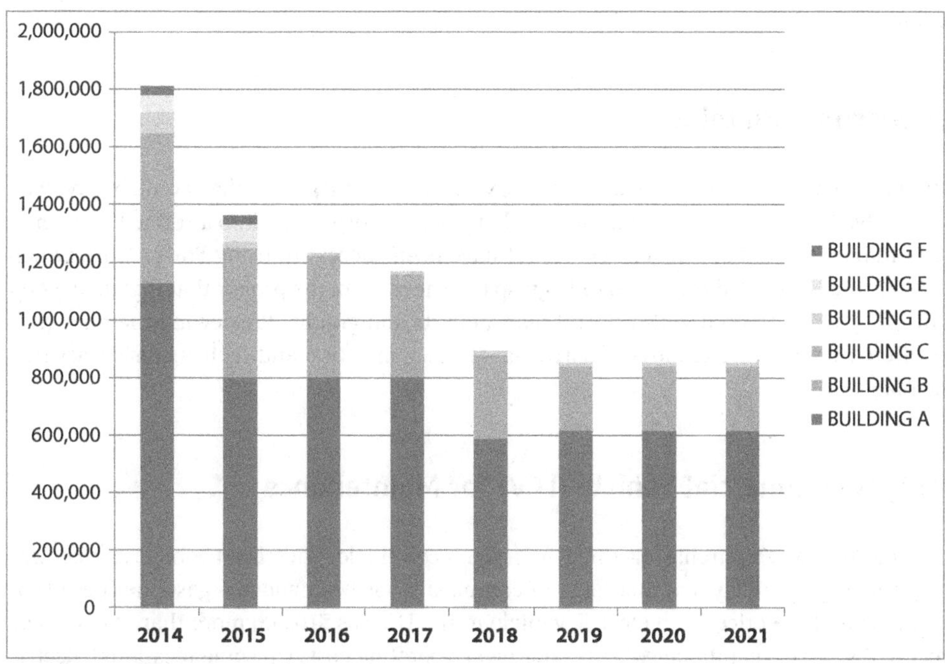

NOTES:
Reductions in cleaning costs achieved through:
* Reductions in lease space
* Contract Renegotiation

Figure 2 Contract cleaning summary. (*Source:* B. Lynch).

Where We're Headed

When the Leveraged Buyout was announced, we developed a plan to vacate lease space and relocate staff to existing company-owned office facilities and converted factory space. By the time the details had been worked out, but before any staff reduction announcements, we presented to the executive committee. "Where did you get these numbers!" was the angry response. Jim Cogdill drawled "we guessed... sounds like we had some good guesses," in a manner that unnerved the executives. After the executives got over the initial shock that a bunch of facility people could guess their plan, they toured the War Room, asking questions, agreed that we were on the right track, and started to get into the details of the capital funding we would need and Cash Flow implications of our plan. It's important to note that our presentation started with expectations about the future business environment and we focused on the company financial implications of facility moves before moving on to the actual facility plan.

How to Get There

The "Stacking Plan" as Jim Cogdill called it, involved three critical paths of departmental moves over a time-period of 63 weeks. Groups in existing space would be consolidated (remember staff reductions) to free up vacant space and groups from lease space would then move in. The plan had been developed for 53 floorplates on tracing paper that was placed over small scale occupancy plans that were hung on walls of the War Room. Executives could look at the tracing paper plan, then lift it to see the current floor plan. This represented an 80% decrease in cycle time from our previous method.

Decision-Making, Consensus Building

Executives and department managers of groups to be moved along with operational groups that would be involved in implementation were involved in presentations and their feedback was incorporated into a refined version of the plan that was actually implemented. I'm not sure we would have been successful using PowerPoint, which only offers one slide at a time, and does not allow one to fully grasp the enormity of the project that is gained when walking in to a room with four walls covered with colorful plans, charts and graphs. Previewing the plan with departmental managers offered them a rare glimpse of corporate strategy in action and their inclusion assured their commitment to the plan.

Purchasing a New Light Commercial Vehicle (LCV) for Maintenance

In years past, the primary question when purchasing a new maintenance vehicle, aside from vehicle choice, was lease or purchase but that's changing. Today, another layer of complexity has been added – gasoline (petrol) or electric? In 2021, the average purchase price of an electric vehicle in the US was $10,000 more than the average for all vehicles. With tax credits, lower maintenance costs and lower operating costs (in some locations) electric vehicles often have a favorable value proposition, but the question is **"are they the best solution for Light Commercial Vehicles (LCV's) used by groups like Facility Managers?"** Much of our focus here will be sorting out a good set of numbers to assist in making a fact-based purchase decision.[10]

Where we've been The worldwide electric vehicle (EV) passenger car stock has expanded from a few thousand in December 2010 to more than 20 million in 2020. China had the most registrations until 2020 when Europe (and particularly Germany) surged to the lead. Sport utility vehicles (SUVs) and light commercial vehicles are the fastest growing market segments with no end in sight to the introduction of new models with significant performance improvements, which bodes well for Facility Managers. Petrol (gasoline) drives today's economy and the distribution network is ubiquitous while electricity is available at every wall outlet, but not necessarily convenient for road trips (yet).

The story of gasoline (petrol) prices is different depending upon where you live. In the US the average price per gallon is $5.19 at midyear 2022 ($1.37 per liter), which is about double the price in the period from 2016 through 2020. In Canada, prices are slightly higher, and the European average for the same period is $7.53 per gallon ($1.99 per Liter). No matter where you live, if you look at how the price of gasoline has changed over time, the graphs are real headscratchers. Prices have fluctuated significantly over the years because gasoline is a commodity and numerous factors influence the price (price movements over time are similar across national boundaries).

10 Greenhouse gas analysis is specific to each geographic area and is influenced by the sources of power generation used by local utilities.

Like gasoline, the price of electricity varies by location, and each source of power generation (coal, natural gas, fuel oil, nuclear, hydroelectric, tidal, wind, and solar) has a unique history. Hydroelectric and nuclear are steady, low-cost sources of power, but are not available everywhere. High-emission coal power plants have been closing and the use of wind and solar has increased in recent years. Understanding the trends and future plans for your local and regional utility providers is essential before any cost analysis can begin, but the lack of volatility in price makes electricity attractive.

Where we're headed – Automakers world-wide have made astonishing EV investments.[11] Just as in consumption of news, you can find articles supporting your point of view and there is really no telling if the EV enthusiasts are correct, or the ICE (Internal Combustion Engine) proponents are right.[12] What is of concern to you is the continued advancement of battery technology and a cost curve that projects that the cost of internal combustion and EVs evening out by 2030. With an operating cost less than half of that of an ICE vehicle, EVs would seem like a slam dunk, but they are not right for every situation, especially commercial applications, as you will see.

Gasoline prices are expected to remain high due to demand outstripping supply. While worldwide economies are expected grow, refinery capacity is not keeping pace; so demand should continue to outpace supply leading to an upward cost curve.[13]

Electric power prices have been less volatile than petroleum-based sources. Be sure to understand the mix of sources your power provider relies upon because each has a separate carbon emissions profile.

How to get there – During this phase, you assemble what you've learned and create financial or emission models to establish the value proposition for your top choice vehicles. Identifying inputs for an accurate apples-to-apples comparison than with gasoline versus electric is an exercise in detective work, because few EV proponents are keen to share the "extra" costs involved with an EV purchase.

To start your analysis, start out by establishing your *maximum* drive distance per day, then start by identifying electric vehicles with a drive range that comfortably meet your criteria. In the US, the sticker on a new internal combustion (gasoline) vehicle states the city and highway mileage per gallon and if you look up the gas tank capacity, you can easily calculate the distance traveled between fueling stops. With electric vehicles, particularly light commercial vehicles, there are multiple factors that impact your **aveats**:

- Electric vehicles are more efficient at city driving than gas-powered vehicles.
- For EVs, range is the most important statistic. Refueling isn't necessarily a quick stop at a filling station, and can takes hours to complete. If you run out of fuel and are pulled over at the side of the road, what do you do?
- On average, EVs can travel less than half the distance of similar, gas-powered vehicles per fill-up. In 2022, the US Environmental Protection Agency of the battery slows the charge rate and degrades battery capacity over time so best practice is to not top off.
- Only 85–90% of the energy that comes from the wall outlet/ charging station makes it to the battery pack.
- There are two measures of efficiency: MPGe, which includes charging losses, and consumption, which measures energy used while driving.
- *Car and Driver* magazine found that only 3 out of 33 tested EVs beat US EPA projections, while they have found that gas-powered vehicles generally exceed EPA estimates.
- Batteries degrade over time, decreasing range.[14]

---

11 Half a trillion US dollars by 2021.

12 "Ambitious Electric Car Sales Targets May Fall Short And Reprieve ICE Power; Report," Neil Winston, *Forbes*, March 17, 2022, https://www.forbes.com/sites/neilwinton/2022/03/17/ambitious-electric-car-sales-targets-may-fall-short-and-reprieve-ice-power-report/?sh=5b08fab25709.

13 "Gasoline Prices Will Stay High" Kiplinger, https://www.kiplinger.com/economic-forecasts/energy.

14 "Understanding the Effects of Payload and towing on Commercial EV Range" Fleet forward Feb. 15, 2021, https://www.fleetforward.com/10136746/understanding-effects-of-payload-and-towing-on-commercial-ev-range.

- Cold weather decreases range dramatically. In one test, using the heater decreased range by 60%.
- Adam Berger, president of Doering Fleet Management, says driving in winter hurts range 20–30% while excessive speed decreases range 10–15%.[12]
- Heavy payload, ladder racks, trailers, and other accessories can decrease range significantly. There are no statistics, but anecdotal evidence suggests towing a load can cut EV range by as much as 80%.[12]
- The impact of traversing steep grades, like the 70-mile (110 km), 7% grade west of Denver on I-70, is not available.
- EVs have half the annual maintenance cost as gasoline-powered counterparts as well as virtually no downtime.
- Charging times are decreasing. Charging rates for commercial EVs like the Ford Lightning Truck can now recharge from 15% to 80% in 45 minutes.

Let's say you and your team have determined that a light commercial van is the best vehicle to replace your department's aging pick-up with a bed cover (see Table 1). The vehicle is used by Facility staff to serve four facilities in a metropolitan area, averaging 239 miles per week (see Table 2). You've identified an ICE (Internal Combustion) and EV (Electric Vehicle) that meet your need and you've come up with the following Net Present

Table 1 Vehicle: Light Commercial Van – ICE.

Vehicle: Light Commercial Van ICE (INTERNAL COMBUSTION ENGINE)							
		Year 0	**Year1**	**Year2**	**Year3**	**Year4**	**Year5**
Miles per year	12,435						
Miles per Gallon	20						
Annual Fuel Use (Gallons)	622						
Gas (Petrol) Price/gal	$5.19						
Annual Fuel Cost	$3,227		($3,226.88)	($3,226.88)	($3,226.88)	($3,226.88)	($3,226.88)
Insurance, license, registration	$1,616		($1,616.00)	($1,616.00)	($1,616.00)	($1,616.00)	($1,616.00)
Repair/Maintenance/Mile	$0.09						
Maintenance	$1,119		($1,119.15)	($1,119.15)	($1,119.15)	($1,119.15)	($1,119.15)
Depreciation(Accounting)	5		($8,000.00)	($8,000.00)	($8,000.00)	($8,000.00)	($8,000.00)
Resale value (% of Purchase Price)	60%						$24,000.00
Income Before Taxes			($13,962.03)	($13,962.03)	($13,962.03)	($13,962.03)	$10,037.97
Tax	39%		$5,445.19	$5,445.19	$5,445.19	$5,445.19	(3,914.81)
Net Income			($8,516.84)	($8,516.84)	($8,516.84)	($8,516.84)	$6,123.16
Purchase Price	$40,000.00		($40,000.00)				
Add Back Depreciation			$8,000.00	$8,000.00	$8,000.00	($8,516.84)	$8,000.00
After Tax Cash Flow			($40,516.84)	($516.84)	($516.84)	($516.84)	$14,123.16
Net Present Value	4%	($28,729.40)					

(*Source:* adopted from EPA).

Table 2 Vehicle: Light Commercial Van – Electric Vehicle.

Vehicle Light commercial Van ELECTRIC VEHICLE								
		Year0	**Year1**	**Year2**	**Year3**	**Year4**	**Year5**
Vehicle Supply Equipment (EVSE)	$1,000		($1,000)				
EVSE installation	$,000		($1,000)				
Cost to install 50 AMP circuit	$625		($625)				
Smart Charger upgrade							
Cost Construct Garage or Car port							
Miles Per year	12,435						
Cost per Mile							
Cost per kWh	$0.12						
Charging loss factor	1.20						
Cost per kWh	$0.14						
kWh per 100 miles	27						
Cost per 100 miles	$3.89						
Cost per mile	0.03888						
Annual Fuel Cost	$483		($483)	($483)	($483)	($483)	($483)
Insurance, license, registration	$1,616		($,616)	($,616)	($,616)	($,616)	($,616)
Repair/Maintenance/ Mile	$0.09						
Maintenance			($261)	($261)	($261)	($261)	($261)
Depreciation (Accounting)	5		($12,000)	($12,000)	($12,000)	($12,000)	($12,000)
Resale Value	25%						$15,000
Income Before Taxes	39%		($16,985)	($14,360)	($14,360)	($14,360)	$528
Tax			$6,624	$5,600	$5,600	$5,600	($206)
Net Income	60,000		($10,361)	($8,760)	($8,760)	($8,760)	$322
Purchase price Add Back Depreciation			($60,000) $12,000	$12,000	$12,000	$12,000	$12,000
After Tax Cash Flow Net Present Value	4%	$37,342.13)	($58,361)	$3,240	$3,240	$3,240	$12,322

(*Source:* adopted from EPA).

Value Calculations for a corporate presentation. The overwhelming majority of literature on EV cost focuses on one item, annual fuel cost, which you can see is only a small part of the total cost of ownership from the corporate perspective.

Consensus Building/Decision-Making

Decisions today aren't as simple as they used to be. In addition to financial considerations, greenhouse emissions and corporate green strategies can be factors. If your enterprise lacks Green Guiding Principles, your purchase will most likely be evaluated using standard corporate financial analysis spreadsheets which may look like the preceding examples.

Knowing how decisions are made within your organization is the first step in planning a presentation, which should be ready at least three days before the big day. Selecting the presentation method is next. The Socratic method lays out a series of arguments and concludes with "and therefore we recommend option 'X'." The recommended *USA Today* Presentation Method is the exact opposite. You state what you want to do, then follow with "here's why." Your presentation should be concisely summarized in a "value proposition" like:

"We recommend a light commercial Electric Vehicle as a replacement for our maintenance pick-up. This purchase supports corporate emission goals and provides a 50% operating cost savings after the five-year depreciation period." See Table 3.

You might even have a table to support your assertion regarding operating cost after year five:[15]

Table 3 Annual Operating Cost – Years 6–15 (costs not inflated).

	INTERNAL COMBUSTION ENGINE	ELECTRIC VEHICLE
Annual Fuel Cost	$3,227	$483
Insurance, License, Registration	$1,616	$1,616
Maintenance	$1,119	$373
Total	$5,962	$2,472

(*Source:* adopted from EPA).

Presenting your value proposition should be followed by a discussion and because you have done your homework you should do well in answering questions and explaining your reasoning.

As you drive down the freeway of life, you may feel there are billboards all around flashing "Danger" because we live in challenging times. You've survived COVID-19 lock downs and hybrid work situations having thrived through your ingenuity. No one has a crystal ball, but it's my guess that many Facility Managers will be faced with the task of developing a reorganization plan or capital purchase presentation similar to the electric vehicle example and I hope that the decision-making process offered here will provide a framework for developing and presenting your plans. Remember "Burton's Rule of No," do your research, put together a solid presentation and you will succeed!

15 You might state that the five-year financial analysis offered in this article is how the finance department will evaluate the proposal, but that your department intends to keep the vehicle for 15 years, while you expect the Internal combustion vehicle to have a much shorter Expected Useful Life.

Barry Lynch, CFM, SFP, NCARB, MBA, IFMA Fellow
Project Manager for the State of Louisiana
Denham Springs, Louisiana, USA

I started my career working for a small design firm in Colorado, where I learned that I loved creating projects out of thin air. Clients had needs and I developed architectural programs and project budgets to fulfill those needs. I soon realized that tying architectural projects to organizational financial performance was a weak spot in my capabilities (all architects are lacking in this area), so I was off to the University of Illinois at Champaign-Urbana, where I earned a Master's in Architecture and a MBA. The combination provided a solid foundation for performing financial analysis for feasibility studies and complex long-term capital plans called Strategic Facilities Plans.

My financial background has allowed me to provide insights that are often lacking in planning. For example, in my role as a Corporate Architect, I noticed that we had pushed off vacating a lease space in a 63-week space consolidation plan because the lease terminated four years in the future and we wanted to minimize the lease penalty. However, after listening to a talk by our CFO, who said the new emphasis was Cash Flow, I re-evaluated the plan and realized we could vacate the space immediately, pay the lease penalty and the financial cash flow benefit would be significant due to depreciation write-offs. I realized the value of training in my role as a Senior Corporate Architect and earned the Certified Facility Manager and Sustainable Facility Professional designations from the International Facility Management Association (IFMA). My active involvement resulted in being elected as one of 127 IFMA Fellows in the 24,000-member Association. As an independent consultant, my most significant projects were developing a Master Plan for the $160 million renovation and underground expansion at the Kansas Capitol and a Feasibility Study that identified how the State of New Mexico could develop the country's oldest frontier fort (Fort Stanton) into an entity that would no longer require continual state operating and capital funds.

As a consultant for a firm that specialized in planning and design for financial institutions, I specialized in Strategic Facility Planning, Long Term Capital Planning, Project Planning and Scoping, Feasibility Studies, Branch Network Analysis and Planning, Branch Network "Upside Down" Analysis, Master Planning, Architectural Programming, Space Use Analysis/Planning, Rentable Area Studies and Lease Reduction Studies.

In my current role as a Project Manager for the State of Louisiana I manage programming, design, construction, and renovation projects for LSU, Southern University, the University of Louisiana and Community Colleges. Projects range from renovating a 10,000-seat basketball arena to renovating lab buildings and everything in between. The State allows me to do occasional small consulting projects, which are primarily Branch feasibility studies for past clients. I've published more than 40 articles and given more than 50 presentations and continue these endeavors in my current role. I like planning because the task always starts out with a smorgasbord of challenges, goals, and constraints. By working through each in a systematic manner, I revel in creating order out of chaos.

2.10

FM Maturity

Research, Standards, and Influence

Kathy O. Roper

In the early 1980s the administration of many companies and government offices fell under the operations, finance, or human resources departments. Each company or each leader of these departments decided how to organize, and few wanted the diverse responsibilities that we have now come to know as "Facility Management" or FM. In one smaller organization where I worked, the "Administrative Services" as the FM department was then known, fell under four different Vice-Presidents within two years.

Apparently, none really wanted a cost center that could not generate income, especially in a Financial Services company. But management of the real estate, leases, space allocation, renovations, moves, telephone services (few computers were allocated to general employees at that time) and mail distribution, along with coordination with the landlord for security (mainly keys for office door locks) maintenance and janitorial services were required to have an accountable employee somewhere. They are all basic but necessary functions.

Over the course of a 20+-year career as a manager of facilities and later for 14 years teaching and researching the field, I was able to participate in and contribute to the growth in Facility Management. Modernization, automation, consolidation, and a new focus on longer-term strategies made FM an exciting and sometimes creative field. The FM department often tried new office designs prior to rolling them out to all employees. Innovative technologies for managing facility data are commonplace now but were new in the 1990s and early 2000s. Focusing on sustainable facilities and services has brought savings and improvements to FM.

Combining experience and research from related professions was an interesting and important component of my teaching at Georgia Tech. As part of a Master's program, students learned how to do basic research and apply that to real life situations. In the early days of the Master's program most students were existing facility managers who wanted to advance their careers. They brought a wealth of experiences from their jobs and shared with students in the classroom and with projects that each course required. Over the course of the program, more and more inexperienced students came into the program directly from undergraduate programs.

They were interested in FM but had little or no real experience to share with classmates. This was perhaps because the local market of potential M.S. degree students was saturated, and the program did not launch an online component due to the extremely high cost to students for that platform. Or perhaps because more and more international students were attracted to the program, the make-up of classes included fewer existing professionals in FM. This made providing examples, cases and projects more difficult, as real-world projects had to be found outside the student body. It did, however, provide a more research-focused basis within the program overall. Now seeing graduate students and PhDs go forward to further expand the field has been one of the most gratifying aspects of my career. There are now higher education programs in FM or related fields across the US

Facilities @ Management, Concept - Realization - Vision, A Global Perspective, First Edition. Edited by Edmond Rondeau and Michaela Hellerforth.

and many countries worldwide. More graduates from these programs are moving into professional management of the built environment, giving it more professionalism and recognition.

As a part of contributing to the growth of the profession, I was asked to co-author the third edition of *The Facility Management Handbook*, and I provided a whole new section focused on sustainability and new educational programs providing FM degrees. The later fourth edition provided the opportunity to enhance and update technologies in FM, as well as provide gender-neutral formatting throughout the text. Building Information Modeling (BIM), augmented reality (AI), and other updates made this edition the go-to text for university programs worldwide, and FM professionals working in many locations around the globe. A second book, *International Facility Management*, was published in the UK. These contributions helped me to see the changes and impact their growth as wide distribution has enhanced knowledge and focus of the facilities function.

After more than 40 years as a defined profession, FM can now be considered a maturing profession. However, more can be done to advance the practice based on research and best practices from other more mature professions. Here I provide an overview of this maturation process, and the new ISO Standard on FM, its development and rollout, as well as examples of how to implement ISO Standards for growth and expansion of the field of Facility Management.

Based on industry standards, FM is now poised to increase its leadership role in guiding organizations toward better decision-making around assets and operations. This strong model will also provide more credibility and value to the profession, enabling FMs to contribute to business strategy in meaningful ways. Discussion of the influence FM has had over its 40-year history will set the stage for the benefits and evolution of new standards and links to research as industry best practices.

FM as a Profession

A profession is defined as having three main qualities that set it above other "jobs":

1) The mastery of a complex set of knowledge and skills through formal education and/or practical experience. "Specialized knowledge and skills in a widely **recognized body of learning derived from research, education and training at a high level**, and is recognized by the public as such." (Australian Council of Professions 2016)
2) Every organized profession (accounting, law, medicine, etc.) is **governed by** its respective **professional body**.
3) A disciplined group of individuals who adhere to proscribed **ethical standards**.

Facility Management was first defined in 1979 and governing bodies (professional associations) were established such as the National Facility Management Association (NFMA) in the US in 1980 and the International Facility Management Association (IFMA) in 1982. Codes of Ethics required by professional associations have been in place for NFMA, IFMA and most associations related to facility management.

Formal education began in 1980s and rapid growth in early 2000s led to academic research in the field.

However, in 40 years, has the body of learning derived from research, education and training been adopted into practice? There is little evidence of this, unfortunately.

An "on-the-job" history of FM has been in place within most organizations since their inception, but this created a number of barriers to research into practice integration:

- Limited TIME!
- "We've always done it this way" or "If it ain't broke, don't fix it"
- Leadership knowledge and history

Now is the time to move beyond these barriers!

Research into Practice

There is general agreement across many industries that effective practice is based on the reflective application and adaptation of research. In turn, practice that includes principled experimentation can raise critical questions to be answered by research (see Figure 1). This feedback loop can create a sustainable strategy for innovation in the nation's education enterprise. However, a US National Research Council (NRC) report on education research observed that transfer of scholarship into practice is "a last frontier," despite the fact that the application of research is paramount for moving education reform from transitory fads to proven strategies (NRC 2011)

This is the area that has received little focus from the FM industry overall. A methodology or approach is needed to align the industry with the research that is being conducted in multiple universities globally. Following the example of other industries meaning that FM does not need to recreate the process but understand and adapt it from others.

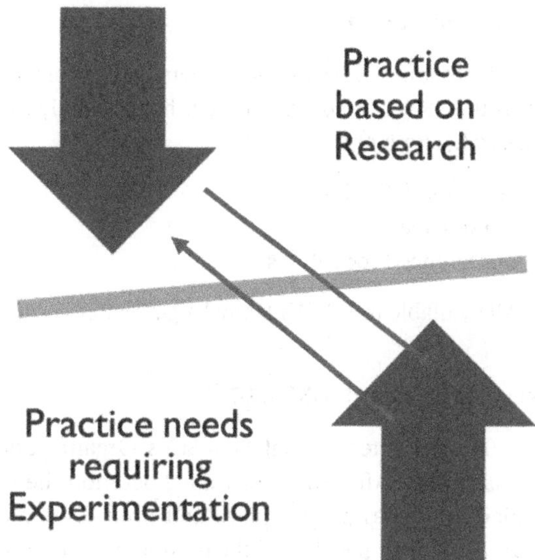

Figure 1 The give and take of research and practice. (*Source:* K. Roper).

How Others Do It

Healthcare professionals have led research into practice, probably based on the academic requirements experienced in gaining medical degrees and continuing research in the field.

The American Medical Association (AMA) requires strict compliance with new standards once new standards are developed, and all of these are based on research.

Doctors and nurses differ in how they deal with the "evidence-practice gap." They even want to expedite the gap in timing between research results and implementation. AMA found that nurses' decisions tended to be problem-oriented and managed on a person-driven basis, whereas doctors' decisions were consensus-oriented and managed by autonomy. All, however, experienced a **knowledge-based execution of the research results**, as the implementation process ended. This means that rather than simply trying different approaches, the research is first consulted to determine where the best path forward might lie. While standards now exist and best practices are shared the professional associations, the use of these research-based results is inconsistent.

Psychiatry utilizes what they term **Intermediaries** to carry research into practice. Requirements and characteristics of intermediaries include:

1) trust
2) neutrality and transparency
3) collegiality and enthusiasm (Biebel et al. 2013)

Intermediaries are people or organizations that help to accumulate and disseminate the research into the practice. FM associations need to consider this role and determine how to function as intermediaries to allow broader take-up of existing research.

How IFMA Does It

The International Facility Management Association (IFMA) currently utilizes the Research and Benchmarking Institute (RBI) to conduct in-depth data analysis, benchmarking and forecasting of key FM trends. Contact: research@ifma.org

- Benchmark Studies
- White Papers
- Conference Proceedings

All available in IFMA's Knowledge Center community.ifma.org/knowledge_library/

New FM Standard – ISO 41001

In 2015 the International Standards Organization (ISO) began development of a new standard for Facility Management. After two years of deliberation the initial definition of FM for ISO is: "An organizational function which integrates people, place and process within the built environment with the purpose of improving the quality of life of people and the productivity of the core business."

Related or similar ISO standards include:

- ISO 9001 for quality management
- ISO 14001 for environmental management
- OHSAS 18000 for occupational health and safety
- ISO 31000 for risk management
- ISO 26000 for social responsibility (Also GRI)
- ISO 20121 – sustainable events
- ISO 55000 for asset management

ISO also has standards for multiple issues in construction.

Overall, the mission of ISO is to provide standards to align businesses across borders and governments (see Figure 2). For ISO 41001 the mission is: To develop standards that articulate and enhance the understanding of Facility Management as the leading professional discipline within the Built Environment profession. (ISO 2016)

Research by ISO has found that implementing standards can provide economic benefits from between 0.5% and 4% of annual sales revenues. Standards are a strategic business issue with a direct impact on new product development. Major global organizations look for standardization prior to considering doing business with an organization (typically asking, are you following ISO Standards?). Therefore, not participating in standardization hands decision-making over to the competition who have aligned with standardization. These benefits also help to maximize organizational streamlining of practices, products and purchasing, which provide additional savings to the organization. This change to ISO standard is part of the professionalization and advancement of FM and will continue to bring evolution and enhancements to our field.

Benefits of Standards and Integration of Research

A primary business focus of standardization is to improve procedures, productivity, and performance. Specifically oriented to FM, standards help to enhance sustainability, mitigating negative environmental impacts and keeping facilities in compliance with local, regional, and international regulations. A safe workplace is also a key benefit. Optimization of life cycle performance and cost controls helps to improve resilience and relevance with provide the supported organization with a better organizational identify and reputation.

Additional benefits accrue to organizations aligned with industry matched research. Finding that research easily (and quickly for most FMs) is required in the rapid reaction arena that FM operates. Too often companies

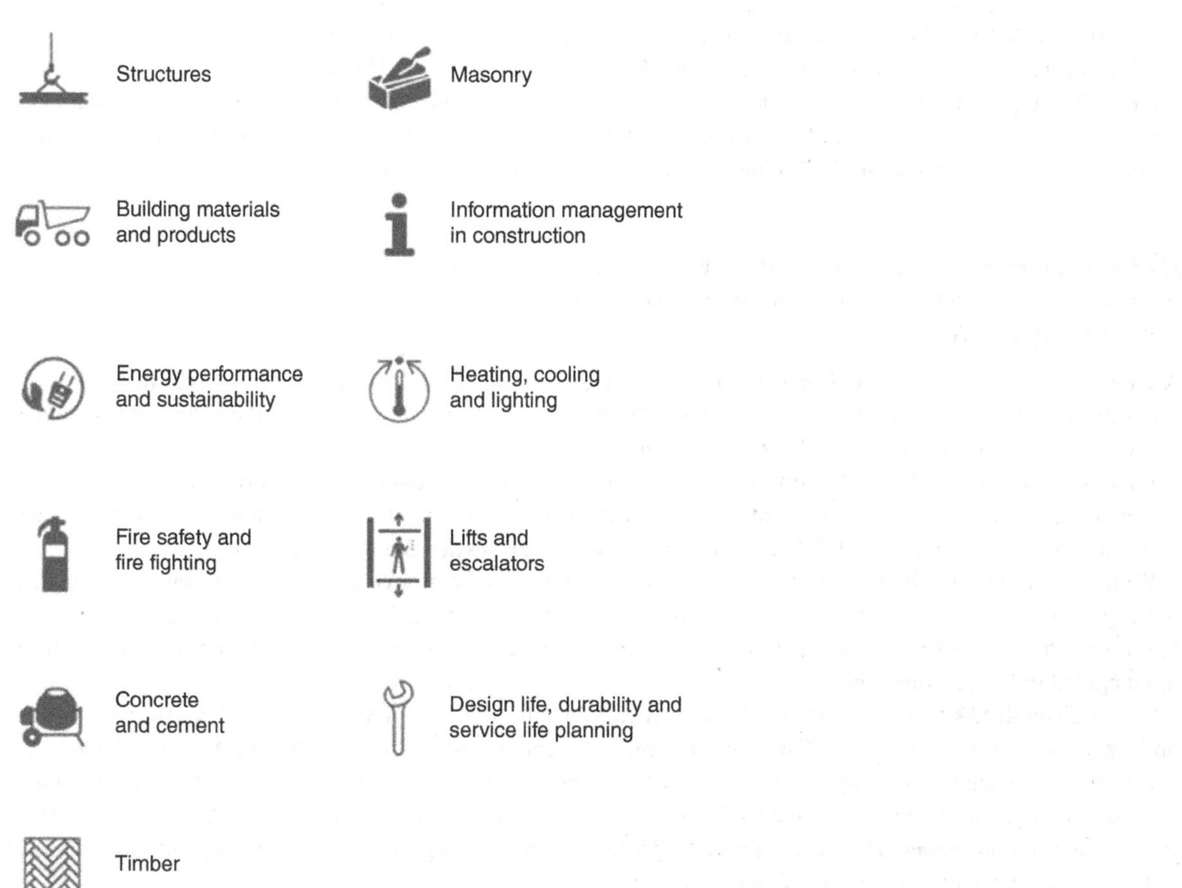

Figure 2 ISO standards related to construction and facilities. (*Source:* ISO / https://www.iso.org/files/live/sites/isoorg/files/store/en/PUB100317.pdf).

try to "recreate the wheel" with internal practices attempting to improve performance, save money and/or increase customer satisfaction. But a lot of work has already been accomplished in these areas. Finding them is the difficulty and could be made easier with a clearinghouse for FM knowledge. My personal recommendation is for the professional FM associations to undertake this task and provide true value to the industry as well as members. Models for this transmission of research into practice already exist and can be adapted to FM.

Over my 40+-year career in FM, there have been dramatic changes, improvements, and somewhat better understanding of what and why facility management is a necessary and beneficial component of any organization. The continuing changes, especially in technologies to assist FM, are expanding exponentially. As John F. Kennedy once said, "Change is the law of life. And those who look only to the past or present are certain to miss the future." FMs know the challenges and looking to the future is an important task to becoming a better FM.

References

Australian Council of Professions (2016). Available at: https://www.professions.org.au/what-is-a-professional.

Biebel, K., Maciolek, S., Nicholson, J. et al. (2013). Intermediaries promote the use of research evidence in children's behavioral health systems change. *Psychiatry Issue Brief*, University of Massachusetts Medical School publication 10 (4): 2.

ISO (2016). ISO/TC 267 – facilities management standards catalogue. Available at: www.iso.org/iso/home/store/catalogue_tc/catalogue_tc_browse.htm?commid=652901&published=on&development=on.

National Research Council (2011). Assessing 21st century skills: summary of a workshop. J.A. Koenig, Rapporteur, Committee on the Assessment of 21st Century Skills, Board on Testing and Assessment, Division of Behavioral and Social Sciences and Education. The National Academies Press, Washington, DC, p. 2.

Kathy O. Roper, RCFM, LEED AP, IFMA Fellow
Associate Professor, Retired, Georgia Institute of Technology
Atlanta, Georgia, USA

After over 20 years in practice, Roper led the Integrated Facility Management graduate program at Georgia Institute of Technology in Atlanta, GA, USA. She retired from teaching in 2016, but still consults and researches and writes about workplace and facility management issues.

Roper is an active member of the International Facility Management Association (IFMA) and served as Chair of the Board of Directors in 2011–2012; she holds the Retired Certified Facility Manager designation and was recognized as an IFMA Fellow in 2007. She is a LEED Accredited Professional with the USGBC.

With over two dozen scholarly publications Roper co-founded the open-access International Journal of Facility Management (www.ijfm.net). She is co-author of *The Facility Management Handbook*, third and fourth editions, and *International Facility Management*. Numerous white papers and presentations at international conferences are included in her accomplishments.

She received the 2005 and 2016 IFMA Distinguished Educator and 2007 IFMA Distinguished Author Awards, and her College awarded her Early Career Achievement Award in 2006 and Faculty Service Award in 2007.

Focusing on sustainability, impacting climate change, and aiding facility management professionals in understanding and implementing improvements, Roper wrote two white papers for IFMA, *Climate Change Fundamentals for Facility Management Professionals* and *Adapting to Climate Change for Facility Management Professionals*. Both are available in the Knowledge Library of IFMA.

2.11

FM Transformation in High-Tech Industries and Global FM Update

John Carrillo

During the 1970s, the impact of high interest rates weakened the economy, slowed construction growth, while jobless numbers approached 10% in the US. The job market for graduating students with a degree in either Construction and or Architecture was not promising. I graduated from California Polytechnical University in California with a BS degree in Architecture in 1975. For approximately one year, I worked as a draftsman for an Electrical Engineering company and also for an Architectural firm. My new career path was not exactly paying the bills while trying to take care of a young family.

In 1976, I was hired on as an Industrial Engineer working for General Dynamics – Pomona Division, an aerospace company building tactical weapons for the US Government which operated at various sites located in the cities of Pomona and San Diego, California. I worked in the Plant Engineering department, and little did I know that this organization along with the rest of industry would transform over the next decade into the Facility Management profession. So, you could say that I was one of those frustrated (Want to be) Architects, who by accident, migrated into the FM industry.

During the 1960s and 1970s aviation companies implemented various "Quality Assurance Programs," to ensure that business procedures were being executed without compromising standard quality requirements, regulations, and compliance guidelines. These programs were extended to support organizations like Plant Engineering which comprised the following functional organizations: Long-range strategic planning, Building infrastructure procurement, Engineering (Office and manufacturing equipment layout), Construction and Maintenance. Our Quality Assurance Program monitored building and infrastructure equipment, including performance metrics to improve organizational effectiveness, systems and processes.

All employees at General Dynamics were expected to continually enhance their skillset with the goal to improve operational effectiveness and reduce cost. I immediately joined AIPE (American Institute for Plant Engineers), which would later in 1995 be named AFE (Association for Facilities Engineering).

In 1978, the General Dynamics Plant Engineering organization was recipient of the prestigious AIPE annual award: "Outstanding Plant Engineering Program in Industry – Number one in USA Large Plant Division."

During the late 1970s, I introduced and transitioned the concept of open bullpen furniture configuration to modular furniture systems at General Dynamics – Pomona Division. We worked with Herman Miller on measuring the effectiveness and productivity increase of the office worker. The aesthetics of this furniture system elevated the importance of the office environment.

With an interest to improve the operation of our facilities organization, I was invited by Cecil Williams and Marty Dugan to participate in several FMI sponsored FM senior management forum discussions. Primary focus of these discussions was to define the role, responsibilities, and value proposition that Facility Managers bring to corporations. It became evident that combining all of the key FM functional core competencies together into one

Facilities @ Management, Concept - Realization - Vision, A Global Perspective, First Edition. Edited by Edmond Rondeau and Michaela Hellerforth.

department helped to align and support with corporate strategic goals. Organization effectiveness improved and with more opportunities to reduce cost demonstrated FM's value to the company's CFO.

In 1981, I was hired by Northrop Corporation – Aircraft Division in Los Angeles, California. My job responsibilities as Strategic and Budgets Manager for the Manufacturing Facilities Department, included 9 million sq. ft. of administration, computer, manufacturing, industrial machining and final aircraft assemble space. The plans that were developed by the FM Department, included a seven-year long-range strategic and four-year operational plan including the deployment of new technological systems into all aspects of support organizations within Northrop Corporation, Aircraft Division.

I joined the Los Angeles Chapter of IFMA in 1983, was a founding sponsor of the Orange County Chapter of IFMA in 1984, and also became a member of IFMA's Manufacturing Council. During the next several years, I participated at several World Workplace conferences, including various leadership programs with the primary focus to determine the most effective way of aligning the Facilities Management organization.

In 1984, at the age of 32, I was promoted as Director of the Capital Asset and Facilities Program (CAFP) for Northrop Corporation – Aircraft Division. The CAFP department's annual budget was US $100 million per year managing 10 million sq. ft. and 1,000 employees – 300 FM professionals and 700 blue collar workers. The CAFP organization comprised the following organizations: Strategic Planning, Real Estate Administration, Design & Construction, Environmental, Plant Engineering, Numerical Control Maintenance and Building Maintenance.

Based on a study conducted by the GAO – US Government Accountability Office (Published in 1989) "Women and Minority Aerospace Industry Profile 1979 – 1986," there was an effort within Northrop Corporation and other Aerospace companies to promote diversity into the workforce and increase women and minority representation in management positions. At the time I was promoted as Director of CAFP, there was no representation of women in management. The aerospace industry in general was very male-dominant, and with the directive to promote qualified women, I was faced with several challenges within the hierarchy of leadership, but during the next couple of years with the help of the Human Resources Department, I was successful in promoting women into key positions within the CAFP department.

During the next eight years that I managed the CAFP organization, we made multiple changes that improved organizational effectiveness while streamlining processes, systems and standards and procedures. Most of the craft/engineering technician personnel were outsourced (primarily thru attrition), which led to improved productivity, reduced cost, and an increase of client satisfaction results.

In 1985, there was no formal FM educational or certification programs in industry that could be used to enhance the skillset levels of Facilities Managers, especially those individuals that did not get a college technical degree in either Architecture, Interior Design, Construction, or other related engineering degrees. I engaged with the Building Owners and Managers Institute International (BOMA) to create a multi-week course for 150 of Northrop's Facilities Managers – "Productive Decision Making for Integrated Offices and Facilities – The Integrand Institute 1985." The seven-module course contents included Course Introduction, Orientation to Facility Integration, Project Planning Phases, Identifying Personnel Needs, Facility Technology Factors for Processing Work, Facility Technology for the Environment, and Applying Your Knowledge. One of the program facilitators who presented this program material was a very respected colleague in the FM industry and a good friend of mine – Nancy Sanquist, IFMA Fellow.

In the early 1980s, companies in the Aerospace & Defense sector in the US recognized the importance of incorporating Total Quality Management (TQM) in manufacturing. TQM emphasizes the improvement of quality and services throughout the companies support organizations by including all management employees, customers, and 3rd party supply chain vendors and contractors into the process. The CAFP department of Northrop Corporation fully engaged with TQM and also implemented the Plan Do Check Act (PDCA) program which tracks performances measures and supports the continual improvement processes. The end result is to increase profitability to the company's bottom line.

Another huge challenge for the CAFP Department, was the management of Northrop's Aircraft Division manufacturing process of aircraft wings and fuselage components which stored large quantities of virgin and used hazardous materials – acidic/alkaline and corrosive/caustic chemicals. Northrop Aircraft Division in Los Angeles, California was classified as a hazardous waste generator. The CAFP department was responsible for all EPA regulatory compliance requirements and frequently met with over 35 local, county, state, and federal enforcement agencies. During the 1980s we invested significant capital asset funds for the containment of hazardous materials storage and the retrofit of all chemical processing vats and tanks, including frequent ground water plume analysis testing.

During the late 1980s, the CAFP department was responsible for the decommissioning and dismantling of a small nuclear reactor which was used for the assemble of long range intercontinental ballistic missile seeker heads. There were extensive requirements and procedures thru every phase of demolition, including removal of spent fuel, as to ensure that there was no release of any radioactive material during the entire decommissioning process. The concrete reactor pool was dismantled into small pieces and after proper disposal of irradiated materials, small concrete rubble was sent to the appropriate waste site. Spent fuel was donated and transported to various university programs located in the US East Coast. This overall process took nearly three years.

In the mid-1980s, Northrop Corporation conducted a 20-state evaluation to identify a green field 1,500-acre parcel of land for the construction of a missile assemble munitions facility. The evaluation, plan, and bid construction process took several years to complete, costing US$150 million.

Over the last 16 years, I was very fortunate to have worked for two Aerospace companies that embraced the TQM concept for best-in-class operations. While at Northrop Corporation which would later become Northrop Grumman, I was able to effectively rearrange the Facility Management organizational which improved operations and contributed to the company's bottom line, while enhancing the skill set of Facility Management professionals.

In 1992, I was recruited to work for another technology company. My family and I moved to the Bay Area located in Northern California, and I worked for Pacific Bell Telecommunications, which would later become AT&T. I worked for the Corporate Real Estate and Facilities Management organization, and my first position was Director of Project Management/Construction.

****

I also transferred my IFMA membership to the East Bay Chapter and became actively involved participating at World Workplace events and I also became a member of NACORE which would later become CORENET. As a senior member of AT&T's CRE/FM leadership team, I participated at various leadership round table forums both with IFMA's leadership and NACORE's best in class training programs. I also served on the board of WCCC (Western Council of Construction Consumers) and became chair of WCCC in 2002/2003.

As president of the East Bay Chapter of IFMA, we implemented a five-year strategy plan that doubled membership and sponsorship while enhancing membership offerings and networking thru education, web site, newsletter, public relations, and golf outings events. During multiple World Workplace Award of excellence programs, the East Bay Chapter of IFMA received the following awards; Membership Marketing in 1999, Web Communications in 2001 and Chapter of the year in 2002. I also participated and was a member of the Utilities Council and received my CFM certification in 2000.

On October 17, 1989, there was a major earthquake – 6.9 on the Richter scale that occurred in the San Francisco Bay area. Pacific Bell's largest telephone switching building CO (Central Office – 15 story, 300 ft high, 650,000 sq. ft.) located in Oakland, California, sustained major damage that supported over a million customers, including critical operational services to Emergency 911, local military, and local airports circuitry. Emergency repairs were underway when I was hired as Director, Construction Management, to manage this four-year critical path (7,000-line items) two building restoration/structural modification programs. While working with structural engineers and performing concrete/steel solutions with multiple stress tests being performed at University of California – Berkeley, we developed two seismic concrete and steel repair techniques – "Passive confinement and column encapsulation." These repair techniques were later used by the State of California to repair freeway concrete overpasses throughout the state of California after various major earthquake events during the 1980s and 1990s.

During the construction process there were no network/business interruptions, and the cost of overall project was US$130 million.

In 1997, my next challenge was to develop and implement an optimization program of retracting office personnel from a dozen leased/owned locations and moving these employees into the Pacific Bell headquarters in SRV – San Ramon, California. We increased the density and population of the building from 5,200 to 8,700 employees. This complete rearrangement of the building required the relocation of up to 500 personnel per month during a two-timeframe. This US$35 million investment included the complete furniture systems upgrade and consolidation of this 2 million sq. ft. building and savings/cost avoidance over time was US$100 million.

In 1994, the entire Pacific Bell Corporate Real Estate group received the ISO 9001 – (DNV) Det Norske Veritas Quality System Certificate. This certificate, issued October 24, 1994, was valid for three years for the following products/service ranges: Portfolio Planning, Acquisition and Disposition, Design, Construction and Maintenance of Corporate Real Estate Facilities. We were the second organization to receive this prestigious award. The first facility management organization to win this award was the US GSA (General Services Administration) Corporate Real Estate organization in Washington DC.

The Pacific Bell Corporate Real Estate organization also spent several years incorporating best in organizational practices by instituting the "Malcolm Baldridge National Quality Practices Procedures." The five key areas of focus were: Product and process outcomes, Customer outcomes, Workforce outcomes, Leadership and Governance outcomes, and Financial and market outcomes. This process assures continuous improvement in overall performance in delivering products and/or services and provides an approach for satisfaction results measurements to customers and stakeholders.

From 1996 to 2019, I moved into various FM Director positions (Portfolio Planning – California, Project Management – California, Property Management – 13 Western states). In 1997, Pacific Bell merged with SBC Communications and ultimately merged with AT&T. In my last position before retirement in 2019, I was Director of Design and Construction for the 13 western states, including Guam.

During the 2018 timeframe, the US Senate was considering the formulation of a group of CRE/FM subject matter experts to evaluate the utilization of 1 billion sq. ft. of US government (GSA – General Services Administration) portfolio. This effort would evaluate the effectiveness and efficient use and operations of Federal Government buildings and space. Various US Senators would select a committee member, and at the time, President Trump would select the chair of this committee. Senator Nancy Pelosi from California recommended me for this committee, but the initiative by the US government was canceled.

In an article featured in September/October 2015 in the IFMA-FMJ magazine, Nancy Sanquist – IFMA Fellow provided a spotlight article called "Closing Achievement Gaps in High Schools and Community Colleges – IFMA Foundation Global Workforce Initiative." This article covered various programs and outreach efforts that reached students from high school and community colleges and talking to students about the Facility Management profession and the benefits of selecting "FM a career of choice."

At this particular time, I was Chair of the IFMA Foundation. The article reads as follows: "In a particular successful presentation to high school students in the Bay Area, IFMA Foundation Chair John Carrillo gave a heartfelt story of his incredibly successful FM career. He shared his beginnings in a Mexican-American neighborhood near East Los Angeles and his use of the community college system to study Architecture. Through superior athletic skills he was able to obtain a bachelor of science through a partial scholarship at California State Polytechnical University, Pomona. Upon graduation, he started learning facility management, project management, and strategic long-range planning at General Dynamics.

At that time, the aerospace industry was one of the leading market segments through which to understand the value of the emerging field of FM. Carrillo later joined Northrop Grumman Corporation at an IFMA chapter and sent some of his key staff to the first IFMA seminar ever taught (note: the author of this article was the creator and instructor of that course). He spent the last 24 years at AT&T where today he is director of corporate real estate for the western region and is responsible for 40 million square feet and a US$250 million budget, in addition to serving as chair of the IFMA Foundation.

One student, Karla Iraheta, emailed Carrillo after the event: "It was a pleasure having you as a guest speaker at Hayward High School. It was an honor meeting you, sir. I must say that I was quite shocked by all your achievements, starting from the low and rising up to the top. From the thing that you said in your speech, I was inspired. inspired to fight for what I believe and that we have to start from the bottom and work our way up in order to achieve our goals. Once again, thank you very much for giving us your time and presenting to us."

****

In 2002/2003, I served on IFMA's Global Board of Directors. I also received the prestigious award of IFMA Fellow at World Workplace in 2006, San Diego, California.

From 2011 to 2016, while serving on the Board of Trustees and becoming Chair for the IFMA Foundation, we created a pipeline and presented to students in high school and community colleges in many locations across the US, providing opportunities for individuals to "Make FM a Career of Choice." Along with the support of IFMA's Chapters, Councils and Communities, the Foundation was able to raise approximately US$150,000 each year providing scholarships to 50–60 students each year. The IFMA Accredited Degree Program offerings (bachelor/master/doctorate) peaked at 31 universities and colleges across the world.

In 2019, I was asked to join the Executive Committee of IFMA's Board of Directors as 1st Vice Chair. Our mission and focus from 2019 to 2022 were to engage, transform, and strengthen IFMA's position and finances. And to institute a Strategic Plan that will consistently enhance IFMA's value proposition offerings to our FM members and practitioners.

In the last IFMA (FMJ) journal dated May/June 2022, Don Gilpin, President & CEO-IFMA, posted some kind words when I left the Board of Directors as Immediate Past Chair: "In 2019, IFMA Fellow, John Carrillo took the helm as chair, bringing a more sophisticated approach to our strategic planning. Instead of changing direction with each new year (and each chair), John initiated a strategic planning process with the help of a strategic consultant, fellow board members and staff, restructuring association decision making to support an overarching goal for the future. We unveiled our strategy last April. As an organization, we know exactly where we are heading. John's fingerprint is the return of decision making that supports IFMA's long-term strategy."

****

I was elected by the Global FM Board of Directors as Chair of Global FM (two-year term) which started on February 1, 2021. I replaced Duncan Waddell, who went on to become chair of ISO/TC267. Duncan replaced Stan Michell, who was chair of ISO/TC267. Both gentlemen are scholars, consultants, and educators of the FM profession.

Founded on July 1, 2006, Global FM – Global Facility Management Association is a worldwide Federation of member-centric organizations, committed to providing leadership in the Facilities Management profession. As a single, unite entity promoting Facilities Management, Global FM is a conduit for furthering the advancement, knowledge, and understanding of Facilities Management and the sharing of best practices, resulting in added value to the individual members of each member organization.

During the last year and one-half, Global FM membership has grown from 8 to 14 member-centric associations. The current list of FM Association members are as follows: ABRAFAC – Brasil, FMA – Australia, FMANZ – New Zealand, HFMS – Hungary, IFMA – Headquarters in Texas, US, IWFM – Great Britain, MEFMA – United Arab Emirates, SAFMA – South Africa, TRFMA – Turkey, APAFAM – Panama, ACFM – Catalana, Spain, EGYFMA – Egypt, AFMPN – Nigeria, SFMA – Saudi Arabia.

During the last 18 months, much has happened in our post-pandemic environment, and facility managers/practitioners have been thrust into the center of the conversation of what do we do next. Public and private sector entities have redefined their business continuity plans into economic survivability strategies. The implementation of back to work health and safety protocols continued to be measured and for the most part, remote hybrid office work environment alternative is here to stay in many regions of the world. Facility managers need to continue providing employees with the most productive, latest technology, cyber security, flexible, and safe work environment. Collaborative workstations at remote site and locations are ideal in order to hire the best talent from anywhere. Many companies and businesses are providing enhanced training for their employees as a way to improve collaborative organizational restructuring to improve productivity and investing in digitization technological solutions that will contribute to the company's profitability bottom line.

Corporations/companies in most regions of the world are investing in 4.0 technologies in order to meet their economic challenges and goals and to improve their place in the market place and leap frog their competition. Process improvement solutions can be achieved thru IOT (Internet of things), smart infrastructure equipment applications, cloud computing analysis with third-party supply chain vendors and contractors, and AI (Artificial Intelligence) solutions.

Facility practitioners need to enhance their knowledge base and leadership skillset. We as individuals need to take charge and be prepared for the challenges that lie ahead in a sustainable resilient environment. And we need to have a basic understanding of viable digitization applications and smart programs that are applicable to our facilities and building infrastructure equipment.

FM Association members of Global FM now have an opportunity to collaborate and share best practices, research, and best-in-class FM educational opportunities from around the world. Global FM is now sharing a common platform with ISO/TC267 FM standards. There are currently (5) 41,000 standards published with (7) future standards under development. It is important to note that the next standard to be published will be 41,019 "The Role of FM in Sustainability and Resilience." Most of the ISO standards are in alignment with the United Nations 17 Sustainable Development Goals.

We just recently finished updating our strategic plan, including our Vision, Mission, Strategies, and Initiatives (see Figure 1).

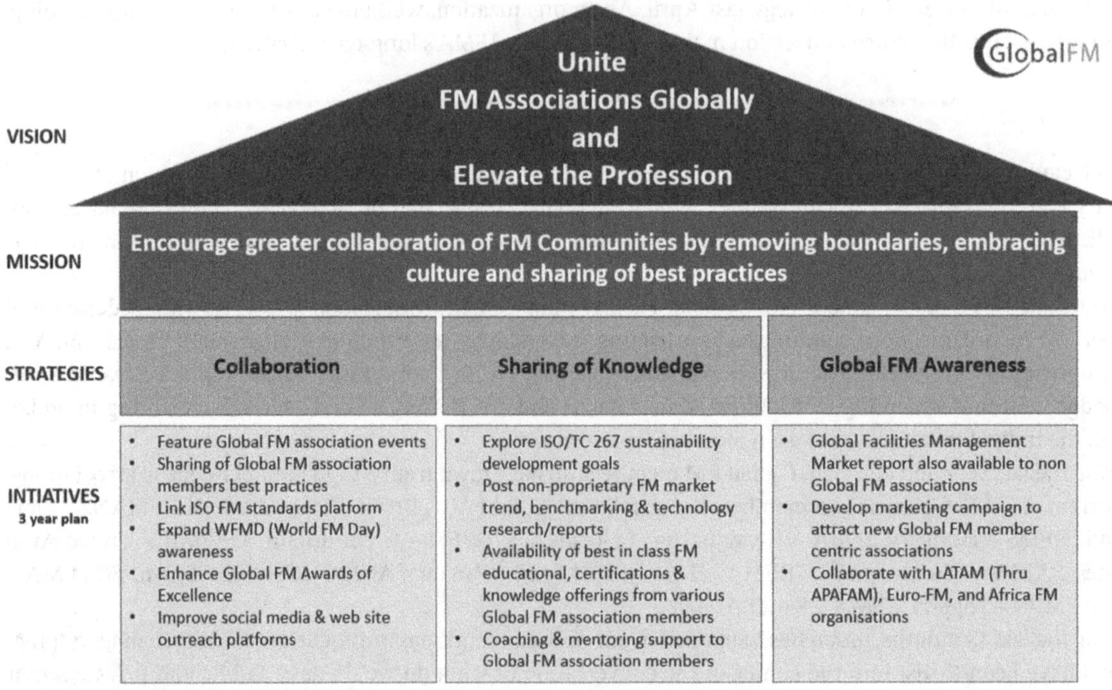

Figure 1 GlobalFM vision, mission, strategies and initiatives. (*Source:* GlobalFM).

Global FM's value proposition includes two significant events each year and a global Facilities Management market report produced by Frost & Sullivan every two years.

Once a year, the Global Facilities Management community comes together on **World FM Day (WFMD)** to celebrate the Facilities Management profession and its successes (see Figure 2). There is an annual posting of WFMD celebrations throughout the FM community. WFMD 2022 theme was "Leading the Sustainable Future." As people from around the world begin to emerge from their homes and return to the built environment to work, learn, worship, and play, we are witnessing an incredible phenomenon for Facility Management with a renewed importance being placed on human health and safety coupled with building sustainability and resilience, the Facility Manager today has been thrust into the center of the "Return to conversation." This year's WFMD is special to me since celebrating the work of FM has never been so important to all peoples of this world. "Happy World FM Day!" – John Carrillo CFM, IFMA Fellow, Chairman – Global FM.

The **Global FM (GFM) Awards of Excellence** was created to enable GFM members to further showcase and communicate achievements of their FM members to an international audience. This Awards program is only open to GFM members and must be nominated by the member association. GFM members nominate "Best in Class" from either GFM member award programs or, from those who don't have an award program, encourage their members to submit suitable projects or achievements via the GFM nomination process. The awards bring entries from around the world each year. These nominations must be submitted in line with our guidelines.

Figure 2 World FM day. (*Source:* GlobalFM).

Every two years, Global FM produces a **Global Facilities Management Market Report** from Frost & Sullivan. Reports were produced in 2018 and 2022. The 2022 Report was delayed due to the extraordinary eco nomic impact of COVID-19. Scope of this study includes FM industry size and significance by country, global strategic imperatives for FM industry, FM industry trends by the countries in scope, FM industry revenues by the countries in scope FM industry workforce by the countries in scope, progress on digital transformation of FM, and progress on sustainability in FM practice.

John Carrillo, CFM, IFMA Fellow
Past Chair, IFMA
Current Chair, Global FM
Acampo, CA USA

John Carrillo has more than 40 years of experience in facility management and corporate real estate at technology-related companies. John just retired as Director, Planning, Design and Construction, West Region for AT&T. Responsibilities included project management oversight of 40M square feet for the 13 western US states and Guam. His team included facility/project managers, alliance architectural and engineering firms, general contractors and technical consultants, who managed several thousand capital and expense infrastructure replacement and repair projects totaling up to US$250M per year. As Director of Property Management for AT&T John developed a preventive maintenance program for 20,000 buildings.

Before AT&T, Carrillo worked in the Corporate Real Estate – Aircraft Division at Northrop Grumman. Over the course of his career, he has managed as many as 1,000 FM employees, developed strategic programs, built a missile factory, decommissioned a nuclear reactor, and managed a large seismic retrofit project in the San Francisco Bay Area.

A frequent presenter at IFMA events, Carrillo has also prepared and presented coursework on FM/real estate, quality processes, use of digitization technology in the workplace and related topics at UC Berkeley and other locations. Mr. Carrillo is currently Past Chair of IFMA Global Board and current Chair of Global FM.

Carrillo earned a Bachelor of Science degree in architecture from California State Polytechnical University in 1975. In 1986, he earned an Advanced Management Certificate from Claremont University Graduate School, and in 1996 he was awarded ISO 9001 Registrar Team Certification. He earned his IFMA Fellowship award in 2006. His significant contributions to the FM industry have been recognized by IFMA and other organizations, such as the Association for Facilities Engineering (AFE), and the Western Construction Consumer Council.

2.12

The Development of Facility Management in Hong Kong

A Personal Reflection

John D. Gilleard

The 1980s – Early days

Visiting the then British territory of Hong Kong for the first time in April 1984 as an academic member of a study group of building and estate management students from the National University of Singapore (where I then worked) proved a memorable trip. Norman Foster's partially constructed HSBC Main Building, located in Hong Kong's "eye of the tiger," was a particular highlight. The site footprint was incredibly tight for such a complex building, with the majority of superstructure and internal systems fabricated off site. Hong Kong's Mass Transit Railway System (now running several similar systems worldwide) was an eye-opening example of how to move large numbers of people around a highly congested city. Ingenuity, resourcefulness, and a can-do attitude have always prevailed in the local Cantonese/ Chinese culture where the willingness to adapt creatively has played into the historical development of facility management as a profession in the city and its environs.

Before returning to Singapore, I also traveled to (then) Portuguese-controlled Macau on China's southern coast, 40 miles west of Hong Kong, staying at the recently built Hyatt Regency Hotel – 11 floors in height and apparently unremarkable. However, later I learned that only the hotel's foundations, ground floor, and external works had been constructed on site. The rest, floors two to eleven, had been fabricated as reinforced volumetric units with all M&E services, interior fixtures, and fitments as well as the furniture installed in Alabama, USA, from where they were subsequently shipped to Macau and assembled. Another example, if perhaps costly, of an innovative solution to the dilemma of high product demand coupled to a shortage of specialist-skilled Asian construction workers.

By 1988 when I was back in Hong Kong traveling from Atlanta where I was, by then, an Associate Professor in the School of Architecture at Georgia Tech, I was able to visit the now-completed HSBC's Main Building. Given its status within the business community and its prime location in the city, it had obviously drawn other new commercial towers to Hong Kong's Central District and eastwards towards the more diverse retail district of Causeway Bay. Across the harbor the Kowloon skyline was also changing – all perhaps a reflection of the Sino-British Joint Declaration that had been signed in 1984 setting the conditions for which Hong Kong was to be transferred to Chinese control post July 1st, 1997. Change was coming, and fast.

Facilities @ Management, Concept - Realization - Vision, A Global Perspective, First Edition. Edited by Edmond Rondeau and Michaela Hellerforth.

The 1990s – The Nascent Development of FM

In September 1990 I commenced a three-month visiting appointment at the University of Hong subsequently extending my stay in the territory thereafter for 25 years (not an unknown phenomenon for many expatriates in the city!) as a Principal lecturer and later as Professor and Head of Department of Building Services Engineering at the Hong Kong Polytechnic University (PolyU).

In November 1992, the Hong Kong Chapter of the International Facility Management Association (IFMA) was established and, if I remember rightly, April 1993 I became its 15th Chapter member. The Chapter quickly grew and by the late 1990s was IFMA's largest Chapter outside of North America with over 150 members. The Chapter's first President, Philip Lo, IFMA Fellow, was a driving force behind its early success. Building on this, the Chapter attracted representatives from both private and public sector organizations, the workplace design community, furniture manufacturers, academics, and FM service providers – and eventually students studying to become professionals with the wider FM and property community.

The Chapter was awarded IFMA's "Small Chapter of the Year" in 1998, and later "Large Chapter of the Year," both a reflection of its wide range of activities and innovative practices. As a member of the Chapter's board of directors, we organized monthly learning events, workshops, visits, etc.; this was a heady period in the history of FM in Hong Kong and I was privileged to be both VP and President (1999) during this exciting period on FM's growth.

In April 1993 the Hong Kong Polytechnic University hosted Hong Kong's first FM conference. Frank Duffy, founder of DEGW, the international architectural and design practice, gave the morning's keynote address to an audience of almost 200. Duffy is best known for office design and workplace strategy, and during his talk he spoke about the worldwide spread of FM, the emergence of professional FM groups and the likely impact that FM would have on future business practices. Duffy's talk also underlined the need for a customer focused, service orientated approach to managing property. At the time I recognized that many in the audience were simply curious to know about this new profession whilst others were skeptical, questioning the need for FM when Hong Kong already had a strong and vocal property management sector. This was perhaps unsurprising given the dominance of the property sector in Hong Kong's economy, retaining property values was and still is considered to be critical to gauging a company's success.

The afternoon keynote speaker, Bruce Forbes, IFMA Fellow, founder of ARCHIBUS, followed-through with his positivity and vision: how the application of technology to FM and computer programs like ARCHIBUS would enhance productivity and modelling of property portfolios. I found Forbes' presentation to be fascinating and, from a personal perspective, one of those moments that helped to change my academic direction as well.

In the early 1990s, the majority of organizations' property portfolios were managed by their own employees focused primarily, particularly in the private sector, on maintaining the portfolio's asset value. These in-house property management teams gave little attention to the occupiers of their buildings. However, as these sections gained in property management capabilities and experience, they were often reconstituted as independent companies with the objective of expanding the organization's business by gaining property management contracts from other clients. For example, Eastpoint Property Management Services Ltd emerged from a property management department within Colliers Jardine (Hong Kong) Ltd., and was, by 2000, providing property and asset management, security, technical, and integrated facility services to a wide range of residential, commercial and industrial units throughout Hong Kong and Macau. The success of Eastpoint, and other similar company off-shoots, is illustrated by ISS (a Danish-based property management services company) subsequently acquiring Eastpoint in 2005 in order "*to concentrate* [on] *becoming the market leader in Hong Kong and in the Chinese mainland in the provision of integrated property and facility services*," according to Marten Thomsen, ISS Executive VP and Regional Director, for Asia.

However, as the 1990s progressed many Hong Kong–based organizations and businesses began to explore ways to reduce costs or headcount, or to gain increased property/facility management expertise by outsourcing to specialist companies. Initially this was often just "out-tasking" one or more property management activities as single contracts, e.g., cleaning, security, catering services. But gradually these single task contracts were bundled together and outsourced to a single services provider as a means to simplify the organization's contract administration. As this approach grew, specialists "managing agents" began to be hired to act as an organization's representative, responsible for the appointment of both single and bundled service providers.

The Impact of Academia

The development of the supply side of FM did not take place in isolation from the wider role of its professionalization within Hong Kong, however. Chinese culture has always highly valued the importance of education, exemplified by the acquisition of appropriate and accredited academic qualifications. Facility management was no exception, and by the mid 1990s, this time had come in Hong Kong.

By the mid-1990s my academic focus was primarily on aspects of FM. Undergraduate and post-graduate degrees in facility management, although offered in the USA and the UK for example, had little consistency in degree content, and perhaps more importantly, were still few in number. Hence to encourage greater awareness of FM as a professional discipline, IFMA, via the IFMA Foundation, established a task force to develop curriculum objectives and learning outcomes for a facility management degree program. As a member of this task force, I worked along side William (Bill) Sims (IFMA Fellow, Cornell University, USA), Keith Alexander (IFMA Fellow, University of Salford, UK), Jeffrey Campbell (IFMA Fellow, Brigham Young University, USA) together with representatives from Ryerson University, Toronto, Canada and East Michigan State University, USA. Around the mid 1990s, the task's force final report was published, clearly identifying, in broad outline, the subject content and knowledge outcomes for a generic FM degree.

FM education and training in Hong Kong were driven primarily by the demand at postgraduate level, drawing from a broad base of existing professional disciplines ranging from engineering [M&E], estate, and property management. The PolyU started to develop the structure and content for a Master of Science (MSc)/Postgraduate Diploma (PgD) in Facility Management in 1995, and as the leader of the FM program development I was able to draw upon my experience with the IFMA task force to guide the eventual structure and syllabus content.

Acceptance onto the part-time MSc program was via a range of building/engineering/surveying bachelor degrees or via professional certification, such as IFMA's CFM award. This last award allowed more senior FM and property management professionals who had, in their earlier careers, been denied the opportunity for a university education to acquire an accredited degree in the discipline – an important credential, especially in Hong Kong, and a significant aspect in gaining better acceptance and status for FM there.

The PolyU MSc/PgD Facility Management program, accredited by IFMA, the Hong Kong Institute of Facilities Management – HKIFM, and the UK's Royal Institute of Surveyors – RICS started in 1996 with the first students graduating in 1999. The program quickly became popular, graduating 40 students per year from the early 2000s to the current date. Besides teaching two core subjects – strategic management and benchmarking studies, I also supervised research projects, some of which formed the basis for scholarly articles in the UK journal Facilities. Guest speakers, both locally and from abroad, were especially popular as they were able to bridge theory with reality. For example, I vividly recall Kit Tuveson, IFMA Fellow, visiting the university for a week to participate in a student FM case study. Kit was quickly given the title "Professor Kit" from which he derived much pleasure. Professor Jan Ake Granath, from Gothenburg University, Sweden, an acknowledged FM academic, also joined the FM program for a number of years as a visiting professor. So too did Professor Keith Alexander from the University of Salford; departmental colleagues also added breadth, knowledge and experience to the program.

The Late 1990s and Early 2000s – Finessing the FM Model

The out-tasking model was further developed with the appointment of "managing contractors" who were paid a fee for providing FM services, typically as a percentage of the value of expenditure managed. By the early 2000s managing contracts began to be superseded by "total facility management" (TFM) contracts, whereby the responsibility for providing FM services and for generally managing the facilities was placed in the hands of a single organization. This could be said to be the natural finessing of a strategy that emerged in 1991 when IBM UK, "spun off" their property/facility management team in a management buyout, creating a new business, Procord (subsequently acquired by Johnson Controls in 1994, then sold to their Global Workplace Solution's business and finally CBRE in 2015). Procord focused on delivery of FM services to their occupier clients and not simply the built assets. Such an idea is commonplace today, but 30 years ago it was revolutionary.

For example, in the late 1990s I recall a visit to a major Hong Kong bank that employed more that 400 staff in their property management department around as security, janitorial, or maintenance personnel. There was also a small professional property management team responsible for project planning, cost control and coordination of the bank's in-house property management staff. However, as time passed, many of the low budget operations – cleaning, catering, mailroom – were out-tasked, initially on a piece meal basis to separate vendors, with high-risk services, such as building services engineering remaining in-house. Later, the bank would bundle their property management services as a total facility management (TFM) contract, retaining only a small internal facility management team to manage the contract. They, in turn, began to place greater emphasis on "customer services" with the bank's business units, and on improving the bank's core activities.

And thus gradually, in Hong Kong, FM providers began to deliver all FM services to client organizations though now, in full turning of the circle, with their own in-house teams and hierarchy of service vendors contracted directly to the FM TFM service provider. However, there is one crucial difference from the original company on-site property management teams of the early 1990s and that is that the TFM model transfers most of the risk to the FM service provider, with the outsourcing organization typically maintaining a relatively small in-house FM team to manage the contract and strategic development issues. Nevertheless, to quote a seasoned Hong Kong–based FM professional, "*to this day* [2022] *there remains a distinction between Property Management and Facility Management in Asia – the former being the delivery of operations & maintenance services to the base building on behalf of a building owner and the latter being delivery of similar services on behalf of an occupant / tenant. The distinction is a little blurred and many service providers typically deliver both Property Management & Facility Management services.*"

Stepping back a little, as the 1990s progressed, the term "facility management" began to be better understood, although when translated into Chinese it typically included the characters for property management. Property management services were usually restricted to the common areas of a building such as its fabric and envelope, grounds and gardens, lift lobbies and vestibules, as well as common services such as cleaning and security; minimizing costs and tenant complaints, and the idea of managing the property for the benefit of tenants was missing. But in Hong Kong, the majority of international companies chose to lease their office space, rather than own, and thus were more receptive to FM.

This approach was crucial. For International property developers (Hongkong Land, Swire Group), institutions (Hospital Authority), educational establishments (American School, City University of Hong Kong), and the nascent Airport Authority, as well as a number of commercial organizations, all showed an early interest in FM services. Banks (HSBC, Standard Chartered, Chase Manhattan, Deutsche Bank) and a few bigger, more progressive companies that owned significant building stock (Hong Kong Telecom [now PCCW]) were no doubt also aware of FM's growing development in both Europe and the US.

2000 to Early 2020s: The FM market Goes Regional / Global

A main issue with the out-tasking/outsourcing model in Hong Kong was that it was not always easy to scale up, although as time passed bundling of contracts became more common. During the early period post-millennium most government offices, parks, hospitals, schools, etc. moved toward some level of FM outsourcing, as did the private sector, the latter arguably led by the financial services sector. Where FM remained in-house, this typically involved landlords managing their own property portfolios with their own in-house property management subsidiary. But with more experience came more contracts and higher profiles that began to attract attention from mainland China.

For example, Synergis Management Services Ltd (SMS but founded as Synergis in 1978) is a leading property and facilities management company. Initially, its early contracts focused on property management for the Hong Kong Housing Association but after rebranding in 2003 business began to expand and in 2008 the company took on FM services for two of Hong Kong's largest organizations: the MTR corporation and China Light and Power (CLP). Later, FM contracts were signed with private residential owners, government organizations, educational establishments, landlords, developers, and investors in both Hong Kong and China. By January 2022, China Resources Management Ltd of the China Resources Group acquired the Synergis property and facilities management business enabling Synergis to enhance its "*overall business strength and long-term development capabilities in both Hong Kong and Mainland China.*"

CLP's outsourcing contract to SMS could be said to have been a landmark for FM – a five-plus-five years' contract across all of its Hong Kong locations across – and was almost unheard of at that time. The proposal process was also innovative, leading the way for future changes to securing FM contracts by other organizations: CLP replaced the traditional request-for-proposal by a series of one-on-one workshops between the client and its prospective FM suppliers, awarding the contract to the supplier who not only provided a model that the client fully bought into but also the one that developed the best working relationships.

As the FM industry in Hong Kong began to mature, the number of suppliers reduced to a few main competitors with JLL, CBRE, Savills, Cushman Wakefield, and ISS being among the main providers. Many client organizations had growing confidence in FM outsourcing and were moving from local to regional contracts. This regionalization facilitated economies of scale, standardization of service levels and sharing of best practice. FM was gradually recognized as a key part of any large business and many organizations created a Director of FM (or similar) represented in the C-suite. In addition, these larger regional outsourced FM contracts recognized they could also benefit from centralizing certain services in lower cost countries, such as, in Asia, delivering regional or even global FM help desks and security operations from India or the Philippines. Also, where previously a three-year contract was common, five-year (or even five-year plus) contracts became more common. This not only favorably impacted procurement costs, but also facilitated sustainable efficiencies, best practice implementation, and the development of long-term business partnerships between FM supplier and client organization.

In the last decade, yet another twist to FM servicing has started to take place arising from the emergence of these large providers. Global contracts are increasingly becoming more common for large organizations, limited only by the number of FM suppliers able to offer worldwide services – currently only JLL and CBRE. With both companies having a long-standing existence in Hong Kong, their global FM remit has now begun to impact FM locally in this market as it is now common, for example, with global banks with a local presence to outsource their FM operations worldwide on a global contract. Generally, there is an overarching global Master Services Agreement (laying out common contract terms) with a Local Services Agreement (stating key performance indicators and savings targets) tailored for each country, including Hong Kong.

Another feature of these larger contracts is a savings glide path where the FM provider commits to year-on-year percentage savings over the first 5–10 years in order to secure the 5-year plus part of the contract – perhaps 5% year in 1, 4% year in year 2, 2% in years 3 and 4, and 1% in year 5. In addition, to secure the contract extension, the FM provider might typically be expected to meet other non-financial goals alongside Key Performance Indicators

(KPIs) such as sustainability targets related to energy or carbon footprint reductions. More recently, the whole area of environment, social and governance (ESG) issues is also now becoming a significant focus for FM as both providers and clients' businesses seek to future-proof their brands with more eco-aware public communities.

This brief sweep through the development of FM in Hong Kong shows, like the region it is based in, that time never stands still. Hong Kong has always been known for its dynamism, its people for their creativity, and its businesses for their adaptability. FM provision and services are a reflective mirror of this overarching leitmotif with, almost certainly, more twists and turns to come as the profession continues to flex and grow amid regional and global transformations.

Acknowledgments

I would like to thank the following people for their generous contributions to this article and for reminding me of the fallibility of memory: Lee Po Wang, Samson; Deric Probst-Wallace; David Rees; Arnald Ng; Brian Crockford; Lai Hung Kit, Joseph.

John D. Gilleard, PhD, SLCR, IFMA Fellow
West Witton, Leyburn
North Yorkshire, England

John D. Gilleard joined the Department of Building Services Engineering at the Hong Kong Polytechnic University (PolyU) in 1991 where he developed and led Asia's first facility management graduate degree program. During his time at the PolyU John also supervised a number of Doctoral and MPhil students researching the FM area and published widely on the formative development of facility management in both Hong Kong and China.

On retiring from the PolyU as Professor and Head of Department in 2007, John accepted a post as visiting professor, Kufstein University of Applied Sciences, Austria, where he lectured on the international development of facility management. From 2010 to 2019, John was Senior Director of Learning for the APAC & EMEA regions of CoreNet Global.

2.13

Forty Years of FM – From a German Perspective

Jochen Abel

I am very pleased to have the opportunity to contribute to this book. When Facility Management was born, I was 12 years old and had other things on my mind than my professional career. Today, I hold the professorship of Facility Management Strategies at Frankfurt University of Applied Sciences. I suppose for this reason, and for the reason that Ed Rondeau and I met during an appointment process at the Georgia Institute of Technology, I was approached. I am very happy to comply and report on facility management from my personal perspective – in the form of a professional biography from facility management.

First Contact with FM

I started my professional career at the Johann Wolfgang Goethe University Hospital in Frankfurt as a freshly qualified engineer, FM was completely unknown to me. I was solidly trained in engineering for healthcare technology. However, starting out as a project engineer was completely different from what I expected. Of course, there were various engineering departments and trade-specific workshops. In addition, though, each engineer had cross-trade responsibility for several buildings. In this function, we were responsible for coordinating all small, medium and large projects. At the same time, we were the central contact person for the users in the relevant building.

This matrix organization was the brainchild of the then head of the Department of Technology and Building Management. The year was 1997, not quite the year FM was born, but in Germany at that time, the term facility management was only known to a small circle of people. I was completely unfamiliar with FM up to that point. The head of department had definitely understood that it was the department's task, as a service provider, to provide a working environment suitable for the primary process (healing and care). At the same time, it was important to him that the value of this working environment – not only from a monetary point of view – be made clear. For example, we had an area-based system for internal cost allocation, detailed data for all of the approximately 15,000 rooms on campus, type of use, and customer (cost center). We had a CAFM system in which this data was processed and via which all service orders were handled and billed, as well as a central incident reporting point for all service orders, including Webprotal. Overall, I would say that we set standards back then. Only I, as a young engineer, was not aware at all that this was FM. That was my first contact with the topic of facility management – and to be quite honest – I was even less aware of how much FM would shape the rest of my life.

Facilities @ Management, Concept - Realization - Vision, A Global Perspective, First Edition. Edited by Edmond Rondeau and Michaela Hellerforth.

My Professional Career in FM

The professional career I was asked to report on here is probably not quite ordinary. I am presenting it quite comprehensively, as this makes it clear that for me it is the basis for my understanding of FM and has significantly shaped my attitude with regard to high-quality training in FM. I will leave out my school years, which I completed with the Abitur.

Professional Training

My professional career began with an apprenticeship as a locksmith, or officially as a metalworker specializing in construction technology. Why am I mentioning this in a book about facility management? Well, in the years of my apprenticeship, I learned three things in addition to the technical stuff that have become relevant to my understanding of FM: First, I was shaped by the communication between people working in practice: clear, direct, goal-oriented, but also very humorous, and almost always appreciative, critical and constructive. Secondly, I learned a lot in the disputes between the workshop and the design office, which were almost like a sporting competition. The comments in the workshop were clear: "What they've drawn again in the design office. That can never be put into practice like that!" Conversely, there were probably similar sayings in the design office. I was impressed by the fact that, despite these disputes, the collaboration on problem solutions was nevertheless also very critical, constructive and appreciative. This made the added value of well-functioning, cross-hierarchical collaboration and communication clear to me. The last part of my lessons learned is respect for the achievements of craftsmen and craftswomen. The statement that I am a very good locksmith in theory sums it up. I passed the theoretical journeyman's exam very well. My practical exam was – to put it kindly – not quite "very good." For me, I took away from it a great respect for practical, craftsmanship skills. Being a gifted craftsman or craftswoman is something wonderful and anything but a matter of course. All in all, this training as a start to my FM career – of which I knew nothing at the time – showed me how important skills such as communication, cooperation and appreciation are.

Study of Technology in Health Care

For me, the study of technology in healthcare was essentially a slightly modified study of building services engineering. Energy, refrigeration, heating, air conditioning, gas, water, and control engineering were the common contents, supplemented by medical and hospital operating technology. The only nontechnical exam was in the area of business administration. It is thus obvious that it was a course of study clearly tailored to technical expertise. We were able to calculate everything up and down, had formulas and constants in our heads, were able to calculate pipe networks and heat requirements, and could determine the performance values of a wide variety of technical components literally in our sleep. What was not part of the course at that time, however, was the teaching of how a contract is awarded or how a bill of quantities is structured and what the legal background to it is. Management or social skills were also not explicitly taught or trained. These were somehow taken for granted or it was assumed that these competencies would arise naturally in professional life or would be acquired in self-study while working. All in all, my studies provided me with excellent technical skills and, above all, I had extremely good professors! This has also shaped me for my life's path and I am very grateful for that. English and rhetoric were subjects I devoted myself to in addition to my studies. The offer was there – but only on a voluntary basis. From today's perspective, I would say that I used the contents of these voluntary modules more in my professional practice than the technical know-how.

Another experience that certainly shaped my career was a self-organized stay abroad during my internship semester. Organizing an internship abroad – in my case at St. Mary's Medical Center in San Francisco. Applying for and receiving a scholarship, preparing myself professionally and linguistically for the stay; all this contributed enormously to the development of my personality.

Equipped with these tools, I began my first practical work.

Practical Work in a University Hospital

I have already briefly outlined my practical work at the university hospital in Frankfurt. This is where I acquired my basic attitude toward the customer-supplier relationship in FM. The things I learned here regarding the customer-supplier relationship I later found in the theoretical approach of Kano (Life Cycle and Creation of Attractive Quality 2001) on the subject of customer satisfaction again: In the case of the hospital, the customer is the medical staff and not the patient. Providing the appropriate working environment for medical processes is a particularly challenging task – and yet it remains the task of facility management – even if I didn't call it that myself at the time. As Kano describes it (see Figure 1) large parts of FM services belong to the mandatory or basic elements.

The fulfillment of these basic services leads to dissatisfaction if they are of lower quality – however, it does not per se lead to positive satisfaction if the basic services have been fulfilled to a high standard of quality. On the other hand, there are the elements of attractiveness or enthusiasm which are not expected by the customer and therefore do not lead to dissatisfaction if they are of inferior quality or not available at all, but which make a particular contribution to customer satisfaction if they are provided in high quality. Accordingly, it remains a challenge in FM to constantly search for these elements of attractiveness and to develop new ones so that the customer continues to value the services of FM. The other way to make the services of FM clear to the customer is a charging model, as was used for a time in the university hospital. In this model, space is not just a necessary evil or a status symbol for the customer – it is a significant cost for the FM. For example, in a hospital, about 60% of total FM costs can be allocated by space to the originator. At the same time, the extent to which such a shift in perspective simultaneously changed the customer-supplier relationship was impressive. Thus, before the introduction of the allocation model, the people responsible for the cost centers were quite likely to say with regard to the floor space: "Everything is mine!" The area was considered a status symbol: More space meant more prestige and, of course, every assistant doctor needed his or her own office. After the introduction of the billing model, it was more likely to be denied that a room was ever used by one's own department or even attributed to a colleague.

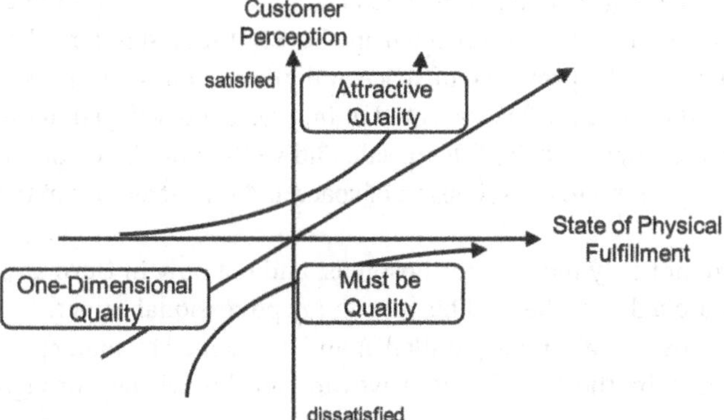

Figure 1 Kano Model of customer satisfaction – own presentation based on (*Source:* Adapted from Hölzing, 2008).

Also, physicians could suddenly share offices. What I learned then is: So, if FM services as mandatory or basic elements tend not to be perceived as a value, then an accounting model that assigns a monetary value to the service is a good way to make FM visible and tangible. At the same time, this form of disclosure of services also leads to a change in the customer's "expectations." The customer expects the agreed quality because he pays for it. This also puts an end to a way of thinking within FM according to the motto: "The customer can be happy that we are doing something for him at all." All in all, I was able to gain very valuable experience during my time at the university hospital with regard to professionalization and customer orientation in FM.

However, at that time I was still firmly convinced that FM (only) encompassed what was the responsibility of the "Technology and Building Management" department. It took a further step in my professional development for me to realize that FM also included areas such as purchasing, security, laundry supply, transport, and IT services.

Doctorate in Karlsruhe

Following my work in Frankfurt, I was offered the opportunity to work on a project on FM in hospitals. At the first university professorship for facility management – my doctoral supervisor Kunibert Lennerts – at the then University of Karlsruhe (TH), which is now known as the Karlsruhe Institute of Technology (KIT), the project on "Optimization and Analysis of Processes in Hospitals" (OPIK) was set up. The aim of the research project was to take a holistic view of all facilitative services and to investigate them using scientific methods. I was a welcome collaborator in the project, as I had both technical know-how and practical experience in hospitals. It was only through scientific involvement with the topic that I became aware that FM does not only consist of the provision and operation of a structural shell and the necessary technical equipment. Through the change of perspective from practitioner to researcher, it became clear to me that almost the entire administrative organization in a hospital can be assigned to facility management. In Germany, hospitals are usually divided into three areas: The medical area, to which the physicians are assigned, represents the "core business" together with the second area, care. The third area comprises all commercial and administrative services. This naturally includes services that are very close to the core business, such as billing for medical services including controlling and patient management, or the human resources department and legal affairs. The rest, however, is the full range of facilitative services. In addition to technology and the provision of the building envelope, these include, for example, the rental of staff apartments, janitorial services, security services, catering, waste disposal, laundry services, transportation services, cleaning and document management.

It became clear to me, by looking closely at the organization, that German hospitals are not that badly organized in terms of facility management. Interestingly, many of the systematizations and elaborations of quality characteristics of hospital-related FM that we designed in the project were reflected in the EN 15221 standard. For example, we had already implemented the subdivision into the provision of space and infrastructure on the one hand and the area of people and organization on the other. The product-oriented billing system developed in the project for FM services in hospitals, which, for example, stipulates the "clean, safe and well-maintained" provision of space as the main product, is largely identical in content to the provision of space and infrastructure in the standard.[1]

The realization, in turn, that these issues of FM are not only relevant for hospitals, and not only in Germany, but also worldwide, and are being discussed and researched, was the next big step in my professional career.

Facility management as a science: Not so long ago, this was still mildly smiled at and questioned by many people. But if you consider that a science is characterized by the fact that it strives to gain knowledge through

1 Compare: DIN EN 15221-4:2011 Page 56 (EN 2011).

methodically guided research, which in turn is communicated and discussed through scientific formats such as publications and conferences as well as through teaching,[2] facility management has definitely established itself as a science: On the subject of FM, knowledge that is orderly and considered safe is sought and produced using recognized methods. This knowledge is also taught at universities around the world. I do not want to deny that in the early days of FM, many things were written that, from today's perspective, cannot be classified as good, i.e. methodically justified or orderly. I also believe that the path towards a stable gain in knowledge, through which knowledge is further consolidated in constructive discourse, can never really come to an end and thus 40 years of FM can only be a good first start.

For me, the time of the doctorate at KIT was the entry into FM as science. But before the topic of science found its continuation, there was another very important change of perspective for me.

Work at a Real Estate Portfolio Company

Before I started working for a real estate portfolio manager, my outlook was, on the one hand, as a result of my studies, that of an engineer who tries to solve technical challenges as best he can with concentrated expertise. In doing so, I have the impression that engineers primarily strive for the best possible technical solution and regard the associated costs as a necessary evil, which, however, is of no further interest to them. From the second perspective of a service-oriented project manager at the university hospital, who tries to satisfy the needs of the customers in the best possible way, other questions suddenly came to the fore. No longer just the question of what is technically feasible, but above all the question of what makes the most sense from the customer's point of view thus moved into focus. In this perspective, the requirements for FM are defined by the customer's work process. The customer thus also specifies which is the most suitable solution. The third perspective in the doctorate consisted of viewing FM as a complex system of "interrelated or interacting activities that use inputs to deliver an intended result,"[3] as DIN EN ISO 9000 puts it so nicely in a nutshell. FM refers to many different processes that need to be coordinated with each other, that are partially dependent on each other, and that as a whole should contribute to customer satisfaction.

With the change to a large real estate portfolio manager, another perspective was to be added: The fact that it is possible and legitimate to make money with real estate changed my view of FM once again. My tasks here included looking after the topics of technical asset management, computer-aided facility management (CAFM) and reporting at the company's headquarters. Incidentally, it is interesting with regard to the topic of 40 years of FM that the department in which I worked had previously – even before I worked there – been renamed from Facility Management to Asset Management. The question arises as to why FM, contrary to its original idea of taking a holistic view of all real estate-related processes over the entire life cycle, was broken up into the terms portfolio, asset, property and facility management, and facility services. I think it was a process of finding an identity that was unavoidable or difficult to avoid. In his lectures, my doctoral advisor Kunibert Lennterts talked about the fact that toward the end of the 1990s, FM figuratively learned to walk, if you compare it to the development of a human being.

I would like to take this further and say that puberty inevitably had to follow after that: a process of self-discovery, combined with many an identity crisis. Since every janitorial service and cleaning company in Germany began to write Facility Management in big letters on the company car in the late 1990s, the name change is obviously an understandable reaction of a large real estate portfolio holder, which was intended to prevent a department that was responsible for close to 100 million euros in rental income from being grouped together with the janitors. At the same time, the value chain was thus hierarchized. There is nothing wrong with this – in the

2 Compare: Definition „Wissenschaft"(science) in Gabler Wirtschaftslexikon (Gabler 2009).

3 See Definition "Prozess" in DIN EN ISO 9000:2015 page 33 (ISO 2015).

sense of dividing up tasks and responsibilities. The very broad spectrum includes so many different tasks that it makes sense to subdivide them further. In some cases, different competencies are required. In any case, the overall area of facility management can be differentiated into different processes with different content emphases and results. Unfortunately, the conceptual bracket that encompasses all these tasks has been lost. For me, the term "facility management" remains this bracket. DIN EN ISO 41011 sums this up very nicely in its definition of FM, when it talks about FM integrating all processes and what belongs to them "within the built environment with the purpose of improving the quality of life of people and the productivity of the core business" (ISO 2019).

Back to my work for the large real estate portfolio holder. The perspective of an owner to achieve a crisis-proof, good return on investment in the long term, if possible, was initially unfamiliar to me. The question now was not primarily what needed to be done to the property and which solution was technically the best and would lead to the longest service life. Rather, the focus was on what needed to be invested in the property to make it usable and to meet the requirements of the market. Although business management considerations have always played a role in my previous positions, they were never so clearly in the foreground as the overriding maxim. Up to that point, I had only marginally dealt with questions of return on investment or valuation, for example. In the specific case of the real estate portfolio holder, it was precisely during the financial crisis that a refinancing of the company was pending, with the cherry on the cake that the financing bank was existentially affected by the financial crisis. This made it all the more necessary for me to focus on the business aspects of my work, and it was during this time that I was able to recognize the influence of lease terms, vacancy rates and the relationship between rental income and market value. I also gained experience of what it takes to manage a portfolio of around 20 million square meters of space spread across the whole of Germany. Divided into four regions and supported by a large number of service providers, the main task was to uniformly define and delineate areas of responsibility, standardize contracts and report on them in a uniform manner, and manage all internal and external stakeholders in such a way that the targets set by the shareholders were achieved.

During this time, I was able to learn an extremely large amount about facilities management, which was also related to the fact that I had a very good and result-oriented boss as well as very many colleagues who were happy to share their knowledge and were always available for a professional discourse. After four years there, an opportunity arose that I had already considered – but not necessarily thought possible.

Appointment to the Frankfurt UAS

Starting at the Frankfurt University of Applied Sciences (UAS) was a bit of a gamble. A temporary position for two years, where it was not entirely certain how things would continue in the long term. Nevertheless, my decision was clear. I well remember the words of a colleague and friend: "What are Karlsruhe or Kaiserslautern when you can be the professor of facility management in Frankfurt?" The two-year fixed term became four, and yet I never regretted the decision. Quite the opposite: I was offered the opportunity to develop and establish the Real Estate and Facility Management course at one of the top real estate locations in Germany. At the beginning, Facility Management at Frankfurt UAS was a niche product. FM was studied as part of a degree program that focused on geoinformation and surveying. Exactly at the time when my time as a substitute professor was coming to an end and my appointment as a full professor was underway, ideas were being sought to further develop the range of courses. I remember well the research on the market situation. The downright funny finding of the research was that almost all of the websites of the respective universities offering facility management degree programs in Germany featured photos of the Frankfurt skyline and/or regularly hosted field trips to Frankfurt, regardless of geographic distance. So we were where the facilities were and had nothing corresponding on offer. Even more important, however, was the research into the need for graduates in the field of facility management.

The question of whether there was a sufficient market to absorb graduates from an FM program was addressed through a ratio calculation. The first step was to try to estimate how many employees with a university degree are needed in facility management. According to a study commissioned by the German Facility Management Association, there are over 4.5 million people working in FM in Germany – as of 2014 – which is close to 11% of the workforce.[4] Assuming that the ratio of people with and without university degrees is similar to that in the construction industry, it is possible to estimate the size of the market to be served. For the construction industry, the figures are available via Bundesagentur für Arbeit. According to these, around 4%[5] of employees in building construction and civil engineering have a university degree. This ratio is quite low compared to other industries and does not take into account the engineers employed in construction planning and supervision. To estimate the number of new students needed, it is assumed that a ratio similar to that between new students and employees with a university degree in the field of architecture is reasonable. This is 7.5%.[6] This results in a required number of over 12,000 students per year in the first semester for Germany. This compares with 19 bachelor's degree programs listed with GEFMA for Germany. Even if 200 students were to start each of these courses each year, not even a third of the necessary students would be covered in the first semester. Thus, the market demand has not been and is still not nearly met.

In summary, it can be said that there was a great need then, as there is today, and that Frankfurt was and is an outstandingly suitable location. In addition, there were higher education policy conditions that led to the establishment of new courses of study at the university being desired.

I took advantage of this extremely favorable starting position and, figuratively speaking, due to the favorable winds, only had to set the sails appropriately in order to pick up speed with the idea.

In terms of my personal professional career, this corresponds to the current status. I teach and do research in Frankfurt. My research focus is on the sustainable operation of high-rise buildings. The development of the courses is described in the following chapter.

In summary, I would say that my career path was never explicitly planned. Ideas and desires were always existent – however, I can only explain many developmental thrusts by very favorable and unexpected encounters or events, to avoid words like coincidence or fate.

Education and Promotion in the Field of FM

My previously described professional career in FM and experiences gained therein have had a significant influence on my perspective regarding the development of a suitable academic education at Frankfurt University of Applied Sciences. A fundamental question in the development of training programs is whether a generalist approach should be taken – i.e. a broad spectrum of knowledge should be imparted; or whether specialized knowledge in a particular area should form the core of the training. The generalist approach is of course obvious at first glance with FM, with its holistic idea that encompasses the entire life cycle of the building. Nevertheless, specialized knowledge is also needed, of course. In my opinion, a facility manager defines, initiates and coordinates the necessary processes on behalf of the client and permanently monitors quality. This person thus fulfills the function of a communicator between clients and users on the one hand and internal or external service providers, specialist planners and other experts on the other.

In my view, people in facility management must therefore have sufficient detailed knowledge to understand an expert and be able to formulate the customer's requirements precisely. Even more important is the willingness to deal with complex contexts and, if necessary, to familiarize oneself with a subject far beyond basic know-how.

4 See FM-Branchenreport 2014 on page 13 (GEFMA 2014).

5 Own calculation based on the ratio of the number of employees in building construction and civil engineering occupations with an academic degree to the total number according to the German Federal Employment Agency. (BA 2021)

6 Own calculation based on the ratio of employees with a university degree in the field of architecture according to the Federal Employment Agency (BA 2021) on the number of first-year students in the winter semester 2020/21 and summer semester 2021 (DESTATIS 2021).

This willingness to engage in lifelong learning has also been described to me many times as a particularly important characteristic of employees.

The two degree programs at Frankfurt UAS dedicated to the operation of buildings, "Real Estate and Facility Management" and "Real Estate and Integral Building Technology," take the more generalist approach. This results not only from my professional career in FM but is a result of a very comprehensive development process.

In 2016, we began developing degree programs at Frankfurt UAS that deal with the operation of buildings. During the development, it was made clear on the part of the companies involved in the development that the need for people who are very familiar with the planning, construction and operation of technical systems in buildings is particularly high. Interestingly, the ability to think and work across trades was seen not only in the operation phase but also quite urgently in the design and construction phase. In various workshops, small working groups and many discussions, the designated two study programs were then developed together with the practice partners. From the first meeting in May 2016 to the first day of lectures in October 2018, just two years and five months passed. From the outside, these are two courses of study that lead to different degrees. A closer look reveals that the degree programs are five-sixths identical (see Tables 1 and 2).

Students start together in the first two semesters and are taught the necessary general fundamentals. The basics relating to architecture, construction technology, technical building equipment, business administration and the management of services are also common topics in both degree programs. In the third and fourth semesters, the two-degree programs differ in a total of six modules. While students in the Real Estate and Integral Building Technology program (Bachelor of Engineering) are introduced in depth to the planning and operation of technical systems such as heating, air conditioning, refrigeration, electrical, plumbing and control engineering, students in the Real Estate and Facility Management program (Bachelor of Science) are introduced intensively to project development, FM strategies, space and energy management, real estate transactions and property management.

Table 1 Module overview Real Estate and Facility Management program.

Real Estate and Facility Management
Bachelor of Science (B.Sc.)

FRANKFURT UNIVERSITY OF APPLIED SCIENCES

7	Elective Module 1	Elective Module 2	Study Project 1	Bachelor-Thesis including Colloquium		
6	Integral Planning	Building Information Modeling		Study Project 2	Key Competencies	Interdisciplinary General Studies
5	Internship Semester					
4	Corporate Governance Principles	Facility Services 2	Strategic Facility Management	Space and Energy Management	Property Management	Real Estate Transactions
3	Planning and Design	Building Services Engineering 2	Project Development	Construction Management	Operator's Responsibility	English for Real Estate
2	Briefing for Buildings	Building Services Engineering 1	Civil Engineering	Project Management	Law	Building Construction and Fire Protection
1	Real Estate Market	Basics of Physics	Computer Science	Facility Services 1	Materials Science and Chemistry	Mathematics

(*Source:* J. Abel).

Table 2 Module overview Real Estate and Integral Building Technology program.

Real Estate and Integral Building Technology
Bachelor of Engineering (B.Eng.)

7	Elective Module 1	Elective Module 2	Study Project 1	Bachelor-Thesis including Colloquium		
6	Integral Planning	Building Information Modeling		Study Project 2	Integral Planning	Building Information Modeling
5	Internship Semester					
4	Corporate Governance Principles	Facility Services 2	Sanitation / Fire Fighting Technology	Refrigeration Technology	Air Conditioning / Ventilation Technology	Heating Technology
3	Planning and Design	Building Services Engineering 2	Electrical Engineering Measurement and Control	Construction Management	Operator's Responsibility	Physics Specialization
2	Briefing for Buildings	Building Services Engineering 1	Civil Engineering	Project Management	Law	Building Construction and Fire Protection
1	Real Estate Market	Basics of Physics	Computer Science	Facility Services 1	Materials Science and Chemistry	Mathematics

(*Source:* J. Abel).

After a practical phase in the fifth semester, both groups come together again and deal with the different perspectives in projects, elective modules and integral planning with BIM. In addition to the broad spectrum, the aim of these courses is to familiarize students with the necessary key competencies so that they are self-critically able to search for the best possible solutions in constructive discussions with all parties involved.

In order to offer such a broad spectrum, it takes a whole team of professors. Of course, the courses are integrated into a department that is absolutely building-oriented with architecture, civil engineering and geodata management. In this respect, there is an expert team of more than sixty professors. In addition, there are other highly qualified personnel at the university from (business) economics and law disciplines, from logistics or computer science, as well as engineering sciences on various technical topics. A total of five professorships are firmly anchored in the developed courses of study. These cover the topics of FM strategies, technical building equipment, FM in the planning and construction process, real estate management and management of facility services.

Fortunately, highly qualified and motivated people were found for all the new professorships to be filled. This was, or is, a great challenge for two reasons. On the one hand, at the time of the appointments, the market for highly qualified specialists was almost empty, as the industry was booming. During such periods, a professorial appointment may not be lucrative enough for some people. The other challenge is that the subject areas to be filled have not yet been established as scientific disciplines, or only marginally so. For example, people who have specifically dealt with topics such as technical building equipment or facility services management in their academic careers are very rare.

Overall, I am pleased that a functioning team was able to emerge from a one-man show, which is able to cover the entire spectrum of real estate and facility management. Two courses of study with over 50 places per year guarantee a continuous supply of specialists and thus ensure further professionalization of facility management. Together with solid laboratory equipment, which is continuously being expanded, this has resulted in the Real Estate Research Lab (ReReL). In my opinion, it is important that the institutions active in the field of real estate and facility management network even better in order to continue working on not only producing and imparting knowledge, but also making it widely visible and making facility management appear as what it has been for 40 years – an independent scientific discipline!

References

BA, Bundesagentur für Arbeit (2021). *Beschäftigte nach Berufen*. Nürnberg: Bundesagentur für Arbeit.

DESTATIS, Statistisches Bundesamt (2021). *Bildung und Kultur – Fachserie 11 Reihe 4.1*. Wiesbaden: Statistisches Bundesamt.

EN, DIN (2011). *15221-3:2011 Facility Management – Teil 4: Taxonomie, Klassifikation und Strukturen im Facility Management*. Berlin: Beuth Verlag.

Gabler (2009 August 2). *Gabler Wirtschaftslexikon*. Wiesbaden: Springer Gabler.

GEFMA (2014). *FM-Branchenreport*. Bonn: GEFMA.

Hölzing, J.A. (2008). *Die Kano-Theorie der Kundenzufriedenheitsmessung*. Wiesbaden: Gabler | GWV Fachverlage GmbH.

ISO, DIN EN (2015). *9000:2015 Qualitätsmanagementsysteme – Grundlagen und Begriffe*. Berlin: Beuth Verlag.

ISO, DIN EN (2019). *41011:2019 – Facility Management – Begriffe*. Berlin: Beuth Verlag.

Kano, N. (2001). Life cycle and creation of attractive quality. *Proceedings of the 4th QMOD Conference*. Linkoping, Sweden: s.n, S. 12–14.

Prof. Dr.-Ing. Jochen Abel
Frankfurt University of Applied Sciences
Frankfurt, Germany.

Prof. Dr.-Ing. Jochen Abel is head of the Real Estate section at the Frankfurt University of Applied Sciences, Frankfurt, Germany. He brings a long-standing experience in Real Estate and Facility Management out of a variety of perspectives.

He teaches Real Estate management, Facility services management, Strategies of Facility Management, CAFM, Sustainable operation of buildings, Property Management, Energy and Space Management. His research focus lies on Building performance measurement, sustainable operation of high-rise buildings, customer-supplier relationship in FM and customer satisfaction.

Prior to his appointment to Frankfurt Dr. Abel was working in the asset management division of Aurelis Real Estate where he was in charge of the asset management reporting system. At the Karlsruhe Institute of Technology (KIT) he led the OPIK (Optimization and Analysis of Processes in Hospitals) research project covering the topic FM in hospitals. At KIT he also wrote his Dissertation with the title: "A Product Oriented Cost Allocation System for FM Services in Hospitals." He also has practical experience in FM out of his time working for the Frankfurt University Clinic. Dr. Abel is author of numerous publications on Facility Management.

2.14

*ATTA GIRL*One Woman's Portfolio Life in the Built Environment

Nancy J. Sanquist

Introduction

Thinking about the last 40 plus years of my life learning, working, traveling, and volunteering, I realize I have had a "portfolio life." This is a term used by Professor Lynda Gratton of London Business School, describing how to "reimagine work" today. It means a departure from the normal linear progression in the stages of life (school – work – retirement) where one takes a more circuitous route by dividing time doing paid and unpaid work, taking time off along the way and indulging one's passions while following multiple career paths.

This sums up my life's journey; while working for salaries in the built environment, including two new professions, one in historic preservation of cities and buildings, and the other in facility management, I also volunteered in various community endeavors. Along the way, I indulged my passions for viewing and collecting art, traveling around the world, working with the latest technologies and visiting and saving great architecture (ATTA). Having insatiable curiosity can lead one to a life of continuous learning and careers that are meaningful, and purpose driven. Plus merging the sciences and the arts are a great combination for a truly fulfilling life where good salaries and opportunities can feed one's passions (see Photo 1).

I am telling my story by dividing it into five sections covering the decades of my life, starting with two decades involved in formal education to prepare me for what followed which were four decades of both paid and unpaid work which continues today. IFMA began its life in the 1980s which is when I got involved (SECTION TWO on), and am still today, as I am an Advisor to the IFMA Foundation actively engaged thanks to the work with my partner in FM, Diane Levine. It is to this Executive Director and all the women and, men, in my life, who have given me the opportunities to experience the life I had and am still participating. If I had to say the one thing our wonderful IFMA community of friends have in common, none of us have stayed in "our swim lane," but preferred the excitement of catching the waves in an ocean and riding them to shore. And we still keep searching for that perfect wave in all sorts of waters.

SECTION ONE: 1960s and 1970s: UCLA-Paris-UCLA-Bryn Mawr-Easton-NYC

It was the 1960s in Los Angeles when "there was a concurrent revolutionary ferment in art, music and movies made into a cultural force" (Russo 2022). I intended to experience it all moving from New Jersey to California after graduating high school and enrolling at UCLA. During that time, I was recruited by a fashion photographer, George Barkinton, to be in a shoot in an issue on the girls of UCLA for ***Mademoiselle*** magazine (Photo 2). Now for a small town Jersey-girl, this was truly amazing (see Photos 2 and 3)!

Facilities @ Management, Concept - Realization - Vision, A Global Perspective, First Edition. Edited by Edmond Rondeau and Michaela Hellerforth.

Photo 1 NJS by Erik Jaspers, Amsterdam 2017.

Having a miserable first semester, I quit school for a while and ran away to Paris with a Madison Ave. ad guy turned jazz drummer and when he took his drums and left me for a gig in Lyons, I stayed in the city to study art and learn French. It was when I walked into the Jeu de Paume's room full of Monet's waterlilies covering the walls that I had one of those "ah ha" moments and realized I wanted to go back to school and learn to understand such incredible works of art. I then came back home, returned to UCLA and graduated with a BA in the history of art. From there I continued to study the same subject at Bryn Mawr College (BMC) Graduate School. The first day, I knew then that this was going to be a whole different academic experience. Across from my simple room in a grad dorm was another student in the History of Art, Katherine Porter Aichele, who already had a Master's in French and was learning German before she went to bed at night "for fun."

Porter and I would spend three years together, taking all the courses for the PhD and we both received grants one summer to go to London to do research. I was now a serious scholar, who would line up with Porter early each morning to elbow our way into the building at the entry doors to ensure we get our favorite carrels at the British Museum where we were conducting our research.

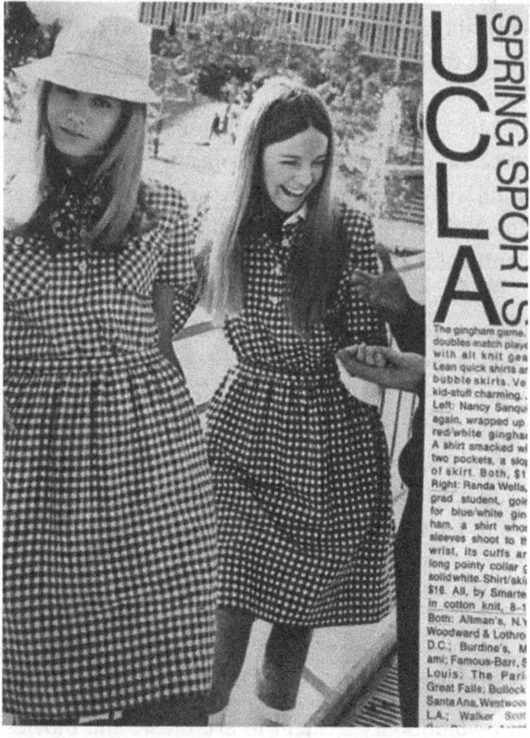

Photo 2 Photo George Barkinton, 1965.

We were known as two of the "golden greyhounds" (Photo 2) who consisted of the other three blond females in our graduate class which had also included three men. Our most memorial experience at Bryn Mawr occurred in Louis Kahn's brutalist dorm, Erdman Hall, where an exhibit of Georgia O'Keefe's paintings hung on the grey concrete walls of the medieval-like building. O'Keefe had loaned them for the occasion of receiving an award from BMC in light of her achievements. Porter and I looked down from a balcony during the reception and below us we saw the diminutive figures of Louis Kahn meeting Ms. O'Keefe. One said "it works" to the other (we can't remember who spoke the words) meaning her paintings "worked" in his building which was a dream scene to two budding art historians.

While I received a Master's of Art, Porter went on to get her PhD while marrying the artist-in-residence, the Austrian painter, sculptor, and graphic artist Fritzl Janschka, whose muses were James Joyce, as well as my friend. As it was time for me to get a real job, I luckily secured a teaching position in the introduction to art and architecture at Lafayette College in Easton, Pennsylvania. My lack of knowledge in architecture, as well as my nonexisting teaching skills, became apparent when my first lecture lasted only fifteen minutes when I dismissed my rather startled, but delighted undergrad class.

Not only did I learn to prepare more material, I met an older professor who was a sculptor and painter and became my first husband, but I also got to know some of my amazing students. One of whom, June Sprigg, a young scholar writing and illustrating a book on the Shakers, had a boyfriend who was applying to a new program on historic preservation at Columbia University School of Architecture and Urban Planning. It was a multidisciplinary program combing courses in architecture, law, finance, science, and urban planning. These courses were preparation to wage a war against the wrecking ball of urban renewal smashing up neighborhoods and downtowns in cities all over the US.

Photo 3 Photo BMC Photographer, 1971.

After Lafayette, I got a job teaching twentieth-century architecture at Muhlenberg College in the next town and, realized, inspired by reading Jane Jacobs and Ada Louise Huxtable, I was ready to stop teaching for a while since I was prepared to actually do something to save the built environment. I had worked as a volunteer in Easton along with a local family, the Mitmans, who owned the oldest jewelry store in America. We established a preservation group, Historic Easton, to help save the valuable fabric of the eighteenth-century city. I applied to the aforementioned Columbia University graduate program concentrating on getting a Master's of Science in historic preservation to gain architecture and urban planning skills.

The time was right as I had to have spinal surgery due to the effects of polio I had as a baby, and I couldn't work for nine months due to a Frieda Khalo-like body cast I had to wear on my torso post-op. I started my New York City classes the straightest student to work at a drafting board my classmates had ever seen. We lived part of the time in Easton and part in grad housing overlooking Riverside Drive in the city where we would hang out with artists in the slowly changing SoHo in those early days before being gentrified. I worked part-time selling jewelry by artists (including my husbands) from a closet sized store in the Plaza Hotel and got my MS in just over a year. My thesis was supervised by Adele Chatfield-Taylor who later became the President of the American Academy in Rome and was the wife of the playwright John Guare ("Six Degrees of Separation").

My fellow students were not just ordinary folks, either. One had a developer boyfriend who lived in The Dakota (where John and Yoko lived) and we could party on one of the top floors, another worked part time on a new TV show called SNL, and still another was plotting to nab Prince Charles when she met him in London over a break. My closest friend, however, was a young woman who came to the program to save people from being thrown out of their neighborhoods volunteering in the Lower East Side. She and I traveled to Savannah to observe how preservationists were rehabilitating their poor neighborhoods and yet keeping the residents. This led me to use the Easton as a thesis project as I had worked, albeit to no avail, to save a poor neighborhood, "Gallows Hill," from the wrecking ball. In spite of that, I got a paying job in the town becoming the first historic preservation planner for the City of Easton upon graduation.

I continued my work in historic preservation when I returned to Los Angeles, where my husband attended UCLA Architecture for a grad degree while I volunteered to work with the newly formed Los Angeles Conservancy to help save the newer landmarks. I got a job being the first to teach this new field at UCLA, while at the same time trying to revitalize Hollywood Boulevard and bringing my new skills to El Pueblo State Historic Park. From there, now divorced, I was hired by the Bixby Ranch Company to create a master plan for Rancho Los Alamitos, the family homestead.

The Bixby family had bought ranches for sheep and cattle grazing from the Mexican landowners which included most of what is known today as the City of Long Beach and 24,000 acres north of Santa Barbara on the Pacific Coast known as the Jalama and Cojo Ranches (see Photo 4). I worked for the President, Preston Hotchkis, a descendant of the original rancher, and in those two years I got to see the life of Southern California movers and shakers (the LA Times Chandlers were friends of theirs) and understand the growth of the state from the fascinating vantage point of descendants of the pioneers of SoCal.

In the class I taught at UCLA, it was once again a "student" Robert Sweeney (an expert on Frank Lloyd Wright, Photo 4), who got me involved with one of the thorniest preservation issues at the time in LA. Rudolph Schindler, an Austrian architect, escaped Europe in the 1920s, went to Chicago to work with Frank Lloyd Wright, and then traveled across the US on his way to join other immigrants in West Hollywood. He bought a plot of land in and built one of the most iconic modern houses which would be known as "The Schindler House." It was in danger of destruction, and Robert and I, with other passionate architectural enthusiasts, convinced the family to preserve it rather than sell the land and building, got it listed on the National Register for Historic Landmarks, and created a board of directors called the Friends of the Schindler House or FOSH. I quote Mark Mack in his description of its importance:

> The Schindler House, as it is known, embodies for me the rare confluence of zeitgeist and timelessness, where aesthetic purity and radical politics were allowed to mingle freely. In a time such as ours today, where ideals are giving way to pragmatism and conformity, where artificial and speculative smoothness overcomes rugged individual expression, the Kings Road house, and the life lived within it, remain a lively marker of a history whose time is sorely needed again. (Sweeney, Sheine, 2012)

We listened to what Schindler's masterpiece told us and it had one time served as the gallery for the first exhibition of the painters, the Blue Four, in this country. At the same time, it held conversations between some of the leading radicals of the time in the flexible studios. It was designed for two couples with an untraditional layout, with

Photo 4 Robert Sweeney, NJS and James Johnson, 1995 Drawing by The Greeter of 72 Market, Venice, CA.

studios for two architects and one each for their musician wives. Fireplaces warmed their bodies both inside and outside where the landscaping embraced the outdoor "living" rooms. The kitchen was centered between the two music studios and the upper roof was where the sleeping porches were designed, but never really functioned. Shoji screens closed the studios to the outside giving the building a simple, yet elegant, Japanese aesthetic. The building would soon be adapted to one of its original functions.

I found out that some remarkable buildings can have a soul, and, as Stewart Brand observed, this building "learned" to take on a new life. During this time that I was teaching preservation at UCLA, the most dedicated students would join me, Robert, soon to become the Director, and the architect and USC Professor Stefanos Polyzoides with his fabulous daughter for a picnic on Saturdays on the grounds and then work parties helping to restore the house to its original state. My group scraped the paint ever so delicately off the redwood interior walls. Nothing made our hearts beat more than seeing the original bare redwood join the concrete walls in a dynamic composition. We continued through the next decades to bring the Schindler House back to life with help from the Austrian government, but more on that later (see Photo 5).

Photo 5 Door, Tel-Aviv on cover of BMC Alumnae Magazine – NJS, 1989.

SECTION TWO: 1980s: IFMA begins: Los Angeles

Meanwhile my parents had returned to LA, bought a condo in an 18-unit building right on the beach in Marina del Rey, adjacent to Venice. One of the other residents, Werner Heumann, a Bauhaus trained architect, owned a firm called Cannell-Heumann located in downtown LA. My father convinced me to give up my poorly paid preservation consulting and I interviewed with Werner to work there and learn the new field of interior architecture. With my degree, and a limited miserable time on "the boards" working for a big LA architectural firm, I was hired and taught to program or plan interiors for building projects. At the same time, I was introduced to a monstrous mainframe DEC computer which was fed software for this planning work created by a team of brilliant architects and computer experts at UCLA.

Our clients were organizations like automobile manufacturers, financial institutions, pharmacies, and other large corporations. In working with them, I got to travel all over the country, work for a while in Germany and finally meet the team who designed and maintained the software. Two of them lived in a unique garden-like housing compound called Village Green where I also owned a condo. One morning Tom Kvan and Bill Mitchell, well known as a pioneer in digital design, visited me to invite me to join their new venture funded company called The Computer-Aided Design Group and Facility Management firm or CADG+FM. This was 1984 and IFMA had just been formed two years before, and our small firm capitalized on supplying this new field with technology.

They had realized that the software that they had created which was a database (DBMS) interfaced to a computer-aided design (CAD) system was an ideal automation tool for the newly coined facility management profession. And since I had already been trained on it, I was the ideal person to create a training program, write the manuals with Bill Mitchell, and work with the salespeople to market it to corporations and government organizations, one of which was the Israeli government in Tel-Aviv (where I took Photo 5 of the ancient door in 1989). And at the same time, I was the right candidate to join the new IFMA organization and use it as a vehicle to market the suite of software products for automating the planning, design, operation, maintenance and leasing of buildings

IFMA's small staff were an amazing group of people. There were four particularly savvy women who gave me life-changing opportunities: Linda Pate (still with IFMA), Nancy Minni, MaryGrace Huber, and Randa Wells. At their suggestion I created the first seminar for IFMA in 1987 entitled "Principles of Facility Management" which I wrote from my education and experience working with corporations' buildings and real estate portfolios. The first session was held in my backyard, Marina del Rey, where two of my initial students were sent by Northrop's John Carillo who understood the new field of FM and encouraged his employees to take the course. Since John was a prospect for CADG+FM software, I got to see his strategic war room at Northrop where he commanded their large portfolio of buildings. John and I have come together on various committees through the years at IFMA and he remains a friend and colleague, where he is now commanding Global FM. This seminar was followed by others in Houston, New York, and San Diego. At the end of this decade, I married for the second time to a young architect whom I met at an exhibit at the Schindler House, James M. Johnson (Photo 4) and I moved from LA to a small coastal community in San Diego, Del Mar, where my new spouse lived, worked, and surfed.

SECTION THREE: 1990s: Los Angeles-Europe-Canada=Mexico-Tokyo-Boston-San Diego

This was the era of Dennis Longworth as CEO of IFMA and my most intense involvement with the growing IFMA organization. You would think I was an employee due to all the work I did in this decade which was why I gained Fellow status by the end of the 1990s (see Photo 6). I spent an inordinate amount of time preparing and delivering seminars, lectures, and writing articles. I delivered 25 seminars and 21 lectures in cities all over the US and abroad. I also wrote numerous articles on such topics as strategic planning, asset management and DBMS (database management systems).

One IFMA presentation was the first given in the US with an international panel including Moto Nakatsu from IBM Tokyo and Einar Venold from Oslo's Det Norske Veritas. When we got the evaluations, a number of the audience members said they had no interest in hearing from non-US participants. That was shocking to us and luckily that attitude has changed as FMs have realized that the US had much to learn from other countries. I traveled to 19 US cities, as well as Tokyo, Mexico City, Calgary, Barcelona, Brussels, Malmo, Oslo, Copenhagen, Amsterdam, and London delivering my presentations and seminars.

One of these trips was truly extraordinary:

1) *The Antonio Gaudi Trip*: I was invited by IBM to give a lecture in France and Madrid, as well as by a FM group in Barcelona and the sponsors knew of my love of Antonio Gaudi's architecture. They put me up in a hotel down the

Photo 6 IFMA LA Chapter Winners at WW Philadelphia 1996. (*Source:* IFMA).

street from Gaudi's Casa Mila and in the evening when I went back to the hotel there was a dumpster in front that when I peeked into it I could barely make out some broken ceramic tiles which I just threw a few in my purse as they looked rather interesting. The next morning as I surveyed my finds, I saw these beautifully designed broken blue tiles that had been dug up from the sidewalk which I believed were the work of Gaudi. I rushed to the window to see if the dumpster was still there, but it was gone. It is one of the best treasures from my travels to this day.

I began this decade still working for CADG+ FM, but in one of my seminars an attendee, Connor Hickey, invited me to meet his boss, Allen Shay. Allen, a young man in his 20s, was the son of Edward Shea who owned one of the earliest outsourcing companies for the US military, Pacific Architects and Engineers (PAE) with 20,000 employees around the world. Mr. Shay had started one of the first outsourcing companies during the Vietnam War supplying people and equipment for everything that the military didn't want to own and he alone grew this into a large global enterprise.

I gave a presentation on SAM (strategic asset management) and afterward was asked how I could help them. I said "hire me" which Allen did. I was brought on to break into the corporate market which I found not so easy to accomplish even though their workforce understood FM better than most private sector companies. It was hard to convince corporations that a defense contractor could handle their facilities. I did a disaster recovery plan for the US Embassy in Tokyo and secured a lucrative contract with the US Navy to conduct a condition assessment for all the buildings at the Yokosuka Naval Base in Japan. I had to sell it to the FM Director, Bill Johnson, and I also ran the project. I needed to fly to Japan to work out of the Tokyo PAE office. It was headed by an American architect, John Diefenbach who had a staff of fabulous Japanese women administrators (who taught me Ikebana in the evenings), as well as teams of male engineers and architects.

I had discovered a company, introduced to me by my IFMA colleague Eric Teicholtz,, which created software for building condition assessments called Vanderweil Facility Advisors (VFA). After visiting them in Boston, one of their architects helped me by modifying the database to do a higher level and quicker condition assessment for the large base as we had to complete it in a matter of weeks. Moto Nakatsu, now a retired CRE exec from IBM, and the PAE Japanese engineers completed the team.

As women were not so common to lead a team like this, the engineers were not only unhappy with me, but also not pleased with the process we used to determine the condition of building components by using our experiences and perusing the maintenance logs. What we also discovered in the process was that preventive maintenance was not being done as it should have been making the deferred maintenance backlog even larger. And we were able to estimate repair costs due to VFA's and Moto's input to the calculations. At the end of the project, which took many phases, we were able to produce a list of prioritized deferred maintenance projects with estimated costs so the Commander of the base could go to Washington DC with a plea for needed funds with backup data to support it to repair the buildings on his base. And I was able to experience living and working in Tokyo and Yokosuka, with a side trip to Kyoto, which was an amazing experience.

I worked for PAE for four years and then joined VFA to go back to the tech world. Since that time, Allen and his sister Anna sold the company, after Ed died, for $1.2 billion to Lockheed Martin. Anna became a celebrity flushed with her $600 million, in the show "Bling Empire" streaming segments on Netflix. And I, less fortunate, started commuting to Boston, one of my favorite cities, every month to work for VFA in their Summer Street office with harbor views. It was here in Boston, in a temporary apartment on Commonwealth Ave, that I received the call from Dennis that I had been admitted to the Fellows group. As my husband said, he imagined them in a room thinking "will she give us more trouble if we let her in her keep her out?" Obviously the first choice won out and I have more or less "behaved" myself ever since.

This is also the period I joined the Industrial Development Research Council (IDRC) and got involved to learn real estate practices. This gave me a whole new network of people and knowledge that I needed. I got involved in the Councils and loved the "swag" from the economic development booths as their salespeople tried to lure CREs planning any new headquarters, office building or factory to their neck of the woods (one of the states or cities). My LL Bean bag, compliments from the State of Maine, is still in use after all these years (see Photo 7).

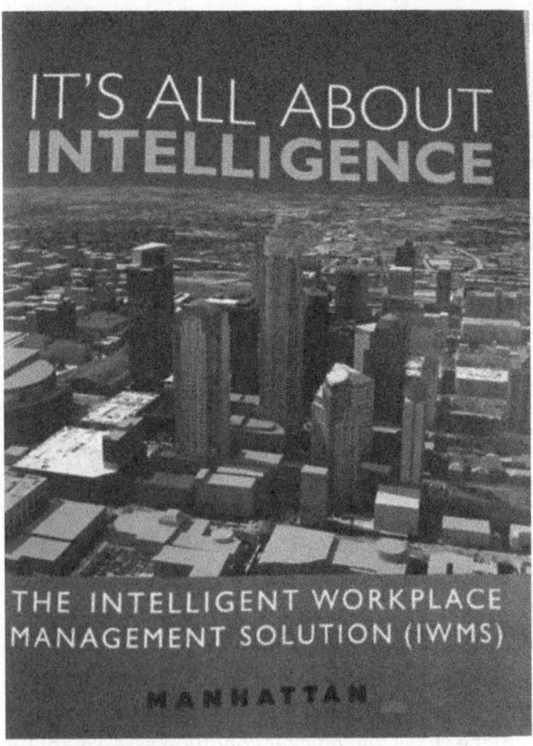

IT'S ALL ABOUT
INTELLIGENCE
THE INTELLIGENT WORKPLACE
MANAGEMENT SOLUTION (IWMS)
MANHATTAN

Photo 7 Manhattan Brochure designed. (*Source:* N. Sanquist)

I traveled to a conference in Hong Kong for IDRC where I proudly talked about my work in Japan and the value of long-term maintenance. One audience member jumped up at the end of my talk, and in a snide British accent, told me "Here in Hong Kong if a building gets old or rundown, we just build another one" and people chuckled. Since that time, maintenance is getting its due appreciation as opinions have changed regarding viewing assets as disposable when no longer useful. We have come to value existing resources in this environmentally conscious age from the clothes on our backs to the existing buildings in our cities.

SECTION FOUR: 2000–2010: World Tour-Del Mar-NYC-London-Boston

It was the beginning of the twenty-first century and although Y2K didn't end the world as some predicted, this new century began with ominous events, a terrorist attack and the downfall of some corporations due to the work of criminal executives. These "Black Swans" would also affect the course of my career. I had drifted away from IFMA as the new CEO and I did not see eye to eye on moving IFMA forward. By this time, I had become more active in IDRC which merged with another RE organization, NACORE to become a new entity, CoreNet Global, and I was viewing our world as FM/CRE. An article I wrote for ***Today's Facility Manager*** in 1998 encapsulated what I thought at the time: "Are You Ready for Infrastructure Management?" The idea came from IDRC which created the concept of Corporate Infrastructure Resource Management (CIRM) which acknowledged that management of real estate needed to be integrated closely with management and strategies for people (human resources), capital (finance), and technology (IT=software, hardware, networks) meaning FM, CRE, HR and IT were coming together to work in the new corporate structures, as well as enabled by the new technology that was developing by companies like the one I was employed by.

This article landed on the desk of a charismatic CEO, Stephen Gardner, of a technology company headquartered in the coastal town I lived in, Del Mar, CA. Peregrine had just acquired a small FM/CRE company, Innovative Tech, for millions of dollars making their CEO, Bill Thompson, his family and a few lucky employees very wealthy. I gave Bill his kudos for making more money in FM tech than anyone in the business to date and wished him good luck on his celebratory trip to Paris on the Concorde. I also asked to meet the executives of this dotcom to set up an interview for me and he gleefully accepted the praise and the request. Next thing I remember is sitting in the waiting room of the Peregrine executive suite reading a *Time* magazine which opened to a story of how difficult it was for women over 50 to get a job, which I was, and this did not seem like a good omen. I met the executives knowing that they did not have anyone amongst their 200 employees who had experience in real estate which was necessary to flesh out their infrastructure management story, so I had a good chance of being hired, ancient though I was (particularly in the tech biz).

Gardner was determined on turning this IT technology company into one that was set on acquiring companies that would round out the automation of the management of a portfolio of all assets. They had software that managed IT assets and now FM and real estate and were searching for technologies that could manage fleets of vehicles and even knowledge assets to add to their portfolio. I was hired in 1999 as the Strategic Director of Real Estate and immediately began working with teams in San Diego and London.

At the same time I joined Peregrine, Mike Bell had become an analyst at Gartner (1998 to 2007), the leading company for IT analysts. This was like the perfect storm, as I had learned how important these analysts were to credibility and sales to a technology company and FM/CRE had never had their own analyst covering our industry. I knew Mike from meetings at IDRC and also by his association with a real estate professor from MIT, Michael Joroff. A project was designed by Bell and Joroff to convince 20 companies to sponsor workshops at MIT to work on new concepts for the workplace with a publication as the output. I represented Peregrine as part of this group which also included AT&T, GSA, and other large organizations. This resulted in the **Agile Workplace** published in 2002 by MIT and Gartner.

Working for this company was the only job where I had an almost limitless budget for the marketing I needed to do internationally. I stitched together a world tour which resulted in a trip from San Diego-Chicago-Oslo-Brussels-Nice-Rome-Bangkok-Sydney-and Hong Kong back to San Diego via LA. I was interviewed, along with a Norwegian Peregrine technologist, by a Oslo financial news reporter on the real estate and FM market in Norway and how it was behind other countries, like the US, the UK, and the Netherlands. And I did a tour of the local airport which was using Peregrine and the GIS system Esri to manage all their facilities. I flew to Nice looking over the snowcapped Alps to attend the MIPEM conference on the coast of France in Cannes. We launched our new real estate portfolio in a hospitality suite with a balcony overlooking the Cannes harbor dotted with yachts leased by the large real estate companies to entertain the executives who had come from all over the world to this huge real estate conference. This was tough duty as we had to attend a IDRC party on the beach and later watched rich fur clad Russian gamblers surrendering their ill-gotten gains at the Monte Carlo Casino. I even found time to visit a Picasso haunt and the Matisse Chapel in Venice.

In Sydney I joined an IDRC sailing party in the harbor past the infamous Opera House and in Hong Kong put my feet in the water on the grounds of the HK Yacht Club where I entertained my friends Alex Lam and Joe Ouye before setting out to have my fortune read by a chicken at the Kowloon market before giving our presentations at a conference.

Later on, my budget supported a formal dinner party for 24 at an IFMA World Workplace Europe in an old Innsbruck hotel. It was an amazing event with some of the Peregrine people in period Austrian costumes and invited IFMA friends. There were jugglers, gymnasts, and even a phony drunk waiter to entertain people as they ate from the lavish menu. IFMA founder George Graves who attended with his lovely wife, said to me many years after the event, that it was the best party he had ever attended and the best one I ever gave!

And then the Black Swans flew in and all hell broke loose. Watching on TV, I saw the terrorists hit the World Trade Center which was one of the most horrendous things I had ever seen. Everybody's world changed. The terrorists had invaded our country killing so many people as we watched two more planes go down, one in the Pentagon. These events totally shook our sense of safety and security. Dennis Longworth was still CEO of IFMA at the time and bravely held the World Workplace 2001 as planned in Kansas City. I wrote an article about it in which I said, At this World Workplace, it had become clear to us that facility management is also about reducing the risk of the loss of life and the suspension of business activities. FM has moved beyond building new structures and maintaining them effectively and safely. The role of FM was redefined. And so, we left Kansas City proud of the organization, which many of us had helped to grow in many cities and countries year after year. It was a conference none of us will ever forget. We came unsure of what we would be feeling, unsure of simply the decision we made to come. We left full of hope, for both our profession and our American way of life and business.

And then another Black Swan flew into our midst damaging our "way of business" when Arthur Andersen (AA) laid off 20,000 employees as some of their unethical staff had destroyed this 89-year-old consulting firm in the fraudulent activities they committed for the sake of their client Enron. Arthur Andersen was not only Peregrine's auditors, but my boss had come from their real estate group as I had introduced an AA real estate expert, Gary Lens, to Steve Gardner after a presentation in Santa Monica earlier that year. Gary had brought along a few other executives to the company before AA's downfall. We should have predicted what happened next, but employees never want to think that their own leaders could actually be crooks. But that is exactly what they were as they reported false numbers as revenue for many quarters. It took a few years for them to come to trial, but the executives of Peregrine had destroyed the company and many of them wound up in jail put there by a federal judge. The once popular leader, Steve Gartner, never saw the light of day outside a prison cell as he had a heart attack before ending his sentence.

After being let go In the third round of layoffs at Peregrine. I was hired as a consultant to Autodesk in San Ramon to bring a concept I had created during my IFMA seminars called SAM to conceptually help bring Autodesk different product lines together. Unfortunately, to get the siloed groups together just couldn't be done (see Photo 8).

After this year long consulting project, Nick Moore, a real estate financial expert, opened a US office in New York City for Manhattan Software and hired me. This was a real estate technology company headquartered in London. Now I had a new commute from Del Mar to New York City and London, perfect for visits to art exhibits and historic properties in the cities. I brought with me my years of experience in the field and also the unique skill I had gained working with Gartner. This would prove invaluable as Manhattan software became the leader in the Gartner Magic Quadrant for Integrated Workplace Management Software (IWMS which was named by Mike Bell) technology increasing their presence and revenue globally.

SECTION FIVE-2010–2020+: Del Mar, Boston, London, San Ramon, CA, Sydney, Nijmegen, Netherlands, Greensboro, North Carolina

We finally come to the last decade plus of this journey. It is certainly one of the most exciting, productive, and fun periods since I met my current "partner in crime" Diane Levine, but it also has had its tragedies. During this time, Manhattan was bought by a publicly traded large tech company, Trimble, to complete their technology portfolio. They had software for design and construction, but none for operations and maintenance which made Manhattan perfect for acquisition. Once again, I thought we were in the catbird seat to dominate the AEC/FM/RE marketplace, but once again it was difficult to break the silos down between product groups. I left after a few years, but I did have a great time experimenting with the new Microsoft HoloLens technology in a pilot project in Sydney.

I moved to what would be my last tech company, Planon, headquartered in the Netherlands with the US office located outside of Boston. I worked there until 2018 when my husband lost a 10-year battle with Alzheimer's, then three months later I suffered a stroke which left my left side disabled. So ended my work at Planon which had been great fun working with my IFMA buddies, Erik Jaspers and Dave Karpook.

A few years before this all happened, I had attended an IFMA Foundation cocktail party, and was softened by a few glasses of wine given to me by this incredibly energetic, passionate, brilliant Trustee named Diane Levine. She lured me to the Board to work with her on a Summit at Cornell and help edit a book she was writing and editing, the beginning of the Work on the Move series (now three have been published). The purpose of the Foundation at that time was twofold: the Advanced Degree Program (ADP) for crediting FM programs at colleges and

Photo 8 Chaffey Community College. (*Source:* IFMA / Chaffey College Student).

universities around the world and raising money for scholarships for students who attended these higher ed institutions, two incredibly worthy causes. But Diane and I felt we should be doing more, not really sure what that was. We decided to get an outsider's opinion and invited a CoreNet economic development guy to one of our off-site Board meetings at a Milliken owned residential compound in Georgia. He sat quietly in a corner as he observed the Board's discussions and then we took him aside in a breakout room. He told Tony Keane (CEO of IFMA) and a few others that we were missing a huge opportunity. This involved helping to solve the problem that many companies were facing, hiring a qualified FM workforce. It was at the same time as I had read that the workforce was an issue globally and so we created the Global Workplace Workforce Initiative (GWWI) that got shortened to GWI.

Diane and I put a long PowerPoint presentation together, thoroughly researched to sell the Board on the idea which some members agreed we had made a case for this new direction while others took more time to see our vision. Our idea was that the Foundation would have to form partnerships with IFMA to offer content, economic development groups for workforce expertise, local non-profits that understood the communities we were addressing and community colleges where we would find the underserved communities and bring FM education to them thus changing lives. Luckily through our work with CoreNet, we knew a star player, Mary Jane Ohlasso, serving as a financial executive in the office of San Bernadino County, the poorest county in the State of California. She believed in the idea, and with others we were able to convince a Dean at Chaffee Community College in the County to add the "Essentials of FM" to the business curriculum. Some of these students which also included people looking for a new career established an IFMA student chapter which would later get an award for its excellence (Photo 8 shows Diane in the midst of these students in the first GWI program).

Meanwhile CoreNet Global was going through growing pains. Barry Varco, an IFMA biggie and brilliant gentleman became Chair at CoreNet at one point and tried to merge the two organizations, but that wasn't in the cards. Then CoreNet strangely decided to eliminate the communities and just have SMEs (subject matter experts), and this gave Di and I an idea. We should start a workplace community at IFMA and lure Kate North, the reigning queen of the now defunct CoreNet Workplace Community, to come to our side and bring the fabulous people from that group with her. We convinced Tony Keane on the fact it would increase membership, as well as increase its knowledge base and the rest is history. That was a beginning of the most vibrant of IFMA's communities, the Workplace Evolutionaries or WE. Meanwhile, Diane and I worked to publish more FM books and create and hold a London Summit (see Photo 9).

Conclusion

I write this today living and working in my North Carolina mid-century modern house which I have turned into a library, art gallery and writing studio close to where my dear friend Porter lives. I have listed the latest developments in the projects in my portfolio which I have been lucky enough to serve as a member of a team:

1) Today Easton, PA now has a large historic district protecting the historic character of the city thanks to the continued work of the Mitman family, now involving an entirely new generation of preservationists, and Historic Easton.
2) The Schindler House is now the Viennese MAK Centre for Art and Architecture, a contemporary, experimental, multidisciplined endeavor founded in 1994 as the hub of three houses designed by Rudolph Schindler, the other two now owned by MAK as residences for visiting artists and architects. FOSH negotiated this unusual international

Photo 9 (*Source:* N. Sanquist, 2015).

agreement with MAK to staff and provide programs and remains very much involved with the care and renovation of the historic landmark, still more revered in Europe than the US.

3) Rancho Los Alamitos has been adapted to be an educational center for the understanding of ranch life in early Southern California, following my master plan. But even better (and nothing to do with me), Jack Dangermond and his wife, owners of ESRI, saved the Bixby's Cojo and Jalama ranches (24,000 acres) by purchasing it for $165 m. and turning it over to the Nature Conservancy, thus eliminating any threats of residential or mineral development or oil extraction. Now this 11-mile stretch of the California coastline of rich biodiversity will be studied by the nearby UCSB and public access available to the waters around Pt. Conception, one of the best surfing spots on the coast.

1) The IFMA Foundation was recently granted consultative status by the United Nations making it a NGO, non-government organization, which will open up more international opportunities to expand the Global Workforce Initiative. Diane and I watched with excitement on TV as the Chair of the Foundation Board gave a brief speech in support of our application. GWI with its Talent Pipeline has had great success in the US with programs in New York City, Houston, and Denver supported by grants with more cities in the wings.
2) We successfully completed our first Virtual Summit on FM and the UN's Sustainable Development Goals (SDGs) on a new platform called REMO which allowed participants from all over the world to choose a seat at a table or a couch and have a conversation or work during breakouts. The virtual simulation worked, with a few hiccups, as our incredible slate of speakers gave information on such topics as technology, buildings, green rooftop farms, financial services for a janitorial staff, but I was proudest when Diane explained the incredible progress of GWI for immigrants, minorities, women and veterans in a number of US cities along with jobs waiting for them by the Foundations Advisors like JLL, Sodexo and ABM. Publications will be produced on the findings of the Summit which will also go to the FM Taskforce on SDGs for ISO.
3) I was asked to serve on the IFMA Research Advisory Committee (RAC) to recommend and participate in international research and publications on the future of facility management; the first white paper authored by Professor Mark Mobach, Jeffrey Saunders, and myself should be available this summer and is entitled "Higher Ground" and I will continue to work on Baukultur from the Davos Declaration, and a tech roadmap.
4) I convinced the editor of *FMJ*, Bobby Vasquez, to allow me to do a new column for each issue entitled "FMJ Review of Books," named after the great ones, NY and London. The first one is entitled "The Changing Workplace" in honor of Sir Francis Duffy, on four 2022 books on this now popular topic.
5) A few years ago I received an IFMA Board Chair Citation "for more than 25 years of passionate and visionary work dedicated to the instrumental role of the workplace" and "for inspiring others to imagine the future of the built environment and fostering FMs to start the journey."

Now after explaining this one portfolio life, I can sit back from this story and see how IFMA has grown and prospered, but the definition of FM is the same 40 years later, only much expanded to include the broadened world of ESG, including a new P, Planet: It is about the interdependent relationships of:

1) PEOPLE(S): in the organization, FM team, supply chain, external community and society in general;
2) PROCESS (G): the work, technology and governance of that work
3) PLACE (E): the regions, cities, neighborhoods, campuses, buildings and landscapes
4) PLANET (E): the planet, oceans, forests, land, and the air we breathe.

I use this sculpture by Sangeeta Sandrasegar (Photo 9) as a last image, "To be carried away by the current, to be dissolved into the other," designed for the terrace of the Museum of Contemporary Art (MCA) in Sydney as a symbol for the future. The artist created this as a hybrid embodiment of ideas inspired by such diverse influences as Virginia Woolfe, the Little Mermaid and Leonardo's St John the Baptist. It encapsulates fleeting moments as changing colorful reflections bounce off the highly polished fiberglass expressing the bridge between space and time, male and female, the ocean and culture. I spent a few afternoons under this sculpture on the MCA terrace with a colleague and co-author, Joe Poskie, designing a Building Digital Workplace Innovation Lab (**Work on the Move 2**, Chapter 9, 2016) which turned out well for the client and our company, Trimble. I see this image today as pointing to future directions for both IFMA and my continuing learning and working with the organization. I can now say I had a small part to play in this bigger story as we celebrate 40 years. ATTA IFMA and ATTA GIRL!!

*American slang expression used in a hybrid context, to encourage a person and also the initials for art, travel, technology and architecture, my passions.

Nancy J. Sanquist, IFMA Fellow
Greensboro, NC USA

Nancy J Sanquist, IFMA Fellow, is an author and art collector living in a Mid Century Modern home/studio located in North Carolina and is an Advisor to the IFMA Foundation and to the Greensboro Opera. She researches and writes on the built environment, is a member of the IFMA Research Advisory Committee, and writes a bimonthly "Review of Books" for the IFMA *FMJ* periodical. She is currently writing a book on her experiences in Los Angeles and Rockland, Maine.

2.15

Facility Management Education

Personal Journeys of Facility Management Educators and Researchers

Joint Contribution By Eunhwa Yang, Jun Ha Kim, and Matthew Tucker

Introduction

Over the last 40 years of Facility Management (FM), the definition of FM has evolved. According to the International Facility Management Association (IFMA), the latest definition of FM is a profession that "encompasses multiple disciplines to ensure functionality, comfort, safety, and efficiency of the built environment by integrating people, place, process, and technology" (IFMA 2017). Regardless of the evolving definitions of FM in these previous years, global standardization of the definition and scope of FM is a relatively recent achievement. In 2018, the International Organization for Standards (ISO) defined FM as an "organizational function which integrates people, place, and process within the built environment to improve the quality of life of people and the productivity of the core business" (ISO 2018). The IFMA's latest definition and ISO's definition of FM highlight the human aspects of managing the built environment. As such, the ultimate goal of FM activities and functions is to contribute to the quality of human lives. With the global standardization of the definition and scope of FM, we, as educators, expect a general understanding of FM and awareness of FM functions to become more prominent, yet we are expected to provide the next workforce in FM with localized FM practices as we have seen regional differences of FM functions and practices, such as North America, Asia/Pacific, and Europe. All of us are in higher education with the duties of research, teaching, and service in different regions, such as the United States, South Korea, and the UK. In this chapter, we will share our own stories about why and how we chose to do a doctoral study in Facility Management (FM), why we chose FM education as a career in higher education, and what our research and teaching experiences have been.

Personal Journey for Dr. Eunhwa Yang

My interest in the built environment has evolved, starting from design and construction. My undergraduate study was architectural engineering, with exposure to architecture design studios. I worked as a site superintendent for a general contracting firm right after college in South Korea; then, I quickly realized that I wanted to study more. When I was searching and looking for what subject to study within the realm of the built environment for my master's degree, I learned about facility management, more specifically, the Masters of Science in Building Construction and Facility Management (MSBCFM) at the Georgia Institute of Technology in Atlanta, GA USA. During the MSBCFM at Georgia Tech, I realized that my academic focus has been on the building itself, including engineering and construction management; yet the FM is a broader business function that integrates people and

Facilities @ Management, Concept - Realization - Vision, A Global Perspective, First Edition. Edited by Edmond Rondeau and Michaela Hellerforth.

three other components (place, process, and technology). I found the study of human-environment relations and human-centered design fascinating, so I pursued a doctoral study in the Department of Design and Environmental Analysis (now called the Department of Human Centered Design) at the College of Human Ecology at Cornell University in Ithaca, NY USA. A doctoral study typically requires a long timeframe and much higher dedication and determination for degree completion. I was not an exception. It took me a while, but looking back, it was a wonderful time that I studied without much concern – although I had self-doubts and worries about uncertainty around getting a job – I took classes in various fields and absorbed a lot of different information and knowledge, from environmental psychology, organizational behavior, sociology, city and regional planning, to statistics.

During this time at Cornell, I developed my academic worldview, in which I understand the built environment from the humanities' perspectives. More specifically, I believe in understanding the relationship between humans and environments through the lens of Bronfenbrenner's ecological systems theory to fully reflect the interwoven layers of environments. In human development, Uri Bronfenbrenner (1979) developed the ecological systems theory, which states that human development is affected by different layers of environmental systems, including the microsystem, the mesosystem, the exosystem, the macro system, and the chronosystem (from self, home, school, workplace/job, neighborhoods/communities, to society which change over time). His ecological systems theory adds an emphasis on environmental influences and moderating effects of human attributes on corresponding consequences. The ecological systems theory inspired many social scientists to expand and test the theory not only for children but also for adults who evolve throughout one's lifespan. Five ecological systems highly emphasize the impact of physical environmental factors, including various types of facilities in one's everyday life. Therefore, within the ecological systems theory, findings from multi-disciplines, such as architecture and design, environmental psychology, business and management, human resources, construction management, and project management, can be applied in workplace and FM research and practices.

From here and there, anecdotally, I heard there might be an "aha" moment in the doctoral study. For me, it was a gradual process rather than one distinctive moment when reading other studies with the joy of finding and learning more about the subject with much more critical thinking in research design and methods in mind. And more recently, writing and creating knowledge – contributing to the body of knowledge – has become a genuinely enjoyable daily activity for me. So, I wanted to conduct more research (as I thought my doctoral study was large pilot research for my research career), spread the joy of learning that I realized, and help the next generations find joyful learning experiences. I believe that the joy of learning really translates one's external motivation to intrinsic motivation for learning, and that is the way in which people can be life-long learners in ever-changing professional trends for not only FM but also other broader professional paths in the built environment sector.

At the point of writing this chapter, I have been teaching FM courses for over seven years at Georgia Tech. Within this teaching period, I have witnessed many different cases of students' career paths with diverse backgrounds, such as business administration, real estate, architecture and design, construction management, facility management, and civil engineering. More importantly, I would like to emphasize the importance of having the mindset of being willing to learn topics around the building's life cycle, from the inception of the ideas to the operation and maintenance phase for students. When people start their careers in the building sector, they (often only) focus on the scope or phase of a building project in which they are involved, whether it is development, design, engineering, construction, or operation and maintenance. Yet, the complete picture of those activities in each phase cannot be possible without thinking about the ultimate goal of the project – supporting businesses, people (both immediate users and next generations who might be affected by buildings' carbon footprint), and society. In this sense, I lay the foundations of my courses and teaching on the idea of building life cycle management and human-centered design/practices.

With the general expectations that the future of work, workers, workplace, and technology will evolve and can be drastically different from the current practices, everyone in the field, of course including educators like myself, should not stop exploring and testing new ideas and continue learning the subjects and learning from each other. Furthermore, with the emphasis on environmental, social, and economic sustainability and quality of human life, the role of FM and workplace management/strategies will become more and more important; at the same time, the boundary of different FM functions from human resources-related work to building science-related work may become blurry where they are required to closely work together. This will enhance the quality of strategic facility planning with the human-centered, data-driven FM approach.

Personal Journey for Dr. Jun Ha Kim

Growing up in a family of architects since childhood, I majored in architectural design at the University of Minnesota, Twin Cities, MN, USA, with a minor in management. I minored in management, while many other architecture students minored in landscape architecture because I was fascinated by specific and measurable management techniques. After graduation, I started working for a construction company without any hesitation. However, I came to know about the existence of FM one day, having been looking for something that can combine architecture and management to some extent, and went to Georgia Tech to start a master's in FM. While working for the Georgia Tech FM department, I could learn practical skills. After I obtained my master's in FM, my academic advisor stimulated my intellectual curiosity which was still not resolved. Eventually, I started a PhD in FM which was a bit of an adventure for me at that time. After I got my PhD in the US, after having a hard time abroad for a long time, I wanted to do FM education in Asia. Between the National University of Singapore (NUS) and Kyung Hee University (KHU) in Seoul, Korea. Kyung Hee was my first choice. KHU was one of the most prestigious private universities in South Korea.

In Seoul area, 17 out of 37 universities have FM-related majors, such as architecture, real estate, and architectural engineering; of these, relatively few FM-related courses are being offered. The College of Human Ecology at KHU offers BA, MS, and PhD in Housing and Interior Design with a concentration in property and facility management track (PM & FM track). This is the only FM program in Korea, with six years of full accreditation from the IFMA Foundation in 2016. The mission of our department was to cultivate competent students with specialized knowledge and practical skills in the fields of the built environment, and the PM & FM track aims to support excellence in education and make leaders in the area of PM & FM. I have been teaching students here at KYU for about 13 years to prepare students for the current and future trends in the AEC-FM industry. My classes, such as "Introduction to FM," "Housing Studies," "Construction Management," "Housing Environment System," and "FM Practice," deal with the interrelationships between people and housing in multiple dimensions and provides students with such learning opportunities as extensive curricular activities, internships for acquiring professional experiences in the AEC-FM industry, and international programs for obtaining a variety of advanced experiences in a global era. My courses are designed to train students as property and facility managers who will become leaders in the area of PM & FM. Specifically, the specific objectives of my courses are to build up firm foundations leading to progressive solutions to issues in the FM industry. Additionally, my lecture puts its goal to raise specialists needed in today's business environment by carrying out systematic curriculums based on the basic theory and the current practice. Students will learn the necessary knowledge and skills through lectures ranging from PM & FM theory to individual tasks, group tasks, and solving problems through various hands-on experiences to become a PM & FM industry leader.

All professors have control of having guest lecturers from the AEC-FM industry. At least once per course per semester, special guest lectures are provided in our program. Additionally, field trips are also provided as needed. Students are encouraged to participate in Summer Abroad programs. An internship is required for all students as it provides opportunities to apply knowledge to industry problems. After 160 hours or more internship, students must be confirmed by submitting working diaries, performance evaluations, reports, practice journals for submission, and credit approval slip in order to complete the internship course.

Additionally, as a part of the requirements of the course "Facility Management Practice" taught by me, the other type of internship program is being held within the FM department of Kyung Hee University through the agreement with the school office. The internship program runs for about four weeks and gives students opportunities to apply theory and knowledge in practice. Students are not only practicing but also becoming part of the research team to provide solutions to real-world problems based on actual cases and data, and these solutions are applied to the campus through consultations with the FM department. Furthermore, students can meet invited alumni at the "Homecoming Day" event held by our department. Through this event, students can consult with alums, which helps them start thinking about their future careers after graduation.

My research laboratory (Facility Management Lab.) combines various disciplines such as architectural engineering, business administration, and industrial engineering for strategic FM facility condition assessment (FCA), remote workplaces, benchmarking, outsourcing strategy development, key performance indicators and decision support system, Korean-style social housing development, and facility management. Recently, BIM-based smart university campus FM system development and smart troop barrack research have been conducted.

Personal Journey for Professor Matthew Tucker

The one constant in my professional career is that I have always been a researcher; however, I have not always worked within the FM field. Prior to my life in academia, I worked in the affordable housing sector in Liverpool, UK, as a research analyst. I was responsible for coordinating various research projects across a parent group company. However, without knowing it at the time, I was increasingly getting involved in FM, for example, through the production of planned and scheduled maintenance surveys and customer surveys on various aspects of housing services.

After working in the affordable housing sector for a few years, I stumbled across a vacancy as a Research Associate at Liverpool John Moores University (LJMU), to work on a fixed-term project in FM. As such, you could say that at this point, I fell into FM – a common phrase that many people I am sure can relate to. My introduction to FM is familiar with many in the industry who fell into it from different walks of life, from office administration, property management, engineering, construction, and beyond.

During this early part of my academic life, I was given the opportunity to study a PhD. I was then faced with the daunting prospect of "what will my PhD be about?" I commenced my PhD journey in 2007, at a time when FM was rapidly growing and becoming increasingly more prevalent to businesses who were faced with understanding their entire cost structure in a time of global recession in 2008. This presented an opportunity to explore the added value of FM through a lens that I was familiar with from my housing days: customer satisfaction and performance measurement. At the time, it was fair to argue that FM's increasing importance in the workplace, coupled with its rapid growth, meant that there was a need for FM organizations to prove the added value of FM service delivery, meaning that performance measurement had become a vital and increasingly reliant element in the effective management of facilities.

I argued that the delivery of FM services is naturally centered on people and the customers receiving FM services. Measuring customer satisfaction should therefore be a vital component within an FM organization's performance measurement framework. In 2004, the British Institute of Facilities Management (BIFM) – now Institute of Workplace and Facilities Management (a landmark moment that I will come on to later) – issued a report (now unavailable) titled "Rethinking Facilities Management: Accelerating change through best practice." The report highlighted the importance of customer performance measurement in FM, illustrating the importance now and in the future across various FM functions. The report provided an opportunity to explore how customer performance measurement is strategically used to improve core service delivery, and moreover, how the use of customer satisfaction benchmarks can be incorporated into this process.

My PhD created a significant contribution to knowledge of the subject by understanding how to strategically use a theoretical framework, termed the customer performance measurement system (CPMS), in order to provide FM organizations in the industry the opportunity to enhance their existing customer performance measurement processes, to assist in innovating and improving the provision of FM services. During my PhD study, I also started delivering lectures on FM and research methods. It was the first time in my professional career I felt that I had found my calling. I loved my subject area, and I loved teaching others about it. I thrived on the interconnection between academia and industry. I felt that I could slide to either side and create a bridge between them, with the goal of progressing the knowledge and maturity of the FM discipline. This passion was rewarded by being promoted to a Senior Lecturer in Facilities Management in 2008.

The next defining moment in my research career would come in 2012 when I was proud to win a Fulbright award, providing an invaluable opportunity to expand my PhD research in the United States. The research applied concepts of the balanced scorecard in order to analyze how to maximize customer feedback in facilities management. This was a crucial moment in my academic career, being able to expand FM research internationally, by sharing and exchanging knowledge across continents.

For the next few years, I worked in the School of Built Environment at LJMU, teaching students primarily studying construction-related degree courses. The more I grew in confidence and experience, the more I realized that I was an outlier here, consuming a lot of time convincing students that buildings only exist to serve the organizations who use them, for the sole purpose to work. Moreover, the extent organizations would be able to do that work, would rely on a holistic definition of the workplace, that encompasses cultural, digital, social, and physical spaces.

This frustration would drive my research agenda, to gradually take a more targeted focus on exploring the concept of workplace, that had outreached from contemporary views of FM. For myself, and others in my inner circle, it was argued that FM had moved away from a heritage in what early pioneers outlined as "expert workplace management" (Price 2003) which recognizes workplaces as the *social* and *distributed* rather than purely *physical* spaces where people use the tools available to them to get their work done. This is further evidenced in isolated past articles charting either FM's history or proposed futures at the time, with FM being described during the 1990s as "a belief in potential to improve processes by which workplaces can be managed to inspire people to give their best, to support their effectiveness and ultimately to make a positive contribution to economic growth and organizational success" (Alexander 1994).

This eventually led to the next seminal moment in my career, moving to Liverpool Business School in 2018 and becoming a Reader (Associate Professor) in Workplace and Facilities Management. A similar transition was also occurring in the professional sphere, with the BIFM repositioning as the Institute of Workplace and Facilities Management (IWFM). This landmark change allowed my research to be emancipated from the transactional culture of FM, driven by procurement, delivery, and contracts to embrace the discipline as the one outlined by Price (2003) and Alexander (1994). But not only that, in 2020 my research would find a meaning and pertinence beyond anything I had ever anticipated: to understand the significance of workplace during and beyond COVID-19.

At the time of writing this chapter in 2022, I am a Professor of Workplace and Facilities Management, and despite the immense challenges and life-changing experiences we have all faced since 2020, I am truly excited about the future of workplace and FM research: that embraces expert workplace management, beyond a physical space, and into a socially connected and distributed one. I would like to conclude with a personal remark to anyone reading this who is interested in a research career in workplace and facilities management: embrace the experiences handed to you, and always keep searching for what you believe in.

Looking Ahead: 40 Years and Beyond

All three of us have a career as researchers and educators after acquiring doctoral degrees. However, at the same time, we also witnessed students pursuing their careers in the industry upon doctoral degree completion in more recent years. We do not believe this is a coincidence; rather, this shows that the industry values highly skilled researchers more and more as there is higher demand and potential to improve the built environments and FM functions with advanced technologies and data analytics. We think this trend will persist in the future.

Second, we expect that the importance of FM functions will increase as the work itself is changing with the economy, automation, multi-generations in the workforce (from newly entering generation to never retire population), and the lasting effects of COVID-19, such as a desire for higher workplace flexibility (i.e., where and when to work for how long). In fact, workplace flexibility, environmental conditions of home offices, and organizational supports were found to be positively associated with individuals' productivity, satisfaction with working from home, and work-life balance during the COVID-19 pandemic; yet, certain work-related activities, such as getting trained and training others, could be better performed in corporate offices (Yang et al. 2021). As organizations cautiously explore what a new "normal" may look like, a socio-technically aware profession, like workplace management and strategies, can contribute to shaping the future world of work (Moriarty et al. 2020).

Third, the old view that considers FM as a cost center that merely operates and maintains the physical facility in a reactive manner will be drastically shifted. Instead, multidisciplinary approaches truly combining the practices of management/human resources, design (physical and digital workplaces), psychology (psychological health and wellbeing), and engineering (highly efficient, sophisticated building systems and fixtures) will be essential for organizations' success. As a result, the FM functions will evolve with immersive technologies, high importance on employee health and wellbeing, and the revolution of work, collaboration work, and work modes.

Last, we advocate for a strong, synergetic relationship between academia and industry to reduce the gap between the two and for strong support from professional organizations for FM higher education programs (Lai et al. 2019). Academic institutions are there to produce the next generation of employment/workforce as well as bring innovative collaboration through novel approaches to solve perplexing problems/questions to advance the field of FM. As the nature of applied sciences, research in workplace and facility management can be best performed in real settings, which is only possible when these academic research endeavors are supported by the industry and vice versa.

References

Alexander, K. (1994). A strategy for facilities management. *Facilities* 12 (11): 6–10.

Bronfenbrenner, U. (1979). *The Ecology of Human Development: Experiments by Nature and Design*. Cambridge, MA: Harvard University Press.

IFMA (2017). What is facility management? Retrieved October 25, 2022, from https://www.ifma.org/about/what-is-facility-management.

ISO (2018). *ISO 41001:2018(en)* Facility management — management systems — requirements with guidance for use. Retrieved October 25, 2022, from https://www.iso.org/obp/ui/#iso:std:iso:41001:ed-1:v1:en.

Lai, J., Tu, K., Lian, J., and Kim, J. (2019). Facilities Management education in the Four Asian Dragons: a review. *Facilities* 37 (11): 723–742. https://doi.org/10.1108/F-06-2018-0066.

Moriarty, C., Tucker, M., Ellison, I. et al. (2020). Recognising the socio-technical opportunity of workplace: an analysis of early responses to COVID-19. *Corporate Real Estate Journal* 10 (1): 51–62.

Price, I. (2003). Facility management as an emerging discipline. In: *Workplace Strategies and Facilities Management* (ed. R. Best, C. Langston, and G. De Valence), 30–48. Oxford: Butterworth-Heinemann.

Yang, E., Kim, Y., and Hong, S. (2021). Does working from homework? Experience of working from home and the value of hybrid workplace post-COVID-19. *Journal of Corporate Real Estate*, Vol. ahead-of-print No. ahead-of-print. https://doi.org/10.1108/JCRE-04-2021-0015.

Eunhwa Yang, PhD.
Assistant Professor, School of Building Construction, College of Design
Georgia Institute of Technology
Atlanta, GA USA

Eunhwa Yang, PhD, is the director of the Workplace Ecology Lab and an assistant professor in the School of Building Construction, College of Design, at the Georgia Institute of Technology. Eunhwa has received a PhD in Human Behavior and Design from Cornell University and a Master's in Building Construction and Facility Management from Georgi Tech. Her scholarly passion lies in the area of sustainable and healthy built environments. Her current research focuses on three main areas: (1) understanding healthy workplaces and the future of workplaces, (2) understanding the built environment and cumulative environmental impacts on older adults experiencing cognitive decline, and (3) supporting data-driven decision-making processes during the operation and maintenance of the built environment. Eunhwa brings a unique set of multi-dimensional perspectives toward the idea of a human-centered built environment, with academic backgrounds in human behavior and design, architectural engineering, facility management, and workplace strategies and management, across a building's life cycle. She is especially interested in translational research utilizing mixed methods and has been actively involved with the International Facility Management Association (IFMA) and CoreNet Global. She has developed/redesigned and taught graduate courses for the Master of Science in Building Construction and Facility Management and Professional Masters in Occupational Safety and Health. Her students have been recognized in the field by winning many scholarships and competitions, such as the IFMA Foundation Scholarships and CoreNet Global Academic Challenges.

Jun Ha Kim, PhD, FMP, KCFM
Professor, Department of Housing & Interior Design
Kyung Hee University
Seoul, South Korea

Jun Ha Kim, PhD, FMP, KCFM, is a professor and the director of the FM Lab in the Department of Housing & Interior Design at Kyung Hee University. He got his Master's (2003) and PhD (2009) specializing in FM from the School of Building Construction, Georgia Institute of Technology. As a five-time IFMA Foundation scholarship winner (2003, 2004, 2005, 2008, 2009), Jun Ha Kim has been an active IFMA member. Prior to his current position, he worked for Georgia Tech's facilities department and Turner Broadcasting System (CNN). After returning to South Korea, as the director of the Property Management & Facility Management (PM & FM) track and the FM Lab, he developed the FM curriculum based on the IFMA guideline (core competencies) and advised PM & FM track students. Additionally, he developed and coordinated the FM exchange program five times between Kyung Hee University and Georgia Tech in 2013, 2014, 2017, 2019, and 2022. He was a chief editor for the Journal of Korea Facility Management Association (KFMA), and he has been an active board member and joined in the development of the Korean FM certificate called Korea Certified Facility Manager (KCFM) for the Korea Facility Management Association (KFMA). He is now a vice president of KFMA, and his continuous efforts have helped to establish Kyung Hee University as a leader in Korea in the area of Property and Facility Management research and teaching.

Matthew Tucker, PhD
Professor of Workplace and Facilities Management
Liverpool Business School
Liverpool John Moores University
Liverpool, England

Matthew Tucker, PhD, is a professor of Workplace and Facilities Management at Liverpool Business School. Matthew adopts a multi-disciplinary approach to his workplace and facilities management research, combining concepts and techniques from the built environment, environmental psychology, and business and management. Matthew's specialist areas include workplace strategy, workplace productivity, the role of facilities management in building information modeling, performance measurement, and sustainable facilities management. Matthew also works externally in different international roles, most recently as an Executive Board member of the Executive Doctorate in Business Administration Council (EDBAC) and a Facilities Management Expert for the British Standards Institution (BSI).

Matthew is a Fulbright Scholar, winning the first ever RICS-Fulbright award, enabling him to undertake a pioneering research project on customer performance measurement in the USA. Matthew has over 70 publications that include peer-reviewed journals, book chapters, refereed conference papers, and industry reports. He has vast editorial experience and sits on various Editorial Advisory Boards. Matthew is currently the Associate Editor for 'Facilities' (Emerald Publications). Matthew maintains close industry links as a Certified Member of the Institute of Workplace and Facilities Management (CIWFM); and a Member of the Royal Institution of Chartered Surveyors (MRICS) and a Chartered Facilities Management Surveyor.

2.16

The Acceptance of FM and IFMA

Doug Aldrich

Those remarkable people who founded IFMA in 1982 never realized how their initial meetings, deliberations, and decisions would come full-flower 40 years later. With my high regard for their vision and dedication, I believe it took some years for the seeds to take hold, initially in North America and then expand to Europe, Asia, South America, "Down Under" and elsewhere. FM jobs and responsibilities were known by a range of names in other countries, which turned out to be sticking points in global expansion, implementation of the profession, and acceptance of its new presence. That happens whenever a new concept or organization bursts upon the scene locally and even more so when it grows around the world. There were no real surprises when IFMA began to become visible and its story continues.

Becoming an FM

I would like to weave in personal history with observations and experiences that I had during two decades in this new profession and its flagship association. After my Army stint, I worked in private industry as a chemical engineer and then department manager in a variety of positions. At the time, these were logical progressions of what I foresaw in college after graduation and even more so after a year to obtain my master's degree. I didn't realize during those initial 18 years in industry that working in research, product, and process labs, supplemented by jobs in human resources (HR), new product-market development, business management and planning would give me such expansive and realistic views of our large company. Little did I realize at the time how my new SKEs (skills, knowledge, and experiences) would enable me to create and build an FM role and later become a facility director for our global R&D operations.

A second increment of my personal growth during these years was in team-building, leadership and communications, especially with people from different backgrounds and missions. It is one thing to talk with other engineers about projects and technology; it is totally different when engaging employees from marketing, quality assurance (QA), and finance. Whether I was participating on a multifunctional team or leading it, the use of speaking, graphics, listening, and conflict resolution showed me how critical those tools could be in my technical future (never guessing a different career track would turn out to demand those same ones).

There were also special assignments along the way, as well as "job enrichment" tasks in safety audits and training, that gave me perspectives how our company had many roles and goals. All these encounters helped me to better understand where our organization was headed, dominated by innovations from new products. On average, we invested ~7% of our sales in R&D and reaped 25% of our total growth from new products and improved processes. Later I used that perspective to develop strategic plans for how our buildings and specifically labs could make those objectives happen.

Facilities @ Management, Concept - Realization - Vision, A Global Perspective, First Edition. Edited by Edmond Rondeau and Michaela Hellerforth.

A third aspect of my technical and management years was in communicating with executives. This came from such activities as participating on business task forces to evaluate new technologies, creating a "dual ladder" of professional advancement for scientists in parallel with managers, leading tour groups of important visitors (especially from outside the country), and being a visible representative with government agencies. In each case, the interactions with our top people were insights as to how they thought, perceived, and decided. These made me a strong advocate of "elevator speeches." (Imagine you are with the CEO going up four floors, and he asks how the new building project is going. Being long-winded doesn't work!)

Other than an HR position, my real career surprise occurred when the VP R&D asked me to stop by. He had a $6 million renovation project write-up on his desk, but wanted a critical review before signing it. He was talking to me because I had worked in many different types of labs and was known for contributing solutions to sticky problems. (The VP added privately that my reputation for candor was important because he was being pressured to authorize the work.) After the usual "Why me?" discussion, I accepted the temporary job, which turned out to be one of my life's pivotal decisions.

At the time, we had one individual who "cared for" our Michigan R&D labs, with roles of fixing building problems, answering people's questions, and overseeing our lab safety program. He was a good administrator of past traditions, and well known for helping lab people with a host of answers. Although we had worked together in the past, my new assignment was clearly a radical departure for him as the safety guru for labs. When I outlined the 3–5 goals needed to accomplish my special assignment, he was surprised by my desire to hire a lab programming consultant, visit other labs to assess their best practices and lessons learned, and relook at everything we were doing in our current facilities. He was most supportive since I included him right away on my initial team.

I did just that, spending energies attending seminars, talking to architects, reading all the articles and visiting labs that were publicized as "best in class." After eight months of discussions, visitations and calculations, I took my final report to the VP and said, "The renovation project is a dog; we should build a new center that consolidates our R&D departments and prepares them for the future." His reply (literally) was, "Doug, why don't you put money where your mouth is and build it." I agreed, although never trained in design and construction, and accepted the challenge to complete this project. This was truly a huge change in my career expectations and later life's experiences.

The purpose of this history is not to laud my first project, but rather to point out there was no such position like mine in the company. Our engineers built chemical plants, not buildings; even then, our QA labs were quite vanilla. So my first lesson was to grasp this new opportunity with energy (even when I had trepidations), willingness to learn, and an open mind. This was the beginning of my role as a visible, emerging FM leader in the company and ultimately establishing that role and responsibility worldwide.

When running this project, I used principles that IFMA contains in its core competencies today; to me, they were what a good leader should have. In this case, I applied them to project management and then subsequently operating buildings. At the end of this first journey, during its dedication, I felt a great sense of accomplishment, but even more importantly, I was openly acknowledged as the person who took care of lab buildings.

The bottom line to my reflections is how this project prepared me to assume an FM role in my organization, and how that led to a strategic presence and operational excellence in our laboratories globally. My special assignment in 1985 then led to networking and facility interactions with others, a prelude to hearing about and joining IFMA.

My Journey with IFMA

I was at a conference in Boston, Massachusetts, chatting with Erik Lund, IFMA Chair, who asked if I had ever considered joining his organization. I replied I had never heard of it, but asked him to tell me more. So I signed on without a clue what the association did or stood for. It was the first time I connected with FM as in Facility Management. My horizons were broadened with my initial membership dues. Four events occurred to build my interest in and support of FM's and IFMA in North America.

First, Lund called me in 1987 and asked if I had any interest in forming an R&D Council (Only a Utilities one existed at the time.) to promote networking among FM's heading up the care of complex facilities across a variety of industries. An FM at Westinghouse and I conferred and "clicked," agreeing to spearhead the formation of this special-interest group. In short order, we accumulated a list of 100–125 names of labs and their FMs, and I sent introductory letters to all of them. The letter cited the benefits of sharing information, data, and best practices among facility leaders with common interests of helping R&D people be productive. Our approach was to sponsor a spring meeting at Steelcase, the furniture manufacturer, for three reasons: It had unusual types of labs, new approaches to officing, and liked to sponsor tours, presentations and a networking dinner for potential customers. It was a great success with ~75 in attendance.

The formula we leaders decided upon was to ask for a host company to provide those activities at an annual spring event. These meetings were at various locations around the United States, across a variety of industry types, with special speakers, member presentations, and work sessions. Between meetings, stimulated by our member lists, there were frequent contacts among those of us with common interests such as vendor recommendations, developing technical criteria, benchmarking, and design/engineering processes. The meetings became very popular with informative lab tours, discussions, and problem-solving topics. We averaged 100 attendees at each spring conference.

The R&D Council took off, bringing in many new IFMA members as well as broadening an active network of FMs. There were also a few new members from Canada, which expanded our horizons to North America. Then those attending World Workplace each fall gathered for a discussion meeting, as well as any relevant tours in the nearby area. This increased the number of R&D FM's attending the autumn event and stimulated more networking until the following spring meetings.

Second, publicity blossomed with articles in several magazines, both within the building world and in industry-specific publications. Many were written by prominent IFMA members who, in some cases, found themselves to be on magazine covers. These were usually profiles of that person at his or her organization, talking about FM job responsibilities, challenges, and accomplishments. These articles "seeded" follow-up inquiries for more interviews and writings. It was rare that three months went by when an industry magazine didn't feature an IFMA member talking about the real world. Many articles included hard data on project costs, productivity gains, and step changes in FM performance. In my case, media connections were in magazines featuring science labs with a chemical bent and further promoted the profession and IFMA.

Additionally, *FMJ* (*Facility Management Journal*) began featuring articles on different types of facilities, showing the breadth of building types that FM's covered. IFMA members were asked to contribute, and I had the opportunity to write about labs, an unusual environment when compared to offices. The most fun part of that request was educating non-technical people about labs, their purposes, complexities, and value to business and country. After my first major project, my headquarters R&D facility was a photo-opportunity with me on two covers, with my articles and pictures featured inside.

Another visible manifestation of FM and IFMA coming on the scene was the number of seminars, panels and discussions that brought together RE (real estate), service providers, and FM people. These forums covered both mutual and different roles of our associations and professions, and often their involvement with developers and A/E (architect/engineering) firms. It was a great time to extol the breadth and contributions of FM's just new to the built environment field, just 10–12 years from IFMA's formation. It was during these types of activities, supplemented by sessions at World Workplace, that collaborations among future association leaders flourished.

After the completion of my company's R&D headquarters in 1990, I had the opportunity to run it and correct my mistakes. I was asked if my design/engineering techniques and, more importantly, philosophies could be applied to our other domestic labs. Thus FM was becoming recognized at my company for bringing a different set of tools and practices to labs and its valuable users. My written articles resulted in my being asked to join five editorial advisory boards, to bring IFMA and FM to the forefront of our profession. We were becoming truly visible!

Third, I was asked in 1993 by Bill Gregory, IFMA Chair, if I would join the Board with the purpose of representing networking councils and growing their numbers for IFMA. He was hoping that, using lessons learned from my co-founding the R&D Council, our FM network could be spread to other industries. I accepted the challenge. During Board meetings, every time Chapters were mentioned, I added the words "and Councils" so often that eventually I just had to raise my hand to have them acknowledged!

Interestingly, early members of the R&D group from university labs split off to form a separate Council for Academic Facilities, to focus on their opportunities and challenges. During the next two years, we grew from two to 11 Councils, promoting the notion that FM permeated many industries and businesses, with a common set of skills, knowledge, and contributions. Every Council developed a regimen for meetings and networking, and Council of the Year became a sought-after award. In 1993, R&D was named as such, and Councils, as well as the award, have thrived since then.

Councils were a key part of promoting FM across many industries previously untouched by IFMA. In later years Communities of Practice were developed, further expanding the reach of networking and utilization of professional skills and FM knowledge. It was a most successful two years for me to lead IFMA into a strong network of Councils.

During this time, two other key events by IFMA elevated the profession and association to a higher level of awareness and visibility. The first was the establishment of the CFM (Certified Facility Manager) in 1992. The basis of the certification was by passing an examination on a number of questions relevant to FM knowledge and practice. The result was many members sought that distinction, with hard work in preparing for the exam. At World Workplace, a breakfast was hosted to recognize those who had earned their CFM. It was such a distinctive honor when I received it that my leader, VP of R&D, personally wrote the notice for our company newsletter. He was so pleased with this first CFM award in the company that he promised to give me more work!

The other occurrence, also in 1992, was the creation of the honor of IFMA Fellow. This was to identify candidates through a nomination committee, and then selection by a jury of Fellows, IFMA Chair, and Staff CEO. The critical distinction is the number of Fellows is strictly limited to 0.5% of the membership. The process is formidable with nominators working with prospective nominees to ensure a high-quality submission to the jury. That group then reviewed all documents with numerical scoring, qualitative descriptions, and open discussion of the packages "between the lines." At the fall World Workplace, those chosen are recognized on stage and each receives a coveted pyramid. IFMA Fellows are linked with quarterly networking sessions and frequent conversations for task forces and committees during the year. Fellows have now been selected from many parts of the world, with she or he focused on mentoring others in the association.

Fourth, I was asked to run for the IFMA Board's Executive Committee (EC) following a great cadre of FM leaders before me. When treasurer, the Foundation took hold with raising funds, promoting FM college programs, and recognizing outstanding students with scholarships. During 1995–1997, the EC and Board redefined membership grades to be more inviting, and opened the door to substantial growth. We developed a strategic planning process to elevate association goals and highlight the "I" in IFMA. Our Board meetings were well planned, vibrant, and productive. My five years as Council Director and ascending EC positions were not only hard work but also just plain fun!

One of the proudest moments in my professional life was moving up on the EC to become IFMA's Chair. Never had I envisioned becoming an FM and then joining IFMA, would lead to this honor and responsibility. I remember my acceptance address was based on the acronym CHANGE (Communications, Humor, Action, Networking, Growth and Energy). Those words became the theme of my tenure and our Board's actions. I culminated my term by restructuring the Board for global awareness and participation, inviting the "best" to be in our Board leadership roles regardless of postal code. Finally, we created the first FM Megatrends study and report, which has been updated three times since, as a stimulus to thinking about major changes affecting us in the future.

At the same time, my corporate position changed from a domestic focus to Europe, with executives anticipating that laboratories abroad should become as well managed and supportive as the ones in the United States. Because there was such a close fit and synergy between my Chair role and company position, an exciting new phase in my FM life was about to happen. The stage was set in North America to grow and optimize our association and profession, and now it was time to exploit that internationally.

Spreading FM Globally

As a Board and EC (Executive Committee), we were eager to develop meaningful relationships outside North America. Those of us with IFMA leadership roles and similar business positions often journeyed to other countries in Europe and Asia for projects and O&M (operations and maintenance) of existing facilities. From these efforts, connections were made to FM's and later chapters or networks forming first in Europe, then Asia, Brazil, and "Down Under" in Australia and New Zealand. New places are coming forth now too.

The IFMA staff began developing communication vehicles with key contacts in other countries, so that special invitations could be given to come to World Workplace. There we had great opportunities to meet and dialogue with people in our profession, but often practiced quite differently. This was most fruitful as we explored different approaches and best practices for serving our "customers" and supporting our organizations. The common-interest conversations were truly stimulating, and relationships were born!

After my global headquarters project and the consolidation of our US labs under my O&M wing, I was asked to evaluate the position of my company in Europe with respect to TS&D (technical service and development) labs. Having three years of business experience, I was able to examine our five-year strategic plans and how our European labs leveraged them. From this analysis, I developed plans to close three local labs and build two new major facilities. The first was a renovation and expansion double goal in Germany with construction and relocations. A Belgian project was a basis of my article that promoted FM's role in strategic thinking and tactical actions. Our C-Suite liked the results, as did our European customers, and the post-occupancy audit scored well.

It was during these years that those of us with foreign responsibilities strategized how we could work with people in the UK and on the Continent who led their organizations and countries in developing FM as a viable profession. Out of these efforts were born country chapters or similar organizations as IFMA manifested FM as a viable force in their businesses and industries. There were IFMA meetings in Europe that served as a forum for building relationships and promoting the profession, with minor differences due to local laws and customs. It was also an opportunity to realize that there were many ways to manage facilities and services in other environments.

With my experiences in Europe, our executives asked me to assemble global plans so corporate investments and realized productivity gains were consistently measured and evaluated. In addition to R&D (research and development) labs, we also placed small ones for regional technical service at key locations around the world. As the global FM for the company, I was asked to assemble a comprehensive picture of facilities for international business. This assignment led to many interactions with FM practitioners, while I was building major labs in Brazil and Shanghai, and fitting out small ones in Singapore and Australia. There were also O&M roles in Wales, Taiwan, and South Korea to help educate me about doing business around the world.

An interesting aspect and part of my learnings in spreading IFMA and doing projects internationally were the many aspects of culture, customs, and history of each country. When speaking to college classes, I always recommended they did their homework in preparing for trips (and maybe work) outside the country. For my ventures to other parts of the world, I did as much research about getting along while leading the design, engineering, construction, and start-up of new labs. One of the most important roles for an FM is to learn how to do business elsewhere. A/E processes are often different, and certainly legalities must prevail in them. There were differences in bidding, contracts, environmental approvals, and economics than I was used to in America. Discussions with my local hosts were essential to do them well as I listened intently to their help.

Thus, by combining my IFMA EC role with my corporate responsibilities, I was able to expand IFMA's visibility while just doing my job. Others with similar positions offered several opportunities for me to speak at FM conferences, give advice to local groups and Chapters, and exchange good friendships. Some of my finest memories are from speaking in Finland, Sweden, the UK, Hong Kong, and Chongqing. Similar meetings were held in Asia that broadened facility management understandings with excellent support from various active chapters in places like Australia and Hong Kong. Conferences were held with speakers and presenters that exemplified how FM could contribute to many enterprises and their successes. IFMA Fellows were emblematic in carrying the image of what FM was all about. (There were occasions in which my talks were publicized in local languages as well as English, thank heavens! It was always courteous, however, for me to begin and end my seminars with words from the local language.)

My FM and IFMA Coda

From my first involvement with "facility leadership" in 1985 and IFMA a few years later, I am amazed at the expansion of our profession and the growth of our association. True, there have been bumps, adjustments and learnings along the way, but today IFMA is the focus of what our profession is all about.

This happened in just 40 years since our profession was articulated. After my Board service, there were IFMA extensions into new areas: sustainability, technology, training, surveys, knowledge libraries, and new core competencies. The megatrends report has now highlighted the implications of FM demographics that portend another set of issues in replacing those who are retiring. Our creators never imagined what their first meeting would lead to; IFMA will continue to grow and evolve in its coming decades.

It's critical to note that IFMA has the attributes of 21,000 members worldwide who are practicing their SKEs for the benefit of their organizations. There are numerous stories, memories, anecdotes, laughs, hellos, and good-byes in our meetings, networking, phone calls, publications, and emails. The FM bond is indeed strong through our efforts.

The real contribution to be noted is how IFMA and its members have benefited their countries, organizations, and facilities. They span industries, businesses, churches, universities, suppliers, transporters, service providers, consultants, utilities, and, of course, R&D labs. The clear mandate of FM's is to serve their clientele respectfully, openly, creatively and strategically.

So many times, I've thought about how much my life has changed since I told our R&D VP that this "Renovation project is a dog." He told me later I put my money where my mouth was very well. This was the genesis of my personal trip with FM and IFMA.

Dr. Doug Aldrich, IFMA Fellow, CFM, Past IFMA Chair
President and Owner
Aldrich & Associates LLC
Arvada, CO USA

As president of Aldrich & Associates, Doug leads key processes for lab facilities that ensure a strong link with organizational objectives and strategic intent. His advisory services range from broad concepts to detailed designs, from physical lab adaptabilities to soft people behaviors, and from visionary ideas to tangible tactics. He has built and/or operated labs in Australia, Belgium, Brazil, China, Germany, Japan, South Korea, Singapore, the United Kingdom, and the United States.

Aldrich spent 18 years as a chemical engineer before shifting into the FM side at Dow Corning. As global laboratory director, he was accountable for strategic planning, design and engineering, construction, and operation of one million SF (93K M2). This included governance of laboratory EHS activities as well.

His focus is on the integration of people and technical spaces, for on-going work and new ventures which endure in impact. He has been a noted persona about global FM, strategic planning, and how real-life projects impact R&D. Doug has been on five editorial boards and two magazine covers, was profiled in an FM book chapter, and written several articles about laboratories.

Aldrich co-founded IFMA's R&D Council and served on the Board to represent all Councils. He was elected IFMA Chair in 1998, to lead 18,000 global members. He earned his CFM in 1993 and was named an IFMA Fellow in 2003.

Doug has BS and MS degrees from South Dakota Mines who named him the first Distinguished Alumnus, gave him the prestigious March Medal, and awarded him an Honorary Doctorate in Humane Letters. Theta Tau elected him to its Alumni Hall of Fame, and Mines named its Student Learning Center after him.

An active participant in the university environment, he was a campus recruiter for 30 years, developed a technical communications curriculum, and wrote a guide for student interviewing. In addition to class talks, he served on both department and institution advisory boards, and as president of the alumni association.

2.17

Academic Career Paths in Real Estate and Facility Management – a Personal Experience Report

Antje Junghans

"The future is not some place we are going to, but one we are creating. The paths to it are not found but made. The making of those pathways changes both the maker and the destination." Australian Commission for the Future (Jopling, John 1999).

With great pleasure I accepted the invitation by Edmond Rondeau to contribute to the concept of this new Facility Management (FM) book, which he described as follows. "Our concept envisioned not another textbook, but a gathering and displaying of the reasons, the motivation, the impressions, and the experience provided by eyewitnesses, and participants of the birth and growth of this phenomenal FM concept, which found its way into associations, academia, and corporations worldwide in these four decades" (Edmond Rondeau, January 2023).

I am looking forward to sharing my personal academic career path toward Facility Management (FM) since its beginnings in Germany in the early 1990s. In my studies, working life as well as in tertiary education and science, I have been involved in the fields of architecture, construction, real estate, and facility management since 1985. Initially, the later career path was still open. I received inspiration for the academic direction during my studies, which awakened my curiosity and enthusiasm for scientific work. Working as a professor became both a profession and a vocation. My ambition is to explore and improve the interaction between living and working conditions and the built environment. In the new academic field of Facility Management, I had the opportunity to shape this path.

More than 30 years ago, I started to develop a deeper understanding of the international state of the art in FM by living and working in different universities and countries. The choice of topic for the diploma thesis "FM in the context of the planning and use of buildings" shaped my path`s direction and has been a constant theme in my practical and university working life. Facility Management met me at the end of the 1990s with new contents in the consecutive diploma course in construction management at the University of Wuppertal. At the same time, I received a request from the Chamber of Architects of North Rhine-Westphalia to improve the management of the growing real estate portfolio in the fixed assets of the architects' pension fund. At this time, I started to deal with the international scene in facility management. In 2006 my appointment as professor at the Frankfurt University of Applied Sciences further paved my way into the realm of academia. In 2011, my path led me to Trondheim as a university professor at the Norwegian University of Science and Technology (NTNU). The journey is the reward, so to speak, and I was drawn further to Switzerland. Since 2016, I have been working as a professor in strategic facility management and director of the Institute of Facility Management (IFM) at the Zurich University of Applied Sciences (ZHAW).

Facilities @ Management, Concept - Realization - Vision, A Global Perspective, First Edition. Edited by Edmond Rondeau and Michaela Hellerforth.

Education and Career Start

In 1985, I completed secondary school in Dortmund, Germany, and gained entrance to university. In my parents' house there were no academics whose career paths I could take as a model, and I decided to choose a course of study that followed my artistic interests. At the same time, I wanted to become financially independent from my parents. Architecture or civil engineering? Traditional university or university of applied sciences? What were the differences? I didn't know and opted for architecture. In the autumn semester of 1985/86, I started studying architecture at the Dortmund University of Applied Sciences. At the same time, I took on student jobs to be able to finance my studies and cover my living expenses.

The Dortmund University of Applied Sciences enrolment process required demonstration of practical experience. For secondary school graduates, this requirement could be satisfied with a three-month construction internship. In the summer of 1985, I started a paid internship with a local construction company. The focus of the work was on modernizing old buildings in an inner-city residential block. Apartment layouts were adapted to the new requirements, and facades and roofs were renewed. I gained insight into all areas of construction work. After successfully completing the construction site internship, I was admitted to study architecture at the Dortmund University of Applied Sciences. The name of the course was Architecture – specialisation Building Construction.

I remember the speech welcoming the 100 or so first-year students. We learned that only 10 – 20% would make a successful start to their careers. In the mid-1980s was downturn in the construction industry, and many architects and civil engineers lost their jobs. However, with the reunification of Germany, this situation changed abruptly. Starting in the fall of 1990, highly qualified experts and engineers were once again in great demand. A highlight of my studies was also the second internship, which I completed in an architectural office, DOMINO in Rotterdam in the Netherlands. Here, I got to know the professional practice as an architect. The tasks included the preparation of construction drawings, taking measurements and drawing design plans, especially for inner-city residential projects. Now, with the three months of work experience from the Netherlands under my belt, I returned to Germany. There I found a job in an architectural office called "Gesundes Stadthaus" (English: healthy townhouse) in Bochum. Sustainable building in an urban setting fitted very well with my main interests in my architectural studies, which I completed in 1990 with a diploma thesis on the subject of "Office and commercial building at the Amiens square in Dortmund" as a graduate engineer (Dipl.-Ing.).

I started my career in an architectural and urban planning office in Dortmund. Based on the professional experience gained there, I was admitted to the Chamber of Architects of North Rhine-Westphalia in 1992 and was entitled to use the professional title "Architect." The projects I worked on included building permits, design plans, rural development plans, and urban design plans in Dortmund and the surrounding area, as well as in the new federal states of Germany, e.g., in Schönebeck/Elbe and Rostock.

From 1993 to 1995, I accepted a new challenge as a client representative for a company group in the field of hotel investments. I was responsible for project development, planning, construction, and commissioning of hotels. My area of responsibility included various hotel projects throughout Germany. After two years of intensive project work and a lot of travel and overtime, I became self-employed in 1995 as an architect with my own office operating in architecture, consulting, planning and construction management. The focus of my work was the time schedule and cost control for a brewery reconstruction in Düsseldorf as well as invoice reviews for the final accounting of various hotel projects, building applications and urban design plans. In addition, I carried out teaching activities at the TOP CAD School in Dortmund. I also worked at the Dortmund University of Applied Sciences in the Faculty of Architecture in the research area Women and Society of Professor Dipl. Ing. Sigrun Dechêne.

The financial freedom generated by my freelancing enabled me to take up a consecutive diploma course in construction Management at the Faculty of Architecture at the University of Wuppertal in 1997 and to complete it by the summer of 1999. The study program in Wuppertal was offered jointly by the Faculties of Architecture and Civil Engineering. Theories and methods of sustainability assessment had already been researched at the University of Wuppertal since the late 1990s.

There was also fruitful collaboration with the Wuppertal Institute, where the MIPS concept (Material Input per Service Unit) was being developed at that time. The studies in Wuppertal also included lectures in real estate management, project development, project management and facility management, which were offered in cooperation with the teaching and research area of construction management. The topic of my diploma thesis was "Facility Management in the context of planning and utilisation of buildings." The university degree of Diplom-Ingenieurin (Dipl.-Ing.) qualified me then pursue a doctorate.

I applied to the Institute of Construction Management (IQ-Bau) under the leadership of Prof. Dr.-Ing. Claus Jürgen Diederichs with a research project in cooperation with the Real Estate and Building Department in Dortmund. I was hired as a research staff member and became a doctoral candidate.

After completion of the project assignment, I got an opportunity to work in the engineering and consulting company DU Diederichs. My main task was the time and cost control for the reconstruction and modernization of the Münchner Kammerspiele, so I moved to Munich in 2001. After the successful completion of the project in 2004, I was assigned to the renovation of the State Theatre in Darmstadt and commuted between Munich and Darmstadt. From 2005 I worked as a research and teaching staff member at the TU Munich at the Chair of Building Realization and Robotics under the leadership of Prof. Dr.-Ing. Thomas Bock. At the TUM I was able to resume and further develop the work on my doctoral thesis and focus it on the field of energy efficiency in existing buildings. In parallel, I qualified as an energy consultant and became a member of the Bavarian Chamber of Architects. In the summer of 2006, I became Professor in Facility Management at the Frankfurt University of Applied Sciences. At the same time, I further developed my doctoral thesis with the topic "Evaluation and increase of energy efficiency of municipal inventory buildings" and completed it in 2009 at the University of Wuppertal.

The Academic Career Begins as Professor in Facility Management in Frankfurt am Main

During my time at TUM, I developed my first contacts with international facility management academics. I nurtured these contacts further during the European Facility Management Conference EFMC 2005 in Frankfurt am Main. At that time, the Chair of Building Realization and Robotics, Prof. Dr.-Ing. Thomas Bock, was also becoming active in the field of Facility Management, and I took the opportunity to participate at the EFMC 2005 as a TUM scientist.

At the EFMC 2005 in Frankfurt I met Prof. Dr.-Ing. Tore Haugen, who was Dean of the Faculty of Architecture and Fine Art at the Norwegian University of Science and Technology (NTNU) at that time. I also met Prof. Dr.-Ing. Kunibert Lennerts, who had been appointed as Germany's first Professor in Facility Management at the University of Karlsruhe in 2000.

I established further contacts with the European Facility Management Network (EuroFM), the International Facility Management Association (IFMA) and the German Facility Management Association (GEFMA), which I knew from my diploma studies at the University of Wuppertal. Through TUM, I received an invitation to an international symposium in Trondheim.

The 2006 "Trondheim International Symposium – Changing user Demands on buildings, needs for lifecycle planning and management" took place at NTNU (Norwegian University of Science and Technology) from June 12 to 14, 2006. The symposium was organized in collaboration with Nordic FM and the CIB W70 commission.

CIB W 70 is the commission for Facilities Management and Maintenance of the International Council for Research and Innovation in Building and Construction (CIB). The CIBW70 2006 Trondheim International Symposium's theme on "Changing User Demands on Buildings – Needs for lifecycle planning and management." This linked together the different elements of FM and Maintenance reaching from the hard side with a focus on technical questions in maintenance and modernization to the softer side with strong focus on social, cultural, and organizational issues related to FM and Maintenance. (Haugen Tore I., Moum Anita, and Bröchner Jan 2006: Trondheim international symposium, Changing User Demands on Buildings – Needs for lifecycle planning and management 12–14 June 2006, Proceedings, NTNU).

The conference program included the following main topics:

1) Nordic FM – Meeting the future of FM
2) Building conservation and refurbishment
3) Changing user demands on adaptability and flexibility
4) Workshop Hospitals of tomorrow
5) Nordic FM – Operation and Service management
6) Sustainability in FM and Design
7) Workplace Management
8) EuroFM in Healthcare Network meeting
9) Usability of Workplaces
10) EuroFM Research Network Group meeting

In February 2006, I became the first woman to receive a Professorship in Facility Management in Germany. I became Professor at the Frankfurt University of Applied Sciences. The professorship was to be filled in the field of Facility Management at the Faculty of Architecture, Civil Engineering and Geomatics. The focus of my new position was to teach courses in all degree programs of the faculty, especially in the new degree program Geoinformation and Municipal Engineering. I was expected to have in-depth specialist knowledge and professional experience in various areas of facility management, including project development, project management, utilization management and construction management. At that time, I was not aware that at the age of 40, I had become the first woman to hold a professorship in facility management in Germany. From the winter semester 2006/2007 onward, I took on courses in facility management and project management, in particular building studies, management of municipal real estate, building management, and technical development, as well as supervising project work.

In December 2006, I gave my inaugural lecture "Blue overalls with pinstripes – Improving buildings in facility management." In it, I quoted, among other works, from the second edition of the book *Facility Management* (Edmond P. Rondeau, Robert Kevin Brown, Paul D. Lapides, Wiley, Jan 23, 2006 - Architecture - 624 pages), Chapter 10 "Successful FM" in paragraph "Moving toward the future" as follows: "In a number of organisations, facility management has moved from the boiler room to the board room. The facility professional, who was stuck with whatever design, building, furniture, or system he or she was handed, now leads the team that solves complex corporate facility problems." At the beginning of 2000, FM in Germany was still understood in technical terms, especially in building services and facility management in the German standard DIN 32736: Facility management (building management) comprises "the totality of all services for the operation and management of buildings, including structural and technical facilities, based on holistic strategies" (DIN 32736, 2000–08 edition (baunormenlexikon.de)). Later, the entire real estate life cycle was considered, for example, in the GEFMA 100 guidelines.

The overlapping areas of responsibility of real estate and facility management as well as architecture, construction, and engineering were documented in the service specifications of the Fee Structure for Architects and Engineers (HOAI) as well as in the publication "Facility Management-Consulting" of the AHO (Committee of the Associations of Engineers and Chambers of Engineers for the Fee Schedule e.V.) (AHO 2001).

Other leading publications from that time were:

- *Real Estate Management in the Life Cycle* (Diederichs 2006) as the second edition of the 1999 book "Führungswissen für Bau- und Immobilienfachleute" (Diederichs 1999),
- *Facility Management – Basics, computer support, system implementation, application examples* (Nävy 2006) as the fourth updated and supplemented edition of the first edition already published in 1997 (Nävy 1998),
- *The Facility Management Handbook* second edition (Cotts 1999), and the third edition *The Facility Management Handbook* by David Cotts, Cathy Roper, and Richard P. Payant (Cotts et al. 2010) reinforced my desire to gain international professional experience.

These books accompanied me as a source of inspiration in the early years of my professorship at Frankfurt UAS and are still stored on my home office's bookshelf. The development of facility management in Germany was closely linked to international developments and roots in the USA. I am very grateful that I was able to meet personalities and authors from the early days. In the numerous conferences and networking events, as well as international Scientific Committee, teaching, and research, I was able to intensify the scientific exchange and collaboration. The professorship at the Frankfurt UAS gave me new opportunities to work internationally.

My Research Semester at the Norwegian University of Science and Technology

In the academic year 2007, I organized a study trip to Norway. Students presented the results of their building science project "Frankfurt Office Guide FROG" in the context of the international facility management master's program at NTNU from September 17 to 23, 2007, in Trondheim. After the successful planning and execution of the study trip to Trondheim, my interest in fostering collaboration with the Norwegian colleagues grew, especially in the Centre for Real Estate and Facility Management at NTNU, and to further develop FM internationally.

Facility management students from Frankfurt UAS qualified with the project work "The Janitors job in the course of time" for the participation in the international "Student Poster Award in Facility Management" at the European Facility Management Conference 2007. The EFMC 2007 was organized by EuroFM and IFMA in collaboration with local universities and industry representatives and took place at the ETH (Swiss Federal Institute of Technology from June 26 to 27 in Zurich, Switzerland). The Conference Proceedings included the following ten main topics (EUROFORUM 2007: EFMC2007 – European Facility Management Conference 2007, Conference Proceedings, EUROFORUM Handelszeitung Konferenz AG, Zurich)

1) European Best Practice in the private sector
2) European Best Practice in the public sector
3) IT in FM – Space and occupancy costs
4) IT in FM – Implementation and Integration
5) FM Supporting Business continuity and risk management
6) Defining Core Business
7) Customer Care and Communication in FM
8) Trends and Innovation in FM – Standards, Performance and Measurements
9) Trends and Innovation in FM – workplace
10) FM Providing sustainable Buildings and a healthy environment

In 2010, I was the first German to be elected to the board of the European Facility Management Network (EuroFM).

In the autumn semester 2010/11, I received an invitation to spend a research semester from SINTEF and the NTNU Faculty of Architecture at the Centre for Real Estate and Facility Management. During the research semester I took the opportunity to get to know the Norwegian network, teaching and research projects intensively. The Centre for Real Estate and Facility Management was initiated and managed by Prof. Tore Haugen and team from 2002 to 2006 (The Centre for Real Estate and Facilities Management "Metamorphosis" – NTNU).

Public developers and real estate managers were interested in professionalizing the provision, maintenance and efficient use and management of the built environment. "Well-maintained buildings give added value for all" (NOU no. 22, 2004). A public sector publication and cooperating partners in the building administration as well as consultancies, FM industry associations, construction companies and international research partners in Scandinavia, the Netherlands and Germany worked together. A platform for the supervision of doctoral students was established at NTNU, as well as the establishment of two master's degree programs in real estate and facility management. When I came to NTNU in 2010, the establishment of the MSc Real Estate and FM was still in its infancy. From 2010, a professorship was advertised to strengthen the team.

Qualification for a University Professorship at the Norwegian University of Science and Technology

In December 2010, I became Associate Professor in Facilities Management at the Faculty of Architecture and Fine Art at the Norwegian University of Science and Technology and started to learn the Norwegian language.

The goal of the position, as stated in the job description, was to emphasize FM from a strategic and tactical management and organizational perspective. The focus was on the support and service functions important to running an organization, including planning, and managing operations and maintenance services as well as assessing user needs and quality levels. Facilities Management was considered a new academic discipline, and the professor would be involved in the further development of the field in collaboration with other partners. The professorship was at the Department of Architecture and Management, and the activities were associated with the Centre for Real Estate and Facilities Management.

At NTNU I took the opportunity to qualify for a full university professorship within two years. This path was supported by my international scientific engagement. In September 2010 I became chairperson of the research network group (RNG) in the European Facility Management Network (EuroFM). As this leadership role would be instrumental in building a research agenda in Facility Management recognized throughout Europe.

In May 2011 I assumed responsibility for the development and organization of the internationally recognized 11th EuroFM Research Symposium, held at the European Facilities Management Conference (EFMC) in Copenhagen May 24–25, 2012, and attended by more than 500 participants. As Chairperson of the Scientific Committee, I was responsible for the call for papers and managing the double-blind peer review process. In addition, I co-edited the research symposium conference proceedings.

My work as RNG Chairman included responsibility for the development and organization of six RNG meetings within two years, held in Brussels, Vienna, Munich, Kufstein, Copenhagen, and Trondheim.

Moreover, I have been an active member of scientific committees at international conferences. For example, the 12th Research Symposium at the European Facility Management Conference EFMC 2013 in Prague May 23–24, 2013, Member of the Scientific Committee, and referee for double-blind peer review for the 7th Nordic Conference on Construction Economics and Organisation, June 13–14, 2013 at NTNU, Trondheim, and the 10th EuroFM Research Symposium at EFMC 2011 at the TU Vienna. As of November 1, 2012, I have been University Professor in Real Estate and Facility Management at NTNU. The appointment was confirmed by letter dated June 26, 2013, and became effective retroactively. I had reached my career goal at the age of 46 – I thought – however the way is the goal.

The qualification was assessed by an international review panel. I had worked for two years in the Faculty of Architecture at NTNU, in particular developing and teaching the FM and sustainable facility management modules in the Master of Sciences in Real Estate and Facility Management (MSc REFM), as well as supervising project and thesis work of two PhD students and serving on PhD review panels at NTNU and internationally.

In addition, a research project in pertaining to energy efficiency improvement of non-residential buildings (MINDER) was successfully acquired. Furthermore, I used the time to intensify my international network, especially CIB, Passive and Low Energy Architecture (PLEA) and EuroFM. Travel and international experience as well as employment of a research assistant were supported by NTNU and the Faculty of Architecture and Fine Art.

The MINDER research project focused on energy efficiency issues and was interdisciplinary in nature. Researchers from three faculties – Architecture, Social Sciences, and Product Design – worked together. PhD candidates were supervised within the framework of MINDER. There was also good cooperation with SINTEF (The Norwegian SINTEF is one of Europe's largest independent research organizations), the Centre for Zero Emission Building (ZEB Centre), and other research projects in the field of sustainability, e.g., the Broeset project and the work of NTNU Sustainable. Common to all was the search for energy and resource saving solutions for the built environment. In ZEB, business partners were represented alongside SINTEF and NTNU. Sustainability was also studied in 2016 during the research semester at the UC Berkeley, where I worked on sustainability rating systems at the Department of Civil and Environmental Engineering at the Project Production Systems Laboratory (P2SL) directed by Prof. Dr.-Ing. Iris Tommelein. The P2SL at UC Berkeley is a research institute dedicated to developing and deploying knowledge and tools for project management.

Review and Outlook and the Creating of the Ways into the Future

My journey toward becoming an academic in Facility Management required deep understanding of the international state of the art. It has been enormously rewarding to enrich my own context of experience in science and practice with inspirations from other cultures and languages. Learning foreign languages and integrating into other cultures and countries provided deep insights into both the facility management industry and the associated scientific discipline.

I grew up in Dortmund, studied and worked in Dortmund and Wuppertal, Germany. I gained my first experience abroad in 1987 during my architectural internship while studying in Rotterdam, the Netherlands. Within Germany I moved from Dortmund to Munich for professional reasons and then commuted between Darmstadt and Munich and from 2005/6 between Frankfurt am Main and Munich. I was tenured Professor in Facility Management at the Frankfurt University of Applied Sciences. In Norway, I was hired as Associate Professor and after two years, with appropriate scientific achievements, I was promoted to University Professor in Real Estate and Facility Management.

Facility Management was still little known in Germany at the end of the 1990s. The German Facility Management Association (GEFMA) was one of the first German-speaking FM associations. In a time of cost- and space-saving construction at the end of the 1990s, FM became known in German-speaking countries. FM received a lot of attention with respect to operating cost optimization and outsourcing of internal FM services. The new approach to public administration management supported efficiency and cost effectiveness efforts. The municipal office for administrative simplification (KGST) was founded in North-Rhine Westphalia. With the Energy Saving Ordinance, European efforts to save energy and increase efficiency were added. Modernization and maintenance requirements for existing buildings resulted from the energy consultant training courses conducted by the Chambers of Architects in the Federal States in Germany.

The first FM course in Germany was established at the Albstadt Sigmaringen University of Applied Sciences. This was followed by around 16 other GEFMA-certified universities of applied sciences and universities. In 2000, a first university professorship in FM was filled at the University of Karlsruhe. GEFMA established a professors' working group. The European FM Network (EuroFM) had around 45 university members in 2010. As part of a research project on the "European FM Research Agenda," the universities with the strongest research were identified in Scandinavia, the Netherlands, and Switzerland.

Today, the impression is that technical solutions "Smart City" or "Digital Transformation" and process optimizations are in the foreground. The next wave is already on the horizon and will have to bring people back together with a technically optimized environment in order to develop, manage, and use healthy living and working environments. To this end, psychology, sociology, and environmental sciences will gain in importance.

From 2005 onwards, I worked on combining technical, commercial and management competencies. I had gained relevant professional experience in user requirements planning and client representation. I was able to use this scientifically and develop it in teaching and research projects. Since my appointment as professor at the Frankfurt University of Applied Sciences, I have consistently pursued an academic career path. In real estate and facility management, I was appointed Professor at NTNU and qualified there from Associate Professor to full University Professor. In the Teaching and Research Centre for Real Estate and Facility Management at NTNU, task focuses pertaining to public sector responsibilities had to be solved. It was expected that technical solutions, holistic strategies and the methods for budget planning and control for the building and infrastructure would be developed and implemented.

Facility management goes beyond the traditional scope of services provided by architects and civil engineers in the planning and realization of buildings. In FM, needs are continuously identified, analyzed and improved with foresight in order to maintain a high quality of use and availability.

In laying the stepping stones to my future and in doing so I have built my personal academic career path, which has led me to international universities in different countries. Financial conditions were difficult paving my academic way in the beginning, but became firmer and more settled over time and took a clear direction since becoming Professor at the Frankfurt University of Applied Sciences. The path-building has changed me; I have become stronger and grown from the demanding nature of my experiences. The direction has changed from Germany to Norway to Switzerland. In 2016, an opportunity arose to combine the professional and family location, and to take up a new challenge. I became Director at the Institute of Facility Management (IFM) at the Zurich University of Applied Sciences (ZHAW) and moved to Zurich.

My sincere thanks to Edmond Rondeau for inviting me to describe and helped me to understand my way toward FM and calling to mind an old Scandinavian wisdom "Life is lived forward and understood backward." Soren Aabye Kierkegaard (1813 – 1855) Danish philosopher, essayist, and theologian.

References

Diederichs, C.J. (2006). *Real Estate Management in the Life Cycle*, 2e. Springer.

Cotts, D., Roper, C., and Payant, R.P. (2010). *The Facility Management Handbook*, 3e. American Management Association. ISBN: 978-0-8144-1380-7.

Jopling, J. (1999). *Pathways to the Future – Thinking Differently*. London: Sustainable London Trust, 136.

Nävy, J. (2006 January). *Facility Management – Basics, Computer Support, System Implementation, Application Examples*.https://doi.org/10.1007/3-540-36670-9. ISBN: 978-3-540-25164-4.

Edmond P. Rondeau, Robert Kevin Brown, Paul D. Lapides. (2006 January). Wiley – Architecture, 624.

Antje Junghans, Professor Dr.-Ing.
Director, Institute of Facility Management (IFM) at the
School of Life Sciences and Facility Management (LSFM)
Zurich University of Applied Sciences (ZHAW).
Zurich, Switzerland

Professor Dr.-Ing. Antje Junghans has been involved in the construction and real estate industry for over 30 years. Her current focus is managing the Institute of Facility Management (IFM). As director of the institute, she has scientific, personnel and financial management responsibility, and is a member of the School of Life Sciences and Facility Management (LSFM) at the Zurich University of Applied Sciences (ZHAW).

Antje Junghans has many years of professional experience in university teaching and research. She is and has been professionally active particularly at traditional universities and universities of applied sciences in Switzerland, Germany, and Norway. In addition, she has established herself nationally and internationally as a consultant, reviewer, and project manager. Consulting and expert activities include, for example, work for government departments and agencies with respect to accreditation and appointment procedures.

Professor Junghans many years of experience includes architecture and urban planning, construction project management, real estate, and facility management. Her professional interests are the analysis and improvement of functionality, design quality, and economic efficiency of the built environment. She has worked as a researcher and lecturer at TU Munich, RWTH Aachen, and the University of Wuppertal. At the University of Wuppertal, she completed her doctorate while working, and received her PhD (Dr.-Ing.) with her dissertation entitled "Evaluation and enhancement of energy efficiency of municipal inventory buildings." In 2006, she was appointed professor at the Frankfurt University of Applied Sciences (Frankfurt UAS).

At the Frankfurt UAS, she taught and researched in the Faculty of Architecture, Civil Engineering, and Geomatics, especially in the teaching and research areas of facility management, computer aided facility management (CAFM) and sustainable building construction. In 2011, she was appointed professor at the Norwegian University of Science and Technology (NTNU). In this role she engaged in teaching and research on sustainable real estate project development and management in the Centre for Real Estate and Facility Management at the Faculty of Architecture and Fine Art at NTNU. She delved deeper into research work in sustainable facility management in a research semester at UC Berkeley and since 2016, Prof. Dr.-Ing. Antje Junghans has been institute director and professor at the Zurich University of Applied Sciences (ZHAW).

2.18

Review of FM in LatAm

Joint Contribution By Norma Pleitez and Eduardo Becerril

Norma Pleitez

Latin-American market is always challenging because of their different countries, languages, laws; and although commercial needs are quite similar the way to do business is very particular in each country, something that I as a Business Administration professional learned to managed and prepare myself to get a postgraduate in project management. In the upcoming years I served different types of corporations giving and receiving learning to implement and properly improve the Facilities Management model.

With more than 400 million square feet under formal Facilities Management in the market in the region provides a diverse portfolio such as commercial, technology, corporate, retail and logistics, managed by a small group of Real Estate firms. The consolidation of FM is giving us the opportunity to expand the business and orchestrate the adaptation of Facilities Management in a well-prepared premises managed by qualified professionals.

For 13 years I managed the Microsoft Central America and the Caribbean portfolio to create Facilities Management processes including: setting Service Level Agreement's and Key Performance Indicator's, measuring performance and analyzing the data to implement budget control while satisfying customer experience and act as a Subject Matter Expert in Occupancy and Project Management for the LatAm (Latin American) region. Based on that experience, I act as a private consultant for micro and small businesses including ISO certification processes and becoming an Internal Auditor for the ISO 9001:2015

During the past five years I have developed the roles of: Senior Occupancy Planning, Regional Facilities & Workplace Leader, and Real Estate Portfolio Management for the corporate premises of multinational companies across the Latin America region. The management of these roles helped me to challenge myself and be part of the IFMA Mexico Chapter and a founding member and active board representative with the commitment to help promote the FM industry in Mexico. During this time, I decided to obtain the IFMA CFM certification, being the first woman in Latin America to have achieved it in February 2022.

During more than 18 years of experience in FM, I have faced the evolution from being an Office Manager more focused in administrative tasks to become a specialized Facilities Manager that as the years go by, new technologies, competencies, and specialized hard services training were developed. Actually, we are facing due to the changes in the ways of working, sustainability commitments and new FM tendencies, a need to improve in quality and performance, learn more about being more efficient with less resources and special focus in sustainable actions. It's a challenge that all FM professionals must develop to be successful in this new era.

Facilities @ Management, Concept - Realization - Vision, A Global Perspective, First Edition. Edited by Edmond Rondeau and Michaela Hellerforth.

Another challenge that we have is to make the FM women population grow, at this moment women represent only a 20% of labor in Facilities Management globally and around 12% in Latin America. (According the Global Salary and Compensation report the IFMA)

We have a commitment to help and support women to learn and grow in FM, Project Management, Sustainability, and other relevant competencies of our Industry. From technical to executive skills opening a broader spectrum of women in business of FM across the region, this is not just a new trend but a reality to embrace the diverse inclusion of women in FM.

I perceive a future of opportunities based on our experience, background and adapting ourselves in a constant disruptive market in LatAm that will involve more prepared women to deliver the highest level of service and innovative solutions to owners, occupiers, and users in a variety of different type of portfolios. We'll see in the upcoming years a strengthened workforce of women in FM, a group of young professionals and experienced woman will mark the difference on how Facilities Management are tailored, sold, and implemented by networking and educating in professional organization. A good example of that is the IFMA Mexico Chapter which is helping professionals to enhance, elevate, and increase their industry knowledge, networking and exposure to the market proposing to the FM world a better result.

For example, in El Salvador when I was with Microsoft, I always worked to have the best-in-class suppliers and I've seen other companies like MS, a high-level company that seeks that their suppliers have certification of the ISO 9001 Quality certification to ensure similar processes, systems and staff training. This consistency is valuable for existing local or global companies and for business expansion as well. This demonstrates that quality is a basic service that comes with certification, processes, systems, and people coordinated on deliverables measurement; once the data is accurate and properly analyzed, Facilities Management professionals can take appropriate decisions.

Eduardo Becerril

As an Architect during my early steps in designing, I dedicated all my efforts to be a renowned designer after being granted the design for residential houses, apartment buildings, manufacturing facilities, furniture including a church, most of the designs I planned were built during the nineties and remains as a recollection of a great era, at least for myself and the happy clients that trusted my inexperience but bold ideas. One comes to my mind now. I had a very peculiar request as the design for my grandmother's gravestone as she passed away before I finished college. Certainly, this was a challenging request that opened my eyes to have a more human touch on the specific designs I'd worked on during those years and remains nowadays as the most relevant actor of every space and that is people.

I was focused on servicing the feeling of aesthetics in designing from the outside and the fresh look of simple, functional, and not so often used shapes, colors, lights, and shadows effects and a comfortable sense to form as part of the buildings.

Designing days turned into construction days, taking the risk, and endeavoring at my young age I opened a contractor business. As a 21-year-old architect I began to build houses and furniture, made great relationships with customers and in some cases I learned from mistakes on all the complexity of having several workers, with the responsibility to pay wages and collect payments while dealing with several unimaginable things, then the unfavorable economic environment in Mexico in 1994 hit my small business but not my great world.

I learned the three main areas of expertise an architect can develop and serve as interlocutor between buildings and people, Design, Construction, and certainly Administration, as I had to find a way of keeping and maintaining my involvement professionally in the real estate realm. I was doing some small renovation projects but looking at premises administration, that's the name I knew during those days for what became the formal Facilities Management.

Life opportunities really happen either personally or professionally. Then during a lunch I was invited to, I met a Sr Director of a really small bank with the ambition to immensely growth from 150 to 2,000 branches in less than three years. During the first years that I worked for this bank, I was looking for land, retail spaces to rent, coordinating designs and construction, preparing reports and once in operation find at each branch the best way to maintain each branch under specific conditions. Eureka, the whole package is actually Facilities Management. In our own understanding all the Real Estate group of the bank were under the same goal, challenging, and organized to deliver results.

Early in 2000 globalization hit us with a new reality, a huge global bank acquired the formerly small bank that grew exponentially in a few years in Mexico. With this change, new policies, regulations, and protocols began to spread across the organization; one in particular represented what I was with or without expecting it, the great opportunity to professionalize all I was doing. Fortunately learning from other regions and apporting my experience the global bank gave me an assignment: Let's find a FM expert firm to manage the bank's portfolio to outsource all services as Real Estate is not the core business of the bank. Well said I thought as we must get the experts to obtain the most value of this huge portfolio.

Long story short, I participated in developing a very interesting RFP to select the expert in FM to manage the bank's portfolio. Well after getting all the eye-opening details and elevating my ambitions, I raised my hand to participate as one of the service provider executives instead of being the client receiving the service. I started my career at JLL, one of my greatest decisions in my professional life. I started knowing the true Facilities Management world, attended seminars, meetings, and expanding this for us, a brand-new business of FM in LatAm.

I met IFMA early in 2003, and soon after I went through the endeavor of getting my CFM which I got the credential in the same year and keeping recertified every three years. Surprisingly I was the only CFM in all LatAm for almost 20 years. One of the goals I set to myself was to increase the members, create an IFMA chapter in Mexico, and get more professionals certified. This turned out to be a long way with a many challenges until I was able to gather a group of pioneers, and 2019 was the great opening of IFMA Mexico Chapter. We held a celebration and planned for 2020 activities, with many challenges in front of us as the pandemics hit globally and many things happened.

Other professionals from different groups of global and local companies were actively involved in the Chapter, and I was elected President of the chapter. We had great news about two professionals getting their CFM certification: one independent contractor and consultant, and another was Norma Pleitez the first woman in all LatAm to get CFM certification. What a way to start moving the needle.

Academic efforts began as we were having webinars, training sessions, professional gatherings, sharing experiences, and growing the interaction with the industry of Facilities Management in Mexico and LatAm.

The decade of 2000 was mainly the process of understanding and educating ourselves and building owners, users, and occupiers about FM. We called this the evangelization period which gave us the opportunity to grow from almost scratch to a larger group of These clients were receiving services and professionals were providing FM in a very organized manner. Larger Real Estate firms and local Facilities Services companies embraced the concept and implemented the model. We were growing and making noise, that good noise was based on consistency in service deliverables, normative compliance, human resources organization, and specialization, budget control, and operational consistency.

The 2010 decade began with a boom of FM services growth across 21 countries in LatAm. Ten years after consolidating and gaining experience prepared us for the pandemic's challenges, well-being, health and safety, protocols, space utilization, and flexible work. Keeping essential business up and running was possible because we were very well prepared to embrace the challenge at the same time uncertainty became the day-to-day activities. With decisions based on best trends understanding we aimed to properly utilize building space. The year 2020 and beyond is here, and once we thought about the future of work, we were imagining how the space utilization, technology, flexibility, and sustainable operations will become more the path to follow at all types of corporations, the use of space and how it is managed and is changing.

We as FM professionals, practitioners, academics, or students must change our mindset to aggressively adapt ourselves to this disruptive change. Again, flexibility is required in a constantly changing world.

We concur with the experience and results achieved during all these years in the FM Latin America region based on the strong platform created. We are ready to oversee and embrace the new tendencies in our industry, with special focus for the needs in our region has:

- Sustainability commitments.
- Alternative energies usage.
- Smart buildings implementation.
- ESG awareness.
- Improvement on quality and performance.
- Elevating FM's education and certification.

The most difficult task is to obtain accurate projections on what is expected to happen in the short-, mid-, and long-term due to the uncertainty that actually we're facing post pandemic plus the challenges that all companies are having environmentally, economic and social complexity, and at the same time the easiest task is to be prepared and adapt to change as Facilities Managers.

Norma Pleitez, CFM
Real Estate Portfolio Management Leader
JLL
Work Dynamics, Mexico

Norma Pleitez joined JLL in July 2009 and is currently serving as the Real Estate Portfolio Management Leader Latin America for Procter & Gamble account based in Mexico City. Norma is responsible to drive, formulate and the delivery of Real Estate portfolio strategy; develop workplace solutions and leading client-facing project engagements around the future of work, portfolio optimization, workplace planning and execution.

Norma has wide experience in Facilities and Project Management. She was responsible for the LatAm IFM and Operations Team, supporting client facilities of 44 sites in 14 countries across the Latin America Region.

She was responsible for the planning and execution of the annual operating budget, supporting the projects, transactions, space optimization, and portfolio strategy to achieve the client's strategic plan in Latin America. She also managed the Caribbean and Central America Region for the Microsoft Account where she develops the SLAs, new FM processes, KPIs' implementation, budgeting control, and the Client Satisfaction program for the account.

Norma holds a master's degree in business administration and a postgraduate in project management. Has more than 15 years' experience in Corporate Real Estate in Latin America. Recently has received the Certification in Facilities Management (CFM) from IFMA; being the first woman in Latin America to obtain the certification.

Eduardo Becerril, CFM
Executive Vice President, Country head
JLL
Work Dynamics, Mexico

As Director for FM Services for JLL, Mr. Becerril is responsible of overseeing 45 plus accounts in Mexico with an operating budget more than US$200 million and a team over 2000 service professionals. His main duties include ensuring financial and operating performance, as well as client management and team development functions. He brings over 30 years of relevant experience, including architectural projects, real estate administration, and facilities management for retail, corporate, and industrial portfolios.

Prior to his current position, Mr. Becerril acted as account director for a large retail base portfolio encompassing more than 2,400 sites throughout the country; also, he has held similar positions for Philips, Solectron, Honeywell, and Motorola covering more than 12 countries and a diverse team of multicultural professionals. In this capacity Mr. Becerril has been accountable for delivering and exceeding KPIs' performance, budget management, and savings delivery.

Before joining JLL Mr. Becerril headed HSBC's Corporate Real Estate for Mexico's Northern Region, where he oversaw facilities management, appraisals, property and project management, new branch site selection and construction, also the Health Fire and Safety department (code implementation). Mr. Becerril has developed architectural projects for Hotels, Housing, and Pharmaceutical facilities.

Mr. Becerril holds a bachelor's degree with Honors in Architecture from the Escuela Nacional de Estudios Profesionales Acatlan, UNAM in Mexico City, where he graduated in 1993. He is a recipient of the "Alberto J. Pani" award, a prestigious architectural design accolade. In 2000 he received from the Instituto Tecnologico de Estudios Superiores de Monterrey a Real Estate Administration Diploma. Mr. Becerril is also an active member of the International Facilities Management Association and a Certified Facility Manager since 2006, and also serves as IFMA Mexico Chapter founder and president.

Mr. Becerril is a retired Mexican Red Cross volunteer, he is active member of the Mexican Scouts Association, and his general interests are outdoor activities and aviation. He manages English as a second language.

2.19

About the Many Human Factors in Facilities Management

Christine Sasse

People make facility management. Facility management makes people.

Two banal sentences? On no account. They reflect the experience we have had in the Sasse Group since our company was founded almost 50 years ago. At the very beginning, the founder, Dr. Eberhard Sasse, and his team went to a Munich bank branch to clean there. Overnight, they removed the dirt and waste left there by customers and employees. A work that is measured almost exclusively by the result. Those who carry them out remain invisible. When the night is over, they are gone.

That sounds like "interchangeable." After "low value." After "any." None of these attributes, however, strike at the core of this activity at the lower end of what we now call facility management. The young entrepreneur Sasse learned this at an early age. He saw that the good business opportunity he had seen for the leap into self-employment had to evolve. That she needed added value.

One of these added values was the alert mind of his team (and himself). With open eyes and curiosity about the business of the first customer, they discovered opportunities to offer additional services. At the same time, they recognized that the newly acquired application knowledge from the bank can also be transferred to other sectors of the economy. Today, we would call this a "learning organization" – a characteristic that is generally not expected of companies in such a simple service sector as building cleaning. Except for the people who work there. They have to and want to prove that their performance is not arbitrary, interchangeable or inferior.

What has made the difference at the Sasse Group from the very beginning: We are a family business. With their name, their assets and their understanding of public spirit, the owners are right in the middle of day-to-day business. Families, we know, look at the people who belong to it with a different view. This is true in private; this is just as true in business. You adapt to individual characteristics and abilities. You learn to listen to others and communicate with them in their own personal way. You can see where guidance and help are needed and where there is maturity for someone to take on their own responsibility in the next step. And, most importantly, looking at the reactions of relatives teaches you what goes down well and what doesn't.

In the almost five decades of the Sasse Group, it has been proven that these various "human factors" are ideally suited to developing an efficient workforce, building mutual trust and convincingly advancing new ideas. This is best seen where employees have been "on board" for 10, 20, or even 30 years and have developed from simple specialists to highly motivated managers. In an industry where many people pay attention to technology first, then target the methods, and finally negotiate bills of quantities and prices, people make the difference between "being there" and "designing." People make FM – and FM causes these people to make something of themselves.

Facilities @ Management, Concept - Realization - Vision, A Global Perspective, First Edition. Edited by Edmond Rondeau and Michaela Hellerforth.

Family Businesses, of Course, Focus on People

For the Sasse Group, the characteristics "family business" and "service provider" are a single unit. There are 8800 people working in Germany and Europe who feel they belong to the Sasse Group. This makes it clear at first glance that family and service are closely linked in our self-image. This is particularly evident in our people-focused attitude at all levels. The owners see themselves as partners and service providers for customers and employees. The employees support each other as a team. And everything we do for our customers should help people to feel safe and in good hands at their workplaces in the long term. This principle of "people business" with its many dialogues and the continuous exchange of knowledge is at the same time the foundation for our distinctive innovative spirit. We reinvent FM practically every day.

Enthusiasm Through Mutual Participation

This not only offers us opportunities to compete again and again. This also makes us innovators. This is a quality with which, among other things, we want to inspire young people for FM.

Creating this impact on the outside world is not an easy task in view of the many prejudices against our industry. Undoubtedly, it is easy for us to feel enthusiasm for FM ourselves. After all, deep down we are convinced of what a varied, exciting, innovative industry we are. But how do we get this message out there?

Let's take a look at the situation on the personnel market: Today, it's no longer just about getting young people or experienced lane changers excited about FM. Especially if we are looking for someone for leadership roles. For newcomers today, completely different things play a role than the industry in which they work. They look at the company culture because they are looking for a say and responsibility. Because, as beginners, they also want to be insiders, not followers. They take a close look at how to reconcile private and work life. A family that succeeds in doing this itself is seen in a special light. She is recognizable as a role model.

What we as owners, but also all our managers, pay attention to is more than a well-calculated "work-life balance." We are already a few steps further. Strictly speaking, four:

1) Attractiveness of the location, including opportunities for life partners and children.
2) Affordability of attractive housing.
3) Because employment relationships today are mainly considered temporary, salary also plays a very important role.
4) And, as resigned as that may sound, we are keeping an eye on the prospects for the next change of employer at the FM employer's location. After all, poaching has become a reality of business life. We have sometimes seen this in tenders.

In a market where employees decide where they want to go, we can't stand still rooted to the ground. Optimistically, we operate a very fast-spinning merry-go-round, where someone jumps up and someone jumps off again and again. Conclusion: There will be winners and losers among employers.

But, fatalism is the wrong attitude. Because it is precisely when it comes to the framework conditions that we have immense creative possibilities. We have to use them and not bunker them "in stock." For all the factors mentioned – schools, housing, partner career – we can and must activate our network. We have to contribute our know-how. We are most convincing where we live honest values and make them visible.

Just think of the current discussions on conserving resources and energy efficiency: Who, if not FM, has the know-how? Who, if not FM, has the data on how our customers can solve the tasks at hand? Which brings us back to the enthusiasm: FMFF – Facility Management For the Future, that is our message.

Is There a Difference to Other Industries?

At this point, some may ask themselves: What is the difference between FM and other industries that have to make just as much of an effort to recruit junior staff? If you look closely, you will see that we have developed a plan to manage the "FM" facility.

Wanting to Make the World a Little Better is Not Only a Concern for the Younger Generation

Let's stay with the people. Let's stay with the enthusiasm for FM. Let's stay with the question of which specialists we want to use to provide our services in the future.

When it comes to the shortage of skilled workers, our image does not depend solely on the fact that there is no apprenticeship for our service. This does not matter, if only because the number of trainees and junior staff is currently shrinking in every industry. So, the task is to attract young people in the first place. Young people for whom the consideration of "study or training?" does not necessarily lead to university.

Both material arguments and idealistic values count. It's not enough to post a few fancy videos on social media. Or to find well-balanced words at public events and to announce a corporate culture that may not be as flawless as we would like it to be on a daily basis.

Rather, it is important for this target group that there are companies that act credibly and verifiably in such a way that the world becomes a little better. It's not just the young people who feel it, we feel it ourselves. That's what we have to work on, we have to connect it with our brands – and with our work as an FM service provider. Ideally, we turn each of our employees into brand and value ambassadors. This is also worthwhile in the business relationship. This is because it benefits our clients' CSR and ESG balance sheets.

Let's take a look at our technicians. They make the most important contribution to our work. You adjust the air conditioning system so that consumption and effect are in balance. They prevent wear and tear and thus waste. They recognize when there is a gap between aspiration and reality. This is a strong asset of our industry in the skilled labor market. Especially because we can offer work in alternating and diverse scenarios.

That's why we can once again positively charge the keyword "flexibility." It is no longer a pure "must," but also a "may and can."

We see that before we talk about markets and technologies, about strategies and perspectives, we have to think about the people with whom we want to implement all this. Family businesses have the experience on their side here, how their "own business" has been designed and managed over generations. Families are no less demanding than facilities.

We Work for the Long-term Preservation of Values

Families have internalized another challenge of our time: sustainability. Especially when they are entrepreneurial, the focus has always been on the opportunities of future generations. The question arises no differently for a mechanical engineering company than for an FM service provider: What can, and must we do today so that our children and grandchildren can do well with what we have built?

This is particularly evident in one detail: whether it is machinery, systems, plants or real estate: Facility management, as the Sasse Group understands it, ensures long-term value retention for the investor whose properties we manage. At the same time, they directly contribute to the creation of value in every workplace for our customers. Our services are therefore always holistically oriented. They integrate the economic goals and social responsibility of our clients. That is why they cover the entire life of an investment as well as all the factors associated with it.

Sustainability Asks for Application Knowledge

Under these circumstances, the principle of sustainability shapes our thoughts and actions. In our own work, we use resources in an environmentally friendly and economical manner. We share the experience gained with our customers. We are there to support their own sustainability strategy with data and analysis. This also includes sustainable dialogue and exchange of ideas. From this, we work together to develop better solutions for ever new challenges on the market.

In connection with the energy transition, the increased pursuit of resilience and the outlook for the EU's "Green Deal," we see an increased need for our extensive know-how in this field. Practical examples of this can be found in the preventive maintenance of systems, in the continuous optimization of building services or in the avoidance of waste or emissions. The flexible adaptation of processes as well as the short-term expansion of the range of services – as we implemented them for our customers in resilient hygiene concepts during the pandemic – also contribute to creating long-term sustainability that is responsible for our business.

It is worth repeating at this point: As a family business, this view beyond the here and now is a matter of course for us. In this way, we create the basis for the competitiveness and acceptance of the work of future generations. In line with this, we naturally use innovative technologies and processes to the greatest possible extent. With the help of digitalization, we transform data into knowledge, knowledge into innovation, innovation into communication, communication into service – and everyday life into sustainability.

How Does Digitization Fit into Our Concept?

How do the focus on digitization and the appreciation of human performance fit together?

Digital transformation has been an issue for our company for a long time. In facility management, we take care of plants and objects in which the digital components have been becoming more and more prevalent for years. We have also integrated digital tools into our own organization since they became available. This is anything but an end in itself: we see them as up-to-date and efficient tools that we can use to support our employees and increase the quality of our work.

That's why the digital transformation in the Sasse Group has a particular impact on our work processes. We use digital tools to examine and analyze these processes, then use other tools to redesign them – and still others to communicate the change to our teams and customers. When it comes to education and training, we as a company, but also our employees, appreciate the additional flexibility and intensity that is possible digitally in education and training.

Digital technology is a service, it is a tool to make it easier for our employees to do their daily work.

The central aspect – not least in terms of sustainability and resource efficiency – is the data cycle with which we work. We collect and evaluate the data we receive, for example, from plants. From this, we determine potentials for energy efficiency. We share these with our customers. They then adapt their systems. This, in turn, changes our processes. Then new data is created again.

Nowhere else is it so clear that digitization is not a goal, but a process that always begins anew. This is precisely why we are participating in a project of KIT Karlsruhe that aims to evaluate the degree of building intelligence. These are aspects that will influence our day-to-day business as well as our strategic orientation much more in the future.

For all employees who have these insights, this progress lives and in full at hand, this means an enormous increase in the quality of the workplace. Technologically, the distance to our customers is hardly perceptible.

Facility Management Must Be Understood as a Driver for the Circular Economy

In the field of tension of technological change, FM finds itself in a situation that not only needs highly qualified employees – as described above. As an employer, we are also called upon to establish employee acceptance for digital tools, for dealing with sensors and robotics, for helping to shape change that may make proven knowledge and skills obsolete. In this way, we achieve the subdiscipline of Corporate Social Responsibility (CSR) that is not textbook and for which there are hardly any blueprints.

Circular economy is an age-old topic for us as an FM service provider – and as a family business, it has already been done twice. Every work we do, every decision we make, is designed to ensure that we take responsibility for it in the long term. We have been working on this basis for almost 50 years since the Sasse Group was founded.

It is also a question of profitability to work sustainably and to take responsibility for the social, natural and economic environment. This is how we protect the people and markets to whom we owe our business success. Once again, reference should be made to the KIT project and the Smart Readiness Indicator (SRI) of the European Union. This is an approach to reduce the digitization deficit in the real estate industry and to pave the way for sustainable cooperation between owners, users, and service providers. We need such approaches much more in order to anchor the topic with all stakeholders.

The fact that there is an attitude behind "sustainability" and not the use of cryptic abbreviations in PowerPoint presentations is part of it: Experience in sustainable work is an ongoing process – and if the company management exemplifies it – a regenerative energy for all employees. How comprehensive this is and how deep it is can be seen at the latest when working on the CSR report. There is no aspect where FM is not visible somewhere.

How Social Responsibility Becomes Systemic Relevance

FM, as the Sasse Group understands it, has a social function. This starts where we contribute to "well-being" at the workplace of the employees on behalf of our customers. This can be felt where we increase hygiene and safety in vehicles for public transport companies. And we can prove this where we contribute to the conservation of resources and more economical energy consumption with our work, keyword "sustainability."

Whoever exerts influence in so many places is relevant for the whole system.

The FM industry has been able to directly demonstrate its systemic relevance in the past pandemic years, as its employees have ensured that the relevant infrastructure – from hospitals to office buildings – has continued to function smoothly. But there is often still a lack of understanding of what is necessary for buildings not only to function, but also to offer added value for users.

However, we are also socially relevant when it comes to integration and diversity. If we did not know how to leverage the potential offered to us by integrating motivated and enthusiastic people – then we would have more than deserved the gaps that arise from the shortage of skilled workers. For an FM service provider that wants to stay on the ball and develop further, these are no longer points of discussion, but crystal-clear standards for action.

The (socio-)political task is to facilitate the step over the threshold from unemployment into the individual opportunity of a (initially) low-wage job. Hurdles such as the immediate entry into compulsory social security and tax liability must be cleared out of the way.

The FM industry is a tailor-made professional field to promote social integration and personal development. Regardless of the personal background of employment, an FM job is never the "last resort," but always the "first step" to find a place in economic and social life in the long term and to shape it with one's own resources.

Our great advantage is that the FM industry can bring its employees to a relatively high level of qualification even without dual training. This is because it has a wide range of tasks in which individual talent can develop in a narrow-band manner – within small and manageable teams in which qualifications complement each other and not numbers, but people who work together. This makes us attractive employers beyond the 08/15 spectrum. Comparable to team sports, it is the "supplementary players" and "specialists" who ensure a balanced quality of performance. Once the first step "onto the pitch" has been taken by entering the career ladder, individual opportunities for advancement open up in FM as well as in sport from ongoing training and in-depth qualifications.

Where and How Do We Find the Ways to the FM of Tomorrow Today?

What is the consequence of all these analyses and considerations for us? We work with deep conviction to transform our company within the transformation of our industry.

Our first step is to get facility management out of a mental corner. Too many clients (still) see our services as a colorful and arbitrary sequence of individual activities. In day-to-day operations, on the other hand, a different situation has long prevailed: everything is linked to everything else. For example, in procurement processes for production or in supply chains, there are direct dependencies between a wide variety of trades.

Making these dependencies visible is the task of "FM 21" – a holistic and transparent facility management for the twenty-first century. In which automation and digitization play the role they deserve as accelerators and communication platforms. We need them to bring about desirable positive interactions. Just as we need them to ward off undesirable negative consequences.

We have seen convincing examples of this new way of looking at things in the recent past where we have dealt with the pandemic and its consequences. The short-term stop-and-go, the political and scientific U-turns: they have brought us to the edge of the predictable and predictable world. We were able to hold our own in this situation because we were able to pass on existing knowledge within the company, across business units, directly to the teams. Firstly, this has enabled us to find innovative answers to new challenges quickly and flexibly. And it has provided our customers with reliable solutions that would not have been possible with a narrowed view of individual trades or tasks.

So, what is important when nothing goes as usual, as planned and as trained? On the willingness to change and on the ability to adapt the existing application knowledge to new tasks. We have found convincing answers to crises wherever these forces have come into play – and where we have developed holistic solutions with customers on an equal footing.

We Continue to Develop in Parallel with Our Customers

We have used this knowledge to review and realign our structures and processes. Automation and digitization have matured into helpful and resilient tools. Without them, we would not have been able to make many changes as quickly and as convincingly as we did. This applies equally to production, control and communication. In this way, we are moving in parallel with the developments that are also occupying all our customers today. So, it makes sense to recognize parallels and merge like with like. This is all the more successful the fewer separating interfaces there are: a client here, a service provider there, a common goal on top of that – productivity, value retention, sustainability.

A logical development in this context resulted from the question: Are the contracts we have with our customers actually still able to reflect the everyday life we live together? And – above all: Are they suitable for countering rapid change with something that has a lasting effect? No. If, for example, there is no app in the bill of quantities, its benefits cannot be evaluated and priced. No passage in the contract is suitable for evaluating another example, the use of cleaning robots. Especially not if better, more efficient and more environmentally friendly devices come onto the market during the term. We are in a fluid process here that is faster than the paper we have signed with our customers.

Contract texts are therefore too rigid and too simple to keep up with ongoing improvement and change processes. We see this above all where facility management has developed into a value-added service. It is precisely the fragmentation of bills of quantities into individual trades that ignores interactions and thus decisively restricts the scope for design – on both sides.

Can this be changed? For example, by creating a comprehensive platform that integrates the entire range of services? There are already plenty of examples of this in other sectors of the economy. Think of ordering a festive menu for a company party: Who would order peas or ice cubes separately? Think of buying a car: Who would put the start-stop button and central locking out to tender individually? Think of the wellness holiday: Who would take a sauna in one hotel and take a shower in another?

Three examples that illustrate what makes the relationship between customers and their service providers strong: Some are looking for "well being" in the package, others are putting it together. Thanks to their technical know-how, service providers like us have the talent to adequately address and win different customer segments with adapted package content. This makes it possible to find solutions for a wide variety of strategies or objectives – for those who want to combine employer attractiveness with New Work and Mobile Office, as well as for those for whom sustainability and ESG are worth increasing their commitment at this level. And last but not least, those who want to regain entrepreneurial joy because they certainly can't imagine that the "trappings" run perfectly.

What Are the Contracts that We Will Conclude in the Future?

We have dealt with the question: can contract negotiations be conducted and contracts concluded in such a way that they meet this objective? The answer is a resounding "yes." Certain requirements must be met for the intention and result to match.

- Requirement 1: The two partners must speak the same language and have a comparable level of knowledge about each other's abilities and desires. First of all, this is a task of communication and trusting knowledge transfer, less calculation and controlling. For example, "ESG" is not an absolute value in itself but needs to be weighted and prioritized. The same applies to factors such as safety, value retention or innovativeness.
- Prerequisite 2: The term "facility" must be clarified. Is it all about pure functionality? Or do atmosphere, charisma or interaction with other entrepreneurial factors play a role? Bus depots and concert halls may have more in common than a tractor production hall and a power plant.
- Requirement 3: Times and frequencies of the services must run at the same frequency as the use of the respective facility. In view of the rapidly increasing flexibility and the growing freedom of choice of users, we are moving away at high speed from the norms that dictate "weekend" or "closing time." The mobility of the FM must be able to reflect this volatility.
- Prerequisite 4: Complex structures need holistic support and operation. Otherwise, there is a risk of friction losses and gaps in knowledge about the actual status quo. As a consequence of individual considerations, we find that every cent saved now generates at least one cent of damage if you only pay attention to the individual trade and not to the entire system.

From our point of view, these considerations inevitably lead to a "new work" of partnership and contract negotiations. With tenders in the previous sense, in which each individual item is decidedly predetermined, neither productivity nor efficiency nor satisfaction can be established. Rather, both sides are well advised to work together to develop a viable concept. A concept that incorporates possible changes in the medium and longer term because this is the only way to fully realize the return on investment on applied know-how.

Broken down into a simple sentence, this means that contract negotiations will begin with this sentence in the future: "We want to spend the amount X. What can you offer us in return?"

From this idea, the following steps can be taken directly for the clients, which make up this new way of contract negotiations.

- First of all, a pre-selection of the service providers takes place, whereby the matching corporate cultures play a key role.
- The second step is to define the desired goals – beyond the pure result. At the same time, the results of this process step provide the starting points for what to do if change processes appear necessary or possible during the contract phase.

- After that, a price range is determined, which also includes possible alternatives (e.g., weighting, term) as well as the option for possible cancellations and repeat orders during the term.
- The most complex and at the same time most valuable step concerns the subsequent definition of the process for power control. Here, digitization opens new windows for both the customer and the service provider to check results and goals without the need for high personnel costs. Every hour invested in process development pays off in days during the term of the contract. Especially since the right service providers deliver the components for this process ready for use due to their self-image and their holistic FM way of thinking.
- Here the circle closes under the sign of "trust," which is based on the choice of partner. End-to-end transparency reduces the verification of what has been agreed to random samples – as is the case with calibrated machines or DIN-tested tools.

Everything Changes, People Stay

FM is changing at high speed because the environment of our clients has changed dramatically and continues to change at a rapid pace. In this context, let's take a look at the online dashboard that our teams are now working with – and so are our customers. It is a result of changes in work content and processes in FM. At the same time, however, it is also a contribution to a redesigned workflow and improved, faster communication.

We are convinced that this path makes it easier for all parties involved to grasp the complexity of modern facility management. It has been proven to do justice to a productive, value-adding partnership. It helps to ward off irritations, delays, disruptions and economically harmful consequences that arise from counting the uncountable and measuring the unmeasurable.

The time has come to provide FM with the added value that comes from this perspective. And let's not forget that "management" includes the terms "human hand" (manus) and "lead" (agree).

Dr. Christine Sasse
CEO of the Sasse Group
Munich, Germany

Dr. Christine Sasse is a physician and family entrepreneur. While still studying medicine, she married the founder Dr. Eberhard Sasse of the Sasse Group, and thus gained her first insights into building cleaning, from which today's group of companies emerged. After her first years as a doctor, she moved to her own company in 1996 as a member of the board of directors of Human Relations in order to work with her husband to drive forward the internationalization and development of the group of companies as a full-service provider of integrated facility services. The entrepreneurial couple has two daughters (born in 1989 and 1993), who have succeeded the company in the second generation since the beginning of 2022.

Today, 8,800 people have found their professional home in the Sasse Group and its holdings.

Christine Sasse holds various honorary positions. Among other things, she is Honorary Consul of the Republic of Armenia in Bavaria, Deputy Chairwoman of the Supervisory Board of Bayerische Host AG, member of the Executive Committee of the Bavarian Economic Advisory Council and on the board of several charities that promote the integration of children and young people into work and society. Since 2008, she has been committed to the visibility and recognition of the FM industry in society as part of the initiative "FM-the Enablers."

2.20

How Did I Get Into Facility Management?

Jürgen Schneider

As with most facility management executives of my generation, an FM career was not the original master plan I had set out to do as a young academic. This was not so much due to the image of facility management, but rather to the perception of the industry and its importance, which was still insufficient for me at the time. At that time, I probably simply wouldn't have been able to do anything with the term facility management. But even today, I still meet people who mistakenly associate the term with "janitorial" rather than the management of buildings, properties and operational processes in the operational phase.

"What you can earn with your head, you don't have to do with your hands" was what my father, who was a craftsman, had given me on my training path at the time. At that point in time, it was not foreseeable that the work would catch up with me and that I would enjoy the management of technical facility services. Today, more than 30 years later, I am the managing director of gefma, the largest association for facility management in Germany with over 1,000 member companies. What happened?

Like many university graduates with a commercial degree, I chose a renowned international consulting firm to start my career. Good listening and analysis and a high degree of flexibility characterized the way I worked, and I noticed that I particularly liked the interaction with the people in my job. Translating customer requests into services and finding solutions without looking so closely at the clock were the order of the day and I got the chance to gain insights into a wide variety of industries and international companies.

But what drew me to my next position on the board of directors of a large German telecommunications group was the desire for a "professional home" at the time. Project management of any kind and often in the context of IT was exciting, but I had the feeling that I wanted to answer the question "What do you do for a living?" with a clearly defined activity (today, by the way, I would clearly answer this question with facility management – i.e. with the industry), which led me (back) to the focus of my academic training in the field of human resources. It was a real challenge for employees and management to help develop solutions for complex personnel strategy tasks for an entire corporation that – once state-run – now increasingly had to face free competition in the private sector. It was here that I first came into contact with the real estate department, for whose personnel strategy I was responsible. This was due in particular to the fact that the internal real estate division reported in its entirety to the Group's Head of Human Resources as a cost center. Later after the outsourcing, the responsibility was to change to the finance area.

From this function, I was then offered my first management position in human resources strategy as a young family man in the real estate division of the group, which included facility management, property management and construction in existing buildings, in which I was allowed to take on more and more HR responsibilities almost exactly every two years. Now that I have arrived in facility management, as an HR and organizational

Facilities @ Management, Concept - Realization - Vision, A Global Perspective, First Edition. Edited by Edmond Rondeau and Michaela Hellerforth.

manager, I supported, among other things, outsourcing, i.e., the sale of my own company to an international construction group, which thus expanded its depth of service in the real estate life cycle. In addition to an extensive restructuring program, I particularly remember the moment when, with the switch to the buyer side, the business case of the sale became my own planning. Subsequently, I led numerous reorganization measures of the company as well as the negotiating team regarding the company collective agreements for over 5,000 employees.

What happened in 2013 can perhaps be explained on the one hand with a broad education and my background as a consultant, but I am sure that such permeability within an organization is particularly present in facility management: The management offered me – coming from a cross-section and protected by my then supervisor managing director – to be responsible for operational business. In general, it was the relationship to individual people and the trust in me that made it possible and helped me to perform a wide variety of functions in facility management. But this is also what distinguishes the industry as a whole: FM is a people business, both externally in the service relationship and within the individual organizations. Later, I was to take over this focus with the strategy of the association's work.

Under the circumstances of the loss of the largest customer at the time, I then took over responsibility for the division's international mergers and acquisitions activities in order to drive inorganic growth in the slowly but steadily consolidating market. At the same time, the topic of digitization was already hotly debated in the industry. It was clear that innovation was no longer limited to the further development of conventional technical functionalities, but that processes and services had to be rethought against the background of digitalization. Initially in addition to my M&A tasks, later exclusively, we explored the possibilities that digitization with the available artificial intelligence and IoT approaches now offered us in terms of facility and property management in a newly assembled team and in cooperation with universities and presented them to the customer in a specially built showroom.

The cooperation with young, mostly IT-savvy employees as well as the exchange with multipliers from all areas of the company sparked a great dynamism and spirit of optimism. In addition to the task of bringing this mindset into operations, another challenge was to convince the client side, which was comparatively cautious and often only tested smaller pilot projects, of the advantages. At the same time, I was already looking for integrative solutions within the construction group with the planning and construction stages upstream of the real estate value chain. The fact that at that time there was little interest within the group in products such as "construction-accompanying FM" or the development of life-cycle-oriented and digitally supported holistic and sustainable management in the use phase was due in particular to the market situation: Due to the high demand on the market, what was built was bought and energy was comparatively cheap at that time, so that simply no need was seen to take the long period of use into account in the development of a building for reasons of sustainability and efficiency.

During this time, I received an inquiry from the CEO, who was also chairman of the gefma board, whether I could imagine joining the management of the association. For the managing directors retiring for reasons of age, a management with a broad skillset from facility management was sought and obviously I covered one of the future topics of our industry with the acquired know-how around the topic of digitization.

Admittedly, I didn't know exactly whether I was the right person for association work at first, as I came from corporate structures and functions that always focused on EBIT, sales and costs. But why shouldn't I bring an economic and growth-oriented perspective to a voluntary association? In terms of content, I also saw the opportunity to take a holistic approach to the life cycle of a property through lobbying for the industry in an overarching function and thus to promote the goal of early involvement of facility management in the planning phase. In the end, the decision to work full-time in an association and thus to abandon the classic career and hierarchical system of the economy is a question of purpose.

After coordinating with the other board members with regard to future tasks and challenges and after lunch with my future deputy Annelie Casper, who is experienced in the association, I agreed. It is also true – and I had not mentioned this before – that up to now the distance between home and work of about 380 kilometers took up a large part of the day, which was only interrupted during the pandemic. So, from now on I should also regain a piece of quality of life for myself.

However, the pandemic and the associated lockdown phases proved to be a real challenge for entering a network job. While the FM industry was able to distinguish itself as a systemically relevant and crisis-proof employer with significantly higher appreciation during the pandemic due to its high resilience, the new gefma management looked into small cameras instead of shaking hands in person during their introductory rounds. But still: We had arrived.

Importance of the Industry in Germany

Although the area of facility management is still not included in the figures of the Federal Statistical Office, the latest industry report from 2022 shows a production value of more than 240 billion euros (US$261 billion), a gross value added of approx. 152 billion euros, (US$165 billion), a share of 4.52% of gross domestic product (GDP), and an employment of a good 5 million employees. Out of a total of just under 45 million employees, 11.2% of employees in Germany are employed in internal and external secondary processes of facility services. With this economic importance, the industry has defended its position among the most important sectors in Germany on an equal footing with vehicle construction and is thus a noteworthy part of the German economy.

History and Focus of the Association's Work

Stakeholders have commented on the emergence or definition of the term facility management in Germany long before my time, but in December 1996 the *GEFMA guideline 100 Terms, Structure, Contents* and at the same time the starting point of the gefma guideline system was published for the first time with today 113 guidelines and white papers.

A recent member survey of the association has once again made it clear how important reliable and recognized standards in the form of guidelines are for quality in our industry – a circumstance to which the association ultimately owes its existence. In doing so, the expertise and experience of the association's provider, user and consulting companies are incorporated. The development of standards also includes the certification of software, sustainability (SustainFM) as well as vocational education and training. In 2000, the first gefma guideline on training, which was published in 1997, became the cornerstone for the gefma education pyramid and the certification system for students, business administrators, and service staff in FM (*gefma guidelines 600 ff*). In addition to the re-certification of universities and educational institutions, the working group "Education and Knowledge" focuses on the further development of the Bachelor's and Master's degree programs in terms of form and content, as well as the development of content, framework study and curricula, examination regulations, but also the inclusion of new topics (e.g., digitalization, smart buildings/homes/cities). In the last 25 years, 150 award winners have been awarded for 44 university locations, an initiative that could only be made possible by the sponsorship of member companies. During the same period, more than 4,000 business administrators and service staff trained at private educational institutions were trained.

But not everything can succeed right away. For many years, gefma has been striving to establish the training profession in facility management, a tough process in which each pitfall has to be circumnavigated individually.

In addition to setting standards based on knowledge and experience, gefma would like to open up spaces for professional exchange and thus make its contribution to the (further) development of the market. In addition to the various working groups, current challenges and best practices will be presented and discussed in regional lounges. At the organization's own events and conferences, which are co-organized or in non-material cooperation, a large number of current trend topics are brought to the stage in addition to basic topics such as operator responsibility. In addition to the advancing user-centricity, which is directly reflected in New Work and in which incentives to come to the office regularly in the future instead of working from home are being discussed, we believe that the focus topics of digitization and sustainability in particular will be the central topics of the coming years. In particular, the topic of sustainability offers a special potential for facility management to highlight the existing know-how and thus the importance of facility management. This can and should be used both in terms of political commitment and in terms of status within the real estate industry. In this context, gefma enters into many cooperations, which in particular address the requirements of sustainability.

As a result of the EU's Green Deal, the pricing of CO2 increases the need to equip existing buildings in particular for the future. The package of measures ranges from quick wins such as space optimization and hydraulic balancing of the radiators to extensive refurbishment. In addition, supply bottlenecks and rising interest rates are hampering investment in new construction, so that portfolio holders will have to deal intensively with a sustainable investment strategy to avoid so-called stranded assets in the future.

Facility management is on site, knows the plant conditions, and can optimally assess the requirements as well as the necessary investment volume. In addition, the necessary data to determine the status quo but also for future ESG reporting can be collected directly on site. Digitalization will play a major role as an enabler in this. Although there will not be only one unique data model in FM, there are already initiatives today to allow the different models and systems to communicate with each other via API and thus enable data exchange between all stakeholders involved. The buildings of the future can provide the latest information in the models of the digital twin or the updated BIM model in order to efficiently support the longest phase in the life cycle – that of operation. Because one thing is clear: the battle for CO2 neutrality in the real estate sector will be fought in the existing buildings and thus in the management.

The closer integration with the client's processes in the future (also with regard to data), the expansion of contracts with regard to ESG and the progressive shortage of skilled workers will require the anchoring of contract elements, some of which are output related. In order to have the most effective effect here, facility management must increasingly take on the role of a (strategic) solution partner in order to be able to influence real estate strategy not only in an executing but also advisory capacity in partnership at eye level. This would be nothing less than a cultural change.

View

The main task of the association's work is and remains to increase the public perception of facility management as an industry of the future, but also to achieve the goal of upgrading within the real estate industry itself.

The challenges of climate change, sustainability and energy crises certainly offer a good starting point for this. Facility management makes a decisive contribution to solving the major social tasks as well as the integration of immigrants with a wide variety of qualification profiles and undoubtedly offers future professions that will withstand the challenges of digitalization.

The many different career paths in this book show the opportunities for personal and professional fulfillment in a variety of fields of activity, which will continue to exist in the future.

Jürgen Schneider
Managing Director
gefma (German Association for Facility Management e.V.)
Bonn, Germany

In 2020, Jürgen Schneider became the Managing Director of gefma, the largest association for facility management in Germany with over 1000 member companies, institutions.

After studying business administration at the University of Cologne, he worked in the telecommunications consulting division of KPMG Consulting and later moved to the Deutsche Telekom Group. As part of the Human Resources Management Staff, he was responsible for the HR strategy of the entire Group before moving to the real estate sector (DeTeImmobilien) as a senior executive. There he took over the personnel management including conditions/labor law as well as the area of organization. Subsequently, it supported the sales process to STRABAG AG and assumed regional responsibility for technical facility management for 1200 employees at STRABAG Property and Facility Services GmbH, as well as for international mergers and acquisition activities. From his responsibility for digitization, he moved to the management of the industry association.

2.21

Data Structuring of CAD Drawings – My First Contact with FM

Alexander Redlein

My first contact with FM was when the governmental institutions of Austria set up a project to gain data from the design and construction phase to optimize the utilization phase. This was in the early 1990s. My team and I supported the chamber of architects to establish an open standard based on DXF Data format of Autodesk with layer and block definitions. Architects should structure their CAD drawing in a way that the different elements like carrying walls, windows, and lights should be delivered in the defined structure. My big advantage at this time was my interdisciplinary education as I studied electrical and control engineering at Vienna University of Technology as technical foundation and business administration at Vienna University for Business Administration, which gave me a broad economic but also legal background.

This legal background was also the reason we could change the data structure from being a copy of the data structure of the CAFM system used by the governmental institutions to a more general structure, that most of the used Computer Aided Architectural Design CADD systems could provide.

Why? In the beginning the public body presented the structure of their program and asked all architects and civil engineers to provide the final drawings of the building in this format and data structure. But only few CADD programs could deliver the structure in general, or the users had to work totally different in the programs. Some products could not deliver the structure at all. The governmental body persisted on the structure as it would provide them with the data needed for operation more efficiently. What they did not take into consideration was the European procurement laws, that asks for competition to secure effective and efficient procurement and to prevent distortion of competition. Due to the data structure a big distortion of competition took place, as only few planners using the application of the government or Autodesk itself could deliver the requested data structure. This was at that time less than half of all architects in Austria. As this was against the law, the governmental bodies accepted a data structure that was more open so that the majority of CADD system could automatically provide it.

This project was a really learning experience for me, not only about the demand for data in the utilization phase, but that in FM mainly technicians are looking for good solutions, not always considering the economic and especially the legal framework. Most of the participants in the discussion were technicians, looking for a solution that optimizes the exchange of data necessary for the utilization phase. They were very knowledgeable in design and CADD system and also in data structuring. The architects always depicted the enormous additional workload they would have to bear to deliver the necessary data. Others stated that they cannot deliver the required data at all and were therefore excluded from governmental bidding. These arguments did not really convince the governmental institutions. The legal framework of the procurement regulations in the EU caused a change in the mindset and was the real turning point of the discussion. In most of the projects I conducted later on, this interdisciplinary mix was also very important to gain an optimized solution. So, I always selected the participants in a way that not

Facilities @ Management, Concept - Realization - Vision, A Global Perspective, First Edition. Edited by Edmond Rondeau and Michaela Hellerforth.

only technicians were part but also representatives of the legal and economic departments were included. In the 1990s most of my customers were not used to this procedure, FM was a domain of the architects not so much an interdisciplinary approach.

From Master of Business Administration to Professor for Real Estate and Facility Management – A Bridge between Research and Practice

With my background – a master of electronic and control engineering and a Master of Business Administration – I was quite alone. Most of the people I learned from were architects. My first two academic experts Dries van Wagenberg and Keith Alexander, had this background. But Dries already told me about the service approach, which was already very prominent in the Netherlands in those days. The Dutch Facility Management experts pointed out that a building is not to be regarded as bricks but as a service we are looking for. And that other services make it a useful support for our work tasks. So, a seminar room is not a room we are looking for, but it is an environment including services like cleaning and catering that supports us in an optimal way that we can carry out a meeting and only have to concentrate on our meeting not on the infrastructure. This was an important idea, as it fades away from the pure and simple cost reduction approach to the enablement of people and processes.

That was also the focus of Alfred Taudes, Professor at the University of Business Administration in Vienna. He guided me my whole academic life as a teacher, mentor, coach and friend.

Processes and IT Support as Starting Point

I learned a lot from him about process management, business process reengineering and about the proper IT support. He opened my eyes to the fact that there is more than Computer Aided Drawing Software and that Enterprise Resource planning systems are especially important for Facility Management. Enterprise Resource Planning systems (ERPs) integrate all data and processes of an organization into a unified system. Several processes like procurement, cost accounting but also maintenance management can be used also in our discipline. The advantage of these systems is that they are already in use. The customizing to the needs of Facility Management can easily be done within a limited budget, whereas the necessary functionality of CAFM tools first of all has to be specified, a procurement has to be carried out and then the implementation can take place. These higher costs still lead to the fact that only 40% of large users, or better demand side companies, are not using this type of tool.

This leads me to my second research area: optimized IT support for FM. Analyzing the processes I recognized not only breaks due to the life cycle of buildings but also the changes to the responsible persons are happening:

- Design: architects, civil engineers
- Construction: e.g., construction specialists, electricians, plumbers, carpenters.
- Operation: Facility Managers

In addition, the ownership of the building is also changing in these phases. All those transitions caused a loss of data and the need for different tools.

Out of several case studies and an intensive literature review I realized that it is not one software we are looking for to support FM processes. A combination of systems like ERP and CAFM tools can cover the whole variety of technical and economically oriented processes much better without double entries and using systems as they were planned for. This also asks for data, structured data. In more than 50 case studies with different companies I specified the processes and the data necessary based on the defined processes. This set of process models was used to set up reference process models.

Again, Alfred Taudes was a source of knowledge. Based on his publications in the area of IT support within production, I defined the necessary IT tools and an IT architecture to integrate these tools. The core tools of the solution are CAFM and ERP systems. In addition, several tools were specified, mainly as data deliverers for the core IT tools like CADD and building automation tools

Middleware was used to integrate these tools and to synchronize the data. The middleware enabled the user to carry out the necessary processes across the different tools but had the impression to use only one IT tool. In this way all process steps could be carried out completely and effectively by the different tools: Although different IT tools were engaged for the users the system acted as one solution. In addition, the different data sources were also synchronized.

International Experience – Science Meets Practice

This theoretical concept was implemented by my team together with several CAFM software companies. In the next step my university worked together with these IT companies and consultants worldwide to implement this middleware and to role the solution out to the FM industry. In most of the projects I did the analysis and led the implementation project, in several cases I also wrote the blueprint for the SAP system which acted as ERP system and supervised the SAP customizing. At this time my department consisted of more than twenty people defining reference processes, assigned process steps to the different tools in use, defined the functionality and data exchange done by the middleware.

I personally learned a lot during these years as I traveled to so many countries not only in Europe but also in the USA and Africa. It was about project management but also about different approaches within FM, different cultures and teamwork across borders. I also gained a lot of international REAL friends. I really want to say thank you to all of them as they are still part of my life and network. Together we made a lot of things happen.

During these years I also found in Christoph Achammer another academic mentor and coach. He and Alfred Taudes supported me to become a professor at the TU Vienna in 2003, and I achieved the position of staff unit of the chancellor. As staff unit I was responsible for the SAP system of my university and could follow up with my research activities.

FM Is More – Value Added

At this time, I started with my next research topic. I wanted to get FM away from pure cost reduction to value generation. As there were not a lot of publications on the value added of FM we used theories and approaches from IT support in the area of IT support within office and production processes. My team and I applied these concepts and methods to FM.

We identified three areas:

1) Pure cost reduction
2) Increase of productivity
3) Strategic advantages

To capture the necessary data in the area of FM we started to analyze the demand side. This was done to get away from the marketing publication suggesting too high potential of cost reduction of more than 30%. Most of these articles were based on single projects and were not scientifically valid. Therefore, we started our yearly demand side study in 2005, that we still carry out to get a clear indication of the trends that are in the focus of the demand side and its core business. For this purpose, every year we interrogated a valid sample out of the 500 largest companies in serval countries in Europe about:

- their internal organization of FM,
- their demand for services and their way of sourcing

- their IT Support and
- their cost savings, increase of productivity and strategic goals within FM.

Based on the data gathered by a PhD thesis, we could scientifically validate, that the implementation of an FM department provides a positive impact on cost savings and on the productivity of companies.

In addition, this long-term study gives us a great picture about the development of the behavior, the demand and strategic goals of the user or better the demand side of the industry. These results enabled me to define additional research projects:

1) Workplace Management
2) Sustainability and ESG
3) Digital Transformation of Real Estate and Facility Management

Sustainability and FM a Great Liaison

Since 2010 we have analyzed how FM and sustainability are connected with each other. One source is the Global Reporting Initiative (GRI) which set up a framework for sustainability reporting. On their Web site they published which areas companies focus on in their reports. By comparing this to the list of FM services of the EN15221-4 Taxonometry we could prove that several management tasks of Facility Managers go in line with the strategic goals of the companies. Already in 2014, 96% want to reduce their CO_2 emission, 93% want to increase energy efficiency, 80% want to reduce water and paper usage, and 40% want to reduce their waste. All these areas are part of the job description of FM, but at this time FM was not really included in the internal projects. This was the reason why we put a focus on this topic and published several articles, master theses and books to change this perception. In addition, based on the scientific insights we supported several companies in setting up and conducting projects to reach the goals mentioned above.

Workplace Management – the New Trend

The lack of perception was the same in the area of workplace management. The projects were mainly conducted by interior designers and FM was not included, although it resulted in major changes in the operation. Together with researchers of the Netherlands and the USA we set up projects in Austria to set up Activity based Working environments in 2008. This led to other projects and the publication of several articles and the book *Work on the Move* together with the IFMA foundation. I had the honor to be the only German speaking author. COVID-19 changed the perspective on this subject, as this largest field study on the possibility of working from home depicted that a lot of tasks can be carried out productively from home. But companies have to take care that the binding and engagement of their employees is kept.

Digital Transformation – the Game Changer

In parallel to this I was asked what digital transformation means for our industry by several industry partners in 2016. As I could not give them a valid answer, I started the project "Use Case Database." My team have been collecting international use cases in emerging technologies. Use cases were primarily found in publications like *IEEE Spectrum*, *MIT Technology Review*, *Harvard Business Review*, and strategy documents of well-established consultancies like McKinsey, CBRE, JLL, Gartner Group, EY, PwC, and Deloitte. Many of these publications referenced scientific journal articles, which were used as further input. Overall, 1200 use cases have been collected since 2017. These cases were primarily published in the years from 2015 to 2021.

The use cases collected were not limited to usage in the area of real estate and facility management in order to include innovation from other industries. By doing this, the researchers were able to broaden the scope of the research considerably, while simultaneously ensuring validity of the results. The goal was to collect as many meaningful use cases as possible to predict the impact of the technologies on real estate as well as the facility service and management sector.

The published use cases were analyzed, using grounded theory to ensure that the results were replicable and meaningful. Using a database. the cases were analyzed with regard to which services were affected and which technologies were used. This enabled my team to analyze the relationship between services and technologies and the changes in real estate and facility management. According to this research, digitalization has impacts in two main areas:

1) The changes in the core business, like new ways of working, modify the demand for infrastructure and services changes dramatically.
2) Emerging technologies like Internet of Things (IoT), Big Data, and Artificial Intelligence (AI) allow for disruptive and much more efficient ways of service provisioning. Therefore, the service provision itself is changed by digitalization.

This research was the basis for an invited presentation at Stanford university. As the team led by Larry Leifer – the founder of the methodology design thinking found this knowledge helpful for his design thinking course, since 2019 I have the honor to be part of the ME310 Design Thinking teaching team of Stanford university. Every year we conduct a real estate innovation challenge sponsored by a team of industry partners from Austria. They put forward the topic of the challenge and students at the Vienna University of Technology and Stanford university use design thinking to find innovative user centric solutions. The topics of the last years were:

1) Office of the future – combing Well Being and Resource Efficiency
2) Living of the Future – combing flexible homes supporting working from home with Well Being and Energy Efficiency
3) ESG in practice – what can be Environmental, Social Key Performance Indicators and how can we measure and optimize them

The challenges were very successful, as the depicted new approaches and delivered innovative prototypes for services and IT solutions, that were successfully taken up by the participating companies and implemented. One idea of a scalable IoT-based ESG monitoring and KPI software tool led to the foundation of a successful start-up.

This shows another lesson I had to learn. Several of the concepts and ideas our research came up with were too early for their practical application. In the end of the 1990s we suggested linking fieldbus devices and CAFM tools for visualization purposes and with ERP tools to trigger maintenance processes. This is nowadays common practice and is called IoT Eco system, where IoT sensors deliver status data that is used for heat maps and visualization and for triggering processes or even predictive maintenance. When we came up with the idea most of our partners did not see the potential of the concept, as fieldbuses and their use were quite new in those days.

On the other hand, from a cooperation of industry with scientific institutions, both sides can profit. As researchers, we test new concepts and therefore are "allowed" to make errors if we learn from them. Therefore, we as researchers can show up validated new ways and concepts and set up prototypes to prove feasibility. These approaches can be taken up by industry and put into practice so that they can differentiate themselves and ensure their innovation power and long-term survival. Therefore, this bridge between Research and Practice became the basic principle of my work and life as a researcher. My team and I get input from the demand side, what are interesting new topics for research. We as researchers get additional input from other colleagues, sometimes even from other fields of expertise. Based on this input we can look into the future and develop new concepts. These ideas are challenged in a feedback loop with our demand side and industry cooperation. In that way we can shape the future TOGETHER.

My most important learning was that a team is stronger and more innovative than any individual. A team has more knowledge and more experience than any individual. Not only to recognize the experience and ideas of the team members but to show that that you appreciate them makes the difference. I always see my success as a success of my team. That is also the reason why I love to teach at Stanford, as this university incorporates the team idea. Larry always asked us to put together diverse teams, to gain as many insights and perspectives as possible. And I also see this approach in the way the colleagues interact with each other and with the students. This is very valuable, trustful and respectful. This is how I also act in my business and private life. I think the number of real friends like Pat Turnbull, Karin Schaad, Diane Coles Levine, Geza Richard Horn, Peter Ankerstjerne, and Wolfgang Gleissner shows the rightness of this behavior.

This concept and my work as a scientist are not only recognized by the students, as I really love to teach and my lectures on IT support are taken by around 400 students every year, but also by the industry and by academia. The Academy of Science of Austria recognized our work so that I was decorated by the Austrian government as "academic founder of a new academic area."

Education Is More than Training and Lecturing

Based on the research and experience described before I established a *Master of Business Administration FM* at my university. I selected this format on purpose to foster the management skills of the students. The program includes beside a solid management education the modulus

- Legal & Technical basics and Interdisciplinary Approach including buildings physics and technology, IT support;
- Facility Management: Strategic – Tactical – Operative;
- Workplace Management

A special focus is to put on legal compliance and how FM works together with asset, real estate and property management. The teaching concepts fosters the practical application of the knowledge in projects. We also include new topics like workplace management and emerging technologies promptly in the curriculum to prepare our students for the new developments and be ready for the future.

Most of the topics have to be applied directly after the lecture in workshops or on small projects. To show the interdependences the students have to fulfil an interdisciplinary project where they have to analyze the users' demand for infrastructure and services, reconfigure the building and its equipment, define the services and operation and make the make or buy decision. All investments and costs have to be included in an investment budget and the economic feasibility and efficiency has to be proved.

This approach should prepare the students for their business life. The success of the approach is visible in the career paths of our students. The majority has made a step forward in their career, being head of strategy, COO, or even MA of large service providers or head the FM department within large user organizations.

Besides the MBA my PhD students supported me in my research to a high degree. Horst Pichlmüller for example, proved the feasibility of integrated service provision combining technical and infrastructural services thereby increasing productivity. Others enabled us to cover interesting topics like value added of FM, reference processes, workplace management, emerging technologies, IoT ecosystem. Eva Stopajnik covers the whole macroeconomic field, analyzing how the value added, the turnover and the number of employees within the outsourced facility service industry developed over the time in the EU and the USA based on official statistical data of the EUROStat and the consensus office of the States.

As my PhD students needed a forum to exchange with each other and the scientific community, I established in 2007 the "International Facility Management" Congress. The slogan is- Science meets Practice. This yearly congress covers the topics raised by the "Real Estate User Group." This association is an international mirror group consisting of around 3000 expert members from all over the world like Australia, India, almost all European countries, South and North America. They raise the topics of interest and then my department sets up a scientific call for papers to get scientific papers to represent the current research on the topic. In parallel my team and I are looking for practitioners to provide a holistic picture and innovative approaches from practice. The congress is not only a source of knowledge but also a real networking event with over 200 visitors from more than 15 countries. By these different learning offerings my university, my team and I hope to strengthen the exchange of knowledge and learning but also the networking between the individuals.

FM – Quo Vadis?

When we look to the future, we recognize that several trends that have their origin in the early 2000s are getting more importance. The most important trends are according to our scientific studies in four areas:

1) Workplace Management as the combination of all services to an optimized People-Place-Process approach
2) Sustainability with a focus on the 17 Sustainable development goals of the UN and Environment, Social Governance (ESG) targets of the EU
3) Digital Transformation changing the way we work and the work environment but also the way we operate buildings
4) Circle Economy: urban mining

COVID-19 dramatically changed the way we work. Beforehand only few companies applied working from home and even activity-based working was a very innovative approach and therefore not widely spread. During the pandemic working from home boomed not only in the DACH (German-speaking countries) region, but around the whole world. Now, almost two years after first decisions on sending a broad range of people home to work from there, many employees still remain working from home, even though the pandemic is officially over. According to the IFM home office study of 2023 there is a clear preference of performing concentrated work tasks at home. Routine work is equally done from home or the office. Regarding teamwork the picture is different. The study and also other publications show a clear preference to come to the office to exchange with their colleagues (86.4%). According to the responses, almost two thirds (59.1%) answered that the coordination of work tasks with colleagues and supervisors was better at the office than from home. Direct, immediate contact between colleagues without distance seems to be a clear advantage of office spaces.

It seems that the future is a hybrid office where the employees decide where and when to perform their tasks. Concentrated work seems to happen at the home office whereas the traditional office will be mainly used for meetings and formal and informal communication. This will lead to a higher demand for flexibility and especially communication and recreation areas like lounges at the office. The traditional office has to use workplace experience to "convince" the employees to come back to the office. This experience is also to increase the engagement and binding with the employee and foster employer branding. I totally agree with Leesman's statement: "People are voting with their feet. Employees consider investment in the workplace as an investment in the employees." Therefore, management has to acknowledge that modern workplaces can increase the binding to the company and make acquiring new workforces easier. This is especially important in times of workforce shortage to succeed in the war for talents. Therefore. workplace management is one of the areas Facility Managers shall be acknowledged and lead the change management in a holistic way.

The next trend to follow is sustainability. The topic was always in the focus of the European Union. The laws became more demandable due to the ESG legislation asking for sustainability reports as part of the non-financial reporting. Companies have to state their goals in the area of Environment, Social and Governance, define KPIs and depict the process they make reaching these goals. This development can be seen as a follow up of the GRI reporting mentioned above with the main goals of CO_2 energy, water, and waste reduction staring in the 2000s. Now the goal is to be carbon neutral in 2040 and this reporting became a legal requirement which has to be attested by the company's auditors from 2026 on.

The facility manager is the enabler to reach these goals as buildings are responsible for around 40% of the CO_2 emissions. Optimization of the design, the construction, and especially the operation can reduce the impact dramatically. A prerequisite is to have valid data about the usage in time to set proper actions to make and keep buildings "Green." Emerging technologies can help. IoT ecosystems consisting of IoT sensors for data capturing, big data to store the data and analytics or even machine learning to define proper actions are the enabler for data driven management decisions. Therefore, FMs have not only to show that they can manage these areas, but also to have a knowledge of the emerging technologies and how they can be applied properly. In this respect the Facility Managers are helping to counteract climate change and keep this planet livable. FM should also be a job of choice for those who do not only want to talk about climate change but to set steps to improve the world.

This leads me to the next point: the digital transformation. FM was always seen as janitors and cleaning people. The shortage of people and the emerging technologies are changing the picture. Due to the people shortage and the new possibilities a lot of tasks get automated either by robots taking over cleaning and risky inspections are smart technology usage. IoT data enables predictive maintenance, forecasting failures before they happen and rescheduling the workforce. Augmented reality can be used to support local workers by specialists working remotely. This enables a totally different operation of buildings. A more customized service is offered according to the needs of the final customers. The well-being, comfort, and enablement of people is in the focus.

Technology Is Not the Focus but the Customer Experience

In addition, this industry is very local as buildings can only be serviced where they are situated. Our macroeconomic studies prove that this industry is very stable. All of the crises, either the economic one after 2008 or the COVID pandemic, did not really impact the number of people employed. This industry offers secure jobs and applies future technologies.

The last trend is quite a new one: cycle economy. We saw the first roots of it in the goal to reduce waste. Now we want to go further. We want to reduce the waste to zero by reusing everything. One attempt is the cradle-to-cradle approach leading to urban mining. Existing buildings are the "Mines" of the future. The resources used like steel, concrete, and glass is to be reused in the new buildings. Therefore, we have to record the materials used to be able to find them again and to reuse them. Several projects, especially in Scandinavia, are already in place. Marketplaces for the materials are established. This approach will be the next step to make our industry more sustainable and to save the resources of the world.

In general, it can be stated, whoever wants to really change the world for the better, make it greener and more sustainable should consider a career in FM as it really makes the difference.

Dr. Alexander Redlein
University Professor
Real Estate and Facility Management
Vienna University of Technology
Vienna, Austria

Dr. Alexander Redlein is university professor for Real Estate and Facility Management at the Vienna University of Technology, president of REUG and past president of IFMA Austria. After his interdisciplinary studies at the Vienna University of Technology and at the Vienna University of Economics and Business Administration he is now engaged in research, education, and consultancy in the area of FM for more than 25 years. He is head of the research group "Real Estate and Facility Management" (IFM) at the Vienna University of Technology which consists of 15 researchers. Besides his research activities he acts as a strategic advisor, setting up FM concepts for international companies and municipalities. He optimized their FM processes, ICT and workplace management.

As a researcher he conducts international studies about macro- and micro-economic importance of FM. His workplace research concentrates on the influence of workplace on the productivity and well-being of the employees. He also analysis the impact of digitalization and automation on Workplace Management, FM and Facility services. In addition, he heads an MBA for FM at the Vienna University of Technology and several FM certification courses in CEE and also in India. Since September 2019 he is a member of the ME310 international teaching staff at Stanford University.

Affiliation: Institute for Real Estate and Facility Management (IFM), Vienna University of Technology.

2.22

How I Became Aware of and Became Active in Facility Management (FM) in Brazil

Aleksander C. Gomes

Celebrating the 40 years of FM internationally (1982–2022) and looking at the future of FM, I will tell you a little about the trajectory in this industry in Brazil and the Latin America region working in a large organization with RE/FM global experience.

I learned about facilities management just when I started at IBM, 23 years ago. Because I was hired to work as a project assistant for an important project to expand the datacenter located in Hortolandia/Sao Paulo – Brazil, I worked very closely with a construction management company and facility management processes that supported these DC operations. My firsthand experience was in 2001 conducting a bid to purchase a computerized maintenance management system (CMMS) to standardize all asset management across Brazil locations and later, in 2012, implementing the Maximo CMMS software in Brazil and Argentina.

In 2004, I collaborated in the development of an RFP that for a first time, we went to market to find a unique solution to support maintenance and cleaning for the main Brazil locations.

And the plans to consolidate more services continued, my leadership defined next steps around FM and in 2009 we've reached a new level of maturity in FM optimizing the service providers and streamlining the soft services under a single supplier.

During many years, we worked optimizing the portfolio, adjusting FM services, and growing with the challenge of not impacting expenses in the same proportion, this was and continues to be the challenge of all organizations.

All these changes helped to move my organization forward, so workplace experience starting emerging as a one of the newest topics in FM and significant decision was taken, as crossing borders regardless of language or cultural differences, I started a promote the standardization of the process across Latam (Latin America), mediating the bid process, adapting this process, and galvanizing best practices.

As I always had an interest in participating on corporate RE/FM in a more embracing, I became a leader of the facilities management program in Latam, collaborating on global projects related to safety, overhauling asset management practices, financial controls and formulating the bid process working with the global team helping make decisions from technical requirements to the procurement process.

Most recently, in 2017, a global project was launched to go to market to find a FM integrated provider, outsourcing FM to one provider when I had an opportunity to work again with global team on this pioneered project in Latam, aligning 10 countries with the same statements and service level agreements, supported by a single solution across this region.

Certainly, this was a great learning experience in FM during the pandemic as the FM team worked hard to provide adequate spaces, implementing social distancing, new cleaning protocols, and other several adjustments, and then worked on reopening activities.

Facilities @ Management, Concept - Realization - Vision, A Global Perspective, First Edition. Edited by Edmond Rondeau and Michaela Hellerforth.

In 2020 IBM announced a spin-off of its managed infrastructure service business to kyndryl, and this separation was an incredible and the most recent experience I had in FM, where one of the important actions was related to the assets. I participated in executing lease contracts splits, building separations and all FM services to help the company archive its objectives, also shaping our internal organization of RE/FM to fit with the new demands.

More about My Career in FM and Other Interested Participants in Brazil

In the beginning of my career, I acted for some years providing engineering services as part of a facility management scope. After a while providing these services, I moved from providing this service to contracting FM services, delivering internally for our clients, but always with great interest in all parts of the process. This was very gratifying because it allowed me to stay working very close to the market through various companies.

I always had an interest in FM conferences and workshops, participating actively. Most recently I decided to share my experience to help people that had an interest in FM, companies, and service providers also, launching a group of FM discussion on LinkedIn and a social media on Instagram to promote benchmarking between FM professionals. Recently, as a guest speaker in a university, I had a great opportunity to introduce facility management for a graduation class. This was an important professional opportunity to promote FM for people who are starting in the job market.

The Making of the Term FM in Brazil

The Facility Management (FM) term is continuously being advanced in Brazil and in all Latin America countries. The term was introduced and shaped by global companies that consider the use of the advantages and brought this for the region (see Figure 1).

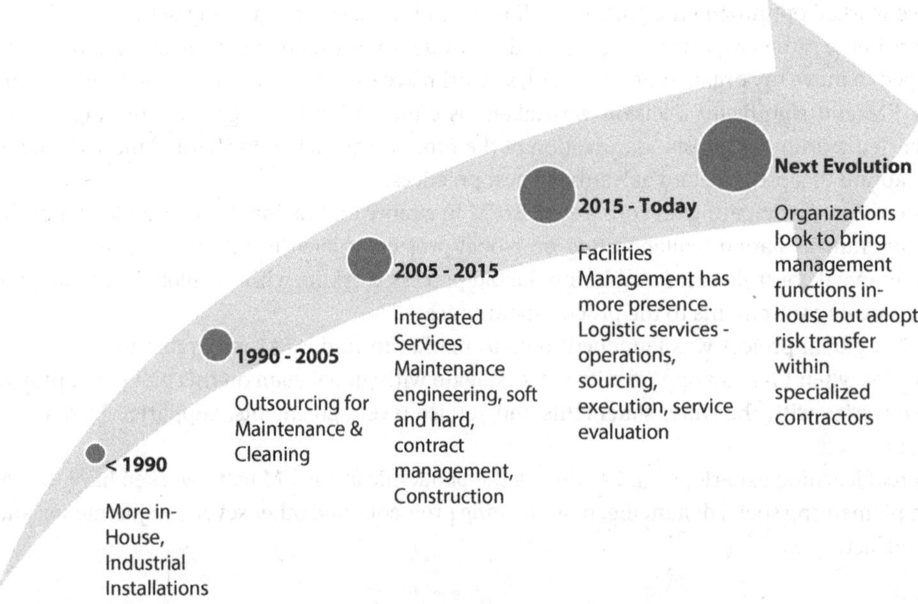
< 1990
More in-House, Industrial Installations
1990 - 2005
Outsourcing for Maintenance & Cleaning
2005 - 2015
Integrated Services Maintenance engineering, soft and hard, contract management, Construction
2015 - Today
Facilities Management has more presence. Logistic services - operations, sourcing, execution, service evaluation
Next Evolution
Organizations look to bring management functions in-house but adopt risk transfer within specialized contractors

Figure 1 According to my experience and participation in the FM industry in Brazil and Latin America countries, the term raised with local companies and some building users are demanding a workplace that satisfies needs, following multinational companies. (*Source:* Adopted from IFMA Brazil).

I see a positive evolution in a post-pandemic world, where corporates are focusing more on workplace and employee well-being than ever. Brazil, among other significant countries in our region, is experiencing an increase of commercial buildings that requires a management of a set of services to deliver a good experience for the users.

In spite of the market and companies focused on providing a better employability and talent attraction, other factors are not favorable to find efficiency in building operations through a common type of facilities management which considers the use of specialized companies – the outsourced management of FM services like housekeeping, maintenance, and security. This happens because the local legislation does not encourage outsourcing and in other countries like Peru or Mexico, this is more restrictive, and management facilities companies mostly have to work as an agency with limited delivery.

I see a more in-house management model for the future of FM in Latam, as we move forward with learning the concept. Despite FM not being a core mission, FM is critical to the business. Currently Brazil and Latam are being served by a few non-regional companies that have highly consolidated the market. Because, as I already mentioned, this market is still predominantly demanded by multinationals, this leaves a large market space for FM to grow among suppliers and local demanders.

Facilities Management is still a new business area that continues to grow in the Brazilian territory. This segment has been showing great interest from the investors and industries and today we already have several publications in this area.

In the same way this market is growing, the offer of courses increased and the organization of events too, making Brazil a hub of FM initiatives for Latin America. After the pandemic of COVID-19, I saw an increase in FM events and participants, in special discussions about workplace and space management processes. Related associations and dedicated magazines are recurring sharing present events, webinar, and offering different courses in FM.

I think the market still lacks maturity to move forward to the integrated facilities management model and it depends on service providers, and companies that are seeking to acquire and offer FM service integration.

The Growth of Women in FM and Women as FM Leaders

The FM industry has large opportunities for growth and involves different groups, and this is an aspect that makes me delighted by the sector. I have not yet had access to precise data on the participation of women in this market, but I see growth, especially in the last ten years.

Organizations are recognizing the value of diverse teams, and in this context, facilities companies seem to be hiring more and more women, including for leadership positions and this diversity brings to the sector an enhancement of the quality of services (see Figure 2).

I am the owner of a social account about workplace and corporate facilities and one thing that calls my attention is the gender balance between the followers:

I think we do not have more women working in facilities jobs due to education/graduate courses, as this area requires technical skills and these courses have predominantly more men than women, so there are significant challenges to overcome.

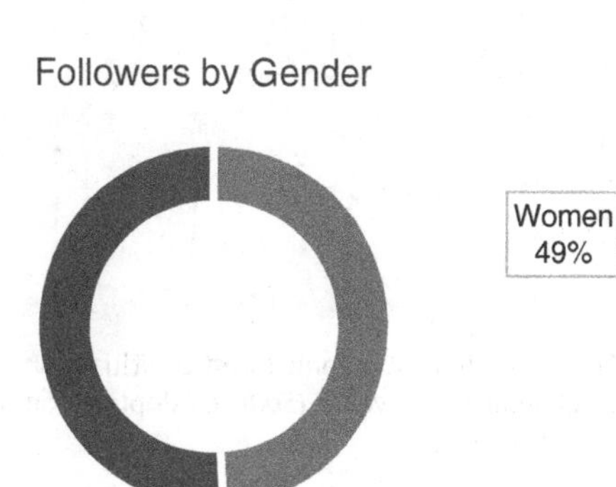

Figure 2 Genda Data from (at Feb,2023). (*Source:* Instagram https://www.instagram.com/workplace.facilities and LinkedIn https://www.linkedin.com/groups/12504953).

The Challenges in the Future for the Built Environment, Especially Post Pandemic and the General Impact to Offices

COVID-19 impacted the FM industry as we had to shut down the workplaces, especially in office spaces, on the other hand, as an exception, healthcare facilities have been operating overcapacity.

Facility Managers supported by his FM service providers worked hard and fast to shutdown many offices (temporarily or permanently), to avoid expenses with unoccupied space. However, companies are inviting employees "back to the office."

In December 2022 I led a roundtable about the workplace perspective and what success looks like for 2023 with a Global FM player and FM team members from different countries in the Latam region to review the factors that companies and service providers can look at to create a more attractive workplace.

Of the many points discussed, already implemented, or planned, the most commented were:

- New Workplace at offices:
  - Ways to measure building utilization
  - Occupancy status and attendance analysis
  - Structured workplace to meet different needs
- Returning to the office – employee demands:
  - This is not only about workplace. The main concern of employees is about urban mobility
  - Accessibility
  - Parking lot available with EV charger
  - Smart buildings to save energy, not a talent attraction
- Companies Action plan
  - Re-think layouts and space planning
  - Weel and FitWeel certificate, also LEED
  - ESG practices

Companies are working on many ways to bring personal back to the office, and they are requesting more competency, and know-how from Facility providers to help improve office spaces to be attractive and comfortable as employee homes, to provide a different level of experience and using more appealing terms such as avoiding the loss of culture (see Figure 3).

The workplace also needs to be planned to meet all types of persons. For many years, several studies have been made to match the workplace with the generations and now, considering the hybrid environmental, it is important to have places that support productivity working together (co-creation, collaboration, and innovation when we are together, face-to-face) and places that provides opportunities to contribute ideas outside of meetings, most preferred for introverts.

Hybrid work is fantastic and brings numerous advantages, as I mentioned above. I think the main one is the time saved on commuting. But in my opinion the great advantage of the present work is the interaction when the

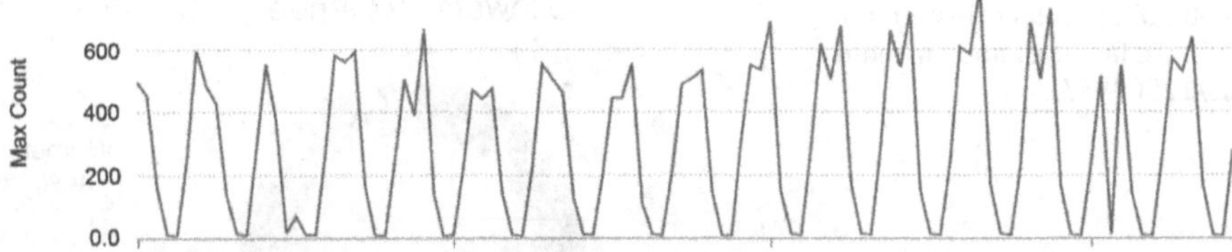

Figure 3 Typical graphic of office utilization after pandemic: low occupancy on Mondays and Fridays, some peak during the week indicating events. (*Source:* Adopted from IFMA Brazil).

company has their team all together – it's cooler, lighter, and something new on exchanging ideas and producing together what you don't see every day.

Aleksander C. Gomes
Facilities and Engineering Manager
Creating innovative and high-performance workplace
Manager, GRE Latin America
Facility Management, IBM Global Real Estate
Sao Paulo, Brazil

A Facilities & Real Estate professional with a degree in Engineering, working for over 25 years, specializing in service contract management, working on global projects and RFP execution for facilities services, implementation of new technologies, safety standards and environmental projects. Always focusing on the workplace experience.

People know me for creating a high-performance culture during project performance, for developing standards and methodologies to support business unit processes, maintaining compliance with global standards, and improving delivery.

Currently my position at IBM is Facilities and Engineering Manager for Latin America (10 countries), working collaboratively with all Global Real Estate team members to contribute to the implementation of the IBM GRE Strategy, deploying our global operating procedures and service level agreements.

I started my career as a maintenance electrician, always working in the service area and being responsible for the maintenance & fitting up of the technical team and being responsible for implementing processes and quality. At IBM, I worked on energy engineering in the free market and on energy conservation projects.

Up-to-date training and good global knowledge of business strategies and management tools with an MBA in Business Management completed in 2020 at Fundação Getúlio Vargas. Graduated in Electrical Engineering with emphasis in electrotechnics from Universidade Paulista.

2.23

Irene's FM Journey "Life Begins at the End of Your Comfort Zone." Neale Donald Walsch

Irene Thomas-Johnson

There is a growing diversity of individuals entering the workforce, bringing different perspectives to the workplace, and dramatically altering how and with whom business gets done. Beyond demographics, diversity is about leveraging strengths and varying ideas, points of view and experiences. Coupled with inclusion – incorporating what makes each person unique to create an environment that engages, welcomes, and values diverse characteristics – diversity provides equal opportunities for career development and success, powering the relationships and experiences that will become more valuable in a changing world.

The facility management profession has evolved beyond buildings to touch nearly every aspect of a company, from operations and human resources to finance and technology. Facility Managers work across business functions, making strategic and tactical decisions that impact space, services, costs, and risk. Yet Facilities Management does not necessarily require an advanced educational degree or years-long training to get started or a prescribed progression from one job title to the next to advance. This presents incumbent workers, the unemployed and underemployed, and youth seeking career direction with an opportunity to enter an industry facing a significant labor shortage.

The gap between available FM talent and job demand is widening. The IFMA Foundation's Global Workforce Initiative (GWI) focuses on attracting diverse talent to FM and supporting their long-term growth through continuous training and education. The GWI program connects with local communities to inform students, parents, teachers, guidance counselors, community organizations, economic development, and government agencies about educational and employment opportunities in facility management. GWI partners, such as JLL, support the initiative to ensure the continued sustainability of the FM profession.

What follows is my journey. A woman with a diverse background who never intended to work in Facilities Management; yet now I hold an executive-level position. I discovered that the field not only opened doors beyond what I ever thought possible, but also fulfilled a deeply held commitment to excellence, instilled by my family (see Figure 1). While I like so many typify the "accidental FM" – inadvertently landing in a role that proved perfectly suited to their interests, skills, and goals – I am dedicated to finding and mentoring the next generation of FM professionals, helping them take a deliberate path toward a rewarding career. My career sojourns also presented an opportunity for my company to become a GWI Advisor which gives affords them one of the foremost chances of hiring the best FM talent for their growth in the industry.

My father grew up poor. As a young man, he enlisted in the US Army and was deployed to Germany, where he met my mother, also from meager beginnings. They fell in love, married, had three girls, and were stationed in Kansas in 1962 (see Figure 2).

It was a difficult time in the US for an interracial couple. Apartment landlords would claim "no vacancies" once they realized the family was multiracial. Living in a perpetually flooded basement apartment, the family had little money;

Facilities @ Management, Concept - Realization - Vision, A Global Perspective, First Edition. Edited by Edmond Rondeau and Michaela Hellerforth.

Figure 1 Family photo. (*Source:* I. Thompson-Johnson).

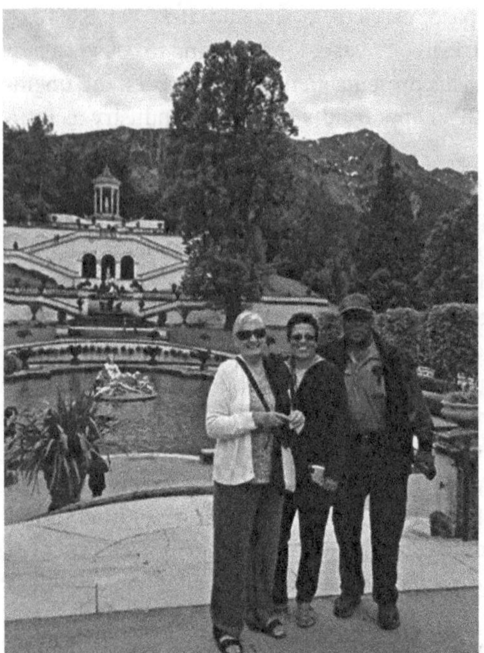

Figure 2 Family photo. (*Source:* I. Thompson-Johnson).

but my parents worked hard, kept their children safe and taught them important life lessons: take challenges head-on, do a job to the best of your abilities, be honest, and always respect others. They made a wonderful life together and are now living their best years.

Thanks in part to a two-year scholarship and Pell Grants, I attended the University of California Los Angeles (UCLA). I got a part-time job at the UCLA store, working in the payroll department doing special projects. I was good with numbers and started helping in the finance department with statements, budgets, and invoices.

When my scholarship expired and the grant money dwindled, I had to work full time to cover my tuition and living expenses. I accepted a position as an administrative assistant at Associated Students UCLA, and it wasn't long before I was promoted to project coordinator, training managers on computers, inventory, budgets, and profit-and-loss statements.

I was then promoted to project manager, completing tenant improvement projects for restaurants, the UCLA bookstore and office space allocations, working directly with architects and contractors. I taught myself the AutoCAD program to design floor plans that would help visualize office moves, retail space and conference rooms as the UCLA store and student union were reconfiguring and expanding spaces. What I am most proud of is the interior remodeling of the UCLA Store and the base isolation construction of Kerckhoff Hall. Both would never have been possible without the leadership, mentorship, and the faith in my ability of all of those who helped me throughout my career at UCLA (see Figure 3).

Over time, I became efficient at project management and was asked to take on the facilities and maintenance department. Although I had no idea what "facilities" were, I took the challenge head-on, as my parents advised. I spent 21 years on the university campus, going to school fulltime and then working fulltime. Now I am a Certified Facility Manager® and a global account executive for the one of the largest real estate companies on the globe, Jones Lang LaSalle (JLL).

When I first started with this global real estate company, I was asked to take on a role in sales. My first response was ok, "but a salesperson you mean like a used car salesman"? Boy, was I ever wrong about the field? I was able to use all my years of experience in facilities management and bring this knowledge to every engagement. I had the experience of being the owner and now working as an outsource provider, I had the opportunity to ensure that the dream that was being sold was one set. A unique and rare circumstance, having experienced both sides and having the opportunity to put the skills gained into reality. Given my experience of owner and operator, I was able to understand fully what it took to manage the environment, operate in the

Figure 3 Irene at UCLA. (*Source:* I. Thompson-Johnson).

environment, and be successful in that same environment. Working directly with companies to help them envision their dreams and aspirations and assisting them in bringing those dreams to life.

Currently as a Global Account Executive, coupled with my many years as a sales executive, along with all the fundamental years in the profession has allowed me to thrive in this field. I now manage and develop successful collaborative partnerships with clients on a global scale. I am accountable for working with them on their corporate real estate objectives including cost reduction, diverse talent improvement, increased service integration driving client occupant productivity and ability to drive consistent real estate delivery locally and globally.

Over the course of my journey, I have grown my breadth of knowledge in facilities (see Figure 4). I worked tirelessly for diversity and inclusion in the workplace; and through programs like GWI, I pay it forward by sharing my experiences with those who can benefit from a nudge toward an industry described as "one of the best-kept secrets in the job market."

There are four nuggets that I have taken away from this amazing career that I fell into so many years ago. First, never, ever turn an assignment down. No matter how small or trivial, accept it, embrace it, and do it to the best of your ability. You will be surprised how this will open doors, and new possibilities. Second, always ask questions and seek knowledge. When I started my career, I knew nothing, but I was not shy to ask. When I had to work with a vendor whether an electrician, plumber I would get to the job site with them, watch, ask questions, and learn. You would be surprised how much you can learn by just being interested in the trade. Third is to network as much as you can and when doing so never be shy about asking for assistance. Maybe you are interested in a particular position, interested in a certain field, ask for them to help you open a door, get an interview, get an introduction. The worst they can say is no. So, it never, ever hurts to ask. Finally, and most importantly always pay it forward. It takes less than a minute for you to make an introduction for someone thru email, LinkedIn, or a conversation. This small gesture can change someone's life!

Work hard, play hard. There was instilled in me a work ethic that came to me thru my parents and was reinforced thru mentors and good friends throughout my career journey. The most important lesson I have learned albeit a little late in my career journey, is to balance by work with my life. So, I work very hard, and I balance that with playing hard. It keeps me healthy, grounded in what is important which is my family and myself, and keeps me motivated to keep engaging in my career (see Figure 5).

When I joined my first cycling club one of my now best friends told me on our very first journey that life begins at the end of your comfort zone (see Figure 6). That became one of my favorite quotes and how I lead my life today (see Figures 7, 8 and 9).

My story is only one small one in the brilliant careers that are possible through FM. I will be forever grateful for the mentors, both sought and happenstance, that shared their wisdom, worth ethic, and most of all their time in helping pay it forward.

Figure 4 Irene at Save the Planet gathering. (*Source:* I. Thompson-Johnson).

Figure 5 Irene Climbing to Everest Basecamp, Nepal. (*Source:* I. Thompson-Johnson).

Figure 6 Irene Climbing to Everest Basecamp, Nepal. (*Source:* I. Thompson-Johnson).

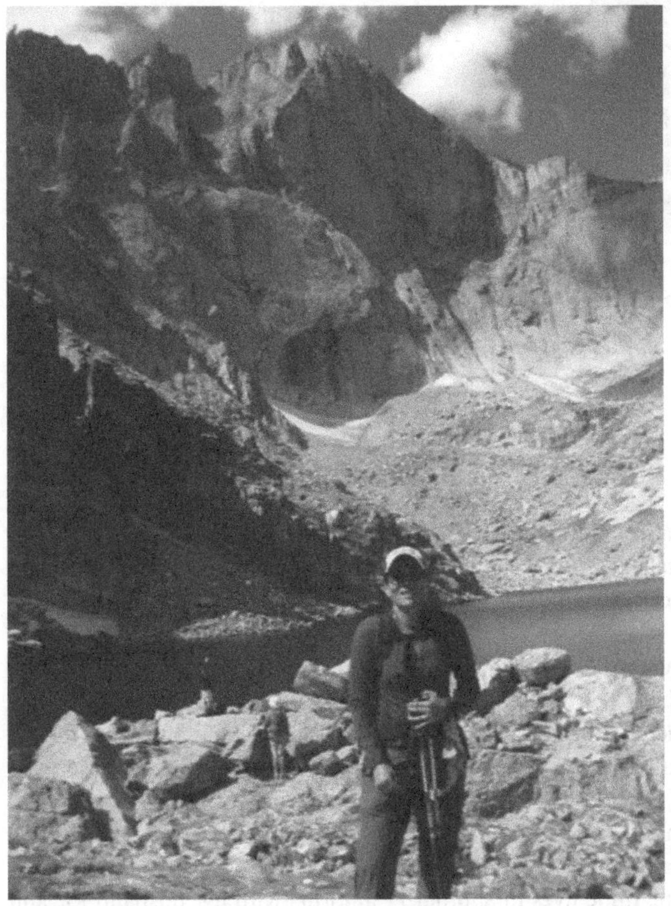

Figure 7 Irebe at Long's Peak, Colorado. (*Source:* I. Thompson-Johnson).

Figure 8 Irene in Iceland. (*Source:* I. Thompson-Johnson).

Figure 9 Irene at the Grand Canyon. (*Source:* I. Thompson-Johnson).

Irene Thomas-Johnson, CFM
JLL Global Account Director / Exec. Director
Solutions Development / Work Dynamics
Los Angeles, CA USA

With over 25 years of consistent leadership advancement, Irene has earned a reputation as a commercial real estate (CRE) industry expert in corporate real estate and integrated facilities management (IFM) outsourcing on a global scale. Currently, as a Global Account Executive with JLL Irene manages and develops successful collaborative partnerships with our clients.

Throughout her career, Irene has been dedicated to delivering excellence through innovation, diversity and inclusion and empowering high-value talent. She is the recipient of the 2021 Globe St. Women of Influence Humanitarian award and in 2022 she is the recipient of the JLL Diversity and Inclusion Champion Award for her work on the Foundation's Global Workforce Initiative. She is a long-time member of IFMA and currently serve as Chair of the IFMA Foundation Board Trustees, the IFMA Foundation Executive Advisor & Chair of Global Workplace Initiative programs, and Past President of the IFMA San Fernando Valley Chapter. Her mission along with the IFMA Foundation is to make Facility Management a career of choice.

A University of California Los Angeles graduate, Irene lives in Southern California with her husband Ric and enjoys spending time with her family and friends. In her spare time, she is an avid outdoor enthusiast and loves to hike, backpack, and bike. Her greatest recent accomplishment, in November 2017, was hiking to Mount Everest Basecamp in Nepal.

2.24

"My Way" – My Way to Facility Management

Annelie Casper

Entertainer Frank Sinatra may not have immortalized my professional life in a song, but the title of his hit "My way" is perfect for my career path in Facility Management.

As we all know, the journey is the reward. Mine was unexpected, but all the more exciting and enriching. When I thought of real estate at the beginning of my professional life, it was for me places of encounter and living spaces for people. Architecturally pleasing or just functional. Today I know that real estate is much more than that. In order to fill these places with life for people, to meet their needs, demands, and desires for living spaces, processes are needed. This soul, real estate, only arises when the human being is at the center of all actions. This is exactly the focus of facility management. An agile and multifaceted industry, an always challenging task and a reward that has a very special value: satisfied, healthy people.

What would happen if all facility management employees suddenly stopped work for 24 hours? Our country would stand still! Nothing would work anymore, because FM is in all areas of our private and professional everyday life. When shopping, in the football stadium, in the office or among modern production facilities, when traveling on vacation or our health care. Facility management is not always in our daily consciousness, because it is especially good when you hardly notice it, because everything runs smoothly. This makes it all the more important to focus on the people who play a significant role in the success of our economy. This is one of the central tasks of gefma – German Association for Facility Management and as Deputy Managing Director, I can contribute to giving FM a face, a voice, and countless names.

Facility management is a creative industry that deals with the processes within a property in a wide variety of ways. Before I embarked on my path to FM, I experienced again and again in my work as an architect that design quality was only seen as a cost factor on the client's side, but not as a contribution to the benefits or cost savings and thus to the success of a property. The perspective usually ended at the masonry and rarely reached the phase that represents the longest and most essential in the life of a property: the operation.

I, on the other hand, am still fascinated by the 360-degree perspective of a building. The life cycle of planning, construction, and operation is not a series of disciplines, but a breathing organism in which all key players with an understanding of their own strengths and the strengths of others must work hand in hand to achieve one goal: to plan, build, and operate the perfect property for the needs of its users. In the processual life cycle approach, I recognized an instrument that turns the pure price competition of the construction industry into a quality competition.

On my way into the facility management industry, I had support from many sides, especially from my former companion and head of the HSG, Otto-Kajetan Weixler. It was he who taught me a lot for my current path with his understanding of values, his appreciation, his appreciative treatment of people and his forward-looking

Facilities @ Management, Concept - Realization - Vision, A Global Perspective, First Edition. Edited by Edmond Rondeau and Michaela Hellerforth.

view of the industry. So, he has become my role model and my orientation in my daily commitment to the FM industry.

Let's turn back my personal career clock a little: It wasn't a direct path that led me to facility management. My path began with the study of architecture, classically focused on the conception and planning of buildings. The implementation of the building and certainly not the phases after its completion were rarely considered. This type of teaching was based on the fact that only individual actors are responsible for certain sub-areas of a property. The overarching area, the so-called construction project management, in which the entire life cycle of a project was mapped, and which covered all areas, from the project idea, the planning, the actual construction execution, the management, i.e., the facility management and the dismantling, was missing from the training.

It was clear to me that this perspective was a one-way street. I wanted to do my part to broaden the focus within my industry. In addition to my profession as a graduate architect, I began a further education course in "Construction Management," in which the approaches of control, economic efficiency and the quality of a building for optimal use were taught, considering the aesthetic, energetic and ecological requirements as well as the current legislation. An important module was the subject of facility management. It was here that I heard for the first time that there was a compass through this almost endless variety of topics: the groundbreaking guidelines of the industry association gefma. I also encountered the "Three Pillars of FM" for the first time: Technical Facility Management (TGM), Infrastructural Facility Management (IGM) and Commercial Facility Management (KGM). Admittedly, this wasn't really the integrative focus I had in mind. A coexistence of the disciplines instead of a cooperation of the actors involved. It was only with the approach of the public-private partnership (PPP) models that I became aware of the influence of facility management on usability, resource efficiency, information availability and environmental impact of PPP models. I was enthusiastic about the concerns and problems involved in the procurement and financing of public infrastructure and the consideration of the entire value chain, as this focus had an enormous added value.

A short time later, I got the chance to bring this approach into the training of young talents in the real estate industry. As a lecturer at HFH – Hamburg Distance Learning University for the module "Strategic and Cross-Life Cycle Facility Management" in the Facility Management program, I was able to teach students the ability to think strategically and at the same time turn theoretical ideas into practical solutions.

For me as a lecturer and an expert convinced of the life cycle approach, one thing was of very important significance: After studying this module, the students knew about the necessity of incorporating facility management in early conceptual planning phases of real estate development.

Facility Management: A Pioneering Role in the Life Cycle

But I didn't just want to convey my conviction theoretically, I wanted to act practically. So, the teaching assignment was the impulse for me to apply to HSG Zander GmbH. A company that had its roots in construction and was just growing the first shoots of facility management.

What fascinated me from the very beginning of the HSG was that even though this unit was part of a group and thus integrated into corresponding structures, the corporate culture was familiar and employee-oriented – high value orientation, strong identification, and close cohesion within the company. This spoke in favor of this FM company and was a completely new experience for me, who had worked exclusively in small architectural offices, because the HSG combined the values of a family business with the strength of a corporation. Our Chief Executive Officer, Otto-Kajetan Weixler, led the company with a high level of personal commitment and contagious verve, leading us to the top of the facility management companies on the German market.

Even before I joined HSG Zander GmbH, the company was part of the listed Bilfinger Berger Group. For my work, this meant that we actually mapped the entire life cycle in a property from a single source, without major friction losses between the individual actors. Perfect for the birth of a product: we developed the "ONE" standard, a life cycle product based on the life cycle approach. The fact that there was little demand for such a product on the market at that time was mainly due to the fact that market participants were not yet ready for this type of consideration, and, in addition, the added value was difficult to measure and not clearly visible.

Facility Management: When Buildings Age Sustainably

At the same time, the principle of sustainability has become more and more influential in all sectors of the economy, including the real estate industry. Aligning economic activities with a long-term time horizon is quite in line with the concept of life cycle analysis for real estate, which often has a service life of more than 50 years and the expectation that this can lead to optimization potential through a cross–life cycle view. This leads to greater effectiveness and efficiency in the use of resources in real estate. In our department, we dealt with this approach on a daily basis, and we wanted to show that in the understanding of building operation in an environmentally and resource-saving manner, work is a matter of course. After all, who, if not corporate management, has the know-how for more sustainability in the real estate industry?

At that time, I met gefma for the second time and this time not only out of teaching. I was actively involved in the association's work. Otto-Kajetan Weixler was not only my boss at the HSG, but also chairman of the board of the gefma association. Looking ahead and with an eagle's eye for the development of the FM industry, he launched a definition and standard for sustainability in FM as chairman of the association. Very important: This development should come from the industry itself. An idea that is still trend-setting today was born and a short time later I sat together with some of our competitors at the platform and we worked hand in hand on the standard for sustainable FM in Germany. Such a committed approach by an entire industry strengthened me very much that the step into facility management was the right one. Here I was able to make a lasting difference.

This meeting of the major FM service providers was the birth of our gefma working group "Sustainability," which still exists today, which I have the privilege of chairing with Prof. Dr. Andrea Pelzeter. It was the starting signal for a whole series of approaches, white papers and guidelines – far beyond sustainability for all areas of facility management.

Facility management is people business. This was particularly evident in the association's work – without the good and honest cooperation with the then managing director Dr. Elke Kuhlmann, I would never have had such good access to the tasks, the people and also the possibilities of association work. Through their support, I understood the added value of representing an industry and volunteering – especially for myself.

gefma – an Association for the Future

Looking back, it was not only the start of my involvement in the Sustainability Working Group. Through my association work, I began to actively shape our industry, to set impulses and to make ideas visible. Over the years, I gained more and more insight into the life of the association and took part in the various network meetings of the regional lounges and in the working group meetings. In 2017, I took over the lounge management Hesse myself. In the meantime, I had two hats on, an entrepreneurial HSG hat and an industry-shaping association hat – an optimal position for me. I was fascinated by the combination of practical, entrepreneurial solutions and the network of an association in which new ideas emerge through the intensive dialogue of experts from companies in the industry.

Up to this point, a multifaceted path, always with the aim of seeing real estate in its entirety and thus making it more sustainable. To this day, I still like the creative added value of my profession. I didn't have to think long for my "yes" when the gefma board asked me if I wanted to take over the deputy management of gefma after the previous management had retired. In the meantime, I was so well integrated into the association and its member network that I saw exactly the right space for success for myself in this position and still see it today to actively shape facility management in Germany hand in hand with many other talents and experts.

From a Wallflower to a Pillar of the National Economy

Facility management is a relatively young and dynamic industry in Germany, which has established itself among users in recent decades, primarily through its positioning as a secondary process designer that supports core business. Despite this obvious importance, FM has long been ridiculed in Germany. This was mainly due to the fact that there was no reliable assessment of the economic relevance of facility management. The challenge: create facts!

In order to gain transparency in the diversity of FM within our society, gefma had an industry report prepared for the first time in 2010. The key figures surveyed exceeded expectations: The added value of facility management proved to be more significant than many a "classic" successful industry. Since then, gefma has had the figures verified in a current industry report every four years.

With a share of 4.52 percent of gross domestic product (GDP), facility management today confirms its stabilizing factor for the German economy. Currently, more than five million people work in facility management. With a gross value added of 152 billion euros, corporate management is one of the top 6 sectors of the German economy and thus ranks just behind the automotive industry and ahead of mechanical engineering.

In short, the industry has long since outgrown the "wallflower existence" and has become a stable pillar of the German economy.

Facility Management – Quo Vadis?

There are some indications that the FM industry will see more changes in the next three years than in the last 30 years combined. On the one hand, external influences are influencing the industry: a fading pandemic, a war in the middle of Europe, climate change, the energy transition, and increased inflation. Germany is in the midst of a comprehensive transformation process towards a climate-neutral and sustainable economy and society. Both the achievement of climate targets and adaptation to the consequences of climate change require rapid, consistent, and coordinated action in many sectors and areas of life. The associated increasing importance of facility management is leading to an increasing perception within the real estate industry.

More than 30 years ago, gefma set out with the mission to intensify the consolidation and promotion of all activities in the field of facility management in Germany as well as the cooperation with the corresponding associations at home and abroad – in short, the professionalization of facility management visibly to make. The objective from its beginnings in 1989 still applies to the association today. In addition, there is now a clear focus on key topics that are crucial for a society and an industry in transition: sustainability, digitization as well as education and training.

In the coming years, industry will have to face many challenges for the built environment. The digital transformation as the dominant technological and social change will have a great influence on people in professional, private and public life – Smart Building, Smart Home or Smart City are just a few buzzwords. Andin this way, digitization also has an impact on facility management. New technologies such as the Internet of Things (IoT), artificial intelligence (AI) and machine learning enable the networking of building infrastructure and systems. Facility management will increasingly adopt data-driven solutions that improve service efficiency, reduce costs, and support decision-making in individual real estate operations. Building on these changed requirements, there are increasing demands on vocational and academic education and training. Digitalization as a key driver will open up new fields of application, such as: Measures to optimize sustainability and climate protection as well as the design of new workplaces in workplace management.

The further path of facility management can only be a sustainable one. This is a way in which all stakeholders in FM are anxious to ensure that the buildings whose usage processes they design no longer have (almost) any impact on the environment. Due to the European Green Deal, CO2 pricing increases the need to go down this path, especially for existing buildings. Facility management is becoming enormously important for the implementation of ESG strategies. Through a rich number of measures ranging from quick wins in energy and resource efficiency to land management and disposal, it will play a key role in integrating sustainability goals. Facility management is clearly the key on the way to a climate-neutral real estate industry.

The topic of workplace management will also become more and more important. In the future, the importance of a positive user experience in the workplace will become more and more important for people, as this is seen as a success factor of the core business: A high level of employee satisfaction has a significant impact on innovation cycles, customer satisfaction, and thus profitability. FM will increasingly focus on creating attractive and efficient working environments by bundling processes as an interface between HR, IT and the core business of building users, serving as a direct point of contact on site and creating attractive working environments. This includes flexible workplace design, technologies to improve employee productivity and well-being, and tailored services that align with the individual needs of companies. The daily experience in the workplace can be significantly positively influenced by intelligent facility management, and thus modern FM becomes a value lever in the core business.

The qualitative and quantitative growth of the FM market will be accompanied by a corresponding increase in the demand for specialists and managers. Here, the industry will find its role as a strategic partner for the client in the future through new ways of training and further education, interlocking of the different processes and the increase in further fields of activity.

gefma – an Active Creator of the Market

In order to successfully solve the tasks described, the swarm intelligence and the connecting nature of gefma are required. gefma is the industry association whose approximately 1100 member companies are heterogeneous and thus represent a colorful mix: FM service providers, asset and property managers, consultants and IT service providers, investors, representatives of the public sector and universities. It is this mixture of people, experts and designers who, through their voluntary work, contribute to the fact that gefma is a functioning network and continues to grow. Above all, however, gefma is a navigator in an increasingly complex real estate industry. The facility guideline system – a firm basis of the association – with more than 110 guidelines and white papers sets standards from the industry for the industry and provides orientation and, above all, security in high quality. Thanks to this expertise, which has grown over almost three decades, gefma's guidelines still represent added value for the success of the member companies and for the "Ready for the future" of the industry. In addition to these standards, the certification of CAFM software, sustainability (Sustain FM), the certification system for students, business administrators and service staff in FM as a quality label, the professors' network, and much more are also part of gefma's future books.

Looking to the Future – My Conclusion

The construction industry, and in particular the profession of architect, is always described as a creative and therefore exciting part of the real estate industry. But this is only a small aspect of the big picture. More important and innovative are the processes that take place in a building and are managed by the FM. They make a significant contribution to the fact that real estate is a place to live, work and succeed. If you were to compare a building with a computer, then the planning and construction of the building would be the hardware. The software, and thus the core of it, are the processes running in it. In the sense of a building: facility management!

In the future, the main task of gefma's association work will be to strengthen the importance of facility management in the public perception and to make a clear statement within the real estate industry that the future of the real estate industry and our society can only be successfully shaped with FM. To this end, the discussion about the challenges of ESG and sustainability in combination with digitization and anchoring in education and training can provide great added value.

I have the privilege of helping to shape a young, exciting, and innovative industry. It makes me happy to make a small contribution to the association and the industry. The different paths of the authors in this book are intended to inspire enthusiasm for the opportunities of a varied and creative industry – facility management.

Annelie Casper
Deputy Managing Director
German Association for Facility Management (gefma)
Bonn, Germany

Dipl.-Ing. Annelie Casper joined the management of gefma, the largest association for facility management in Germany with over 1000 member companies, in 2020.

She studied architecture at the University of Karlsruhe (TH) and business administration at the University of Hagen, where she graduated with a degree in business administration (IWW). Through her many years of work as an architect, Ms. Casper has experience in all phases of construction. In addition, she worked as a lecturer in the field of facility management at the Hamburg Distance Learning University. In 2011, she joined the Public Private Partnership / Lifecycle Projects (PPP / LCP) department at HSG Zander, now Apleona GmbH. Since 2013, she has been advising on the topics of sustainability and climate protection, including co-developer and auditor of the sustainability standard Sustain FM (gefma 160). She is also trained as a DGNB Consultant and BREEAM Associate.

Thanks to her many years of experience in the various areas of the real estate industry, she has in-depth knowledge in the areas of project development, project management, construction, and facility management. She is currently involved in various research projects such as carbon management as deputy head of the sustainability working group at the leading association for facility management gefma e.V.

2.25

Journey into FM

Stan Mitchell

The Beginning

In hindsight my journey into Facilities Management (FM) probably started when I opted out of school. I left before the term ended on the basis that by the time, they found me (I had escaped to the Highlands of Scotland) it would be too late to get me back! My escape route was via a job in farming, which while introducing me to a workplace that was different every day with challenges from equipment, animals, and people, introduced me to some of the skills I apparently had and would serve me extremely well later in my career.

Engineering

Following several years of life on the farm, my longer-term thinking brought me back to Glasgow to serve an engineering apprenticeship and map out a career leveraging an apparent natural trait for getting things working. As would have been predicted by those wiser than me, that move took me back to education which became a prerequisite if I was going to get anywhere in my chosen vocation. While learning the practicalities of my trade building equipment for the nuclear industry I was also progressing formal educational qualifications. During that period of my career, I worked alongside old seadogs who were always reminiscing about the time that they spent at sea in the merchant navy, visiting glamorous parts of the world. However, they did tend to omit elaborating quite so much about the 24/7 lifestyle that was a necessity between ports of call. I was smitten and soon left to take my career to the next level. a life on-board ocean-going ships travelling worldwide.

As an engineering officer on-board ocean-going ships, I had found my natural home and vocation. This effectively was my introduction to FM although it was not called that. I worked in an environment where you have total responsibility for all aspects of FM and more with one significant difference, While at sea you have no supply chain to support you, therefore when something stopped working; when systems failed to fulfil their purpose; when emergencies occurred; everyone on that vessel relied upon you to solve the problem and ensure their safety and wellbeing.

It was during this period I married Valerie who shared the trials and tribulations of life at sea for some five years, following which there was the arrival of Gordon and Sarah. For me It was time to think about my future and ability to support my growing family but importantly also spending time supporting them on dry land and not from a distance via the odd radio or telephone call from some distant port somewhere in the world.

Facilities @ Management, Concept - Realization - Vision, A Global Perspective, First Edition. Edited by Edmond Rondeau and Michaela Hellerforth.

I therefore left my life at sea some years later having again been back to "school" to study and qualify as a Chief Engineer. I did not realize it at the time, but I had during those years been working in Facilities Management, the main difference being that the "Facility" moved quite a lot.

I had, without knowing, experienced what in hindsight was perhaps the greatest preparation for what we know as FM today. What I had gained beyond what every engineer would hopefully gain during their career, was an insight to people and what makes them tick including all the cultural and personal idiosyncrasies that exist but which you seldom fully appreciate when working in the "normal" workplace. When you live, work, socialize, eat, in a confined environment for 3 to 9 months at a time with a very small group of people you learn about every aspect of human nature, the good the bad and the ugly I am afraid. This was a true education and one that has been of considerable benefit as I furthered my career in FM proper.

Manufacturing

Upon leaving life at sea, I joined a manufacturing organization building computers which at the time was delivering leading-edge technology solutions during the early days of the personal computer.

I was extremely lucky that I joined the company before there was a factory; therefore, my first role within an FM context was to work as part of the design team to design, construct, fit out, and establish a Facilities Management organization starting from scratch. It was certainly a new challenge and tested my ability to be agile when involved in commenting, challenging, and critiquing subject matter in the construction industry that I knew absolutely nothing about. It was in that role that I learned what I believe to be one of the most important attributes of a successful Facilities Manager, which is that it is perhaps more important to know what you are not, than opposed to knowing what you are.

I was working for an American Corporation which had recruited its management team from across the USA, the UK, and Ireland. It had a typical approach to the development of its senior management team where they wanted their leadership to undertake responsibilities across all aspects of the business. This resulted for me in going from an FM role once the facility was in a steady state, into warehousing, goods receiving, planning, and production control. I had actually refused twice to transfer out of my FM role when I was invited by the President of the company for a coffee. He spent 30 minutes chatting to me about my family and what it was I found exciting and enjoyable about my job in FM. He then took everything that I enjoyed about the job in FM to explain that it was these skills and attributes that were needed to transform the warehouse operation. That was a true lesson about managing people in a clever, persuasive but positive manner and in the end, he was proven correct. I took the warehousing job, transformed how it operated, rationalized how we store the parts that it housed, introduced "just in time" material into the factory, got myself a formal qualification in production and inventory control, all of which resulted in a reduction of circa $5 million inventory on the shelf at any one time and more efficient processes and procedures related to material handling.

However, one of the most interesting experiences during this deviation from FM was, for the first time in my career, I became a customer of FM and that experience allowed me to view FM in a completely different way. To be the customer was an experience that has influenced what I believe to be true FM ever since.

When looking back on my time and experiences within this organization the most significant learning that I took away from it was what I learned from the President of the company, an Irishman, who was a leader way ahead of his time. Much of which I have seen appear in terms of leadership and good management practice as well as how to interact, motivate, and gain respect from the people for whom you are responsible, I learned from him indirectly. He introduced a single status company, there were no management offices; there were no executive standards that differed from our staff in the company; there was minimal barriers between manufacturing personnel and office-based personnel; this company encouraged people to take risks. When the management team required off-site management discussions all managers attended on the basis that if your team was unable to keep the business fully functional during your absence you had failed in your duty as a manager. He was a wonderful role model and someone who taught me much by example that I was able to apply in my role within that particular organization and in every role, I have undertaken since. That experience transformed me as a Facilities Manager from considering FM to be about buildings and engineering services to the realization FM is primarily all about people.

Networking in Facilities Management

It was during my time when I was mobilizing and establishing the FM organization within the manufacturing organization that I reached out to a number of colleagues who were working with similar manufacturing organizations to encourage a proactive networking community. While we were in some circumstances working for competing companies, from an FM perspective we were not. Therefore, I quickly managed to pull together a small group of like-minded individuals to have regular get-togethers to share challenges and opportunities in what we were all doing within FM. We were all based in the central belt of Scotland where there were many other similar technology and electronic manufacturing organizations. At the time this was a boom industry in Scotland and the area was given the title "Silicon Glen," the little brother of "Silicon Valley."

Having created this network I heard about a fledgling organization that had been established in London called the Association of Facilities Management (AFM). I reached out to them as they had effectively done the same thing but formalized it and our networking group in Silicon Glen became the first region of the AFM. As a result, I became a non-exec officer of the AFM and subsequently a board member.

It was during this period we became aware of similar organization called the Institute of Facilities Management which was an offshoot of an organization called the Institute of Administrative Management. These two organizations subsequently merged, and the British Institute of Facilities Management was born in 1993. Having been involved as a member of the board prior to the merger and thereafter I was involved in developing what was a professional body that was new within the UK and trying to gain credibility alongside the long-standing professional bodies within the built environment sector who were in many cases over a hundred years old. While we were challenged with demonstrating that FM was a strategic and distinctly different professional discipline, we set about doing so creating the building blocks, primarily education programs and qualification, to justify our existence.

While having never intended doing so, I found myself as National Chairman of the BIFM in 2004. While in office, with the support of many others, we made significant changes to the status of the BIFM which was fast evolving. Some of those activities included establishing the "Partners in FM Excellence" agreement following the concept of the networking initiative I had created within central Scotland, by bringing together, for mutual benefit, the US-based International Facility Management Association (IFMA) and the Facility Management Association of Australia (FMAA), both organizations at that time being led by Sheila Sheridan and Steve Gladwin respectively. The logic being that working together we had a greater opportunity to further the cause of professionalizing FM worldwide. This initiative was successful in breaking down many of the barriers that existed at the time.

In 2006 it was decided that if we were able to get these three organizations to collaborate with each other than could we escalate this initiative to a truly international level. With the support of the leadership at the time within IFMA and FMAA we established Global FM, a Brussels-based international federation of FM member organizations with the intent to support FMs setting up their own National Associations. This created an ever-stronger worldwide community of talent and awareness of FM as that strategic professional discipline worldwide. Why Brussels, at the time we felt it important that we avoided any accusation of nepotism so by establishing the head office in a neutral country beyond those of the founding organizations it would hopefully encourage others to join. Given that at that time the emergence of other national associations was most prominent within Europe, Brussels seemed to be a good choice.

In the early days of Global FM, we created the "Hot Topics" that we felt were worthy of our collective focus through further collaborative discussion and actions with the growing number of national associations that were members, in broad terms they were described as follows:

1) **Sustainability** – sustainable communities, business case (model) for sustainability, life cycle sustainability
2) **Innovation** – encourage Facilities Management action from continuous improvement, step change or strategic change of direction
3) **Education and Training** – compare and contrast current member Facilities Management qualifications; develop an international set of core competencies
4) **Risk Management** – develop international glossary of Facilities Management terms and best practice guidelines
5) **Facilities Management now and in the future** – coordinate a greater understanding and access to developments across the world
6) **Economy and Cultures Across Borders** – understand and develop tactics to address the issues and effects of the application of Global standards within diverse cultural environments; consider and contrast both in-house and globally and/ or regionally outsourced service provision.

The above represented our thinking over 20 years ago, and it is interesting to consider them today in 2023 and how successful have they been!

Global FM at that time, as it does today, relies upon the respective leadership of the growing number of FM Associations around the world to adopt and support the mutually beneficial collaborative intent that Global FM was designed to achieve. That remains both the opportunity as well as the difficulty for Global FM today, it needs its leadership to proactively promote and create the organization that can achieve more together than each of its individual members.

Foray into the Standards World

While Chairman of the BIFM I was invited to speak at a Real Estate/ Project Management conference in Spain to introduce the topic of FM to that particular audience. I was the only one there that was talking about FM, so realizing I was literally talking to an audience who probably at the time considered FM to be about janitorial services but not as a professional discipline, I thought I would try to engage with them through a tongue-in-cheek comment about the relationship between the FM and the Real Estate sector. I therefore opened my speech with a smile by stating Real Estate/Property Management is actually a subset of Facilities Management! I was the only one smiling amongst the stony silence of the room.

The interesting comment that became another one of those memorable and influential moments in my career, was the comment from another speaker at the conference who was the Global Real Estate Director for a major engineering corporate organization whose stated during the Q&A session that my comment at the beginning of my speech was correct! He subsequently, however, followed on to say, "until you guys in FM get your act together and have international standards that can be understood and applied worldwide no one is going to take you seriously." That comment was the bell that rang in my head at the time with the realization that he was right. It did not really matter how important we thought we were, what did matter was that there were educational programs with respected qualifications; professional bodies that shared and encouraged best practice; and most of all those international standards that would be accepted as credible on the global stage. My focus on the standards world had just commenced!

My first step into the standards world was to approach the British Standards Institution (BSI) to establish an FM committee. I was able to justify such an initiative due to the fact that my Dutch friends and colleagues had initiated a European wide project within an organization called CEN to establish European standards in FM. CEN was an organization in the standards world, similar in concept to Global FM, which was a European-wide coming together of national standards bodies (NSBs) to collectively collaborate and create European standards that would be adopted by its member countries.

This enabled me to convince BSI that the UK needed a voice, and the BSI FM committee was established. I subsequently served as its Chair for 19+ years, and despite the lack of enthusiasm for many of those years by the more established built environment institutions in the UK to participate, those volunteers who did participate not only managed to contribute to the project within CEN but laid the foundations to what today is a very healthy BSI FM Committee with full representation that is very active having published over 16 standards in FM and with many more in the pipeline.

Within the CEN project I became the Convener of the working group that developed EN 15221-2 standard for the development of FM Agreements. During this period, I thought that my big challenge would be communication in the English language. This became another one of those step change moments in my career. I actually benefited from the fact that my colleagues were good if not better in understanding the English language than I was (I was a native of Scotland after all!), the big challenge was not the understanding of English it was the definition and context of the words that were used that was not always clear. While I knew exactly what I meant it did not always translate appropriately for my colleagues. A very important lesson to learn given that we all work in the so-called global village with colleagues, clients, and suppliers from many different cultures and understanding of the English language as well as some of the colloquial nuances that come as part of the message!

Having both worked with and very much enjoyed the collaborative community that we created within the CEN project I, perhaps inevitably, but with the support of the CEN Committee reached out to the International Standards Organization (ISO) to seek permission for the establishment of an ISO FM committee. This proved slightly more difficult than had been the case with the BSI due to there not being any similar project currently underway. Also, in the opinion of ISO at the time, there already was an FM committee, it was called Asset Management! Following a third attempt at convincing the powers that be within ISO that FM was something significantly different than Asset Management they eventually agreed, and ISO Technical Committee 267 for Facilities Management was officially established in 2012.

This committee took the two initial standards created by CEN TC 348 and elevated them to the first two standards in the ISO 41000 family of standards. They became ISO 41011 Vocabulary, and 41012 Strategic Sourcing and FM Agreements which were the first ISO international standards in FM to be published in 2017.

Having been its chair for the initial nine years of ISO TC 267 Facility Management, it is with some satisfaction that today there are 52 member countries registered as participating in the development of FM standards. This for me personally bodes well in facilitating and responding to that comment way back in the early 2000s that until we had standards that were internationally recognized, we would not be taken seriously. I think I am safe in saying while there is still much to do in the promotion, recognition, and understanding about the value that we offer within FM, we are certainly being taken more seriously now than we were back then.

It should be noted, however, that the success of ISO TC 267 over the years since its establishment has little to do with myself and much to do with the volunteers who give their time in each of the national standard's bodies as well as within ISO TC 267 itself to create these standards. Stan Mitchell was the Chair and had the easiest job of all!

FM Influences Worldwide

Over the years of my involvement within this wonderful discipline of Facilities Management I have attempted to support and encourage others to accept the challenge of developing FM as the strategic professional discipline that I believe it is. My efforts in this regard, in addition to striving to ensure that as a company Key Facilities Management delivers high quality true FM advice services and support, I continue to support my international colleagues where proper in the establishment of FM associations.

While I retain my membership of the Institute of Workplace and Facilities Management (IWFM) (the new name of the BIFM), and IFMA in which I became a member in those early days of AFM, I have also been given honorary memberships with several FM Associations having supported their establishment and development. I am therefore proud to include my professional membership with the Egyptian Facilities Management Association, Polish Facility Management Council, **Panamanian Association** of **Facility Management, and** Middle East Facilities Management Association.

For my current efforts within ISO TC 267 Facility Management Committee, I have been given the task of the review of ISO 41001:2018 Management System Standard, which is the flagship standard of the ISO 41000 series of standards, that under ISO rules needs to review every 5-year years and this work will commence in 2023.

Going It Alone

During these many years of voluntary activity in the BIFM, the BSI and the ISO FM Committees I actually had a day job! I justified my investment in voluntary time promoting the strategic value of FM on the basis of the greater awareness and value-add potential that FM offers, Key Facilities Management will benefit indirectly. While never mixing the two roles with my enthusiasm to promote collaboration and unity across all aspects of FM within the Professional Associations and development of standards, likewise I have always been supportive of similar collaboration and working with like-minded FM organizations within the market and continue to do so.

Following my unplanned redundancy from the manufacturing organization mentioned above, which I doubt I would ever have ever left had it not closed down completely, I decided that this discipline that I was involved in made so much common sense and logic that I could hopefully make a living from it. Please bear in mind that this was at a time when very few people had ever heard of the term Facilities Management so it was decided that I would give it a go for six months, which was back in 1990.

I first established my own consultancy business called Facilities Support Services; the name chosen was in recognition that very few people had ever heard of the term Facilities Management back then therefore it was important that I sold my services under a more generic name. It worked and within a relatively short period of time I found myself having to bring on board additional resources to support me.

Over time and with the emergence and recognition of the term, due in part to the efforts of everyone in establishing UK and European standards as described above, as well as the maturing of the FM Associations around the world, the name of the business and indeed its structure was changed and today its name, as it has been known now for many years, is Key Facilities Management.

The business, which is the longest established FM business in the UK, also claims several other plaudits during its time in business including being the first FM business in UK to gain the "Investor in People" award. When all others were gaining the ISO 9001 standard accreditation, people were considered as being the more important focus for us. Key Facilities Management continues to be a niche player in the market and while we operate internationally and having done so for many years now, our ethos continues to be very much quality rather than quantity. In other words, we are not driven by scale of the business but by the quality of what we deliver; the support and development of our people; the loyalty of the demand organizations that we serve; and enjoying what we do.

This approach has resulted if Key Facilities Management having been appointed FM Consultants as part of the design team on the project that was the first zero carbon city in the world by Fosters & Partners; being appointed as auditors on behalf of the biggest food and beverage company in the world to review their outsourced FM service providers with subsequent recommended improvements, which were eventually adopted by the service provider, enabled them to retain multi-million value contracts; advising some high profile companies regarding contingency and security processes and procedures following 9/11; supporting one of the biggest pharma organizations through the management of their facilities across 19 countries; and perhaps the most enjoyable one in supporting many organizations worldwide with alignment to ISO 41001 Management System Standard, something we continue to do today. What all of the above has taught me is that none of us have all the answers to anything in FM. We have so much to learn from each other and the more that we take that approach the better will be the FM that we deliver.

From a personal point of view over this period of time I have had several acknowledgements for the work I have done on behalf of the sector in general but for me by far the greatest acknowledgement is the small part that I have played in that today the FM market is forecast as being valued at just over $2 trillion and that, I feel, justifies some of the time I have invested into the long-term credibility and future of FM.

While currently supporting the development of FM as well as what continues to be an extremely active involvement in a consulting capacity with clients worldwide, I very much appreciate having had the opportunity to work with quality clients in business; with those that I have met via professional associations and within the standards community; all of which has resulted in my having a truly international network of friends and colleagues which I have viewed as being of considerable value to myself personally and privilege to having had the opportunity to do so.

Looking into the Future

I think that much of what I have articulated above is all about laying foundations. I believe that FM is on the map as a distinct professional discipline although there remains many in Government, Industry and Commerce yet to realize the opportunity and use it to the full.

So where is FM going over the next 40 years?

I do believe that the future of FM is inextricably linked with Technology. I am not talking about CAFM systems, although they certainly have a part to play, I'm talking about the IR4.0 technologies that provide an opportunity for us to evidence data such as we have seldom had the opportunity to do until relatively recently. Having said that, the challenge with FM is how we harness that data into useful FM knowledge and with that knowledge transform the use of space and influence the operational, environmental, and financial value that we can bring to our demand organizations that we serve.

Tech such as AI, Machine Learning, IoT will serve us well as we develop the knowledge of how to use it. The day of the "Nano Second FM functionality in the Autonomous Building" is already here so what is the future for the Facilities Manager? First of all, we need to recognize the change that is on its way and grasp it as an opportunity rather than a threat, which it certainly will be if we do not recognize it. Such tech is just another tool in our tool bag but one that can evidence and articulate what we do that is strategically impactful to those demand organizations.

However, going back to my realization in my career about FM being all about people, I don't think that will fundamentally change. We will still be required to interpret the data that all this tech provides, we will still need to manage the virtual office that many feel is the future, but most important of all we will need to recognize that people do business with people. We will always need other people to develop and nurture social skills and our social societies, whether at work or at play, will need those skills and competencies, it's in our DNA.

Stanley G. Mitchell
Chairman and CEO
Key Facilities Management
Doune, Perthshire, Scotland, UK

Stan Mitchell brings 35+ years' experience in Facilities Management. As well as co-founding Key Facilities Management as a business in 1990 with Valerie his wife, he has led its development internationally delivering FM related Consulting, Training, Technology Solutions, and Operations.

He is recognized worldwide as a pioneer in the development of FM as a Strategic Professional Discipline. His contribution in this regard has included Chairman of the British Institute of Facilities Management, Chairman of Global Facility Management Organization, UK representative of the European Standards body CEN, Chairman of the British Standards Institution FM Committee, Chairman of the ISO Technical Committee for Facility Management all of which have and continue to create International FM standards.

2.26

How I Came to FM

Karin Schaad

As a good student fluent in several languages, I was lucky enough to attend high school and aim for a baccalaureate with a focus on modern languages. Shortly before graduation, it was time to think about further training steps and career opportunities. After I had admitted to myself that teaching needlework was probably not the right thing for me, it seemed that being a translator or even an interpreter was the most obvious career choice for me due to my love of languages. I had even already tried to simultaneously translate for my siblings in front of the television and felt fairly confident, that this was it. So, I made inquiries and visited appropriate schools. It quickly became obvious that the knowledge I had acquired in French, Italian, and English was not particularly in demand for the translation business and that I would also have to complete a corresponding course of study both for translating and for interpreting specialist topics. However, the career counselor showed another option, which I found a lot more appealing, and which suited my profile even better. It was the training to become a so-called home economics manager, then called "Hausbeamtin." This study course not only included my language skills, but also other strengths I knew I brought to the table, such as management and people skills. Luckily, I wasn't put off by the terribly old-fashioned name and, after an informative trial apprenticeship and a pre-study internship in a hospital in Ticino, I started on this career path in 1984.

Already after the first of four years of training, in Switzerland the profession was renamed to "housekeeping manager," at a time when the job title ecotrophologist was already being used in Germany. Even the training, which combined theoretical input with three six-month internships, was extremely interesting and varied. In addition to practical subjects such as cleaning, textile care and cooking, theoretical subjects such as accounting, personnel management and administration, political science and social skills were also part of the training. Italian was taught, too, because at that time it was considered important to be able to communicate with our foreign workers in their own language. Now, since our blue-collar workers come from all over the world, this is no longer considered helpful.

During the first year of our study course, we also had to attend a "Crafts" class, where we were taught in how to arrange flowers and various other handicrafts. Because of our vehement demand for «EDP», we were then allowed to attend a computer course in the second year.

During our various internships, we were then able to apply what we had learned and gain initial management experience. After six months in the "SV service" in the canteen of the former company ABB (Asea Brown Boveri), my assignments also took me to the University Hospital in Zurich and to a rural home for people with disabilities, all well-organized companies that offered a good insight into the world of today's "Hospitality Management."

At that time, at least in Switzerland, the profession was still considered a female profession only. Our headmaster was extremely proud of the fact that we were the first class to have one male fellow student.

Facilities @ Management, Concept - Realization - Vision, A Global Perspective, First Edition. Edited by Edmond Rondeau and Michaela Hellerforth.

First Professional Experiences

After completing my training, I was allowed to call myself a graduated home economics manager HHF, HHF standing for higher home economics technical school. A brief time later, this school was merged with a technical university at another location and later merged to form the ZHAW, the Zurich University of Applied Sciences. Upon graduation, I received a job offer from the University Hospital Zurich, where I had done one of my internships. At the time, our principal had recommended a starting salary of CHF 2,800 (US$3,125). Because my sister was already earning CHF 3,200 (US$3,572) as a newly qualified kindergarten teacher in a canton with lower living costs, I stated this sum as my desired wage, which two other colleagues and I then received.

In this way, at age 24, I started my first job as a sector manager in a large company in the healthcare sector, with around 80 subordinate employees. At first, my colleagues and I had to learn how to assert ourselves against the male group leaders who reported to us, which wasn't always easy. Almost all of them came from southern Europe and therefore had a slightly different view of leadership. In addition, most of them had been with the hospital for years and, due to their advanced age and seniority, often even earned more than we university graduates. However, at the hospital I had several colleagues at the same hierarchical level and with the same training. It was therefore possible for us to discuss challenges in the team and to work out good solutions together.

At that point in time, we facility managers were not yet responsible for the budget, and a certain sum was set aside in the budget for our activities. If this sum was not used up, there was less budget in the following year. Even then however, there were a variety of topics to work on and countless concepts for increasing efficiency to be worked out and implemented. As an example, in close cooperation with several colleagues in our department, we switched from a laundry distribution system with a big distribution trolley and cupboard system to a system in which the laundry trolleys were commissioned to order in the laundry center and delivered into the newly installed cupboard niches on the patient wards (see Figure 1).

Even at this early stage, in 1999, the order was placed electronically using an Excel preadsheet with macros sent by e-mail.

Around the same time, another project was implemented in the textiles sector, in which new uniforms were introduced for the entire staff of the hospital, which for the first time ever hardly provided any visible distinction between the professional groups, except for the technical professions. Up until then, professional groups were recognized by the color on their sleeves. By eliminating this color coding, we were able to realize notable savings.

Another major project was the putting into operation of a large new wing. Because no new jobs were approved despite a calculated need of 23 jobs for maintenance cleaning, the work schedules of all 350 employees had to be adjusted so that we could save the time to staff the new wing. We were able to create the work plans on one of the first computer programs available for this purpose, a forerunner of today's CAFM tools. This software calculated the time needed for cleaning based on data such as square meters, number of water outlets and room purpose. The system used the specifications we had timed ourselves such as 3 min. for a toilet, 3 min. for a washbasin and time per square meters on different kinds of flooring. The implementation of this project was only possible through a drastic reduction in cleaning in all offices, which did not cause any joy, especially among the higher management of the medical profession and the hygiene department.

FM and Women

In 1992 I followed my husband to a work assignment in London, where unfortunately I could not find a suitable job. When I returned to Switzerland a year later, I no longer had the blemish of a graduate, but because of my age and gender I was seen as a "ticking time bomb" and therefore had trouble finding a suitable job right away. So, I chose to "flight forward" and after a few months in HR I took a family break for several years. In our environment there was little organized childcare at that time, so we had to manage this issue privately. I was lucky that the two grandmothers took turns taking care of our children one day a week, so that I could continue to work at least with

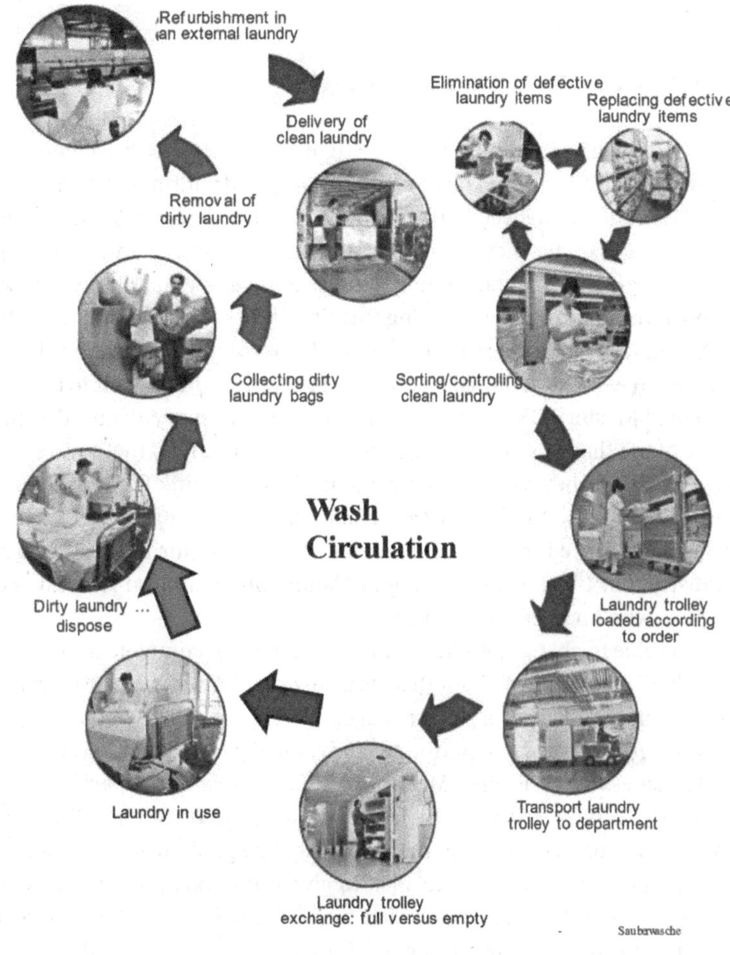

Figure 1 Washing Process at University Hospital Zurich. (*Source:* K. Schaad; USZ).

a very small level of employment, again at the university hospital, but as a support in project management rather than with managerial responsibility.

The topic of motherhood in general seems to be a challenge in our professional field and Facilities Management is probably still "losing" numerous women due to the family phase and the sometimes-difficult conditions that arise for young mothers due to the small number of interesting part-time jobs and insufficient childcare options. This assumption was confirmed by a survey recently launched by the fmpro trade association. In my humble opinion, this issue is not only the problem of women, but a challenge for our society and nothing will change until men also demand and receive more family time.

Nevertheless, there are of course many women who have achieved particularly good management positions in large companies, with support from great bosses or a good network or as a defiant reaction to adverse circumstances. As a board member of EuroFM, I was able to confirm this at a conference in the UK entitled "women in FM." Interestingly, one of the few male listeners, a fellow Swiss, came up to me and vehemently asked me to record this in a video, which we at EuroFM then put into action shortly afterwards. The video can still be found today on U-Tube under the title «women in fm». (See Women in Facility Management – YouTube.) Of course, it is also important that we women support each other, which admittedly is not always easy. I will start the experiment of co-heading a department with two young Facility Managers in my team who have young children and would both like to work part-time. Even nowadays this is not yet quite common. However, I am convinced that this is a good model that will bring great benefits to everyone involved, but especially to the FM industry.

Further Education and Research

After a few years of family life, I thought I was done with Facilities Management and did volunteer work in a variety of ways. Until one day the phone rang, and a former lecturer of mine offered me a part-time job as a course administration employee in the further education department at the Zurich University of applied Sciences, deliberately because she knew I had both a background in FM and knowledge of the booking software they used. This is how I came into contact with the topics of further education and research within the FM industry. I also got to know numerous graduates during this time, which added many valuable contacts to my network.

Again, rather surprisingly, a few years later a lecturer asked me about my FM background and whether I really wanted to work in administration until I retired. Didn't I want to continue my education and complete the newly designed Master of Science (in contrast to the Master of Advanced Studies, for which I worked)? Of course, it was clear to me that this question was not entirely unselfish, there were still too few registrations for the start of this new course at that time. However, I didn't have to think for long about whether I wanted to continue working in the administration office for the remaining 15 years of my career or not. I quickly registered for the master's degree, convinced that I would be able to make up for the further development and professionalization of the industry that I had missed during my family phase and, so to speak, be in the fast lane to catch up with my colleagues who had not taken a break.

I was able to change jobs internally so that I could complete my studies part-time alongside my secretarial work, and that's how it came about that, as a 50-year-old, I was able to receive my master's degree alongside fellow students some of whom were half my age.

During the 2.5 years of part-time study, I had acquired additional knowledge in the areas of Scientific Methods and Business Skills, Facility Management Processes, Understanding Organizations, International FM as well as in Specific FM and Legal Aspects. For my thesis, I chose the topic "Intralogistics in the hospital of the future" advertised by my former employer, the University Hospital Zürich, a very exciting and future-oriented topic that subsequently enabled me, as a specialist, to give numerous presentations on this and related topics. In addition, the sponsor of my thesis recommended me as a delegate for the professional association fmpro for election to the board of EuroFM, the federation of European FM professional associations. Being on that board allowed me to visit numerous FM congresses held by both EuroFM and IFMA, at home and abroad. I also was given the opportunity to write about those conferences in several reports in the specialist journal "fmpro Service," giving me a deeper insight into the research and development topics in the industry such as digitization, robotics, food waste, water hygiene, and NWW. I was also able to expand my network at an international level and meet many well-known FM personalities. After the maximum term of office at EuroFM had expired, I was asked to continue my association related work at fmpro itself, which I was happy to do. Four years ago, in 2019, I was elected to take over the chairman's position at fmpro.

World of Providers, Specifications, and Figures

After completing my master's, it was time to look for a new job where I could put what I had learned into practice. After a few applications, I was able to choose between a conventional FM position in a healthcare company and a position in sales at the service provider ISS. I opted for the unknown, the new. As a senior business consultant, I was introduced

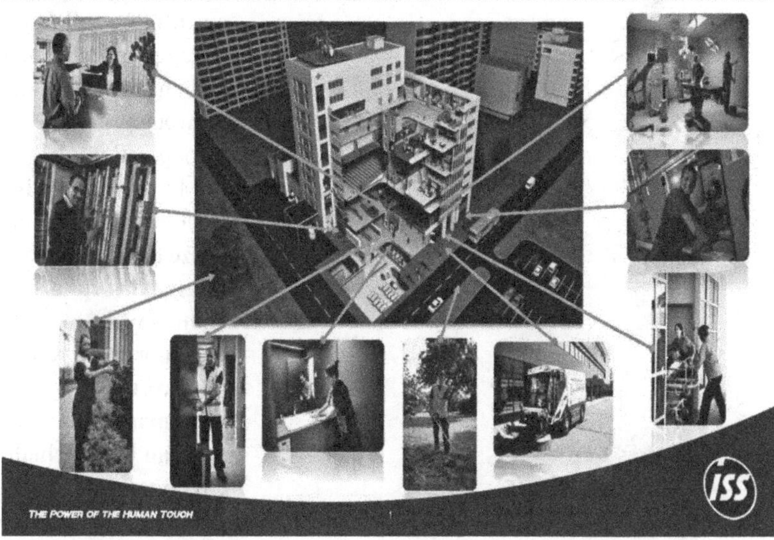

Figure 2 Services in the Health Care Sector. (*Source:* K. Schaad, ISS CH).

to the world of service providers and participated in the calculation of small and then ever larger tenders, most recently in large international projects, where we worked together in big cross-border and interdisciplinary teams.

My tasks included the calculation of services based on a specified list of services as well as the elaboration of service levels and margins, the planning of the handover and implementation and the preparing and holding of correlated customer presentations (See Figure 2). A couple of years later, due to my extended work experience in healthcare, I was given the role of segment sales consultant for the healthcare sector, which was considered by the management to have a particularly great potential for growth.

When, after 4 years, a boss was thrown in front of me who had very little knowledge of healthcare, I happened to find the advertisement for my current position on the job exchange and decided to apply at the University Hospital again after a period of more than 20 years. After several interviews and a professional assessment, I was able to accept the position in January 2019 with a lot of service provider knowledge and a great desire for personnel management and leadership. In between, I have also enjoyed being personally involved in FM training classes at the request of the Vienna University of Technology, the ZHAW and later the IMC Krems.

Head of Facility Services

My new job at the University Hospital was another step up in my career. As head of the Facility Services department in the Real Estate and Operations Directorate, I report directly to the Real Estate and Operations Director, who is placed in the hospital's top management. Together with 350 employees, I ensure that maintenance cleaning and special cleaning are conducted on the USZ campus. This covers an area of approx. 255,000 m2 and approx. 150 different room groups such as patient rooms and offices, laboratories, operating theatres and intensive care units for various medical disciplines.

My tasks also include ensuring that customers are supplied with linen and uniforms, looking after patients and visitors at the information desks, issuing ID cards and renting out staff rooms and apartments. Overall, this means being responsible for a budget of approx. CHF 30 million (US$33,487,750), of which approx. two-thirds are personnel costs and one-third material costs. In the management meeting, which takes place every two weeks, I exchange ideas with my peers and, as a group, work out the strategic direction of our Directorate. Major projects that I am currently working on with my team are the reorganization of the cleaning department, which will lead

Figure 3 UV-Robot "Hero 21." (*Source:* https://www.hero21.ch/home).

to more flexibility, the extended outsourcing of laundry treatment and consignment to order, and the creation of a central bed allocation management for the entire hospital. As opposed to the past, today's projects are much more complex and can only be successfully implemented through greatly improved and agile interdisciplinary cooperation.

By taking part in various FM benchmarks, we support the constant optimization of our processes. Due to the size and influence of our organization as one of five university hospitals, it is also especially important to me and my team to have our finger on the pulse and to support new research work and to advance digitization measures in cooperation with the ZHAW and external companies.

The work schedules of our cleaning staff are created using a modern CAFM system and since 2019 they have been able to display them on an impersonal smartphone and thus acknowledge the work that has been completed. Further modules, such as a module to support quality management and a tool for digital support in planning and executing bed cleaning, are currently being developed. We are also in the process of piloting the use of robots for cleaning floors and are already using one for room disinfection with UV (see Figure 3).

The topic of sustainability is also becoming increasingly important in connection with ESG. We are currently checking the consumption of water and energy, but also of the use and dosing of cleaning chemicals, and support research in the implementation of new solutions, for example the use of ozonated water or hydrochloric acid to clean surfaces. The various FM departments at the USZ are also involved in the planning of new buildings. Our task here is to optimally model the future core processes, to plan our support and to enable an infrastructure suitable for this.

Learnings from COVID

I had hardly gotten a foothold at the USZ and to some extent understood all the processes in my department when COVID came along and presented the healthcare system with major challenges. Our employees at the front could obviously not work from home and were on site day after day. With measures such as protective clothing, various training courses, and sufficient breaks, we tried to protect them as well as possible from infection and dehydration, although we did not have much more information about the disease ourselves. Our employees in the intensive care units caring for COVID patients had to wear protective clothing and goggles over their normal uniforms, and during a visit, I was dripping sweat in all my gear from just watching the cleaning work. Fortunately, facility management was well represented in the relevant task force and was thus able to help develop pragmatic solutions to the many challenges quickly and easily, for example in the procurement and provision of disinfectants and protective masks or in setting up an emergency hospital in the gym of a nearby high school. The hospital later attested to the FM that it was highly systemically relevant during the crisis and that it was the well-functioning backbone of the organization.

Outlook

Due to the variety of topics and the increasing importance of sustainability issues, our association work has a key role to play. At fmpro, this work is based on the three pillars of further training, networking, and services. As an association, we support the organization and implementation of various, partly extra-occupational courses on FM subjects, which are intended for the lower FM management. These are courses for jobs such as "Head of Facility Management and Maintenance," "Operations manager in FM," "Maintenance specialist," and "head of housekeeping department." However, we also conduct courses ourselves, the so-called Academy Courses, which serve to update the knowledge of established professionals. The professional and regional networks, which enable an exchange with specialists from the same industry, are also important in this context. We also offer a specialist magazine, an expert platform, and a job exchange. Together with the trade associations in our German-speaking neighboring countries, we as an association are also working on revising the FM career brochure in order to acquire students for our varied profession and raise our next generation (See Figure 4).

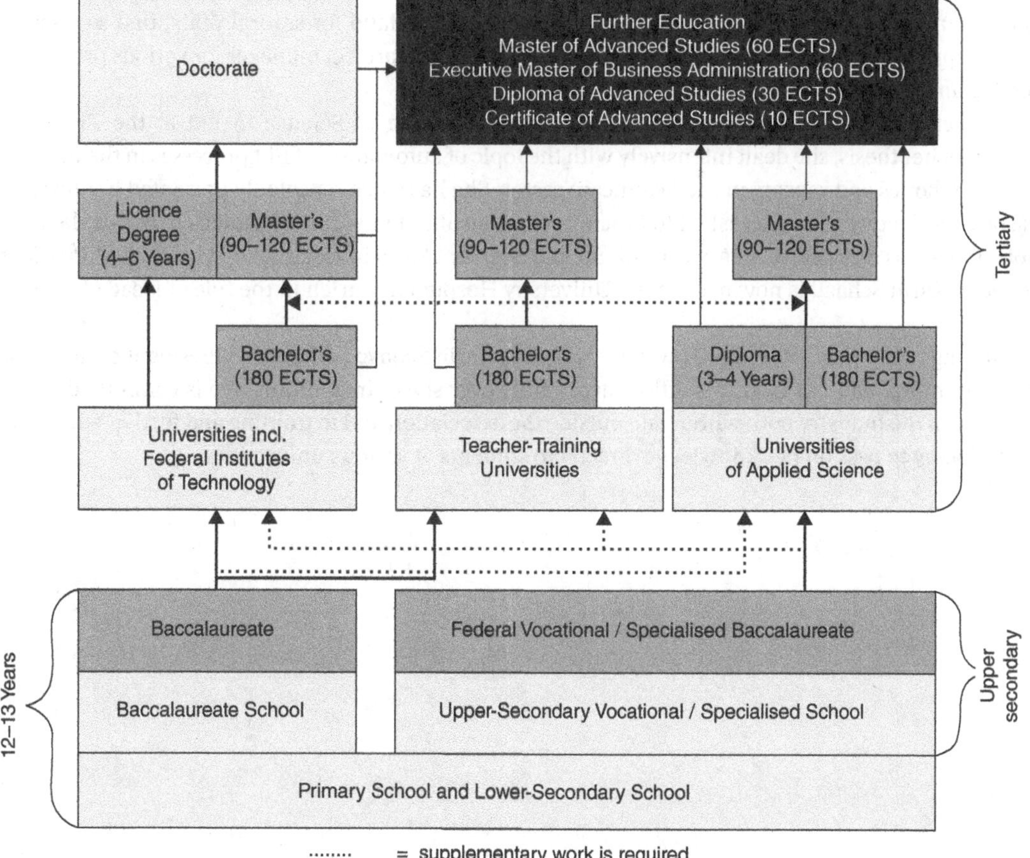

Figure 4 Swiss Higher Education System. (*Source:* Adapted from https://www.swissuniversities.ch/fileadmin/swissuniversities/Dokumente/Kammern/Kammer_FH/Best_practice/Bericht_Diploma_Supplement_2014.pdf).

Not least because of COVID, the concepts of the New Workplace and hybrid working have become established in the working world, which entails new challenges for organizations, but even more so for facility management. Especially when it comes to workplace design, FM can take on a more active role than ever before and thus gain more influence on the advantage of the organization. Good workplace design and the provision of additional FM services can help to improve company culture, encourage return to the office, or alleviate some of the skilled worker shortages. The digitization of many FM processes or the availability of FM-relevant data will also increasingly contribute to the result of a company in the future. This means that the facility managers, e.g., in the role of a CFMO, can and must get involved at C-level, that they need to learn how to discuss with key figures, SLAs, and KPIs and that they do not lose sight of the whole and the life cycle costs.

This is why I consider lifelong learning to be extremely important, especially in the FM industry, be it in the context of constant further development through attending further education and courses or via the exchange in networks and at symposia and congresses. This is how our versatile job is inspiring and fun even after many years of service.

Karin Schaad
Head of Facility Services
University Hospital Zürich
Zurich, Switzerland

Karin Schaad completed her training as a business economist (now BSc) in Facility Management in 1988. She worked in various functions at the University Hospital of Zurich, Switzerland for several years, first as a sector manager, later as Deputy Head of Cleaning Services, followed by tasks as a project manager for various projects in laundry processing and internal logistics.

To complement her FM training, Karin Schaad completed the Master of Science in FM at the ZHAW in Wädenswil. As part of her thesis, she dealt intensively with the topic of automation of FM processes in the hospital and with the use of robotics and sensors in the healthcare sector. She has since repeatedly presented her insights in the form of specialist lectures and was asked to function as a consultant in various hospitals in Switzerland.

After a couple of years as a Senior Business Consultant and Segment Sales in Healthcare with the FM Provider ISS Facility Services, Karin Schaad is now back at the University Hospital in Zurich in the role of Head of Facility Services.

Shortly after starting in her new job at the University Hospital, Karin Schaad was elected President of the Swiss Association for FM, fmpro, and has held this office successfully ever since. In addition, she is committed to the further development of the industry both within and outside the association and to training and further education in FM. She is also happy to pass on her knowledge directly to students at various universities.

2.27

Impact of Global Trends on Real Estate and Facility Management

Sabine Georgi

My interest in cities and architecture led me to start a career in the real estate industry and the built environment. After finishing the Diploma Study, I started working for a housing association where I already realized – based on my Diploma Thesis – that location as well as different aspects, like the availability of social infrastructure, matters most in building sustainable portfolios. I was heavily involved in convincing the government to support the local real estate markets in the eastern part of Berlin and East Germany to secure the housing companies against bankruptcy for avoiding even more negative socioeconomic downturns. Based on that I started in a real estate consultancy company where we implemented portfolio management systems; we gave advise how to structure the portfolio (where to invest, where to demolish, where to sell, etc.) to reduce the risk of the whole portfolio. An integrated part in the commercial real estate properties played a more efficient management – and therefore, to have a perfect Facility Management in place. An integrated approach enables you to gather real good data from the FM operator – if the tenants are happy: How big is the chance that they will stay on the property or will they leave when the contract comes to an end?

I am convinced that the Facility Management industry will play an important role in solving some of the most pressuring issues in our societies based on trends, which became more and more obviously in the last years:

1) New asset classes and operated properties get more important
2) Digitizing
3) Climate Change mitigating risks
4) Transformation of our behavior leads to a need to transform the cities
5) New asset classes and operated properties

For the ZIA I was, for example, responsible for the research of real estate and capital markets. We tried to figure out trends which had an impact on the future developments in the markets – like new asset classes like life science, logistics, data centers and light industry. In these days we also started a transparency initiative for the market, where we clustered the real estate markets and described asset classes. The last version of this report "Taxonomy of Commercial Real Estate" (see Figure 1) – with support of ULI – was published this month, in May 2023 (Zentraler Immobilien Ausschuss (ZIA) e.V.) Together with other organizations like the Deutsche Bundesbank, the bankers Association, Bureau of Statistics, and the ministry for housing, we are convinced that it's important to understand which asset classes exist as they need different approaches in management and also in Facility Management – that's why the clustering of asset classes was introduced. The categorization was also used already by the ECB for their indices.

Furthermore, and in my opinion, this report illustrates that the so-called operated properties, like hotels, coliving or co-working operators, will be more important in the future.

Facilities @ Management, Concept - Realization - Vision, A Global Perspective, First Edition. Edited by Edmond Rondeau and Michaela Hellerforth.

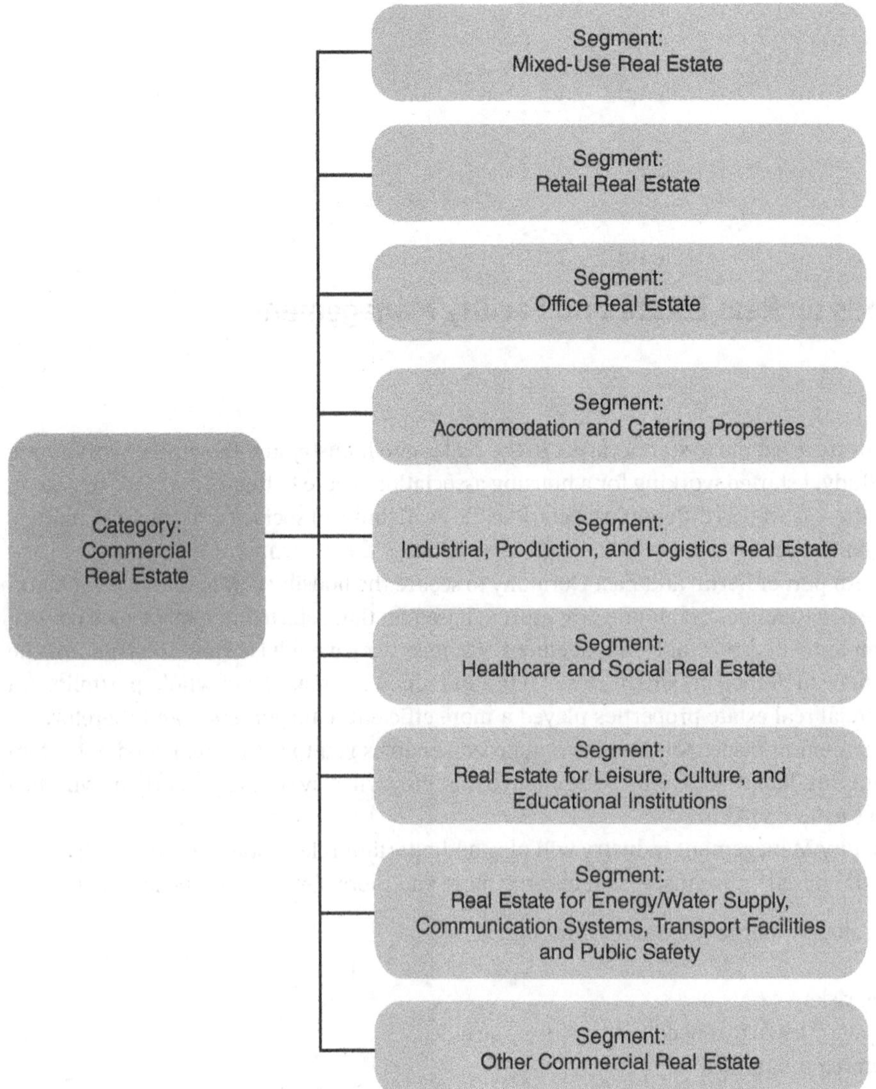

Figure 1 Commercial Real Estate segments. (*Source:* German Property Federation (ZIA) e.V (publisher), 2023, page 25).

In my current position and as Executive Director of ULI Germany/Austria/Switzerland we try to understand how the quarters and cities in the future should be organized. Furthermore, we are publishing research papers for the different trends in real estate, like ULI and PwCs *Emerging trends in real estate Europe*, or reports about new asset classes like life science or co-living properties.

The most recent results point out the great role Facility Management could play in the future (see Figure 2). The last Emerging Trends in Real Estate report showed that the portion of properties which are not fit for purpose will rise; for example, we need new concepts for repositioning offices – workplace solutions could help here. In addition, implementing an FM strategy is needed if offices will transform into mixed-use properties.

From our discussions with retail focused FM contractors, we also learned that they are heavily engaged in support with redevelopments.

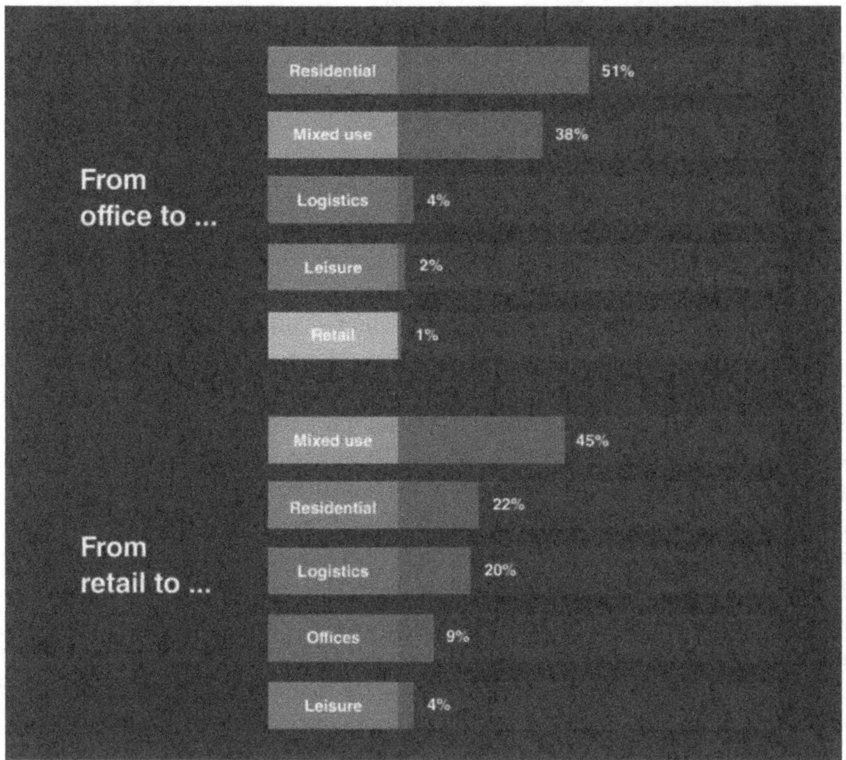

Figure 2 The most common change expected when repurposing an asset over the next three to five years. (*Source:* Urban Land Institute & PwC 2022, page 40).

Transformation of Our Behavior Leads to a Need to Transform the Cities

The pandemic was a big social and economic shock to all European countries and strengthened trends in the change of behaviors.

Firstly, it was a shock for the health system: By April 2022, cumulative COVID-related deaths have risen in consecutive waves to almost 1300 per million inhabitants in the Netherlands, 1565 deaths per million in Germany, roughly 2200 deaths per inhabitants in France and Spain, and almost 2700 deaths per million in Italy (Our World in Data 2022).

Economically, the pandemic produced a significant supply-side shock, as workers were unable to go to their workplaces, many sectors had to limit their services, and health-related absences rose. This also negatively impacted demand, and people experienced a sharp increase of uncertainty in their daily lives. These shocks had multiple implications for real estate and urban life. Production and service facilities were used less, urban amenities could not be visited, and the demand for space both inside and outside of buildings rose as the functions of cities were significantly disrupted. Cities perform three economic functions: they enable economies of scale, a sharp reduction in transaction costs, and significant positive externalities (Just and Plößl 2022).

These principles enable higher productivity and better opportunities for the consumption of both private and public goods within cities. The pandemic has distorted these economic effects and thus depreciated the benefits.

People have had to adjust to these shocks, seeking second-best alternatives for their daily routines at home, at work, and in leisure activities. In some cases, this has required investments in hardware and new know-how, and these physical and mental investments facilitated new institutions and path-dependencies; people have become accustomed to a new normal, gained new experiences, and made new cost–benefit comparisons which are set to impact the overall functionality of cities and thus also buildings within cities.

Impact on Real Estate

In a pan-European survey, Just and Plößl (2022) analyzed the shifts in demand for space expected across various real estate asset classes: clearly, the pandemic has led to a pronounced asymmetry in risk perception for these real estate classes. Residential and logistics are favored by risk-averse investors, while retail and hotels are expected to be watched particularly by investors actively seeking opportunities, risk-conscious investors.

For office properties, the picture is somewhat mixed, as the full impact of the shift to working from home cannot yet be assessed. Eisfeld et al. (2022) have developed seven possible outlooks on the future demand for office space, and even within their comparatively narrow set of assumptions, the scenarios reveal significant and ongoing uncertainties, both in quantity and in quality. Furthermore, investors do not agree whether core city locations or satellite structures, which might be more easily accessed by people living in suburbs, will be required in the future. Nevertheless, real estate market participants broadly agree that, regardless of the exact location, there will be a functional shift in office buildings.

Offices will become a place to bring teams together to co-work, hold meetings, and welcome visitors, with less pure space for performing specific and separate tasks. While uncertainty about future developments remains elevated, a higher degree of flexibility in office space is essential. This holds for both organizational and physical flexibility: how to transform physical space and how to re-organize contracts as well as teams. Space that does not allow for this flexibility will likely be traded at discount.

Interestingly, the implications of these changes on office vacancy rates and office rents are not as straightforward as might be supposed: Morawski (2022) reports mixed results across Europe for the effect of working from home on rents and vacancies.

For residential real estate, many market analysts expected a strong and sustainable shift in demand out of the cities as demand for open spaces and more private spaces increased. Analyzing Google Trend data and comparing market developments for Germany with a counterfactual scenario, Eisfeld and Just (2021) found that search volumes for peripheral space skyrocketed in 2020 and that peripheral prices in particular rose faster during the first waves of the pandemic than in the counterfactual scenario, which assumed no pandemic development.

However, these immediate and strong reactions did not endure, partly because people and administrations learnt that infection risks within cities did not necessarily imply higher fatalities than in peripheral locations (Carozzi et al. 2020). Thus, the sharp increase in internet search volumes for single-family homes and gardens has been fading in recent quarters, and the main cities have regained some of their attraction again

Impact on City Structures and Organization

The pandemic had a strong impact, especially on some city functions such as the provision of services, which resulted in a decrease in the importance of retail and to some extent workspace. Consequently, the future attractiveness of cities was questioned by many market observers throughout the pandemic. In addition, one of the key advantages of dense cities, lower transaction costs, has lost some of its thrust, because many transactions and conversations can now be done online.

After the pandemic, cities will have to replenish their specific sets of advantages by strengthening other functions, such as scale economies and positive externalities, spillovers between people and activities.

This is related to the concept of "good density" (Clark and Moonen 2015): a dense city can both enable and aggravate these positive spillovers. Fostering these effects will lead to a transformational push within the cities. In fact, this transformation was called for before the pandemic, but the pandemic shock has amplified demands on cities and the real estate industry today, so that the transformational process towards better and more successful places is likely to be accelerated.

In the end, cities that focus more on experiences and social interaction, and less on structures based on the division of labor will thrive. However, achieving this focus will take decades rather than years. Citizens, public authorities, and corporations within cities need to prepare for a marathon, not a sprint. The more that this focus is opposed; the stronger that centrifugal forces will tug at cities.

The facilitation of social interaction is key because it also strengthens a city's capacity to find new ideas and to become a hub of innovation. Innovation has always been a function of human interaction and mutual exchange.

In the future, cities will have to concentrate much more on their potential as "consumer cities" beyond mere supply and strengthen the social and interactive aspects of production. A consumer city in the economic sense is not only a retail city, a focal point to purchase a broad variety of goods; rather, a consumer city is a place that offers a broad mix of private and public goods that citizens can consume. This includes parks, theatres, walking paths and museums.

This means that mixed-use uses are desirable both in the neighborhood and in the building. Multiple uses of space within a building are a further development: In addition to considerable space efficiency, they lead to a better economic basis. If, for example, it is possible to offer staggered uses, the rents in particular could be shared, thus benefiting user groups who cannot afford the current market rents in the commercial space (e.g., creative offers).

In the next decade, project development will continue to focus on the further development and transformation of existing buildings, as the development of greenfield projects will continue to be critically scrutinized due to climate change.

The idea is to create curated neighborhoods with a central center in the future. For this purpose, the neighborhood is developed with the local people and on the basis of their needs: Which social and health (supply) facilities are needed?

In particular, there are uses that are complementary and mutually beneficial. In the future, health care facilities and social gatherings will be more important than proximity to retail properties or offices when it comes to choosing where to live. Anticipating this, operators could also be found for uses that are committed to ESG and whose commitment has an immediate positive effect on their own housing stock.

In addition, the other current challenges, such as the availability and costs of building materials, rising energy costs, sustainability requirements for real estate, rapidly changing user requirements and, last but not least, strong upheavals in the demand for space in individual segments, mean that transformation is a central component of the new reality. A new study by buildings and Union Investment (2021) shows that transformation properties have the potential to answer this diversity of topics holistically. And investors are also increasingly recognizing these opportunities.

Meanwhile, 90% of the study participants share the view that mixed-use concepts in transformation properties help to ensure yield security over the entire life cycle. Almost 60% agree with the statement that transformation properties solve many problems caused by rising construction costs, rising energy prices and disrupted supply chains. 57% share the view that this type of real estate is always more sustainable than demolition or new construction – the interest in investing in development to change forms of use is thus given (buildings and Union Investment 2021).

The survey results (bulwiengesa and Union Investment 2021) in the cited study on transformation also suggest a fundamentally strong willingness on the part of investors and developers to make advance payments here as well. When asked about the "willingness to invest" in order to obtain a faster building permit, the real estate players surveyed would clearly focus their transformation on sustainability aspects. These include investments in energy-efficient facades, particularly high-quality outdoor facilities or ecological compensation measures.

Additional costs for recyclable constructions and building materials or for the avoidance of construction waste during conversion are also accepted. Remarkably, 60% of those surveyed would also offer rent-reduced areas for social and cultural areas in exchange for a quick building permit. In contrast, however, only just under 50% would be willing to provide a higher proportion of socially subsidized housing. This, in turn, results in an improved environment and starting position for so-called social impact investments – at least compared to previous years.

All in all, it will be a challenge to rehabilitate vacant spaces. Not only buildings such as churches could be deprived of their current use, but also department stores or shopping centers require a fundamental transformation. This also applies to old railway stations or office properties that no longer meet current requirements.

The majority of those surveyed associate the transformation of an inner-city property with the opportunity to implement a sustainable utilization concept. In second place are the economic goals of "high yield security" and "increasing the value of the property," followed by sustainability goals such as "the lowest possible CO2 emissions for the entire further life cycle" and "architectural and urban environment improvement" (bulwiengesa and Union Investment 2021).

As the study shows, residential uses in the forms of privately financed, subsidized and assisted living are the preferred use offer after a transformation, as are retail of periodic needs. Surprising result: Almost 90% of those surveyed are also willing to accept a certain degree of cost uncertainty when realizing transformation real estate projects (bulwiengesa and Union Investment 2021).

Currently, transformation properties are caught between high construction costs, rising interest rates, high inflation and high energy prices, as well as sustainability requirements imposed by the EU taxonomy. The construction industry also came to an abrupt halt in autumn 2022. It is therefore hardly surprising that, according to the survey participants (77%), the scarce availability of raw materials in particular is a strong driver that favors the transformation of real estate instead of new construction or demolition and new construction. National and EU sustainability requirements are also strong drivers that, according to the respondents, favor the transformation of real estate instead of new construction or demolition and new construction. A majority (59%) also believes that the implementation of transformation properties will cause fewer problems due to rising construction and energy costs and disrupted supply chains than new construction projects (buildings & Union Investment 2021).

There are successful conversions of former industrial monuments into new urban quarters or department stores as well as existing conversion ideas for shopping centers. Here, for example, the planned transformation of Ring Center 1 in Berlin Friedrichshain is shown: The areas for retail and local supply are to be reduced from five to three levels, so that extended office use can be created on the upper floors. The local supply of a grocery store and drugstore is maintained in the basement. In addition, an organic food market is to be relocated – offers from the health sector will be added (GRAFT ARCHITECTS 2023).

A prime example of the further development of an industrial monument is the already advanced success story of the Carlswerk in Cologne. Cables have been produced in the factories of the former factory since 1874, and more than 20,000 people were employed there at last count. The listed halls were sold in 2007 to transform this important industrial area. This proved to be very successful, with the following redesigns taking place on the 125,000 square meters (BEOS AG, n.d.):

Initially, the listed buildings Kupferhütte and Werkstatt were converted into office lofts. The first to move in here was the TV production company Shine Group Germany. Other companies from the media industry, such as Stefan Raab's TV production company Brainpool, are settling in the Carlswerk.

In 2009, BEOS rebuilt the listed Schanzenhaus for the educational company Inbit in just four months. There, as well as in the laboratory and coil factory, which was also quickly rebuilt, Inbit conducts IT and catering training and accommodates its administration.

At the beginning of 2010, the Lübbe publishing group moved its headquarters to the main building of the Carlswerk in Cologne. Approximately 170 employees occupy the fourth to sixth floors. On the ground floor, a former production hall was turned into a foyer, among other things for author readings by the publishing house. In total, Lübbe rents approx. 5,800 m^2. Since April 2010, the theatres of the City of Cologne have been occupying interim quarters for their costume departments and stage workshops in the coil factory and main building for the duration of the renovation work on the opera house. Since September 2010, the copper smelter has also housed the showroom of BEOS AG.

Furthermore, in September 2010, the Italian restaurant Purino opened in the Hypodrom, which BEOS has converted into a loft ensemble, on 1000 m^2 of floor space. In addition, a club and a concert hall, the municipal venues, but also gastronomy and health (bouldering hall, sports clubs, football center) were established in the event halls. An urban quarter has thus been created from a former industrial area. Such new forms of use require correspondingly ambitious facility managers who are not only familiar with one asset class.

However, these examples also make it clear that we need to create more humane, mixed-function and resilient neighborhoods again.

"Good" places, public accessibility and the quality of stay of our places will certainly play a major role in promoting social interaction. Here, too, the study has shown that the importance of outdoor spaces has increased and that a redesign of public spaces is indicated in order to strengthen movement, leisure and recreation as well as social connectedness. All in all, multiple coding of public spaces is indicated, also because in the future – rightly – there will be critical scrutiny as to whom public spaces should be made more accessible. They can be beneficial to health and the climate (pocket parks) but can also serve the health and economic development of the neighborhood and supply (weekly market). Multi-coded spaces thus meet several usage requirements and can – through better integration of attractive outdoor spaces – create added value for the users, which can then also have an impact on the rent. Another advantage of mixed-use districts is the minimization of traffic (keyword: 15-minute district), because the idea of a climate-neutral city/district can only succeed if energy consumption and emissions from traffic are significantly minimized.

Multiple uses of space within a building are a further development: In addition to considerable space efficiency, they lead to a better economic basis. If, for example, it were possible to offer staggered uses, rents in particular could be shared, thus benefiting user groups who cannot afford the current market rents in commercial space, such as cultural workers, providers of creative services, and artists.

Consequently, it will have to be a matter of rethinking social interaction – a challenge that we can only overcome together and in close cooperation between the private real estate industry and the public sector.

Nowadays we are also discussing how to implement placemaking strategies to improve a vibrant ground floor area; therefore, we are looking for solutions to combine non consumer areas in the ground floor areas in addition to more urban and public spaces for the citizens, which are available 24/7. This might also require a joint approach in a quarter where FM could play a role – as facilitator, or even as a quarter manager.

Climate Change

Therefore, it is needed (a) to learn more about the future needs of a city and (b) to find solutions to implement ESG strategies.

At the end of 2021, a group of leading real estate players launched a ULI-led initiative called "C Change" to mobilize the European real estate industry and accelerate decarbonization to limit the impact of climate change. To this end, experts from the entire value chain are brought together, among other things to break down existing barriers.

So far, two points in particular have emerged that are crucial to accelerate decarbonization (Urban Land Institute 2023):

1) An industry-wide uniform methodology for assessing climate risks associated with the necessary transformation and the associated CO2 reduction strategy. Above all, real estate valuations need a strategy to avoid stagnation in investment markets and stranded assets. Without such a uniform methodology, there is a risk that investors will come to "wrong" investment decisions.

While the industry has recognized the need to decarbonize the built environment, it needs a standardized methodology. This is because climate risks are currently not sufficiently included in real estate valuations, so that current building values do not yet sufficiently reflect this. Investors thus come to different price decisions than if these costs were transparent.

2) Practical ways to improve alignment and joint action between tenants and landlords to create common goals for decarbonization and retrofitting.

For the first point, the "Transition Risk Assessment – Guidelines for Consultation" (Urban Land Institute 2022a) is now the first proposal on the table. The guidelines set out a standardized methodology for assessing the costs of decarbonizing buildings and disclosing key transition risks and their impact on value.

One thing is fundamentally certain: all buildings have transition risks and some leading market players have started to consider the costs of decarbonization.

ULI analyses show that decarbonization activities are currently focused on assets, especially in core locations, with correspondingly high assets, where the cost-benefit ratio of retrofitting is lower (Urban Land Institute 2022b). Without cooperation and transparency on the risks of changeover, there is therefore a risk of a two-tier market. However, the goal should be to preserve the value of all buildings in the long term so that our cities remain attractive.

The guidelines (Urban Land Institute 2022a) are therefore intended to help make the transition risks transparent, eliminate them and thus provide guidance on the use of all owners and asset managers. The companies involved, in particular the founding partners of the ULI C Change program, Allianz Real Estate, Arup, Catella, Hines, Immobel, Redevco, and Schroders Capital, are convinced that if all market participants are better informed about these risks, the overall decarbonization objectives can be better achieved.

With the support of the partners as well as technical support from the members of ULI Europe, a proposal for a guideline could be developed in a short time, as already mentioned. In addition, more than 50 individual interviews and a series of consultation workshops with around 100 experts were conducted.

However, the future guidance proposed in the consultation paper is explicitly not intended to replace the industry-standard requirement of a market or fair value valuation, which is carried out as standard when buying/selling/holding a property. Rather, the proposed guidelines must be used in conjunction with this evaluation.

In addition, it is proposed to include certain criteria directly linked to the decarbonization of real estate within and in addition to the already existing valuation methods and tools.

In particular, owners should be informed of the impact of certain risks, namely the intended duration of an investment, in order to assess and prioritize the transition risks that have the greatest potential impact on the safety of the returns sought by owners and managers.

A total of 15 clear transition risks were identified that will have a material impact on real estate values now and in the future. Nine of them can already be financially modeled, standardized and communicated. It is proposed to carry out the assessment of transition risks once a year per property, in line with the regular assessment process. It is also recommended that the assessment of transition risks be carried out by the owner or asset manager with the involvement and support of an internal sustainability expert (Urban Land Institute 2022a).

Once the initial assessment has been made, additional data points related to the transition risk shall be reported and identified in a separate section for the risk-adjusted assessment. For this purpose, a section of discounted cash flow adjusted for transition risk should be prepared and used for all transactions.

Presentation in an Additional Section after Calculation of the DCF

It is recommended that owners and asset managers be able to demonstrate their potential impact on property value when assessing transition risks, even if these risks do not currently affect the actual free cash flows of the asset. To do this, owners and managers should include a so-called shadow area below the bottom line of their free cash flow valuation to explicitly disclose these shadow costs but not include them in the free cash flow measurement (Urban Land Institute 2022a).

ULI has compiled further content in a detailed advisory paper and invites all interested parties to participate in a consultation. This again makes it clear that facility managers, with their know-how in the operation of the systems, are both important data suppliers and are becoming more and more aware of this role, i.e., they can also achieve decisive savings in operation and also transparently present their added value with the evaluation presented.

ULI Europe Young Leader PropTech Innovation Challenge

Another initiative of ULI to address the upcoming challenges facing our cities is ULI Europe's Young Leader PropTech Innovation Challenge (PIC), which will be held for the first time this year. The competition is open to all and no ULI membership is required. Applications are open to European PropTechs that have developed novel, scalable, and competitive solutions. The application categories are based on the three main objectives of the ULI mission:

- Decarbonization and Net Zero
- Improving housing affordability
- Training the next generation of leaders

In each of six European regions, a preliminary decision will be made by a jury of experts. The winner will then take part in the European finals. With the announcement of the competition, ULI has set itself the goal of bringing together innovations from the fields of real estate, technology and start-ups and awarding the most innovative concept to solve the most pressing challenges of our cities.

In the future, it will be particularly important for facility managers to follow the trends described above, to pick up on them, and to further develop their business models based on them. As a result, the FM segment makes a significant contribution to the decarbonization and positive transformation of our cities. However, it will be crucial that all stakeholders involved not only exchange ideas, but also actively work together and jointly develop our urban living spaces, because the complexity of the trends requires increased collaboration. Only together can we make the city of tomorrow fit for the future, so let's get to grips with it!

Bibliography

BEOS AG (n.d.). The Carlswerk: where the first telephone cable was once manufactured, connecting Europe with America. https://www.carlswerk.de/geschichte (Last accessed 30 May 2023).

bulwiengesa & Union Investment (2021). Market study transformation real estate: transformation real estate – a central building block on the way to climate neutrality. https://realestate.union-investment.com/de/im-fokus/transformationsstudie.html.

Carozzi, F., Provenzano, S., and Roth, S. (2020). Urban Density and COVID-19. IZA Discussion Papers 13440, Institute of Labor Economics (IZA). https://www.iza.org/publications/dp/13440/urban-density-and-covid-19 (accessed 8 April 2022).

German Property Federation (ZIA) e.V. (Publisher) (2023). Taxonomy of commercial real estate. Structuring of the factual submarket of economically used real estate for the purpose of market observation and valuation. 3. Results report (2023). https://zia-deutschland.de/wp-content/uploads/2023/05/Taxonomie_der_Wirtschaftsimmobilien.pdf.

Clark, G. and Moonen, T. (2015). The Density Dividend: solutions for growing and shrinking cities. Urban Land Institute (ULI). https://knowledge.uli.org/-/media/files/research-reports/2015/thedensitydividend.pdf?rev=733dd14707c14bf38f5e8aead10ce33e&hash=2F960246FDCBD03F0892A059A1DC9363.

Eisfeld, R., Heinemann, A.-K., Just, T., and Quitzau, J. (2022). Office real estate after Corona: a scenario analysis. Study commissioned by BERENBERG Joh. Berenberg, Gossler & Co. KG. Contributions to the real estate industry 27th IREBS International Real Estate Business School at the University of Regensburg: Regensburg. https://epub.uni-regensburg.de/51496/2/Heft27.pdf (accessed 8 April 2022).

Eisfeld, R.K. and Just, T. (2021). The impact of the COVID-19 pandemic on German housing markets. A study commissioned by the Hans Böckler Foundation. Contributions to the real estate industry 26th IREBS International Real Estate Business School at the University of Regensburg: Regensburg. https://epub.uni-regensburg.de/49390/1/Heft26.pdf (accessed 8 April 2022).

GRAFT ARCHITECTS (2023). RINGCENTER. https://graftlab.com/projects/ringcenter (Last accessed 30 May 2023).

Just, T. and Plößl, F. (2022). *European Cities after Covid-19. Strategies for Resilient Cities and Real Estate*. Springer. Cham.

Morawski, J. (2022). Impact of working from home on European office rents and vacancy rates. In Zeitschrift für Immobilienökonomie. Ahead of print. https://link.springer.com/content/pdf/10.1365/s41056-022-00057-z.pdf (accessed 8 April 2022).

Our World in Data (2022). Covid-19 data explorer. https://ourworldindata.org/coronavirus#explore-the-global-situation (accessed 8 April 2022).

Urban Land Institute (2022a). Transition Risk Assessment. Guidelines for Consultation. https://knowledge.uli.org/-/media/files/research-reports/2022/transition-risk-assessment-guidelines-for-consultation.pdf.

Urban Land Institute (2022b). Breaking the value deadlock: enabling action on decarbonisation. https://knowledge.uli.org/-/media/files/research-reports/2022/breaking-the-value-deadlock-enabling-action-on-decarbonisation.pdf.

Urban Land Institute (2023). C Change. https://europe.uli.org/research/c-change (Last accessed 30 May 2023).

Urban Land Institute & PwC (2022). Emerging Trends in Real Estate® Europe 2023. In the Eye of the Storm. https://knowledge.uli.org/-/media/files/emerging-trends/2023/emerging-trends-in-real-estate-europe-2023–low-resolution-3.pdf.

Table of Figures

- Figure 1: "Commercial real estate segments"; German Property Federation (ZIA) e.V. (Publisher) (2023). Taxonomy of commercial real estate. Structuring of the factual submarket of economically used real estate for the purpose of market observation and valuation. 3. Results report (2023). Page 25. https://zia-deutschland.de/wp-content/uploads/2023/05/Taxonomie_der_Wirtschaftsimmobilien.pdf
- Figure 2: "The most common change expected when repurposing an asset over the next three to five years"; Urban Land Institute & PwC, 2022, "Emerging Trends in Real Estate® Europe 2023. In the Eye of the Storm," Seite40.https://knowledge.uli.org/-/media/files/emerging-trends/2023/emerging-trends-in-real-estate-europe-2023–low-resolution-3.pdf

Sabine Georgi
Executive Director
ULI Germany/Austria/Switzerland
Frankfurt am Main. Germany

Sabine Georgi is Executive Director of ULI Germany/Austria/Switzerland. A BA graduate in business administration specializing in real estate, and a Certified Real Estate Investment Analyst, Sabine has worked in the industry for 25 years. Before joining ULI, she was Country Manager at RICS in Frankfurt. Other roles have included Head of Real Estate and Capital Markets at ZIA Zentraler Immobilien Ausschuss. At management consultancy BBT Group, she advised companies operating in housing and real estate, and also headed the marketing department.

Sabine began her professional career as a consultant to the board at the Association for Berlin-Brandenburg Housing Companies (BBU). While studying, she held professional positions at various companies in the real estate industry.

Sabine lives with her family in Berlin.

2.28

My Episode with Facility Management

Rainer Fischbach

Models have always been tools of central importance for planning – not alone in building but in all kinds of endeavors, particularly technical ones. Traditional ones have been two-dimensional-like schematic sketches and drawings as well as three-dimensional made from wood, paper or clay, supplemented by verbal descriptions, lists of materials and tools. It was in the course of the Italian renaissance that the word ›modello‹ assumed the encompassing meaning of a body consisting of all kinds of artefacts created in the process of design to the purpose of specifying a project or product to be realized (Westfehling 1993, pp. 74–97, 124–200). Ensuing was the standardization of a variety of drawing kinds and scales, tailored to the needs of particular technical disciplines.

At the turn from the 1970s to the 1980s this century-old state of affairs was to change significantly: not only that reasonably priced computing equipment based on microelectronics became available, but there happened also a burst of new software development aimed at the expanding markets created thereby. Not less significant was the new environment of digital communication developing in the 1980s due to the same advances in microelectronics and the TCP/IP protocol suite released out of the military research establishment of the Pentagon's *Advanced Research Project Agency* (ARPA) into the public sphere and giving rise to the Internet.

The *IGP Institut für Grundlagen der Planung* (Institute for Fundamentals of Planning) at the University of Stuttgart, headed by Prof. Horst Rittel, already well-known for his multiple contributions to systems thinking and a critical reflection of design and panning practice (Rittel 1992, 2013), at that time a leading center of research on design theory, methods and tools, quickly took on the emerging digital technology based on microelectronics and started to look into its potential for design and planning. The tools of the day were, starting with the legendary Apple II, with choice of software, including the UCSD-Pascal system, self-imported by Rittel from the USA, where he in parallel held the chair for the science of design at the University of California, Berkeley, and quickly moving to networked SUN workstations running UNIX.

From a background in philosophy, some training in the exact sciences and with a strong interest in methodological questions, I was, guided by a friend who already knew the institute, attracted to the IGP exactly at the time when, in the late 1970s, the microelectronic revolution took off. As a research assistant there, I quickly grasped the opportunity of access to the university computing center and, driven by the idea that some of the various models at the center of Rittel's teaching and institute discussions, should be ported to a computing platform and, while taking part in those discussion, an additional position as an associate lecturer at the (in those days) Berufsakademie Baden-Wuerttemberg, today DHBW Duale Hochschule Baden-Wuerttemberg (Cooperative State University), teaching computing science basics, methodology of programming, algorithms and data structures, and as an advisor for diploma theses based on project work to be carried out at some of the leading enterprises of the region, like IBM, Hewlett-Packard, and Daimler, gave more opportunities of diving into the widths depths of

Facilities @ Management, Concept - Realization - Vision, A Global Perspective, First Edition. Edited by Edmond Rondeau and Michaela Hellerforth.

software and its applications. Porting software from the big machines of the computing center to microcomputers was still a challenge, but one that became less difficult quickly. One thematic thread that increasingly took my awareness was the potential of various paradigms of programming – object-oriented, functional and logical – for modeling, particularly in the fields of science, engineering and design. An additional interest that arose in connection with the former was specification of software systems (Fischbach 1992).

Participants in the research at IGP included assistants Jürgen Georgi, Florian Hoyer, and Matthias Mayer. Additional input provided Mihaly Lenart, one of Rittel's PhD students, doing his thesis on graph-theoretical models of buildings and inspiring me to design and implement related computer models and algorithms (Fischbach and Lenart 1982; Lenart and Fischbach 1982). Another was the modeling of thermal properties of building enclosure, calculation of the metrics and check against the applicable standards (Deutsche Industrienorm, DIN 4108) progressively tightened as a consequence of the energy crisis of the 1970s and 1980s. From a broader discussion of the problems of computer aided design in building emerged a number of insights highlighting some of the shortcomings the related tools and practice of those days were suffering from:

1) using the computer only as a tool for drawing, table calculation and text editing not only missed its potential for advanced technical modeling and calculation in architectural design but also as means for connecting and facilitating downstream activities.
2) the 3-D-CAD software for mechanical design, particularly solid modeling, then introduced in many branches of industry did not fit the requirements of building design, the many levels of detail, the wide range of respective scales with varying standards of representation to be covered there;
3) on the other hand, calculation of quantities and a seamless interface to the further engineering disciplines involved required a 3-D volume model taking care of a number of building-specific requirements;
4) the true object of architectural design is not the walls, but the space enclosed by the walls, nor are the walls the only means facilitating the use of this space;
5) technical equipment and furniture, making architectural space habitable and usable, represented a multifaceted challenge of growing importance;
6) the break line between design and biding was still un-surmounted, compiling bills of material remained a cumbersome, tedious and error-prone task;
7) not even addressed by the majority of the profession was the dimension of building life cycle: what should the contribution of designers and planners to the building's maintenance be and what was the role CAD could assume therein?

The latter question was about the interface between planning and the management of facilities. Facility Management (FM) was beginning to become a scientifically advised discipline, that should make use of a systematically compiled and continuously updated information base comprising all relevant detail of architectural space, construction and technical equipment.

But until today, this approach is all too often suffering from exactly a lack of the respective data, due to deficiencies of its interface to planning on the one hand, but also from a lack of continuous updates running in parallel with changes to buildings and their equipment. We did not even know the name of the game until an academic researcher, Dr. Josef R. G. Mack, hit the IGP in the late 1980s, searching for advice on a number of issues regarding the potential of computers, design theory and methods. Rittel, already tired and increasingly suffering from the disease that led to his dead in 1990, referred him to me – which should become the beginning of an interesting cooperation lasting until 1994.

Dr. Mack, as one of the messengers to Europe of the US-born concept of Facility Management, networking mostly in the background, inspiring, facilitating and organizing a considerable volume of research and related publications on Management Consulting, Mergers & Acquisitions, and Facility Management. Pivotal was in his role as managing director of the Facility Management Institute (FMI), responsible for the *Facility Management* book series published with W. Kohlhammer Verlag, Stuttgart.

In multiple roles, as consultant, copy editor, and coach for a number of contributors to the series, I got involved in the activities of the first FMI Facility Management Institute (founded as an AG, a public corporation, in Zurich/ Switzerland). The first volume of the then started publication series on *Facility Management* stood out of its more technical and economic focus: Wolf Reuter, associate professor at IGP, was given the opportunity to contribute to the series his work for postdoctoral lecture qualification, a philosophical and sociological meditation on the power of architects and planners (Reuter 1989).

A new chapter of my involvement with FM was opened when I started to work with the architectural practice Kahlen & Partner at Aachen, after Dr. Mack had introduced me to Hans Kahlen, the head of the firm. Kahlen was determined to make use of the computer's potential and make bold inquiries to uncover it. To this end he established a working group in his office and appointed me to lead it. A further step was the founding, together with Norbert Gerhards of FCA, a German Limited company, founded in 1990 by Norbert Gehards as a "Reaearch Company for Computer Applications in Architecture" active in Aachen ("FCA" Forschungssgsellschaft fur Computeranwendungen in der Archiektur mbH), and took over the development of necessary tools and support. As a co-founder of the first FMI AG, he continued with Günter Neumann et al, and sometime before partner Dr. Mack in computer related issues and continued FMI's research and development for architects and took over assignments within Corporate Real Estate Management (CREM) including concepts and problem solving in the field of commercial real estate projects. FCA was also a business with activities in the distribution of CAD software and related consulting and education in architecture and planning. Out of this context emerged Hans Kahlen's PhD thesis (Kahlen 1989) which made some impression on the architectural CAD community in computers in the German-speaking countries, eventually leading 1993 to his appointment as a professor of *Architekturinformatik* (computing science for architecture) at the Technical University Cottbus, then in foundation (where FMI – now as a German Ltd. – achieved the status of an An-Institute, an affiliated Institute of the University, under the auspices of representatives of IFMA International Facility Association, Houston, Texas, USA, integrated in the inauguration ceremony of Kahlen on June 27, 1993). Besides this academic achievement a number of publications resulted from the working group and from the CAD experiences made in the architectural practice (Fischbach 1989, 1994a, 1994b, 1994c, 1994d, 1994e; Fischbach et al. 1993; Fischbach et al. 1995).

In 1994, I left FMI and the Kahlen group, after my interests had diverged too much from their course. Besides some minor consulting jobs and continued teaching at the DHBW my next significant role in conjunction to FM was as an advisor on engineering software and IT infrastructure to Dr. Michael Peltzer, head of IT at Ruhrkohle AG (RAG), in those days the corporation where German coal mining and some related engineering and consulting business have been consolidated. A coal mine is a very special kind of facility: it is highly equipped, and it is constantly changing. Its model should, as a base, comprise the geology of crust to be mined and the structure of the headings and faces advanced to this end, the type of buildup employed thereby and additionally all equipment and, of exceeding importance, the mine air as actively conditioned by specific aggregates and conduits.

At RAG, a group headed by Rudolf Dann had, since the late 1970s, developed the *Digitale, integrierte geologisch-markscheiderische Analyse- und Planungssystem* (DIGMAP, digital, integrated system for geological analysis, mine survey and planning), a system based on a model of stratigraphy, headings and faces, stored in a relational database. Under the issues to which my advisory opinion was requested, the future of DIGMAP stood out through its technical complexity and strategic weight. DIGMAP was under pressure from demands for a more intuitive, graphical user interface, driven by the growing spread of personal computing, and suffered from the only one-way interface to the downstream engineering disciplines like mine air and technical equipment. My advice was to undertake a bold refactoring of the system, concentrating own manpower on the extension of the data model to include the downstream engineering disciplines and to rely on some of the emerging frameworks for 3D-geometry, graphical representation and user interaction – a step the management was too fainthearted to do. So, a software platform easily extendable, based on an integrated data model was never realized, while the sound, forward-looking conceptual approach embodied by DIGMAP went to waste.

This was counterproductive regarding the cost situation of coal mining in Germany because the critical factor for this already heavily mechanized and automated industry was, in order to avoid running underground into costly dead ends, improved prediction and planning. Particularly mining in horizons 1500 meters underground, facing a heavily faulted stratigraphy, made more elaborate models compulsory – a situation where today's machine learning techniques could provide matching tools. Suggestions from researchers in the geology of the European carbon to address these problems, which would have benefited from a more versatile software environment, were equally ignored. Aggravating the situation was a loss of expertise through the compulsory early retirement of staff. After the leave of Dr. Peltzer and some retirements, particularly that of Gerhard Engels, head of technical computing, there were no partners left for me in the late 1990s.

Probably it is the fate of a consultant that, looking back, he sees most of his advice ignored and, sometimes, only taken up after costly follies. After many years as an independent consultant, in 1999 I joined ECS Engineering Consulting & Solutions at Neumarkt/Oberpfalz in Bavaria, today with additional locations at Stuttgart, Rostock and Pune/India, founded by Wolfgang Dietzler in 1996. Dietzler had left Siemens management with the aim of targeting the emerging field of *product data management* (PDM), which in the meantime has become *product life-cycle management* (PLM). The main object of PDM/PLM is, more than mere automation, the standardization and faithful documentation of the product development process, particularly the release history, making all steps accountable while building up a structured and accessible base of the data defining a product.

Besides a little bit of internal consulting, some conceptual work and little bit more education of young colleagues, interns and students enrolled with DHBW, while my role as a thesis advisor there was supplemented by that of a member of an examination board, the main tasks at ECS had a narrower technical focus, particularly analysis, cleaning and restructuring of large volumes of data for platform migration and import from legacy systems, including paper based ones – all business much less frustrating than consulting proper.

But finally, I came full cycle: the ›L‹ in PLM until now, almost never meant truly ›life cycle‹ – with some minor exception like the turbines at Siemens Energy and the likes of them, which are unique items, which get a life-cycle folder attached in the system, to be updated while there exists a valid maintenance contract to be served. Of course, true PLM would be a good idea for various reasons: parsimony of energy and materials requires the threefold capital ›R‹ to be obeyed: repair of aggregates, reuse of parts and recycling of materials – and in order to make them work, all relevant data have to be available. But, beyond parsimony, product improvement and services tailored to customer's needs would benefit greatly from true life cycle data, giving hints to actual use and wear of products. This idea carries the name *closed loop engineering*, promising not so much rationalization but positional advantages relative to competitors. Products, production lines, and whole factories fitted with a multitude of sensors could deliver such data, but too flood networks and raise some intricate issues of adequate processing as well as, and not the least, with privacy and ownership. Processing abundant data from a multitude of sensors means detecting specific relevant events from it – a task which always is confronted with the trade-off between sensitivity and specificity: being too sensitive means too many false positives and alarms, being too specific too many events escaped, and perils overlooked. After having advised already a number of theses related to such approaches, involving real-time signal processing and machine learning, I found myself, near the end of my career in computing for engineering, in a working group comprising university and industry staff with the aim of crafting a concept for a research project. The project as conceived by the group did not get the funding requested, but anyway, there is a wide and promising field opening for my successors.

Already while at ECS, my interests made another cycle, coming back to my roots in Philosophy: science, technology and medicine are important mediators in the man-nature metabolism central to human future. Recognizing the weight of the related issues, my publishing activity shifted to the task of putting this metabolism and those mediators into a philosophical perspective informed by my experience in science and technology (Fischbach 2005, 2016a, 2017, 2021b, 2023). But I also found out this experience is still required not only to give some historical record, but also to look into the future (Fischbach 2016b, 2020a, 2020b, 2021a, 2022) ... and certainly has validity for Facility Management too!

Resuming my professional life as well as my philosophical excursions, I can assert that my episode with facility management was a fruitful one. Particularly the aspect of life cycle turned out to be helpful and ubiquitous. Taking it up and carefully applying it to the universe of artefacts will benefit life and all activities, practical as well as intellectual ones. It will be one of the keys to the reconfiguration of man–nature metabolism to the end of making it more enduring.

Sources

Fischbach, R. (1989). Strategig objects of computer-aided facility management. *CAD-Forum '89*. SIA Schweizerischer Ingenieur- und Architektenverein and SCGA Swiss Computer Graphics Association, Wintherthur, 26 October 1989.

Fischbach, R. (1992). Programming by contract — does Eiffel fulfill the ideal? In: *Eiffel: Joint symposium* of *the German Chapter of the ACM with the Gesellschaft für Informatik (GI), on 25 and 26 May 1992 in Darmstadt, Proceedings* (ed. H.-J. Hoffmann), 55–68. Stuttgart: Teubner (German Chapter of the ACM, Berichte; 35). https://doi.org/10.1007/978-3-322-86775-9_5 (29.05.2023).

Fischbach, R. (1994a): Modelling: fundamentals and industrial significance. Part 1: basics. *Bauinformatik*, March/April, 68–75.

Fischbach, R. (1994b). Modelling: fundamentals and industrial significance. Part 2: industrial importance. *Bauinformatik*, May/June, 114–119.

Fischbach, R. (1994c): One-of-a-kind: architectural CAD is different. *iX*, May, 60–72 http://www.rainer-fischbach.info/rf_arch_cad_ix_1994_05.pdf (29.05.2023).

Fischbach, R. (1994d). Modelling with data: fundamentals and significance in the production and operation of complex aggregates. *2nd Beckmann Colloquium*, Wismar, 3–4 June 1994.

Fischbach, R. (1994e): Product modelling in construction. *Bauinformatik*, November/December, 256–263.

Fischbach, R. (2005). *The Myth of the Net: Communication beyond Space and Time?* Zurich: Rotpunktverlag.

Fischbach, R. (2016a). *Human–Nature–Metabolism: Essays on Political Technology*. Cologne: PapyRossa.

Fischbach, R. (2016b). Why are software projects difficult? In: *Informatik und Gesellschaft: Festschrift zum 80. Geburtstag von Klaus Fuchs-Kittowski* (ed. F. Fuchs-Kittowski and W. Kriesel), 393–402. Frankfurt am Main: Peter Lang.

Fischbach, R. (2017). *The Beautiful Utopia: Paul Mason, Post-Capitalism and the Dream of Boundless Abundance*. Cologne: PapyRossa.

Fischbach, R. (2020a). Metaphorical and real machines: a proposal for understanding and dealing with information technology. In: *Future of Work: Sociotechnical Design of the World of Work in the Context of »Digitalization« and »Artificial Intelligence«* (ed. P. Brödner and K. Fuchs-Kittowski), 41–59. Conference of the Leibniz Sozietät der Wissenschaften on 13 December 2019 in Berlin, University of Applied Sciences. Berlin: trafo Wissenschafsverlag (Transactions of the Leibniz-Sozietät der Wissenschaften; vol. 67).

Fischbach, R. (2020b). Big data — big confusion: why there is still no artificial intelligence. *Berliner Debate Initial* 1: 136–147. https://www.academia.edu/64910746/Big_Data_Big_Confusion_Weshalb_es_immer_noch_keine_k%C3%BCnstliche_Intelligenz_gibt (29.05.2023).

Fischbach, R. (2021a). Design of socio-technical systems. In: *The Path to the "Digitalization" of Society — What Can We Learn from the History of Computer Science?* (ed. J. Pohle and K. Lenk), 83–97. Münster: Metropolis. https://www.metropolis-verlag.de/Gestaltung-soziotechnischer-Systeme/14807/book.do;jsessionid=E6BB5A800EBDAF06B1A7CDC1E5EA3DD9 (29.05.2023).

Fischbach, R. (2021b). Technological obstinacy and capitalist logic. In: *Digitization and Technology — Progress or Curse? Perspectives of the Development of Productive Forces in Modern Capitalism* (ed. H.-J. Bontrup and J. Daub), 16–48. Cologne: PapyRossa. https://www.academia.edu/97968886/Technologischer_Eigensinn_und_kapitalistische_Logik (29.05.2023).

Fischbach, R. (2022). Thoughts on the epistemic status of models. In: *Cyberscience — Science Studies and Computer Science. Digital Media and the Future of the Culture of Scientific Activity* (ed. G. Banse and K. Fuchs-Kittowski),

161–174. Conference of the working group "Emergent Systems / Computer Science and Society" of the Leibniz Sozietät der Wissenschaften in cooperation with the Gesellschaft für Wissenschaftsforschung, Berlin, November 26, 2021. Berlin: trafo Wissenschaftsverlag (Transactions of the Leibniz-Sozietät der Wissenschaften; vol. 150/151). https://www.academia.edu/97918896/Gedanken_zum_epistemischen_Status_von_Modellen (29.05.2023).

Fischbach, R. (2023). *A Virus, for Example: How A Society Lost Reason and Humanity — And How to Regain Them.* Düren: Shaker Media. https://www.rainer-fischbach.info/virus_rsrc.html (30.05.2023).

Fischbach, R., Gerber, R., Glaser, J., and Oestereich-Rappaport, R. (1993): Animated spaces: computer use in construction projects. *iX*, July, 68–72.

Fischbach, R. and Lenart, M. (1982). *Straightforward Planarity Testing by Generating Basic Cycles in Graphs for Architectural Planning*. Stuttgart: Institut für Grundlagen der Planung (IGP-A-82-1).

Fischbach, R., Lenart, M., and Stolzenberg, B. (1995): Elements of a design modeling language. In: *Computing in Civil and Building Engineering* (Hrsg. P.J. Pahl and H. Werner), 55–60. Proceedings of the Sixth International Conference, Berlin, 12–15 July 1995. Rotterdam: Balkema.

Kahlen, H. (1989). *CAD-Einsatz in der Architektur*. Stuttgart: Kohlhammer (Facility Management; 2).

Lenart, M. and Fischbach, R. (1982). *On Necessary Metrical Conditions of Existence for Floorplans*. Stuttgart: Institut für Grundlagen der Planung (IGP-A-82-2).

Reuter, W. (1989). *The Power of Planners and Architects*. Stuttgart: Kohlhammer (Facility Management; 1).

Rittel, H.W.J. (1992). *Planning, Designing, Design: Selected Writings on Theory and Methodology* (ed. W.D. Reuter). Stuttgart: Kohlhammer (Facility Management; 5).

Rittel, H.W.J. (2013). *Thinking Design: Transdisciplinary Concepts for Planners and Designers* (ed. W.D. Reuter and W. Jonas). Basel: Birkhäuser (Board of International Research in Design).

Westfehling, U. (1993). *Zeichnung in der Renaissance: Entwicklung—Techniken—Formen—Themen*. Cologne: DuMont.

Rainer Fischbach
Berlin, Germany

Rainer was born in 1950, studied in occidental and oriental philosophy, philology, computing and planning science, international relations, and military policy. From 1978 he was a research assistant at the Institute für Grundlagen der Planning (IGP) with Prof. Horst Rittel, where he provided research in planning theory and software modeling; in military technology assessment with the Forschungsgruppe für europäische Sicherheitspolitik (Peace Research Studies AFES-PRESS). He was an Associate lecturer and thesis advisor at the Cooperative State University Baden-Württemberg (DHBW).

From 1987 until 1994 he was a coworker at the Facility Management Institute (FMI) with Dr. Josef R. G. Mack and was a consultant with the architectural practice of Kahlen & Partner, Aacken, Berlin, and Cottbus. From 1994 to 1999 he worked at Ruhrkohle AG (RAG), and then was employed as a Senior Consultant at ECS Engineering Consulting & Solutions at Neumarkt/Oberpfalz, where he was involved in projects with customers like Siemens Energy and Knorr Bremse. He retired in 2021 and has been the author of hundreds of articles and four monographs.

2.29

FM a Summary

Michaela Hellerforth

Introduction

Originally, my approach to Facility Management (FM) was purely a real estate one. While still at school, I worked for real estate companies. Studying activities in associations in the real estate sector followed, first as an assistant to the board of directors in business management issues, and later also as a member of the board of the Federal Association of Independent Housing Companies ("BFW, Bundesverband Freier Wohnungsunternehmen").

From this perspective, I received the first request to write an article about FM just before the turn of the millennium. At that time, the term FM was not really known to many market participants and the focus of most companies was not yet on areas besides their core business.But there were already the first courses of study in facility management, often with a technical background, which, according to curricula derived from the GEFMA guideline, with the aim to introduce students to the subject and develop expertise needed by the market. There was a great demand from companies for specially trained specialists in this field. In the more than 20 years that have passed since then, the profile of the courses has become more and more specialized. Nonetheless, my real estate experience and view of FM shows me the great importance of soft skills and overarching interface thinking in order to be able to carry out FM adequately. Often, FM is still dominated and handicapped by the core business in companies.

Example: A large hospital campus with its own facilities, rented practices, external operators, e.g., the retirement home and the kindergarten and rented apartments to staff. Most of the buildings were in poor condition. This affects both the appearance ("charm of the 1990s" or even before) and the building technology. Facility management is operated exclusively by our own employees. They do not have a clear leadership, but rush from mission to mission. Whether all test cycles are adhered to is not fully comprehensible at any point. Building data is incomplete in front of and in the minds of the technical manager. A conversation with the board about the problem ended with the following statement: "If I have the choice of hiring an assistant doctor or spending budget on buildings, I will always choose the doctor."

This project showed that even after many years of FM, essential findings have not arrived in organizations. In addition to the questions of operator responsibility, it is not uncommon to overlook the fact that working in real estate and visiting properties that are in a visually modern condition is an increasingly important competitive factor, especially in a surrounding where many hospitals are to be closed. As part of this project, satisfaction surveys were also carried out at the rented practices. Here it turned out that the doctors were very satisfied with the proximity to the hospital and other doctors, but very dissatisfied with the condition of the premises or the overall impression of the center, so that some had at least already considered relocations.

Facilities @ Management, Concept - Realization - Vision, A Global Perspective, First Edition. Edited by Edmond Rondeau and Michaela Hellerforth.

On the other hand, there were two perspectives on the building stock, namely the operational one: the provision of real estate to carry out the core business of "hospital operations" and the landlord's view for the rented buildings and parts of buildings, i.e., the real estate perspective. In such cases, the original boundaries between core business and real estate business become blurred and the head of real estate management has to dance at two weddings. In the case described, there were also tensions as a result, as the interests of the core business and the tenant's interests were not always aligned.

However, this example stands for another important aspect: the recognition of the entire area of real estate management or FM as an equal partner by the management and thus a clear strategy and role of this area: The Corporate Real Estate Management (CREM) should build on a real estate strategy corresponding to the goals of the core business, which would meet the objectives of the next 10 years, the vision around the real estate portfolio, fixed.

Big Data and People

FM can only work if the overall organization and its specifics are considered. The first task is usually to collect and record real estate data in order to better work out the status quo of a company's buildings, facilities and real estate. The data problem exists in almost all organizations that involve the reorganization of FM. There are often redundant databases in different places, Excel lists for support, but sometimes also CAFM-systems that were introduced years ago but were not properly maintained and full of data not up to date, so that no one relies on them. So, Excel lists again after all? This means that many FM projects are also IT projects, especially since the focus is not on data acquisition, but on using the data to improve operations. However, it is just as important to take along the employees who have to deal with these or new systems, who are trying to achieve improvements in projects, and to convince them of the vision "That's where we want to be with our FM in 5 years."

However, the important topics of sustainability and ESG cannot be tackled without taking stock of the current situation. Especially companies that supposedly still have time with their reporting obligations are in favor of ESG in principle, but there is a lack of clear ideas and specifications as to which criteria are to be collected and reported to what extent and, in turn, the data.

Example: Industrial area in a larger industrial area, there had been a large investment backlog for many decades. Parallel to the project of "outsourcing the FM into a separate unit," an inventory of the buildings and their technical facilities took place in order to obtain a database. The data was then entered into the equipment in SAP. This was done on a building-by-building basis, as well as the agreement on updated bills of quantities, because the building survey showed that the specifications used so far were incomplete. Only then and taking into account the budget that the management had planned for the ancillary area, could the funds be prioritized and, in particular, the ecological aspect taken into account in future measures.

Life Cycle – ESG – Life Cycle Costing – Life Cycle Analysis

The life cycle view in FM in the broader sense was one of the first models I was confronted with, but this is often lost in the courses of study, but also among the market participants due to the small-scale nature of the processes and the depth of the consideration. The life cycle model, which was initially purely cost-oriented, has been expanded in recent years to include the consideration of material flows for life cycle analysis. When the phase of partial exploitation of real estate begins, the potential for further development, the material passport or the resource passport is revealed. Here, the installed materials are recorded, and an economic value is determined in this respect, which influences the building value in the sense of the cradle-to-cradle idea, but also an ecological value.

To the extent that these developments are also driven forward and point out perspectives that are becoming increasingly important, most non-property companies are only slowly beginning to orient themselves in this direction. A question that is often asked when considering the implementation of a new building as a BIM project is how the BIM system with all the data perfectly maintained at the end of construction can be integrated into the existing CAFM system with as little effort as possible and also whether the added value of the BIM or the building material passport can be calculated directly in euros. If no satisfactory answers are found for the management, this often means the end of such a project.

People and FM

The tasks facing FM are not diminishing. A major problem is still the shortage of skilled workers, especially in the technical field, with the result that technical contracts are carried out by merchants or lawyers or, incidentally, by technicians and often make it difficult to ensure continuity in the provision of services. In the technical but also in the infrastructural area, savings were made for years, the FM services were at the bottom of the service pyramid as the services with high standardization and low profitability. In many respects, this results in poorly repaired and maintained buildings, where a lot of money has to be spent just to realize the necessary needs of operator responsibility.

Roles

As indicated, many companies also have problems with the roles of the individual protagonists integrated into FM. When looking at the roles, many customers are faced with the question of how they should actually occupy and live the individual roles from investor to portfolio management, asset management and property management to FM in the narrower sense, especially in the context of internal service relationships. This means that the question of efficient organization and the best possible structuring of internal service relationships in corporations is still an area on which the focus is on practice.

Example: Large industrial site The real estate management of a large industrial site had the task of improving the service relationships between the real estate management and the technical subsidiary (the technical department) in such a way that the real estate management could live up to its operator responsibility from the owner's point of view. This included adherence to maintenance and inspection cycles, but also continuity of performance in repair and defect elimination. As a rule, contracts are drawn up for this purpose with differentiated malus regulations, in the event of nonperformance or poor performance, which also provide for the possibility of not paying the remuneration in the event of nonperformance, self-performance, or termination of the services not provided. In this company, every internal order automatically led to the release of the invoice in the system, i.e., a defective or non-executed service could only be compensated back to the client via a credit note and the credit note had to be released by the contractor. Since the internal system of the entire group was structured in this way, other ways had to be found to achieve performance security.

The roles in outsourcing or reorganization projects must always be filled with people. The interfaces between the owner side on the one hand and the service provider side on the other are decisive. Although the task catalogues published by various institutions, for example for asset and property management, help, these must then be implemented in existing job profiles and existing know-how.

The tasks are not free of overlaps and therefore the definition of the interfaces between property and asset management is fluid. A common accrual refers to property management costs as real estate-related property costs. The costs of the levels above, i.e., the owner level and the real estate asset management level, are booked as company costs. In operational practice, it is often the case that property management is completely omitted on the owner side and the asset manager has overall responsibility, but some of the property management services are also done by the service providers, such as obtaining offers. However, it is precisely then that the control tasks of asset property management in personal union are particularly important. In this case, the interfaces must be clearly defined.

Result

Furthermore, the field of FM in the further life cycle concept and its effects is an exciting field and will remain so. In daily work, of course, disillusionment remains regarding the discrepancy between what could be achieved with FM and what is actually achieved due to various circumstances. But on the other hand, the possibilities and further developments of the fields around facility management are also fascinating and working in projects makes it possible to implement at least partial improvements, even if the results are far from teaching and best practice examples.

In all projects, it is evident that specialist knowledge alone is not sufficient, but that it is precisely lateral entrants to the field of FM who critically question existing structures and encourage new approaches. In addition, FM is to a large extent people's business. Only by taking into account the ideas and fears of those involved can a team be formed that is up to the future requirements of FM. The optimization concepts that are purely based on the workplace are only partially sufficient insofar as they can only create the framework conditions in the company but also at the workplace at home, but the soft knowledge or soft skills must be contributed by the managers.

In addition to paying attention to the aspects of sustainability, one of the main focuses must be on the question of how to recruit and retain skilled workers, what working atmosphere, benefits, opportunities and flexibilities an employer should offer employees in this area in order to weld together a team that raises the value potential of real estate and drives forward the goals around the building stock.

Prof. Dr. **Michaela Hellerforth**
Westphalian University of Applied Sciences in Facility Management
Luedenscheid, Germany

Michaela Hellerforth studied economics and law at the University at Münster/ Westphalia (Westfaelische Wilhelms Universitaet, Münster) and started to work for a real estate association, the Federal Association of Independent Housing Companies (BFW Bundesverband Freier Immobilien- und Wohnungsunternehmen e. V., Berlin), as well as for a real estate company. Simultaneously she started with her doctoral thesis at the University of Cologne (Universität zu Köln). The thesis was about the relationship between the individual national real estate markets to the overall European market. While working for BFW, she published a lot, and got in touch with a relatively new field in connection with real estate: Facility Management. At this time a subject that didn't matter for real estate companies, while Facility Management developed more and more in non-property companies. Already during her doctoral thesis, she specialized more and more in Facility Management, published around 130 articles and 14 books in this field, and was also teaching for different institutions.

In 2000 she started working as a Professor for Facility Management at the University of Applied Sciences in the Ruhr Area (Westfaelische Hochschule in Gelsenkirchen). There she works in different fields of Facility Management, as well as continues her publishing. She is also active in consulting public and private institutions in general issues of Facility Management, restructuring organizations, especially CREM and Facility Management divisions, with focus on strategic aspects of Facility Management and Service-Provider Management as well as contract management and Controlling FM-Services.

2.30

My Path to Facility Management and Facility Management

Albert Pilger

Our journey through time begins in 1969, at a time when not only the term itself, but, as a result, the performance profile of facility management was also largely unknown.

For 26 years I worked in the field of building measurement and control technology and building automation. That is, at the beginning of my professional career, there was no talk of facility management. It was only in the last few years, when I was managing director of the company in Austria, that I got to know facility management at international management conferences. What I learned on these occasions seemed extremely exciting to me and I began to delve deeper and deeper into facility management.

I took part in worldwide facility management congresses and thus deepened my knowledge. This led me to start my own business in 1996 with the topic of facility management. Soon I was giving lectures and seminars on the subject of facility management myself. Today, 20 years later, there is a broad training landscape for management Model "Facility Management," which is used in many companies and organizations which is already fully integrated and therefore no longer a foreign word.

Even outside of the persons directly involved in facility management there is an increasing understanding of the question of what facility management actually is. The time of missionary activity is largely over, but only largely. Despite all the positive developments and tendencies, there are still major misunderstandings that need to be cleared up.

In 1994 and 1995, a group of personalities came together, to set up a platform in Austria under the name "Facility Management." Some names from this period were Dr. Ekkehard Wunderer, Markus Aschauer, Gerhard Haumer, Karl-Heinz Lehocky, Heinz Richter, Peter Prischl, Alfred Kleedorfer, and Albert Pilger.

On the basis of this platform, the first facility management network in Austria was developed as early as 1995 and the Association "Facility Management Austria" (FMA) was founded. The first objectives included, above all, an active promotion of the general understanding of Facility Management as management as well as its understanding of quality. As a result, almost by itself, the demand for facility management training institutions, which was immediately complied with the first initiatives were developed to standardize the "Facility Management" management model. This led to a change in the understanding of roles.

In 1997, together with some market participants, I co-developed a professional course of 360 teaching units. This course ran for two semesters in Vienna, Graz, Linz, and Innsbruck. I was the course director from the beginning and still am today. Due to the COVID pandemic, we have changed the course from a face-to-face course to a virtual course and as such this course is still running on the FM Academy platform.

As early as 1997, the first facility management course started at the newly founded University of Applied Sciences in Kufstein (Tyrol), Austria. In 2001, the first students successfully completed their studies as graduate engineers (FH). As the core essence at that time, FM was primarily required to have a high level of technical

Facilities @ Management, Concept - Realization - Vision, A Global Perspective, First Edition. Edited by Edmond Rondeau and Michaela Hellerforth.

competence. Soon after, however, it became clear that FM is primarily concerned with a business question, namely, the increase in productivity in the core business processes. As a result, the graduates soon graduated as Master (FH) before graduating, in implementation of the so-called Bologna Process, to the Bachelor's and Master has been renamed.

The Danube University in Krems and the TU in Vienna also offer academic facility management training. As mentioned before, there are at the FM Academy in Graz and in Vienna professional facility management training.

On the occasion of the European Congress WWP-E 1998 (World Workplace Europe) in Maastricht, the Netherlands, a chapter of the IFMA (International Facility Management Association), IFMA Austria was chartered.

Since 1995, Facility Management Austria (FMA) has also been active in the European network, EuroFM. Through IFMA and EuroFM, FM enthusiasts in Austria were also international and very well connected. The fact that the FMA and the IFMA Austria chapter acted on a common platform from the very beginning contributed significantly to this positive development at both associations that are still supported by a joint office, and in some cases, there are still people who are members of both associations.

Taking Responsibility

In 2001, when the first students graduated from the FH in Kufstein, the European Congress WWP-E 2001 was organized in Innsbruck and held. Even today, we (who were there at the time) are still greeted by our friends in America and also in Australia and Japan about this great event. I must mention at this point that Walter Moslehner supported us tirelessly in making this congress so successful. Walter Moslehner did not live to see the congress himself as he unfortunately died on July 1, 2001.

In the same year that the FMA was integrated into the European network, and the FMA took the first initiatives to develop a so-called FM standard. As early as 2000, the Austrian Standards Institute published the first FM standard, ÖNORM A 7000 (Facility Management – Basic Concepts), and shortly afterward the standards ÖNORM A 7001 (Guidelines for the Preparation of Facility Management Agreements for the use phase of an object) and A 7002 (catalogue of Requirements for Facility Managers) followed. A little later, the standards A 7010-1 to 5 Property Management – Data Structures were added. The Austrian FM associations, together with the Austrian Standards Institute, were also involved in the development of the European series of standards on FM, EN 15221-1 to -7.

This European standard provides an appropriate framework to teach and communicate a common understanding of the term and content of facility management. It is also pleasing that many other courses of study, especially at the Universities of Applied Sciences, provide introductory lectures on the subject of facility management. Thus, the topic can be found in the curriculum of degree programs such as International Studies Management, Project Management, Industrial Maintenance Management, Public Management, Hotel Management, Economics and Management, Architecture, Construction Management, and Civil Engineering.

Many things have become clearer and thus more understandable in recent years. Nevertheless, there are still often, and unfortunately also partly within the industry, large differences in presentation to some of the following questions: "What is FM?" "What are the benefits of FM?" "What are the objectives of FM?" "How is FM to be distinguished from facility services?" "What is the difference between FM and building management?" and "Is there a difference between FM and real estate management?"

Here are the key answers to some of these questions:

Topic 1: Productivity versus cost reduction as an objective.

In many places, it is still believed that the primary goals of FM revolve around the factor of cost reduction. This misconception has existed for several years now. In many cases, however, the reduction of costs is associated with a drastic reduction of service levels and thus also of services and qualities. But to what extent does this affect the hygiene factors of our employees and the Infrastructure at and around the workplace negatively influenced employees and lowered their motivation? Wouldn't it be more effective and efficient to optimize the processes of the services, to improve workplaces in order to create an infrastructure that increases the productivity of the employees? Professional FM often leads to a reduction in costs. However, this can only be a positive side effect of a professional FM and never a primary goal.

Topic 2: Facility management goes far beyond building management.
Building management is responsible for the availability of all agreed building functions for one or more buildings and within the framework of a certain useful life. Facility management is responsible for the infrastructure of the respective organization as well as the respective company long before and even after the buildings are in use. This means that from the perspective of the time horizon as well as from the perspective of the organizational horizon, there is a significant difference between facility management and building management.

Topic 3: Facility management cannot be outsourced, but facility services can be outsourced.
The first two levels in facility management are the strategic facility management and the tactical facility management. At the first level, the strategy of facility management is derived from the strategy of core business processes. On the second level, this strategy is implemented at that level. It is important to determine which supporting processes from the core business processes to which quality is expected to increase productivity. It is also important to answer the question of costs and control. Generally, this is referred to as a qualified ordering function. This means that the so-called service levels are fixed and now you can – but don't have to! – have the Facility Services, including any facility services management, can be outsourced.

Topic 4: FM and Real Estate Management.
The objective of FM is to provide a productivity-enhancing infrastructure for the core business processes. The objective of real estate management is called return. And then there are other important issues that have risen in recent years have been given importance, but which also have room for further development.

Topic 5: Using CAFM correctly as a tool.
In almost all areas of work and life, IT has become indispensable as a tool. Of course, this also applies to FM. When it comes to area information, i.e., information on the equipment of the areas, the vacancy rate, the optimization potentials, cost-by-cause billing, management inspection obligations and maintenance management, resource management to name just a few topics, a CAFM system is required. In terms of relocation management and pre-move scenario planning, CAFM is crucial. Even if a tool for cost recording, cost tracking and budgeting in FM is sought. CAFM is often the adequate solution.

CAFM is not only necessary for the management of real estate-related data and building-related and non-building-related facility services, but is needed for a holistic management of the real estate, the tangible and the intangible infrastructure and is also to be regarded as THE controlling instrument in FM.

Topic 6: Economic, ecological and social sustainability is only possible with FM.
Through a facilitary accompaniment of the planning as well as a facilitary accompaniment of the construction, FM represents the basis for economic, ecological, and social sustainability in relation to the property, even in the planning phase. In the planning phase, a decision is made as to which social and communicative requirements the building must meet (for example, from the point of view of cognitive ergonomics). Furthermore, at that stage, the future management costs are also determined (the future Management costs are 75% determined by the planning, considering the total costs of ownership). In the planning as well as the subsequent execution phases these will also focus on the environmental sustainability of the properties clearly defined and defined (see energy consumption, emissions, CO2 footprint, composites, contaminated materials, etc.).

Topic 7: Users as customers of FM.
If we seriously want to implement the goal of "increasing productivity" and not just leave it as a meaningless slogan, FM must be responsible for the productivity of each individual workplace. This is also preceded by the demand for an intensification of communication with users. First and foremost, it is about presenting the benefits

for the customer. To say that we focus on the customer (employee) and that he/she stands in everyone's way will probably be the wrong approach.

Finally, Topic 8: FM is on its way to the board of directors.
Are the topics that have already been briefly touched upon really important? Are these a success factor? YES. If there is a general appreciation of these points, FM must also be positioned accordingly in the organization. Occasionally, but in increasing numbers, you can already find the Head of Facility Management in a board position. At the second level of management, however, FM can already be found in many organizations and companies.

Since 1998 I have been continuously invited by universities, colleges and technical colleges to give lectures as a lecturer on the topic of facility management. In the end, I worked at 13 universities, colleges and universities of applied sciences.

All this in addition to the consulting work, which I still carry out today as a senior consultant in the areas of industry including the IT industry, universities and research institutions, universities of applied sciences, logistics centers, shopping centers, state, states and cities, zoos, railway companies, real estate companies, congress companies, hospitals, banks, theaters, energy supply companies, social organizations in Austria and abroad, as Einstein concludes (see Figure 1).

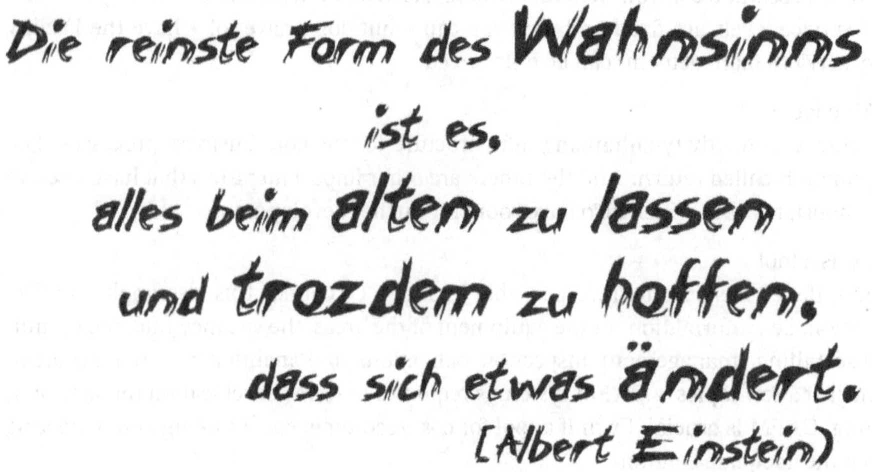

Die reinste Form des Wahnsinns
ist es,
alles beim alten zu lassen
und trozdem zu hoffen,
dass sich etwas ändert.
[Albert Einstein]

"The purest form of madness is to leave everything as it is and still hope that something will change."
Albert Einstein

Figure 1 Quote from Albert Einstein. (*Source:* A. Einstein).

Albert M. M. Pilger, Prof. (FH) Ing. Mag., CFM, IFMA Fellow
Pilger Facility Management GmbH, Senior Consultant
and Head of the FM Akademie
Graz, Austria

Albert Pilger is considered one of the pioneers of facility management in Austria and is still known beyond the borders. He played a key role in shaping one of the first facility management standards (ÖN 7001) and is a sought-after lecturer and consultant in the industry. Albert Pilger was President of IFMA Austria, Chairman of the European Facility Management Association (EuroFM) and the first European to be elected to the Board of Directors of the International Facility Management Association (IFMA).

Despite his many activities – Albert Pilger is also managing director of the Pilger Management Academy – he has been loyal to the Facility Management course for many semesters and also supports it as a member of the development team. Program director Prof. (FH) Dr. Thomas Madritsch honored his commitment with the guest professor's certificate where Albert Pilger was awarded the visiting professorship of the University of Applied Sciences Kufstein Tirol for his services and his many years of involvement in the Facility Management program.

3

Contributors: Vision

This ***Vision*** chapter includes success as well as shortcomings of the realization and transformation in the academic, the institutional, and the commercial world. To create a platform of demands for the immediate future it is required for the built environment to respect social, economic, ecological, and demographic changes under sustainability demands and global climate change.

The following contributors provided their anthology and legacy on the vision and future of facility management:

3.1 Alex Lam
3.2 Parminder Juneja
3.3 Barry Varcoe
3.4 Klaus Zapotoczky
3.5 Fred Kloet
3.6 Mark P. Mobach
3.7 Bernie Gorman
3.8 Joint Contributors: Christin Kuchenbecker and Andreas Kühne
3.9 Diane Levine
3.10 John Adams
3.11 Theo van der Voordt
3.12 Audrey L. Schultz
3.13 Kreon Cyros
3.14 Henning Balck
3.15 Joint Contributors: Volker Hartkopf and Vivian Loftness

Facilities @ Management, Concept - Realization - Vision, A Global Perspective, First Edition. Edited by Edmond Rondeau and Michaela Hellerforth.

3.1

50 Years in the Industry – Leadership Lessons Learned

Alex Lam

In May 2022, I received a congratulatory letter from Don Gilpin, CEO of the International Facility Management Association (IFMA), for my 40th anniversary of being a member. In 1982 I saw an announcement of IFMA. I said to myself, "*This is it! This is what we do! We must belong and support this IFMA.*" Since then, it was a beautiful facility management (FM) journey of learning and networking all over the world. I can now say I practically have friends all over the world. It reminds me of a presentation I made at the 2017 World Workplace Asia in Shanghai: "*50 Years in the Industry – Leadership Lessons Learned.*" It was a retrospective of my professional career with an emphasis on the "M" in FM. I thought I would share my experience and my view of the future of facility management here.

In The Beginning

Back in the 1970s, at the dawn of facility management, I transitioned from private architectural practice to joining a corporation and starting my life as a corporate citizen. Canada's largest telecommunication company Bell Canada (Bell) hired me in 1972 to lead the Interior Space Design department. We were to embark on a process of office redesign and transformation using the open office concept. My introduction from architecture to facility management was through interior design. Reflecting on those early days, my involvement with facility management was just a series of happenstances. These early in-house facilities department was known as the Real Estate department, the Accommodation department, or the Estate department.

The office landscape design concept (*Bürolandschaft*) popularized in the 1950s by the Quickborner Group in Germany was the guiding principle in those revolutionary days. It promoted an open office planning eliminating most enclosed offices. This threatened the North American mindset. The concept fostered workers interaction with free-flowing or organic clusters of workstations. The design on the plan, in some sense, resembled more like an English country garden than a formal French garden.

However, the idea was considered a blow to the North American mentality of the private office culture. Middle management fought tooth and nail to keep their private office because they had earned it as an accomplished status symbol. The result was a constant battle with the middle management for losing their enclosed offices. In the 1960s and 1970s, it was a corporate human resources (HR) strategy to move people around to different positions to receive a more rounded operational experience of the company. At the same time, office renovations to accommodate these moves became expensive. New walls, floors, and ceilings would need to be demolished and reconstructed to meet the new office layout. So, the work involved in office reconfiguration modifying electrical

Facilities @ Management, Concept - Realization - Vision, A Global Perspective, First Edition. Edited by Edmond Rondeau and Michaela Hellerforth.

and mechanical systems was more time-consuming and labor-intensive. At the same time, the rest of the office would be disturbed causing inconvenience to the existing occupants. So, the solution was "*Move People, not Furniture*" and of course, it was much easier to move people within the office landscape concept.

At the time, there was no furniture on the market to address the idea. We worked with a Montreal company called Precision Wood to create our own furniture – freestanding acoustical landscape dividers. The company catered to our needs very well. We spent many hours in their factory to perfect the dividers. The owner is a gentlemanly German and of course, he understood our needs and everything about "*bürolandschaft*." We created a straight and a curved version and the purpose was to provide separations between workstations. The design was an office divider standing on adjustable flat metal feet. The screens were covered with fire-retardant fabric and acoustical insulation inside sandwiching an aluminum foil to reduce sound transmission through the panel. This *primitive* or *generic* design served our needs for a few years until the birth of personal computers which changed the entire office "landscape."

Computers have been around since the 1950s, but it was not until the 1970s that the personal computer (PC) became commercially viable. It posed a new challenge for facility managers. The landscape dividers suddenly became inadequate for the needs of the PC office. The desks now with the PC on them must address ergonomic issues. The placement and the relationship between humans, keyboard, digital input devices, monitors, and CPUs now became critical in the functioning of the workstation. The new ergonomic challenges further ushered in the requirements for integration.

Personal computers now demanded ergonomically designed desks to be adjustable to suit individuals. The dividing screens must be able to support storage with separate built-in channels to carry power and communication cables. A whole new industry was born. Expressions like power panels, integrated systems, and system furniture became our new FM vocabulary. Other elements for the environment to re-consider for the office environment were the new requirements for HVAC, the balance of ambient and task lighting, solar-oriented window shades and power-hungry peripherals. Ideas, innovations, and system adaptability abounded at this stage.

So, "*Move People, Not Furniture*" Entered a New Era

The freestanding dividers now morphed into the new system furniture. They became more functional and interconnectable. Additional loads of worktops, cabinets, and other office accessories incorporated in the design required the frame to be structural. Therefore, metal frames replaced wood frames for stability, connectivity, and durability. The structure became thinner but stronger. The dividers now contained raceways to carry communication and power cables. Independent desks and file cabinets became system furniture hanging on the screens. Some desktops have motorized actions to adjust the height of the work surfaces for the needs of individuals. Integrated lighting and motion sensors are part of workplace integration. New designs of HVAC, lighting, and raised floors continued to make the scene as the modern office evolved.

The office landscape concept in the 1970s had disrupted the office environment opening an opportunity for the new science of integrated office.

Working in a corporate environment allows you to learn facility functions holistically rather than in a fragmented way. Unfortunately, people often tend to be stuck with only one function in their careers. The HR strategy for most corporations in the 1970s was to provide a more diverse business training and experience for their people by sending them to work in various departments. The idea was for them to engage in different aspects of the enterprise and become a more well-rounded manager. I label this learning process in the corporate human capital management system for facility management as the 5-Stage Facility Lifecycle (see Figure 1):

- Office Planning and Strategy,
- Design, Construction & Project Management,
- Planning, Leasing, Acquisition and Disposition,
- Operation & Maintenance, and
- Research and Development.

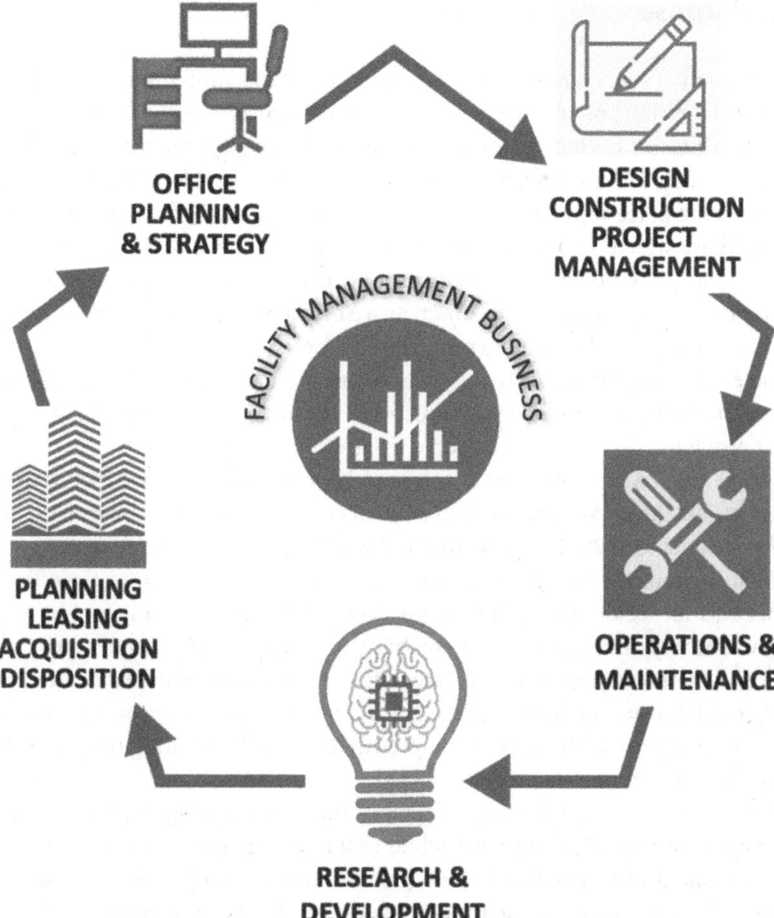

Figure 1 The facility management lifecycle. (*Source:* A. Lam).

It is a lifecycle because of the reality of continuous improvement and timely applications. This process cycles back and forth to adapt to the corporate management and mission changes.

The starting point of this facility lifecycle in most cases is the "Office Planning and Strategy." It sets the motion for office reconfiguration and transformation to meet new workplace requirements. In Bell Canada, we started with a five-year long-range plan to transform the rigid traditional office design into a more flexible and future-adaptable open office design. The workplace must continuously adjust to the ever-changing officing parameters such as business strategies, corporate mission, capital restrictions, environmental impacts, Corporate Social Responsibility (CSR) initiatives, technological advances, adaptation flexibility, and continuous consumer-driven solutions. Once the officing concept was established, the facility department worked with the corporate executives to transform the office together.

The need for a new kind of office planning gave rise to a different way of provisioning and managing. So, the facility management department became critical in partnering with all the stakeholders within the enterprise to meet the corporate goals. In Figure 1, you can see how the "Office Planning and Strategy" stage eventually draws in all the other disciplines to complete the FM cycle.

Facility Management Is About Purpose

A career in facility management in the formative years tended to be the result of chances rather than aspiration. The make-up of a facility team looked like a melting pot of people from various disciplines. We had office administrators, executive assistants, technicians, interior designers, architects, engineers, project managers, English majors, and even nurses. Most corporate citizens had no concept of who we are or what we do. When they came into an office in the morning, they expected the place would be clean, the power was on, the lights were working, the HVAC was running, and most of all, there was toilet paper in the washrooms. No one would ever consider including facility management in their career path, nor anyone would aspire to become a facility manager one day. Once a facility department head remarked: "*If the people do not know we are here, then we must be doing a terrific job*." Perhaps there is some truth in this, but I think it is a wrong attitude for us.

Facility Managers were excellent problem solvers, but we were poor marketers. We had the know-how and the experience to solve difficult and complex situations such as disaster recovery but then, once the problem was resolved, we would move on to the next problem. We failed to tell people what we had done.

In the outsourcing days, facility management was considered a non-core business and was one of the first groups to be outsourced. I once wrote an article titled "We've been sold a Bill of Goods!" that was published in the *Canadian Facility Management + Design* magazine in 2001 arguing that FM was indeed a core business.

The role of a facility manager was not just choosing the right furniture or picking the right color for the office, although many employees thought that was all we do. Our primary function was to ensure employees have a healthy, safe, and productive work environment that could foster creativity and innovation in individuals or teams. The 11 core competencies developed by IFMA in recent years fully reflect the multifaceted qualification of a facility manager. A healthy, safe, and productive workplace supports the core business, and thereby the bottom line. By an error in judgement or action, we might send floors of employees to the hospital or bring the business to a standstill. We could cause a company to prosper or to fail.

In the late 1980s into the early 1990s, the company had a shortage of capital, and expense budget reduction was at the top of everyone's to-do list. Recognizing the corporate need while reducing expenses, we introduced a strategy for office consolidation. We terminated practically all leased quarters and moved people into owned buildings. Then we sold the buildings on a lease-back arrangement with the buyers. We generated over CAD700 million for the corporate capital coffer. Our FM strategy became the corporate strategy.

Yes, We Are CORE

With outsourcing and downsizing, many professionals were leaving the corporation. It means *knowledge* just walked out the door. There was an urgency to ensure continuity and to retain institutional knowledge, especially knowledge related to the design, construction, and operation of specialized facilities such as a datacenter, a production plant, or a research laboratory.

In anticipating changes because of outsourcing, I led a team to establish the core competencies of facility professionals. It was before IFMA developed its 11 core competencies. We worked alongside HR with a consulting psychologist. We created the FM Competency Model and identified the key elements that constituted the basic capabilities and skills of a facility professional. It included business acumen (understanding how a business is run), behavioral attributes (your emotional intelligence), professional knowledge (the skills you bring to the company), and enterprise knowledge (understanding the inner workings of your company). The one unique element that became distinctive for us was the *owner-operator* mindset. In facilities, we did not just do the job, but we do it as if we own the business.

Competency can be defined as any skills, knowledge, behavior, or qualities that are essential to performing a job, or to be differentiating average from superior performance. A competency-based approach to HR planning and development is an important step in building organizations with a competitive advantage.

Based on my work on competencies at Bell, I was invited in 2007 by the National Academies in Washington D.C., founded by Abraham Lincoln in 1863, to participate in a working committee establishing the core competencies required for the next generation of US federal facilities employees. The final document titled *Core Competencies for Federal Facilities Management Through 2020 – Transformational Strategies* served as a hiring guide for the twenty-two sponsoring agencies including the General Services Administration (GSA), Department of Energy, Departments of the Air Force, Army and Navy, and the FBI. They embraced the competency model I developed for Bell and adapted it to identify their unique principal core competencies (see Figure 2). We reached a consensus that the principal core competencies for the future facility manager must have the ability to carry out and generate *integration, alignment,* and *innovation* for the agencies.

In *Competing for the Future*, Gary Hamel and C.K. Prahalad (1996) suggested that "a core competence represents the sum of learning across individual skill sets and individual organization units." Under this definition, core competence is very unlikely to reside in its entirety in a single individual or small team.

Facility Management Is About People

Employee empowerment was the buzzword of the eighties. The slogan is all over the corporate world and on every agenda while cheerleading at corporate motivational rallies. Empowerment is not just a slogan but an actual action. As a leader, you cannot just cheerlead the latest *flavor of the month*, but you must give the "*power*" to the people so that they may act with authority and responsibility.

For example, cleanliness is of paramount importance in telephone switching facilities. Those days, we constantly renovated and upgraded the space to accommodate the rapid growth. Contractors were notorious for not keeping the place clean. I empowered the in-house cleaning staff to be responsible for the facilities they looked after. They were at the lowest level in the facilities organization. I gave them the authority to shut down the job if the contractors did not clean up the place at the end of each day. They exercised their empowered authority, and after a couple of shutdowns, all contractors followed the rules.

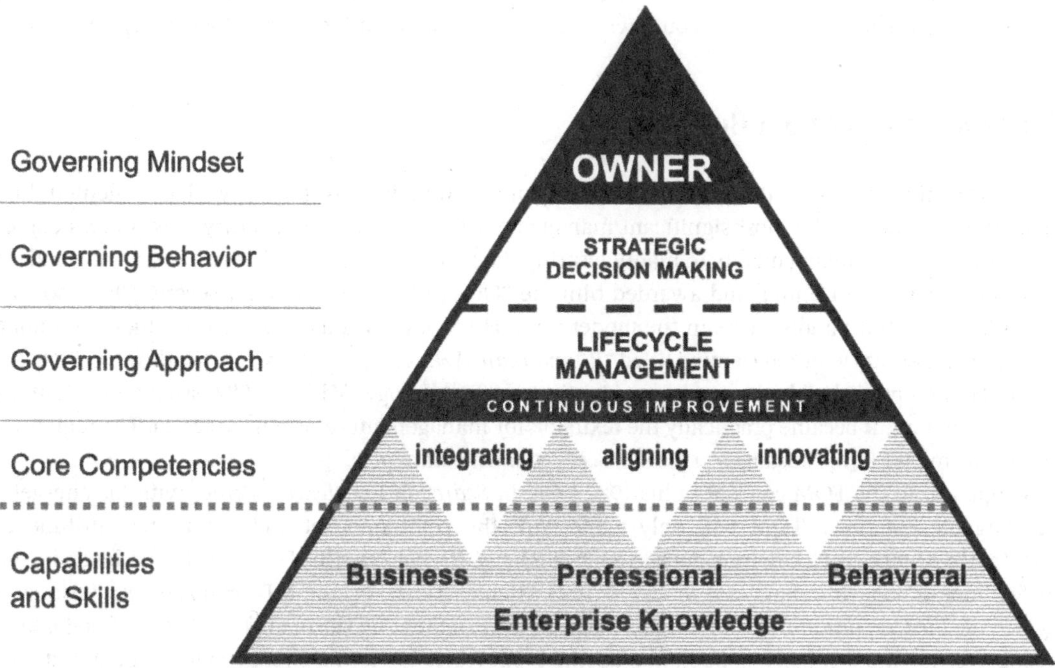

Figure 2 Corporate facility management core competencies. (*Source:* A. Lam).

In 1991 We Introduced the Zero Waste Initiative

The 3Rs of Reduce, Reuse, and Recycle was the early form of corporate Environmental, Social, and Governance (ESG) strategy. We took one step further and created a *Zero Waste* process. We selected a typical 12-story office building and engaged all the building occupants to participate in the program. The facility management team acted as the event leader. The entire team from management to unionized technicians were all hands on. We reduced the use of Styrofoam cups and promoted personal coffee mugs, replaced the ballasts of the fluorescent lighting system with energy-saving devices, allowed the occupants to sort out the three recycling bins themselves, and even introduced "vermi-composting." Yes, real live worms to eat up all the cafeteria organic wastes. We reduced the entire office building's waste from two standard commercial waste disposal bins to two garbage bags. Bruce Simpson, then mayor of Etobicoke, came to congratulate us and offered to present a plaque to the facility management team. But we asked him to change the wording to recognize the employees in the building instead. It was the employees in the building that made the whole event successful. The recognition went to the people.

We rolled out this program successfully across the province in all other office facilities. Our methodology was so well received that it was used as a guide for the province's recycling standards. The result was beyond ESG for the company but more so, the facility team was empowered, respected, and recognized. By giving our facility teams the freedom to innovate, the courage to experiment, and putting them "*outside the box*," we introduced a new creativity paradigm in our company.

Curtis Sittenfeld of the Creative Solution Network remarked in the April 1999 issue of *Fast Company*:

> If you want people to be creative, you have to put them in an environment that let their imagination soar. Most people experience "cubicle creativity" – the size of their ideas is directly proportional to the space they have in which to think.

That was 1991.

Today, the term *zero-waste* is being widely used in different countries by many public and private companies. For us, employee empowerment went beyond our own tuft and had distinct influence on others in a positive way.

Facility Management Is About Business

William Edwards Deming, professor at New York University and Columbia University as well as a talented flute and drum player, is considered the most significant management figure of the 20th century. His success comes from his work in Japan, introducing quality to manufacturing. In 1960, the Prime Minister of Japan, on behalf of Emperor Hirohito, recognized Deming and awarded him the "Order of Sacred Treasure, Second Class" for his contribution to Japan's rebirth and success in the modern era. His work was unknown to the American public until an NBC TV documentary featured Deming titled "*If Japan can... Why can't we?*" He wrote *Quality, Productivity, and Competitive Position* published by Massachusetts Institute of Technology (MIT) in 1982 based on his famous 14 points for management. It became practically the textbook for management reform in America. Through him, total quality management (TQM) became a critical process in most companies.

Tom Peters popularized the TQM concept in his 1982 book *In Search of Excellence* together with his energetic and mesmerizing presentations. TQM will only work with the right mindset and the proper attitude of management. In 1989 Stephen Covey introduced the *7 Habits of Highly Effective People*. It found favor in the boardrooms for human behavior changes. Finally, in 1993 Michael Hammer, the MIT management visionary, wrote *Reengineering the Corporation*. It caught the corporate management by storm and opened the floodgate of corporate reengineering and downsizing. Such management strategies led to the change management and outsourcing trends.

Almost synchronologically successful change management stories from the Japanese auto-manufacturing companies such as Mazda, Toyota, and Honda reached the game rooms of American human resources. A series of Japanese quality-related management techniques found their way to the corporate change management executives. New programs started to flood the management training arena. Programs such as "Just-in-Time" (JIT), Total Quality Management (TQM), Theory Z of Ouchi (meaning "company-for-life"), and Kaizen (meaning "changed for the better") were widely used.

Finally, *outsourcing* topped it all. It was chaotic in the beginning. Companies all scrambled into position. The attitude was "*Everyone is doing it, so why not us?*" Frank Casale founded The Outsourcing Institute in 1993 to help people understand the complexity of outsourcing and to avoid costly errors. I was appointed Senior Advisor on real estate and facility management issues. I conducted several successful outsourcing seminars in China and Asia.

Likewise in Bell, we also went through JIT, TQM and Kaizen, even in facility management, to help people from technicians to senior professionals understand the *Why*, *What*, and *How* of our changing corporation. I implemented outsourcing to an extent that I wanted to "*sell*" our facility management team as an entity to outside companies.

Self-managed teams (SMT) have grown in popularity following the influx of Japanese management techniques. Around 80 percent of companies in the Fortune 1000 and 81 percent of manufacturing companies used self-managed teams within their organizational structure. Companies favored self-managed teams that offered cost savings and increased productivity, when implemented effectively. The key phrase is "*when implemented effectively*." Likewise, we introduced SMT to my facility team. At that time, we had about 400 people in my organization and over half were unionized technicians. This was a new concept for the unionized staff, and it was not an easy group to be convinced.

To self-manage without a "*manager*" did not mean you could do anything you want. We established processes, rules, and procedures so they could manage themselves.

I further modified the concept to incorporate internal outsourcing within the organization. We pioneered the "*insourcing*" concept. In other words, why send the work outside when we had capable and skilled people doing the same job as external contractors. So, we formed seven teams within my geographic territories in the Province of Ontario. Each team behaved like a contractor, and we allowed them to "bid" on jobs against each other. They were organized like an outside contractor with estimators, bookkeepers, and even marketers. They had to manage their cash flow and prepare P&L statements at the end of each quarter.

Surprisingly, the idea met with positive response and the people were enthusiastic and energized to embrace the concept. We started with simple interior alteration jobs and eventually they were allowed to bid on larger projects against outside contractors such as lighting retrofits. When early retirement fell upon us due to corporate downsizing strategies, majority took the incentive and left Bell. They were now equipped with the knowledge and experience to run their own contracting business. It warmed my heart when I heard that some were so successful that they were acquired by larger firms.

We introduced a Facility Management Annual General Meeting (FM-AGM) and issued our *FM Annual Report*. We presented the report to all our stakeholders. At times of corporate budget reductions and spending cutbacks, this report served as a marketing tool to inform (or *convince*) our internal customers and department heads. It demonstrated to them our achievements in expense cost reductions while maintaining high service standards. It was our departmental contribution to the corporate strategy. Our accomplished deliverables included a list of projects completed within budget and time, energy savings in operations, outsourcing savings in maintenance, and project savings using our own FM insourcing. It was all real dollar and cents added to the corporate bottom-line and was part of our marketing effort to show that FM was part of the core business. Facility Management was critical to the success of the company. In the end, it gained their trust and respect, as well as brand recognition for our FM team.

Facility Management, indeed, is about Business.

Facility Management Is About Passion

A unique characteristic of the FM team was that they were good at dealing with the technical aspects of our facilities but lagging in the "M" part of FM. The people who worked in large corporations or government inclined to perform their work according to their daily tasks sheet but would not do more. I failed to see the sparkle in their eyes that were once there when they first joined the company. The "*passion*" was gone. I learned quickly that the most crucial part of managing was ensuring people were motivated to do their work with purpose, recognition, and achievement. That spelled "*job satisfaction.*" When each person felt their contribution was essential, their dedication to their work would grow. The sparkle would return. As David Whyte, an Anglo-Irish poet and philosopher, said in his book *The Heart Aroused: Poetry and the Preservation of the Soul in Corporate America* (1994), "To have a firm persuasion in our work – to feel that what we do is right for ourselves and good for the world at the same time – is one of the great triumphs of human existence."

Once I was waiting for my flight home from New York, I couldn't help noticing five security personnel killing time before spot-checking passengers. One drank coffee, another was lost in thought, and another just dozed off. They looked bored, disinterested in what they were doing. They had no feeling for their work.

I began to think: What's wrong with them? Is it that they see their job as "just a job"?

No job is "just a job." The problem with today's workplace worldwide is that many people are doing "just a job." Most remembered the excitement the day they joined the company when their job was "*the*" job. They remembered "the good times" which, for some peculiar reasons, were always in the past, never today, and definitely not tomorrow. That was the problem. People did not take their jobs seriously, mentally, and morally. Their initial "*passion*" was hijacked.

In a facilities department, it was important for each person to view his or her job as a vital and integral part of a complex organic organization. Each person's contribution was specific, unique, and important regardless of rank.

The late Mihaly Csikszentmihalyi, author of *FLOW*, professor and former chair of the Department of Psychology at I University of Chicago, suggested that true happiness comes from doing a meaningful job and the satisfaction of a job well done.

Values defined people's roles and brought people together for a productive and effective organization. John Keats (1795–1821) once wrote: "*I am certain of nothing but the holiness of the heart's affection.*" Leaders of organizations must translate the poet's notion of "the holiness of the human heart" into what they need to do if they want the heart and energy (*passion*) of their employees to be directed at performing in the workplace.

The question is, where is the heart of the organization?

Since our life journey at work occupies most of our time, we must pay attention to how we walk the path for our own well-being as well as the well-being of those around us. David Whyte again suggested that the human approach "*can be selflessly mature, revelatory and life giving; mature in its long-reaching effects, and life giving in the way it gives back to an individual or a society as much as it has taken.*"

It can be selfless and life-giving, or it can be fatalistic and obsessive. There are plenty of examples of both, but unfortunately the latter often wins out.

Perhaps we should heed the advice of Martin Luther King Jr. to sustain our passion in facility management.

> If it falls your lot to be a street sweeper; sweep streets like Michelangelo painted pictures, like Shakespeare wrote poetry, like Beethoven composed music; sweep streets so well that all the host of Heaven and Earth will have to pause and say, "Here lived a great sweeper; who swept his job well".

We must nurture the *passion* in all our facility team members.

Facility Management Is About Leadership

Daniel Goleman said that the mood of a leader affects the emotions of the people around him or her. In neuroscientific terms, this has to do with the open-loop character of the brain that allows people to come to one another's emotional rescue – enabling a mother, for example, to soothe her crying child. Scientists describe the open-loop as interpersonal limbic regulation which means one person transmits signals that can alter hormone levels, cardiovascular functions, sleep rhythms, even immune functions, inside the body of another. So, in any organization, people take their emotional cues from their leaders, and the result can be beneficial or dysfunctional depending on how well one manages his or her emotions.

This is particularly important because leaders work with people, influence people, and motivate people. Ken Blanchard once said that "In the past, a leader was a boss. Today's leaders must be partners with their people; they can no longer lead solely based on positional power." When suggesting that leaders are also managers, Henry Mintzberg said that managers in a leadership role must help to bring out the energy that exists naturally within people.

Many define leadership as a process of influencing. Followers and constituents become part of this process. In his book *Servant Leadership*, Robert K. Greenleaf discussed the importance of "leadership and followership." Some say that leaders are born, but leadership skills can be developed or honed by training. It is like learning to ride the bicycle. There are intuitive rules that you learn to master almost without knowing it. It becomes an intrinsic part of your body through practice until you begin to be able to keep your balance, ride confidently, and develop more complex skills going forward without thinking about it. If, on the other hand, leadership is like following a procedural manual, then all you see is a series of outward behaviors that exhibit what other leaders exhibit. This does not mean you are a leader; it only means you have demonstrated the appearance of a leader. If all your followers are taken away, are you still a leader? Will you still behave the same way? A real leader is a leader whether there are followers or not because a leader is a certain kind of person. When you do have followers, you must also learn from your followers.

When Abraham Joshua Heschel wrote about teachers and learning, he said: "What we need more than anything else is not *textbooks*, but *textpeople*. It is the personality of the teacher which is the *text* that the pupils read; the text that they will never forget." This kind of leadership will set the right path and will not falter when crises or difficulties appear. This is the type of leader that we need today.

In the corporate world, we always want our leaders to "*walk the talk*" but this is not enough. Is the leader taking the lead to do what he or she advocates? "*Walk the Talk*" relates only to performance. Leaders love to talk big talks, shouting slogans for their "*subjects*" to follow and yet many of them do not follow their own maxim. Hence we have the saying: "*do what I say but don't do what I do.*" This refers to your credibility – the reality of who you are. How well the communication is received depends on how trustworthy and credible the leader is. What we need today is to "*Talk the Walk*" instead. This means you must be the person who demonstrated and exhibits the kind of "*walk*" before you start to "*talk*" about it.

- Leadership is more than just leading or having followers.
- Leadership is about your ontology on life and heartfulness with people.
- Leadership is a matter related to a life-time exercise of thinking and being.
- Leadership is *WHO YOU ARE.*

Facility Management Is About Futuring

There are three basic maintenance processes in facility management. They are scheduled maintenance, preventive maintenance, and predictive maintenance (PdM). A lack of scheduled maintenance will cause premature failure of equipment resulting in downtime and costly financial consequences for the company. Clever folks will use budget acrobatics to switch the funding from the limited "expense budget" to a less restrictive "capital replacement budget" claiming risk management justification. Or, on the other hand, the familiar "*if it ain't broke, don't fix it*" mentality camouflages itself as the "Run-to-Failure" method of maintenance. Unfortunately, it is a reactive process.

The opposite side of the spectrum is predictive maintenance (PdM). The idea of predictive maintenance is to anticipate or predict when a failure might occur. In a sense, it is part of the risk management process. It measures the operational functioning of equipment by monitoring temperature, pressure, lubrication conditions, and noise abnormalities. Studies have shown that PdM could achieve a 25–30% reduction in maintenance costs, 35–45% reduction in downtime, and 70–75% decrease in breakdowns.

In the service industry, there is a similar process of anticipating customer needs, meaning the needs of the customers are taken care of before even asked. We must, therefore, also do a predictive maintenance on our career, capability, and competencies. The 2020 pandemic has led us to question our future and the validity of our profession. Most people work from home or in a hybrid arrangement during the pandemic. The questions many have asked are:

- Will we also become hybrid facility managers?
- What will the post-pandemic workplace look like?
- Will our profession become obsolete?

In the spirit of predictive maintenance, *futuring* facility management can be instructive. It must be considered with all the current social, political, and pandemical influences, to determine the FM future. In like manner, we must consider what our profession may look like and how we may manage our facilities and serve our customers.

The World Future Society acknowledged that speed change is happening in the world: technologically, culturally, and environmentally. They identified six global trends: technological progress, economic growth, improving health, increased mobility, environmental decline and increasing deculturation. In the 2007 November edition of *The Futurists*, several of their projected trends are of interest to facility managers:

- Commerce will become cashless.
- World population will reach 9.2 billion by 2050.
- Water shortage will be a continuing problem. We can now add to it food shortage due to the Ukraine-Russian war of 2022.
- Gen X and Y will likely choose the wrong career path over their predecessors due to unrealistic expectations.
- Privacy will disappear because of cyber-exhibitionism and surveillance technology.

Facility Managers must equip themselves for flexibility and adaptability. I created the Life Applications in Facility Management (LAFM) model as an attempt to provide understanding and clarity to position and future-prove our personal and professional career (see Figure 3).

There are two components in the LAFM model. The first component is the Primary FM Functions, the "*formal*" aspect of our career. It includes the skills required to do our FM tasks of design, planning, operation & maintenance, business acumen, research, and communication. It is, in a sense, why the company hires you. The second component is the Professional Functions, the "*social*" aspect of our career. It includes social networking, professional association participation, mentorship, professional development, and community involvement. Here is where IFMA and other professional organizations interact. It is where the company would like you to engage and contribute to the industry.

Between the formal and the social aspects lies the factor of keeping the work-life balance. Work-life balance must include your understanding of the meaning and purpose of who you are and why you are here. To strike a balance, you need to consider more than just equal time or equal share. The balance within the work–life continuum includes the reward elements of job satisfaction and life fulfillment. Challenges at work and in life stimulate the drive in you.

Stress could become an opportunity. It embraces the relational importance of social responsibility and community participation. The results of a balanced work-life are happiness and the overall well-being of work and family. I always advised my staff, especially the technicians (unionized folks), to go home and be with the

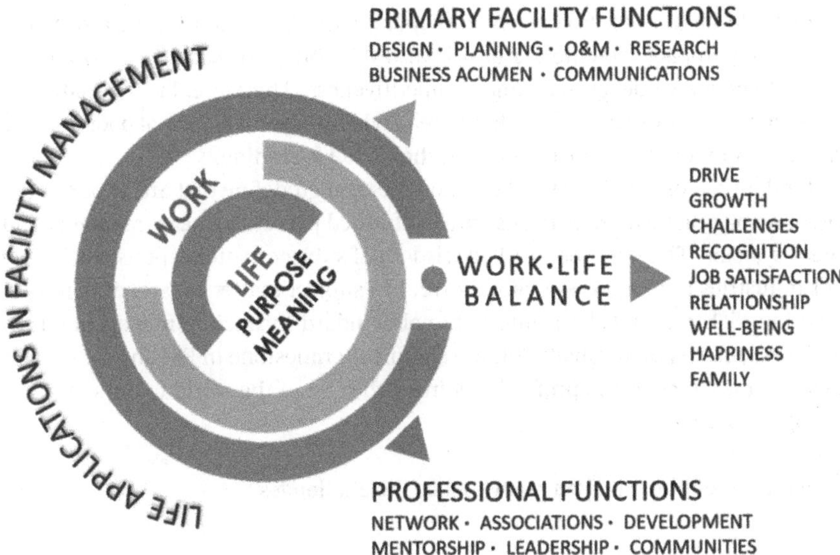

Figure 3 Life applications in facility management. (*Source:* A. Lam).

family when they completed their tasks. I told them there was no need for them to linger in the shop like many who preferred to stay after 7 pm or until their boss had gone home. It was their way to show solidarity but, in my opinion, stupidity. A person with a well-balanced emotion is a person of great value to the company.

Conclusion

To conclude, I thought I would present the highlights of my FM journey:

- Bachelor of Architecture from McGill University (1967) for my technical knowledge, and Master of Theological Studies from Tyndale University (1994) for my pursuit of purpose and meaning in life.
- Joined Bell Canada to begin my FM journey and transformed the real estate department into facility management.
- Took early retirement from Bell and ventured out in a global consultancy while speaking at conferences internationally.
- Served as a member of the Board of Directors at IFMA, ISFE (International Society of Facilities Executives founded through MIT) and SoChange Foundation (a charitable foundation in Canada) and The Beijing Lights Foundation (a registered organization in Beijing to promote quality and wellness social development); Board of Trustees at the BOMI Institute and IFMA Foundation. Introduced the Master of Corporate Real Estate (MCR) designation in Asia with CoreNet Global (2014–2019).
- Founded a new company Aviemore Stirling Inc. focused on designing socio-economic community development for third world countries using FM as a framework for the design models.
- Writing my book on *Human Longevity and your Career* to be published at the end of 2022. I purpose to work and live beyond 100.

It is, I think, appropriate to conclude with the importance of business acumen here. Over the years, I have interviewed and trained many global executives in facility management, property management, and corporate real estate. They have all concurred that our professionals lack business understanding of corporate finance and corporate strategies. Business acumen is the fundamental understanding of how a company operates and makes money. With this knowledge, you should be able to proactively anticipate, participate, and leverage different external and internal influences that may impact the enterprise. It is also the ability to recognize current and future trends so that you will be able to create value, growth, and competitiveness. Therefore, facility managers must maintain an objective business view and worldview. They should read widely beyond technical books. I used to tell my students that if they want a promotion, they must read what their CEO is reading.

International Organization for Standardization (ISO) establishes standards and guideline that are important in today's diverse and complex world of managing businesses to ensure a uniformed process to measure quality and efficiencies in business management systems. They use the distilled wisdom of subject matter experts who know the needs of their organizations or industries they represent. Since 2012, FM subject matter experts drawn from some 45 countries have been pursuing ISO to establish an international standard for facility management. In 2018, the ISO 41001 on facility management was established. This is a significant milestone in FM and for the FM professionals. Now, facility management is a recognized profession joining the rest of the world to pursue excellence, efficiency, and quality in doing business.

Is there a future for facility management?

Yes, there is a future for facility management but, are you ready to face the challenges?

Alex Lam, MRAIC, MTS, IFMA Fellow, Hon. Fellow HKIFM
Chief Design Officer
AVIEMORE STIRLING Canada
Caledon, Ontario, Canada

Alex Lam is Chief Design Officer with AVIEMORE STIRLING in Toronto planning *Campuses for Humanity* for developing countries around the world. With over 50 years in the industry, his career includes 23 years as GM – Facilities with Bell Canada, and over 20 years in his global consultancy in Asia-Pacific and North America.

A frequent speaker at international conferences, his recent presentations brought him to Phoenix, Stockholm, Singapore, Auckland, and San Paulo. He has taught graduate-level master classes on facility management at the University of Hong Kong, the University of Manitoba, and Ryerson University in Canada.

He graduated from McGill University, Canada with a Bachelor of Architecture (1967) and Tyndale University, Canada with a Master of Theological Studies (1995). He was elected IFMA Fellow (2011), Honorary Fellow of the Hong Kong Institute of Facility Management (2004) and a Life Member of the Royal Architectural Institute of Canada (RAIC).

3.2

Facilities Management Industry–Student Partnership – an Exploratory Study

Parminder Juneja

Industry and student partnership is a win-win for both. Research shows that industry engagement with students exist in many forms. Direct partnership where students work on industry projects for their thesis or capstone. Indirect partnership where industry engages with accreditation agencies and sets learning expectations from graduates. The key objective of this partnership is to prepare students as employees of the future who fit right in to fill the talent gap, reduce workforce shortage, and continue to work in the field of facilities management. This also helps the students in finding their interest niche much quicker, sustain their interests, and make a success driven career.

This study was conducted to find the status quo of the partnership between the local construction and facilities management industry and the construction management program, specifically facilities management students, at a public research university (Kennesaw State University – KSU) in the southeastern region of the United States. The study identified the significance of the need to further work on this partnership to create a win-win for the facilities management profession and the students.

Introduction

Partnership between academia, specifically students, and industry, is significant to accomplish a win-win future for both the employers (i.e., the profession) and the students, i.e., the prospective employees. At the core of this partnership is the objective that each stakeholder finds this partnership profitable; students upon graduation are able to seek career options they desire and employers get to hire qualified and well-molded employees (King 2016). The Industry Advisory Board (IAB) of the Construction Management (CM) department at the Kennesaw State University use this core mantra of partnership to stay actively involved in the CM programs' design and development.

Facilities management (FM) program that is housed in the construction management department has a similar objective: get FM students ready for real life and give them the best possible head start for their professional lives. An ongoing challenge for the FM industry is finding FM graduates as "every year fewer and fewer college students are choosing the facility management career path resulting in a lack of qualified personnel to fill the growing number of facility management job positions that are opening yearly" (Hightower and Highsmith 2013). On the contrary, FM students feel that it is challenging to connect with FM companies, professionals or find FM internships. The key question that arises is, what is the root cause of this gap between FM students and the respective industry when both have similar stakes, either win-win or lose-lose, in this partnership?

Facilities @ Management, Concept - Realization - Vision, A Global Perspective, First Edition. Edited by Edmond Rondeau and Michaela Hellerforth.

Research suggests that an optimal approach would be connecting the FM students with the industry at the prior date rather than during the last semester or after graduation (Helyer 2011). This connection is expected to benefit the students and the industry alike. However, to afford college expenses most of the students work while obtaining their higher education. So, the struggle they face is how and where to start. They ask around, but if there is no process in place, they lose it and get back to their daily struggles.

Guided by the above stated benefits and challenges, this study of strengths weakness opportunities and threats (SWOT) was conducted to find the strategic position of the KSU CM and FM student–industry partnership. A SWOT analysis is an effective tool for situational analysis (internal factors), environmental scan (external factors) and developing strategic planning framework. Because of their simplistic framework these analyses are commonly used in the evaluation of an organization, a business activity, a program, a project, or a plan.

In this paper, results of SWOT analysis on FM student–industry partnership are presented. The SWOT questionnaire was filled by the construction management graduate and undergraduate students enrolled in the program at the Kennesaw State University. Conclusion and recommendations follow the analysis.

Literature Review

Industry–student partnership is not a new concept; however, its significance in today's complex knowledge-based economy is vital than ever before. Besides the financial support for education, research and service goals of a university, this partnership can "broaden the experience of students and faculty; identify significant, interesting, and relevant problems; enhance regional economic development; and increase employment opportunities for students" (Prigge and Torraco 2006). Prigge and Torraco (2006) further states that with this partnership employers gain access to: students as potential employees; research opportunities that are otherwise either difficult or impossible; faculty expertise they did not have. The most important fact is that both universities and industry derive benefits from this partnership.

Research on industry–university collaboration suggests that knowledge of current industry practices and needs, attitudes, soft skills, and values must have to achieve and sustain through day-to-day challenges and needs of any industry (Mazhar and Arain 2015). Work Integrated Learning (WIL) is a theory-based pedagogical strategy that has emerged to bridge college education and the work world (Sattler and Peters 2013). WIL is a "pedagogical practice whereby students come to learn from the integration of experiences in educational and workplace settings" (Billett 2009). WIL aims to create "a planned transition from the classroom to the job." (Coco 2000). Sattler (2011) formulated three types of work integrated learning experiences in higher education. One, systematic training, where workplace is the "central piece of the learning" (such as an apprenticeship). Two, structured work experience, where "students are familiarized with the world of work within a postsecondary education program (e.g., field experience, professional practice, co-op, internship"). Three, institutional partnerships, that stands for "education activities [designed] to achieve industry or community goals" (e.g., service learning) (Stirling et al. 2016).

Furthermore, Orrell et al. (2010) identified seven key dimensions of WIL (see Figure 1) as "purpose, context, nature of integration, curriculum issues, learning, institutional partnerships, and support provided to student and workplace." Cantalini-Williams in (2015) further suggested a CANWILL framework to add dimensions of assessment and logistics to the delivery of WIL experiences.

WIL is becoming increasingly popular with construction management undergraduate education (Mazhar and Arain 2015). WIL offers a process to provide students with "transferable skills and to be aware of the changing needs and requirements of employers" (Tucker 2006). Research identifies many types of WIL practices such as "pre-course experience, sandwich courses, job shadowing, joint industry–university courses, new traineeships and apprenticeships, placement or practicum, and post-course internship, work-based learning, vocational learning, experiential education, cooperative education, clinical education, practicum, fieldwork, internship, work experience," and many more. (Sattler and Peters 2013). WIL is indeed an effective means for the industry to

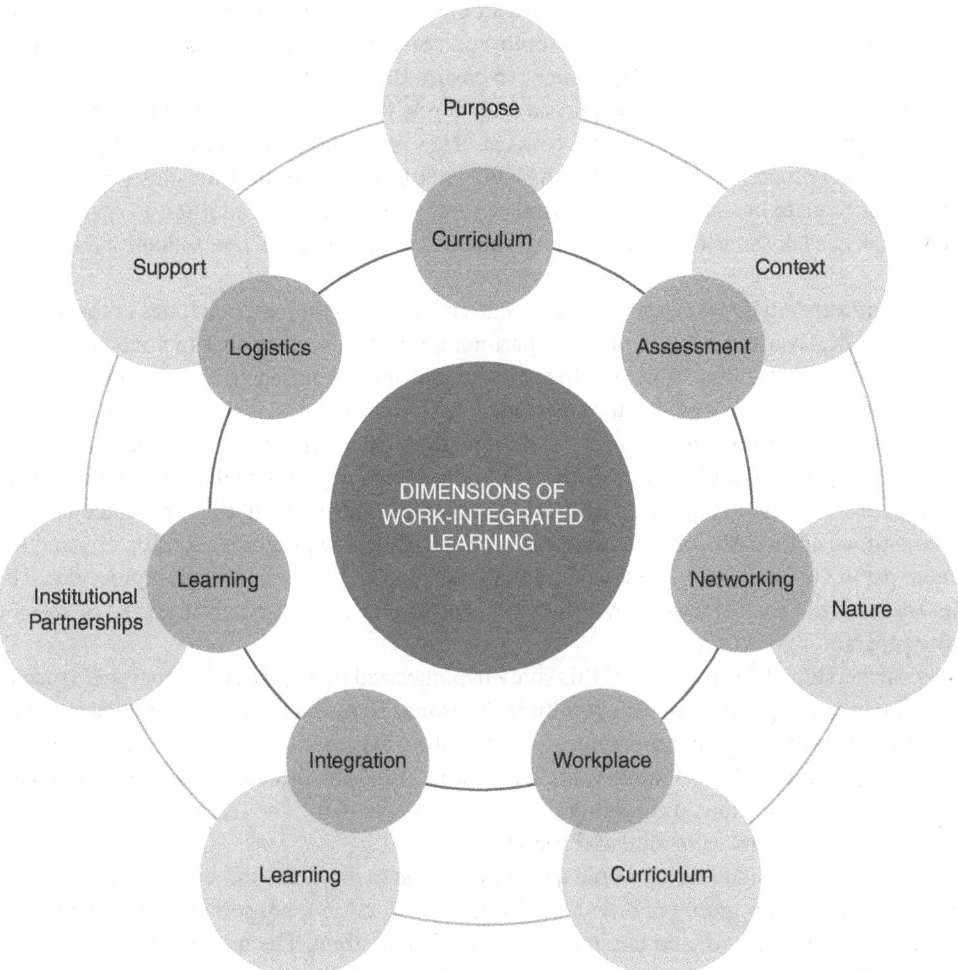

Figure 1 Dimensions of WIL. [*Source:* (Orrell et al. 2010; Cantalini-Williams 2015)].

assess student's competencies and introduce and prepare them for the intricacies of the construction job. Eventually, the objective of the WIL is to create a win-win market for the students, the educators, and the industry.

In view of the significance of the student–industry partnership, a SWOT study was conducted with construction and facilities management students at the Kennesaw State University (KSU). The next section provides the results of the SWOT study. The intent of the SWOT was to study the status quo of the partnership between facilities management (FM) students and the FM industry. The following section provides recommendations to strategically address the weaknesses of this partnership at the KSU and mitigate the threats by capitalizing on the strengths and opportunities.

Methodology

Introduction to Student–Industry Partnership in the Construction Management Department at the KSU

The construction management (CM) department, where the facilities management education is offered, in the College of Architecture and Construction Management at the Kennesaw State University (KSU) has a very strong and active industry advisory board (IAB). The board's objective is to ensure that students meet the educational and career needs of the industry from an employer's perspective. Through their organizational structure, the board recommends actions to the department to ensure that learning outcomes at the program level meet or exceed employer needs. The board also plays a critical role in prescribing state of the art facilities and instructional technologies for teaching different courses. They are diligently engaged with the CM department during the rigorous process of hiring of faculty, the department chair, and the college dean. The board members regularly participate in stakeholder assessment surveys, which are in addition to reviewing of capstone and research projects each semester to benchmark the quality of graduates and to identify areas for improvement.

The KSU CM programs' industry advisory board meets four times in an academic year, two times in the fall and two times in the spring to discuss various aspects of the department and the program that include but are not limited to: infrastructure acquisition, facilitate scholarship programs and alumni outreach, assist in improvement and upgrades of instructional technologies and distance learning infrastructure. Besides these meetings, the IAB participates in yearly fundraising activities that includes, fall foundation gala, fall cook out, spring skeet shoot, and spring cook out. The IAB also holds spring and fall career fair days (elite day in the fall and diamond day in the spring) specific to the field of construction and facilities management. The accreditation and outcomes and capstone assessment committee of the IAB consistently work with the chair, the program coordinators, and the faculty to support and assist in outcomes-based assessment and lend support to re-accreditation processes. The capstone assessments are analyzed each semester by this committee to find areas for improvement in specific courses, or in the entire program.

Guided by this strong partnership, the researcher of this study hypothesized that KSU's CM programs' (specifically FM students) student–industry partnership was significantly strong. To assess the current state of student–industry partnership and to collect ideas for improvement, a SWOT analysis was conducted with the undergraduate and graduate students of the program. A SWOT study was chosen as an initial step because of its simple but powerful framework to investigate organizations' position and develop strategic plan. The framework also allowed formulation of open-ended questions that were vital to the study.

In all 45 students in a group of three students answered questions related to the strengths and weaknesses of the CM program (facilities management program is housed within the construction management department) and commented on the opportunities to improve and possible threats for the program. The method used to generate results was based on focused group technique where each team coordinator led the team by assigning responsibilities to take notes, stay focused, and direct discussions.

SWOT Methodology

A SWOT analysis, that stands for strengths, weaknesses, opportunities, and threats, is a strategic tool that scans the environmental factors to act as a precursor to strategic planning or forecast a solution to an existing problem (Harrison 2010). SWOT analysis is an effective tool to find internal and external factors that can be helpful or harmful for an entity. Internal factors are usually identified via strengths and weakness, whereas threats and opportunities typically focus on external factors. A typical SWOT matrix is shown in Figure 2.

In this study, graduate, and undergraduate students with interest in facilities management concentration completed a SWOT analysis of the student–industry partnership of the CM program at the Kennesaw State University in general and specifically for the facilities management program. The results are presented and discussed below. In all three graduate students and forty-two undergraduate students participated in the SWOT analysis. SWOT analysis was an appropriate methodology to use for this exploratory study as it brings about the root cause of the problem to suggest possible solutions such that threats are mitigated and opportunities are capitalized for improving the partnership between the FM industry and FM students.

Environmental Scan			
Environmental Scan	SWOT Matrix	**Helpful**	**Harmful**
	Internal	**Strengths**	**Weaknesses**
		What are current strengths supporting achievement of goal?	What are current weaknesses impeding achievement of goal?
	External	**Opportunities**	**Threats**
		What opportunities exist in the environment that potentially foster achievement of goal?	What threats exist in the environment that potentially impede achievement of goal

Figure 2 SWOT matrix. (*Source:* P. Juneja).

Strengths of FM Student–Industry Collaboration

The SWOT analysis showed that the student–industry partnership of the KSU CM program possess the following strengths that makes it one of the preferred programs in the southeast region of the United States.

- Smaller FM classes for interactive learning and enriched student/teacher/guest speaker relationships.
- Classes continuously evolve to keep up with industry trends and advances in technology
- STEM career fairs and construction industry-specific diamond day career/internship fair
- In-class guest speakers from the industry. Often, they provide real world examples, videos, images, and industry forms and documents.
- Student invitation to seminars/conferences/training classes exposes students to social etiquette, industry norms, networking, and knowledge of products and services provided by companies in the industry
- Construction site-visits provide exposure to industry environments and relieves anxiety about potentially overwhelming new environments
- Student invitation to industry advisory board meetings helps develop student industry vocabulary, knowledge of current industry issues, and develops relationships and reduction in anxiety related to entering the industry. It is also an effective knowledge transfer mechanism.
- Competition teams and industry–university fundraisers raise money for the university and provide experience and exposure for students.
- Regional facility managers show cases that a facility manager can do practically anything inside the construction field. Individuals with this major are highly sought after in the field and in consulting practices.
- IFMA Atlanta is easy to join and great for networking.
- Career placement statistics help attract more students to the department and industry
- University-industry knowledge transfer benefits the industry.
- Interaction with faculty and students from other universities. (GA Tech)
- Real life projects (Agnes Scott College to LEED certify their library)

Weaknesses of FM Student–Industry Collaboration

The SWOT study suggests the following weaknesses of the student–industry partnership of the KSU CM program.

- Students find it difficult to reach out to the FM industry nationally and internationally
- There is no protocol or process in place to connect students with the FM industry

- On career fair days students meet a lot of professionals from the construction industry but fail to find somebody from the FM industry
- Courses needs more hands-on projects from the FM industry
- None or limited on-campus space for experimental building projects
- There is no internship opportunity database and no commitment from companies in the FM industry
- Incentives or rewards program from internship are not stated clearly; however, they require time commitment, so getting involved is difficult for students that have financial and time obligations to family.
- Lack of FM industry mentorship
- Lack of free access to current FM technologies and management systems.
- Lack of professional seminars, roundtable discussions about FM practices with professors and facility managers from the industry.

Opportunities for FM Students in the Industry

The SWOT study suggests the following opportunities for FM students and student–industry partnership.

- FM industry has many questions to address.
- Each year many jobs become available in the market and get filled.
- Facilities have direct impacts on human health, safety, and overall well-being. It is a profession of responsibility and satisfaction by giving/providing the most appropriate environment to its occupants.
- FM field has wide scope with detail-oriented skills
- Students could work on industry projects for their semester long projects and capstone projects
- Because of the young profession, there is a dire need for new innovative solutions
- Industry could drive the content of the program to fill the technical and knowledge gaps
- Industry could interact with and shape future workforce such that the new employees are ready for the job from day one.

Threats to the FM Industry

The SWOT study discussed the following threats for FM students and thus the FM student–industry partnership.

- FM industry losses its value in the marketplace.
- Challenge to find skilled FM (who are ready for their job from days one), to replace retiring personnel
- Students loose interest in the FM industry and change their field of work to construction or construction-related activities as the construction industry is more heavily involved with students
- Companies do not understand, or underappreciate, the value of having a facility manager
- Use of highly integrated building automated systems may reduce the demand for facility managers, as large complexes can be monitored remotely and with fewer man-hours to manage facilities.
- More contracted services than in-house facilities management services may lower the demand for facility managers.
- Sub-contractors often try to get around rather than going through the FM.

Discussions and Recommendations

In contrast to the study hypothesis that KSU's CM programs' student-industry partnership was significantly strong, the SWOT results showed that there were many areas of weaknesses and many opportunities existed to not only address the weaknesses but mitigate the threats that FM industry and FM students might face if the student–industry partnership is not worked further. In view of the study results, it is explicit that alike other industries, a strategic work integrated learning (WIL) platform is a must have for the FM industry. The FM WIL could take many forms. Some of the recommendations are suggested below.

FM professionals or companies could form a portal to share their FM problem or scenario that keeps them awake at night. FM professionals might want these problems addressed or addressed differently (out of the box solution); but either they do not have the time to consider alternatives, or they do not have the right resources. These problems could be shared with FM students for their semester-long course projects or capstone projects. The expectation could be that students get involved in solving real world problems which could act as a catalytic motivator and industry gets to see the products of the future workforce. For instance, Navon projects provide a repository of mechanical engineering projects for their senior year. Their list contains topics that are identified after research on various mechanical designs and concepts (nevonprojects.com 2000). University of Washington Engineering program under their industry–alumni relations has industry-sponsored capstone project programs. They have industry-led capstone funds to sponsor this program.

According to their industry and alumni relations, "Capstone design projects provide a valuable opportunity for students to apply their education to address real-world engineering challenges. The capstone fund supports engineering students to help move their ideas from design into reality and develop solutions with the potential to save lives." Even during the challenging circumstances of the COVID during the year 2020–2021 their industry capstone program was supported by 51 sponsors and 77 real-world projects and 320 students from across the college of engineering participated in these projects (engr.washington.edu 2021). The Architectural Engineering (ARCE) program at Cal Poly, San Luis Obispo, has employed a student leadership model where students independently establish partnership with industry partners for their capstone projects (Nuttall et al. 2009). In all these cases, the significant value addition is the strong engagement of the industry with academia to establish various programs for the need of the industry rather than programs being offered in isolation or with little interaction.

Another WIL strategy could be FM industry mentorship program for FM students. Such a mentoring program doesn't need to be one-on-one, but one professional to several students to see the students' growth, provide some guides on organizational and day-to-day aspects, and share tacit knowledge acquired from experience. For instance, the "Mentor program" in the college of science and engineering at the University of Minnesota. The program offers a variety of mentor programs for students to connect with industry professionals who provide valuable career information, insights, and strategies. Their different mentor programs include programs such as: North Star STEM alliance peer mentor program, society of women engineers big sis/little sis mentorship program, women in science and engineering (WISE) undergraduate-graduate mentor program to name a few initiatives (Cse.Umn.edu 2022). Kansas State University has an industry mentor program for the STEM fields that provides students with the opportunity to learn from experienced professionals working in STEM fields (Olathe.k-state.edu 2022). They have a student application process to identify a match that in a best fit for the student and the mentor. Department of marketing and professional sales in the Coles college of business at the Kennesaw State University have a professional mentoring program that gives students means and methods to explore their career field for success requirements, build connections between classroom learning and professional world, learn skills for a successful career including leadership, teamwork, critical creative thinking, and networking (Coles.Kennesaw.edu 2022).

The challenges with FM WIL platform include resources in terms of people dedicated to the sole purpose, specific assigned facilities, industry partners that are dedicated to the success of the FM WIL, and not the least, funding sources to support all the activities of the WIL platform. The key to overcome these challenges and form a sustainable FM WIL platform is the commitment to support FM WIL activities. This commitment must come from the institutional leadership, the FM advanced degree programs, and the industry. Tynjala (2008) stated that "employees who cannot network with others to share and construct knowledge will fall visibly behind their peers in the possession of such abilities. Interaction between novices and experts is also of crucial importance in learning at work." Industry must educate academic programs of the current and future practice needs of the profession such that these FM practices are eventually promoted within the academic programs. The FM WIL program is expected to prepare students for their jobs which takes care of the usual lack of confidence and unsureness of students of their abilities to perform a task.

Conclusions

There is a growing demand for skilled and trained facility managers with overall knowledge of all the competencies of facilities management profession and expertise in one or two competencies. Higher education institutions have the responsibility to address these growing needs of the industry. In view, the significance of WIL for FM undergraduate and graduate students cannot be denied. The solution lies in setting up a platform for WIL-related activities. The initiative requires making WIL a core component of the FM academic programs so that learners and educators are motivated to participate and make this program a success. A lot of work must be done to accomplish this goal; however, this study and many other similar studies are an ignition to the future FM WIL satellite that will orbit the FM world nationally and internationally.

References

Billett, S. (2009). Realising the educational worth of integrating work experiences in higher education. *Studies in Higher Education* 34 (7): 827–843.

Cantalini-Williams, M. (ed.) (2015). Teacher candidates' experiences in non-traditional practicum placements: developing dimensions for innovative work-integrated learning models. In: *The Complexity of Hiring, Supporting, and Retaining New Teachers across Canada* (ed. N. Maynes and B. Hatt). Canadian Association for Teacher Education (CATE).

Coco, M. (2000). Internships: a try before you buy arrangement. *SAM Advanced Management Journal* 65 (2).

Coles.Kennesaw.edu (2022). Mentor program. from 10.18260/1-2-5175 https://coles.kennesaw.edu/marketing-sales/mentor-program/index.php.

Cse.Umn.edu (2022). Mentor programs. from https://cse.umn.edu/college/mentor-programs.

engr.washington.edu (2021). Industry-sponsored student capstone project. from https://www.engr.washington.edu/industry/capstone/2020-2021-projects.

Harrison, J.P. (2010). SWOT analysis. In: *Essentials of Strategic Planning in Healthcare (Gateway to Healthcare Management)* (ed. J.P. Harrison). Paperback, March, 2016. Academia.edu.

Helyer, R. (2011). Aligning higher education with the world of work. *Higher Education, Skills and Work-Based Learning* 1 (2): 95–105. https://doi.org/10.1108/20423891111128872.

Hightower, R. and Highsmith, J. (2013). Investigating the facility management professional shortage. *International Journal of Facilities Management* 4.

King, A. (2016). Industry education partnerships in creating career pathways. Techniques. acteonline.org. Association for Career & Technical Education. 91.

Mazhar, N. and Arain, F. (2015). Leveraging on work integrated learning to enhance sustainable design practices in the construction industry. *International Conference on Sustainable Design, Engineering and Construction*, Procedia Engineering.

nevonprojectscom(2000).Mechanicalengineeringmajorprojectslist.https://nevonprojects.com/mechanical-engineering-major-projects-list.

Nuttall, B., Mwangi, J., and Baltimore, C. (2009). Capstone projects: integrating industry through student leadership. *Annual Conference & Exposition*, Austin, Texas, American Society for Engineering Education.

Olathe.k-state.edu (2022). Industry mentor program. https://olathe.k-state.edu/academics/student-resources/mentor-program.

Orrell, J., Cooper, L., and Bowden, M. (2010). *Work Integrated Learning: A Guide to Effective Practice*. New York: Routledge.

Prigge, G.W. and Torraco, R.J. (2006). University-industry partnerships: a study of how top American research universities establish and maintain successful partnerships. *Journal of Higher Education Outreach and Engagement* 11 (2): 89–100.

Sattler, P. (2011). Work-Integrated learning in Ontario's postsecondary sector. Toronto: Higher Education Quality Council of Ontario.

Sattler, P. and Peters, J. (2013). Work-integrated learning in Ontario's postsecondary sector: the experience of Ontario graduates. Higher Education Quality Council of Ontario.

Stirling, A., Kerr, G., Banwell, J. et al. (2016). A practical guide for work-integrated learning: effective practices to enhance the educational quality of structured work experiences offered through colleges and universities, higher education quality Council of Ontario: 192.

Tucker, L.M. (2006). Industry facilitates work-based learning opportunities, NZACE School of Applied Technology Institute, Unitec, New Zealand. 31.

Tynjala, P. (2008). Perspectives into learning at the workplace. *Educational Research Review* 3 (2): 130–154.

Parminder Juneja, Professor, PhD
College of Architecture & Construction Management
Kennesaw State University (KSU)
Marietta, GA, USA

Dr. Parminder Juneja is Associate Professor in the College of Architecture and Construction Management at the Kennesaw State University. Her educational background includes a PhD in Integrated Facility Management from Georgia Institute of Technology; a Masters of Technology in Building Science and Construction Management from Indian Institute of Technology (IIT) Delhi; and a Bachelors of Architecture from Chandigarh College of Architecture, India.

Before joining Kennesaw State University in fall 2014, she possessed 15 years of multi-industry, multi-disciplinary, and international professional experience. As a result, she brings a holistic and integrative perspective to approaching and solving problems which is a key to success in today's complex and transforming education and work environment. She has successfully designed and taught graduate and undergraduate classes, and online, face-to-face, and flip classrooms.

Her research is interdisciplinary and collaborative. She is focused on improving health, safety, and performance in work environments, be it indoors such as, office, classrooms or outdoors such as construction sites. Her overall research objective is to facilitate strategic alignment between work environment and goals of high performance and overall well-being for occupants, thereby, optimizing the whole lifecycle costing of respective environment. She believes that in informed decision making, an informed tradeoff analysis is performed, and the decisions are not restricted due to cognitive limitations of information processing of human brains.

3.3

Four Questions for Forty Years of FM

Barry Varcoe

I started my first student summer job during the early 1980s in the mail room at the UK headquarters of an international chemicals business. I had no idea then that it was the start of a career entwined with "facilities." Neither did I realize that I was witnessing the early gestation of a new industry. It is difficult to remember fully just how different life was then. After all, at that time the first conversations and imaginative ideas about how the "internet" and "world wide web" would change our lives were still at least 10 years distant in the future, if not more. Amongst my daily tasks as a mail room operative was the collection and distribution of the nonstop feed of messages spewing out from the array of telex machines, as well as retrieving and delivering urgent documents fast-tracked around the building in small cylindrical capsules within a pneumatic pipe distribution system.

Fast-forward 40 years and how the world has changed in so many ways. But, as far as the facilities industry is concerned, is it worrying that some of the most important questions facing the early pioneers all those years ago are still pressing and relevant today? Is this evidence that not much has progressed, or is it more the case that the industry has answered the questions well each time they were asked, but the relentless pace of change has soon altered the context and thereby reset the problem? Forty years on, and after a careers worth of experiences, it is now a privilege to be afforded the opportunity to reflect on four of those questions – the ones that have been most significant to me – and how facilities management (FM) has addressed them. Before I start, however, please appreciate that my recollections and observations are compiled with the benefit of 20/20 hindsight, which is always cheap and easy. It is never possible after the event to accurately convey the challenges, pressures, and choices of the time. So I offer this full of respect for the pioneers and leaders of the industry over the years, and what they have achieved.

Question One: Is Facilities Management an Art or a Science?

In the early years Facilities Management drew its talent from a wide spectrum of experience. It was a brave new world of opportunity with no expected or required career path, professional accreditation, or "apprenticeship" to provide a barrier to entry. They came from the industries near-to and supporting the function, such as the building service industries (maintenance, cleaning, security, catering, etc.), surveying (like me) and architecture, engineering, from within organizational operations functions such as finance and HR, and other sectors which especially developed strengths in leadership and administration, such as the military. It was a rich melting pot in which the role developed and evolved through the many and varied influences and experiences of practitioners. Individual strength of character and track record was important to progress, as there wasn't much definitive

Facilities @ Management, Concept - Realization - Vision, A Global Perspective, First Edition. Edited by Edmond Rondeau and Michaela Hellerforth.

knowledge to draw upon regarding what was going on. Finance functions provided clarity about costs, but that was about it. As organizations rapidly grew by taking advantage of a globalizing world, it soon became clear that running buildings was becoming a significant annual expense, and the questions and challenges started.

The industry responded and set about a so far life-long obsession with cutting costs. For a while benchmarking became the cutting edge of the industry, but it quickly turned into more of a sales tool for the emerging service providers, rather than a useful lens on how to improve efficiency. Lies, damn lies, and statistics. The wider problem in my view was at least two-fold. Firstly, there was an inherent resistance to new knowledge across all parts of the industry. After all, by the late 1990s the industry was experiencing rapid growth already, and its prominence was growing too, so why change? Plus, many of its practitioners came with their own views and understanding, were comfortable with them, and didn't want anything to unnecessarily upset the status quo. Secondly, not many understood the whole eco-system that they were a part of, partly because everything was still so new. This changed in time with the introduction of post-graduate education at universities, including Master's degrees, but inevitably this took some time to create enough of an influence to alter prevailing attitudes. As a result of this the body of knowledge that underpinned the industry developed slowly, haphazardly, and unreliably. What development was undertaken tended to be subjective and narrowly focused. Three examples from the late 1990s and early 2000s serve to illustrate the impact of this, and the state of the "art" at that time:

1) Tendering success: in the late 1990s I worked for a prominent global service provider and for a while was fortunate enough to have a wide-ranging role including strategy and innovation. As part of an initiative to build out some reliable performance data to underpin the estimating processes of the day, labor constants and other performance factors were researched, and an overall facilities management pricing model was developed which ended up as an automated calculator for any outsourcing opportunity. Included within this was an understanding of poor, average and good (upper quartile) standards of performance, not just within the management function, but also for all the individual service lines such as maintenance and cleaning as well. The tool was successfully used to set the overall parameters for outsourcing opportunities, and also to analyze historic proposals. One of the more interesting exercises was to correlate the measured performance outcome with contract award success. This showed that there was no correlation between better standards of performance within the proposal and ultimate contract success.

   Furthermore, the average standard of performance embedded within winning proposals was well below the average level. When the difference between that average and consistent good performance was calculated, the difference was a staggering 25%. In what was by then becoming a high-volume low-margin business, the impact for the customer and service provider alike of eliminating this inherent inefficiency would have been transformational. Unfortunately, through the twists and turns of corporate politics, the opportunity to take such a big step forward was lost.
2) Tendering process: a few years later I had moved across to the other side of the industry and was responsible for the facilities and logistics functions of a large international financial services business. It quickly became clear that standards of performance, and the prices we were paying, were inconsistent to say the least. In an attempt to improve the value we were achieving it was decided to reverse-auction all of the main facilities services contracts. At that time it was something of a first to apply the method on such a large scale. Needless to say, it met with huge resistance from the various support services providers, but the contracts on offer were too big to ignore, so they nearly all turned up and bid for the work. The secret to success in such situations is to know your own business model and proposition thoroughly – to know how the level of performance the customer requires and is buying will be reliably delivered, how you as the supplier will make money doing that, as well as a good understanding of the risks involved and how best to manage them. This needs to be a "scientific" model and calculation. Unfortunately, few had this knowledge at hand, and some of the bids were chased down to the point where they were clearly financially suicidal.

   We had to do a lot of due diligence on the final proposals to make sure that they were both financially and operationally viable before we were able to select our preferred service provider partners. After all, there was no benefit to us in selecting a service provider who would lose money on our account, and then possibly be tempted to cut corners to make some money, or perhaps worse, default entirely. We wanted to pay the right price for what we needed but finding that balance was challenging and elusive at best. (As an aside, dealing with the logistics industry in parallel proved to be an eye-opening contrast, and showed what was possible if process quality was embraced at the operational heart of the business.) As subsequent events proved, we didn't get this entirely right! Twice we had a service provider come to us after about 18 months of operation, having concluded that they couldn't make money from the relationship and therefore wanting to negotiate their early exit. On both occasions the reason given for this was our random quality checks to make sure that everything we were paying for had actually been carried out and performed. We routinely found significant portions of work that had been skipped, which we then refused to pay for. It appears that most if not all of the anticipated profit margin was the work billed to us but deliberately not carried out!

3) Cost vs Quality vs Satisfaction: concerned about what I was grappling with as a customer of the facilities industry (and by then real estate as well), I set about a research doctorate to try to establish whether there was any reliable or meaningful correlation between what users of offices paid for them, the quality they received for their money, and the satisfaction of the users within them. To cut a long story short, no meaningful correlations were observed, at either the macro/overall level, nor at an individual service level. In other words, at that time it was clear that, as a facilities customer, you didn't necessarily get what you paid for, and if you did it was more down to chance than anything else. The industry model of service delivery certainly didn't have robust scientific underpinnings.

Fast-forward to today and I hope that this glaring gap in industry capability has been addressed in the intervening years, but I haven't had the chance in recent years to investigate and see for myself whether or not it has. A lot of consolidation within the industry has taken place, though, including across the wider workplace sector, with some of the larger facilities management providers being amalgamated within real estate services provider organizations. So it is to be hoped that this has changed and the facilities industry now really does know what it is doing, and reliably delivers what it sells, not just in terms of input activities, but also the outcomes they should achieve. The days of "facilities" predominantly being an "art" need to be long-gone. "Science" is the platform for reliable and resilient operations, continuous improvement, and sustained growth, upon which a bit of "art" from time to time can be used to conceive new ideas and genuine innovation.

Question Two: Is Facilities Management Really Integrated?

For many years one of the key sales propositions from service providers was their promise of an "integrated" management solution. It still is. But all too often when the surface of this is scratched, it quickly becomes clear that the only integration really on offer is a single point of responsibility for the bundle of services included within the proposition. The services within the remit are not integrated at all. Facilities itself tends to be deeply fragmented across its many massive components – maintenance, cleaning, security, food services, reception, energy, mail, etc. – all of which for the most part perform their separate duties in isolation from each other. True integration would bring them together at a process level in a new way. What we are left with is more a case of "veneer" FM that aims to minimize the loss of value between the parts, rather than a true integrated model that drives a value more than the sum of the parts.

Indeed, when the wider "workplace" eco-system, of which facilities is a part, is considered, there is fragmentation everywhere (see Figure 1).

The fragmentation within facilities is exacerbated at the workplace management level when it is brought together with workplace design, projects, and corporate real estate (CRE&FM) functions. This in turn then suffers a further dimension of fragmentation when the overall workplace environment is considered, and the CRE&FM function is aligned with Technology (IT – that provides the digital workplace) and Human Resources (HR – that

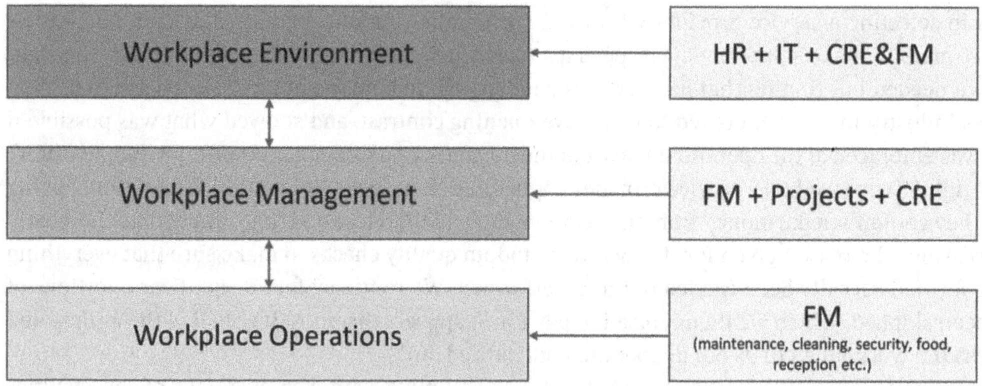

Figure 1 Workplace industry layers of fragmentation. (*Source:* B. Varcoe).

provides the workforce). So, it's not surprising that new service models have emerged over the last two decades or so that seek to address this "value-loss" across the whole of this eco-system, serviced offices being the most obvious.

At least two drivers of change offer hope. This is because each provides challenges that demand an integrated response.

The first is climate change. Commercial buildings are one of the largest contributors to global carbon emissions, and it is clearly critical and urgent that this be addressed. Carbon is produced at every step of a building's life cycle. In its creation, use, adaption, and ultimate destruction. Carbon-neutral operations – facilities – have to be built in from the beginning, so all parts of the wider workplace industry need to work together at the workplace management level to create and shape designs from the outset of any project.

The second is the changing nature of work, which demands a similar integrated response at the workplace environment level if it is to be addressed fully and properly. The forces of the information/digital age, accelerated by the COVID pandemic, have changed the role of the office as a workplace, as well as the expectations of the workforce. The effects of this have to be understood by HR, IT and CRE&FM working together. Interestingly this may change the long-standing perceptions of importance and pre-eminence within the CRE&FM world. Ever since the first green shoots of FM emerged it has usually been third of three in terms of the unofficial rank order, which follows the respective size of the annual budget. The real estate deal makers have always taken precedence (followed within their world by the lease portfolio managers), then the projects professionals, followed by facilities. But now this may be changing. Real estate is no longer an "area per person times headcount" equation. It is far more sophisticated than that and is much more dependent upon a complete understanding as to how the office will be used, and the nature of the work being undertaken within it.

The cultural and experiential aspects are becoming far more important as well, and the amount of space required has to take into account time as a variable if too much of it is to be avoided. This is all the natural domain of facilities. If FM moves fast enough on this, and develops the required expertise, then it becomes the first port of call regarding the workplace. The facilities team also has a unique insight and source of intelligence on any organization. No other function lives so closely with every part of the enterprise every day. Rarely, however, is this latent knowledge captured and used to good effect.

It Is now very encouraging to see facilities service offerings appearing based on the workplace "experience" that is generated. This is a good start in this overall direction. The operating models that underpin these propositions, and the management approach, draw heavily from the hospitality sectors, such as food services and hotels, and have to be integrated if they are to fulfil their promise. This at last is taking the industry away from its relentless cost reduction induced commodity death-spiral of the last couple of decades, as it moves the conversation on to the far more positive proposition of how it can achieve greater value for the enterprises that it supports.

Question Three: What Is the Purpose of Facilities Management?

Beyond providing a span-breaking administrative management function for a range of day-to-day tasks that cause problems if they are not performed properly, what has been the purpose of FM? In the early years it was indeed a convenient amalgamation of responsibilities associated with the running of the building, and the emphasis was on making sure that they stayed operational. Other services then got added, such as mail and catering, which started to add in a service element that touched the user directly. But still the tendency for some time was to stay one part detached from the workforce and interact far more with the edifice and its systems and components. As a result, FM became known by unfair tactical epitaphs such as "from cisterns to sausages, and from soup to nuts." In time though the true value of FM became clear, which was helped by the ongoing advance of education and practice sharing through professional associations such as IFMA, BIFM (now IWFM), EuroFM, FMA, and several universities. The industry's "bench strength" became progressively stronger.

There was a growing appreciation that FM was often significant in its absence i.e., when its services fail, it can stop the organization working, which can ultimately seriously impact profit directly. If a key piece of engineering falters which shuts down a financial services trading floor, or a manufacturing production line, or a service center operation, then that is a significant risk that needs to be properly managed. But this still entrenched the purpose of the function in the avoidance of problems (downtime, unnecessary expenditure, etc.) rather than anything inherently positive.

The relentless progress of the technology revolution gave FM an opportunity to break out of the rut. If the second half of the 20th Century was primarily about the "Taylorist" drive for efficiency, the productivity promises of overwhelming amounts of new information and insights augmented this with a need for creativity and innovation. The previous focus on the efficiency of the individual was replaced with a need to address the effectiveness of the collective as well. Unfortunately, FM and the wider workplace industry were a bit slow in understanding this fully. For too many years there was an ill-informed search for the so-called "holy grail" of productivity, and the workplace's potential role in improving it. This focused too much on improving "efficiency." At first workplace productivity was presented in terms of spatial efficiency, and not even the productivity of the human using it. The simple argument was that if you could get more workers accommodated within a given size of office space, without detrimentally affecting their performance, then the organization becomes more productive.

Strictly speaking this was true, but it completely missed the main opportunity. A greater impact potential was brought into the wider awareness of the industry by appreciation of the simple ratio of the cost of the office to the cost of the people working in it. This was typically 1:5 or 1:6 for most service-based organizations predominantly using offices. The clear argument was that it was all well and good reducing the facility "1" side of the equation to maybe 0.95, but in doing so it had better not increase the risk of negatively impacting the people "5" or "6" side. Furthermore, and more importantly, if the facility "1" could be used more impactfully and could actually help to increase the output of the "5" or "6," then the value or return of the "1" to the organization is much greater.

The focus therefore progressively switched from a sole focus on buildings and space, towards people and how to help them be more productive. Over the years there have been many studies that look at individual facility variables and any people/productivity effects, but none that bring these all together to achieve a reliable understanding of the interactions between each of the factors – the productivity eco-system. They also tend to assess people as individual factors of production, and not consider how they work together. A diagram that attempts to create a logical model of these is shown in Figure 2.

As with a lot of issues that facilities has grappled with and sought to understand over the years, it is complex. This may explain why that elusive "holy grail" has never been found. It simply isn't as easy and straightforward to solve as one magic artefact, insight, process or technique. Added to which, the human factor always has to be

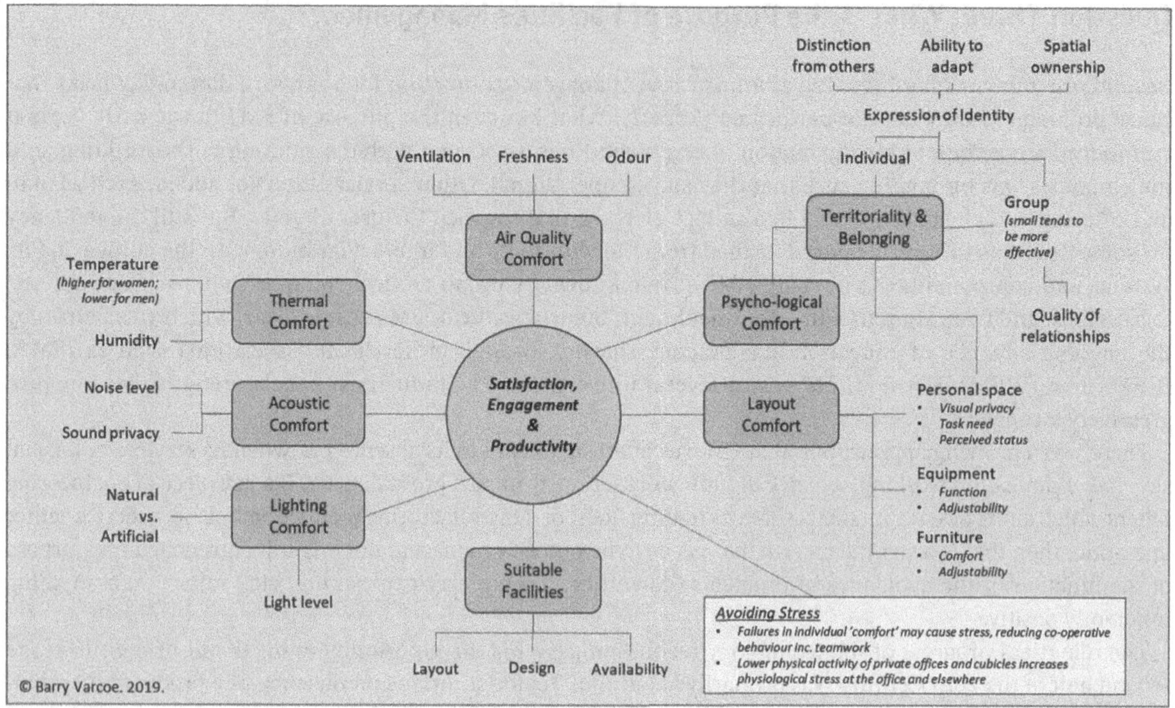

Figure 2 Workplace industry layers of fragmentation. (*Source:* B. Varcoe).

considered. In my view the only claim that anyone involved in creating and operating workplaces can make is that they have done all that they can to encourage more productive work.

There is no way that anyone can guarantee increased productivity from a workplace intervention, whether that be individual or collective. Humans are simply too fickle, and are influenced by far too many factors, within and outside the workplace. The acid test for me would always be to imagine standing in front of the CEO and CFO of an organization and guaranteeing that a certain level of improved productivity could be achieved from a given workforce directly due to what I am proposing to do with the workplace. Could you prove it to a sufficient degree of robustness that they would believe you and "bank it" by reducing the workforce accordingly? I have yet to see any workplace-oriented business case that would come even close to that burden of credibility.

Unfortunately, along the way the industry hasn't helped itself by regularly making many and varied spurious claims that betray its lack of understanding regarding the domain. One of the worst I read was by a leading service provider completely miss-quoting productivity research related to the workplace in a leading management journal. Their headline said that the research proved increased productivity from a particular workplace strategy by up to 30+%! What the research actually said was that around 30% of the workforce using the newly changed environment felt that they were more productive as a result. Even if their own assessment was accurate, and let's say that their productivity had improved by 5%, then that is an overall improvement across the entire workforce of 5% x 30/100 = 1.5%. Also, that assumes that none of the remaining 70% felt that their productivity deteriorated, nor that there was a so-called "Hawthorn effect" meaning that the impact would wear off over time (it being induced more from being the result of attention and a better feeling towards the environment, rather than any underlying and sustainable change in practice).

One aspect of facilities that tended to be overlooked was the potential for causing negative productivity impact through consistently poor standards of service. Not much was made of this, but in my view there was great potential in this over the first 2 or 3 decades of the industry in demonstrating the value of reliable quality and a professional approach. A thorough review of consistently below-average service standards across all aspects of the facilities management remit soon amassed a significant 10% negative impact on the workforce. Consistently poor service went well beyond this, at least in theory. A simple way of considering this was to think of each 5-minute interruption to a working day. This represents 1% of that day's productivity gone. Including this within a facilities or workplace business case would have been a lot easier to support with evidence, even if it only influenced the leaders managing and delivering the service and didn't reach the ears of senior business executives.

Another way that the facilities function has improved its focus on people is fully embrace and to do its part in supporting diversity, equity, and inclusion. This now goes well beyond accessibility and alternative ability accommodations. Neuro-diversity, gender and ethnicity are just a few of the many other dimensions that need to be embraced, so that all staff and visitors walk through the front door of the facility not only feeling equally valued, but also equally able to be "fully themselves" and contribute and give of their best. Health and wellness are other key areas of service that must now be actively supported as well.

Now that FM is learning how to address the needs of people as well as buildings, the requirement scope has moved on and widened once again. As mentioned already, the changing nature of work now has to be understood as much as possible, and a view taken as to how this is best supported, both now and in the future. This is no easy task, as it is another very complex eco-system that is constantly advancing. Just one of many influences serves to illustrate the new challenges. The amount of data available in all industries and markets has exploded, and the "metaverse" has emerged as a way to describe this generically. But how will this change how people work together? The data is only as valuable as the new insights, learnings, products, and services that it inspires, but it will most likely be people working together that will see and realize the potential that lies hidden within. How will work evolve to improve our ability to achieve these breakthroughs together, and how will the design and operation of facilities respond accordingly?

Like so much else, the demands made of FM have moved forward and developed over time, and the industry has had to respond. Looking after the operational effectiveness of buildings is now just the start, the platform for all the rest. At the very least the performance and wellbeing of the people using the facility must be integrated into the service proposition, and now increasingly so too must an understanding and active support of the work that they are trying to undertake, both individually and collectively.

Question Four: How Does Facilities Management See Itself Bringing Value to the Enterprise?

Throughout its 40-year life FM has, in my view, been searching and grasping for its place within the enterprises it supports. I guess this is only natural for the new "kid on the block." Part of the problem, however, has been the other "new kid" – Technology, which quickly and easily established itself within the thoughts and plans of the upper echelons of the organization's leadership. As a result, for many years too many facilities conversations included the frankly implausible proposition that FM too should be at the same level as IT, or better still have a seat at the top table. This betrayed a lack of understanding as to the value the function brings. At one level it can be seen that Technology and FM are similar – both support "work," FM providing the physical workplace, and Technology the digital equivalent. But there the similarities end. Technology doesn't just support work; it is now an integral part of it. There is a total dependency on it – for most enterprises their core propositions to their customers are built around it. So it warrants a place at least near to the top of the organization.

Meanwhile FM, and indeed the whole CRE&FM combined proposition, has as mentioned been relentlessly promoting a one-track benefit–cost reduction. At first this made sense. There was significant inefficiency to be improved, and service providers latched onto this and helped to accelerate the delivery of maybe 10% or more gains. Its significance to the enterprise was emphasized by the repeated assertion that these financial gains would "drop straight to the bottom line," and so make a significant difference to profitability. That may have been true, but organizations more likely just wrapped it up in the overall fight for efficient operations and the need to release money for investment, and it wasn't sustainable in the long term.

As each round of tendering took place after the three- or five-year contract term, sustaining this magnitude of benefit became progressively harder and harder. The law of diminishing returns held true, which lead to something of a perpetual downward spiral. The challenge to drive greater efficiency and lower costs drove service provider margins down, sometimes to wafer thin levels. For a while this was balanced out by scale – contracts got larger and larger, so at least the total amount of profit stayed more or less the same. But with the scale came procurement discipline, and a more sophisticated transfer of risk, as well as even more competition. So down went profit margins again. Further, the increased transfer of risk meant that either unforeseen or uncontrollable events were more likely to happen that would reduce or even eliminate what little profit remained. It is easy to see therefore why suppliers ended up in the position previously described, where their profit came from work tasks deliberately not done in the hope that they were either unnecessary and/or no one would notice. But the real damage done as everything regressed into pure commodity was the impact this had on the industry's collective ability to invest in what it did, and improve its knowledge base, delivery processes, and to innovate new propositions. By and large there wasn't any money left to do much of this at all, even if there was a will.

It has been hard to break out of this vicious loop, but those with foresight started to achieve it by the middle of the last decade, building more robust, quality driven delivery processes that embraced a wider proposition beyond cost, often focusing on the benefits such as improved wellness and quality of life that they could deliver to the workforce via the workplaces they were servicing. This quickly developed further into concepts built on delivering a better work environment experience. The significance of this can't in my view be understated, as it was the much-needed sustained shift away from a focus on cost alone and drew meaningful attention to the important outcomes the money achieved. Here at last were positives that the industry was seeking to achieve, rather than just the avoidance of negatives.

Recent events now make a rapid advance towards improved positive value propositions more important than ever. Society is experiencing forces of crisis and change like never before. As I write the world is still tackling and seeking to tame the global COVID pandemic, is beginning to face up to the imminent catastrophe of climate change, is witnessing a war in Europe that is fracturing the fragile dependencies and networks of globalized trade, and is on the cusp of massive new advances in the digital information revolution that are even beginning to cause some to question the nature of life itself. The nature of work is changing at an unprecedented rate as a result, and so too are the expectations of the workforce. The workplace has never been written about and debated with as much prominence as now. It is high on the enterprise agenda.

So now is the time for facilities as an industry to move forward boldly, alongside its close colleagues in real estate and projects, to redefine the workplace, what it does, and how it does it. There is a ready and willing audience for what it has to say, and the enhanced value it can bring. I can only hope that the industry, after an impressive first four decades, has now reached the maturity and confidence that it will need to make the most of this unique opportunity and moment in time. What it is doing now, and does over the next few years, will likely define its place within the enterprise eco-system for decades to come.

Dr. Barry Varcoe, FRICS
Global Director of Real Estate & Facilities
Open Society Foundations
Winchester, England

As Open Society Foundation's (OSF) Director for Real Estate and Facilities globally, Barry's role embraces responsibility for all aspects of providing and operating OSF's physical places of work around the world. The team are directly responsible for the acquisition, design and fit out of all offices, and facilities management for the main locations. The key to successfully accomplishing this is understanding how best to support and encourage the progressive advancement of a positive culture and a sense of cohesive community across OSF and its extended network. Facilitating productive work and furthering the overall aims and goals of the organization, including minimizing the environmental impact of operations, are also critical.

Prior to joining OSF at the beginning of 2018, Barry was an Associate Fellow at Saïd Business School, University of Oxford. Other appointments include Global Director of Group Services at Zurich Insurance, Group Property Director at Royal Bank of Scotland Group, and Chair of corporate real estate association CoreNet Global in 2009/10. He is currently a William Pitt Fellow at Pembroke College, University of Cambridge, and an editorial board member of Corporate Real Estate Journal.

3.4

Facility Management Yesterday – Today – Tomorrow

Klaus Zapotoczky

Ever since my high school and university years, I was connected to the world of work and all related aspects as a working student, and familiar with facilities of all kinds. After having finished my studies in law and mathematics at the University of Vienna as a Doctor of Law, I decided after having completed the compulsory period as a legal trainee at the district court of Vienna, to broaden my view of actual life and work in society, by complementing my study with social and economic sciences at the Catholic University of Leuven, Belgium. The Licentiate there, with my thesis about "The question of the meaning of work"[1] was subsequently my start in sociology and research in academia.

Finally, as Professor at the Johannes Kepler University Linz (JKU), Austria, I was asked in the late 1980s by colleagues of the business administration department to accompany research projects in the fields of Management Consulting and Mergers and Acquisitions. All areas with particular sociological relevance to working people in systems determined not only by professional work practice but also by the place in which performed, overwhelmingly in facilities of the built environment. I gladly accepted, as the research approach appealed to me, which also included the co-supervision of dissertations from mainly specialists of varied areas, companies of all sizes and types. The results and findings of these cases and field studies in practice promised to be of general scientific interest, eventually to be published in book series, for a wider circle of interested readers. Although I was not directly responsible for economic and business administration issues as head of the Institute for Political and Development Research, my interest was aroused.

Now, decades later, it is a good time for reflections about the outcome of this endeavor, as it includes my personal involvement with the issue Facility Management, even though this was not obvious to me at the beginning. Actually, the research concept started in the field of Management Consultancy. There, above all, two colleagues, Prof. Walter Sertl of JKU in Linz, and Prof. Michael Hofmann from Vienna's University of Economics (WU), had initially in 1983 taken on the project to investigate solutions for problems corporate management was confronted with, when using external advisory services as consultants.

At this time consultancy was mainly performed in Europe by consulting companies from the Anglo-American sphere, having discovered a need for their service especially in the rather undeveloped markets of Germany and Austria. Their still huge industries existed, but the same time a mass of small and medium-sized businesses (in German: "Klein- und Mittelbetriebe, KMU") were actually the predominant drivers of innovation. As this was when the KMU's and the big national industries were faced by the emerging globalization in all fields and felt endangered because of the existing social and economic inequalities.

1 Klaus Zapotoczky, Die Frage nach dem Sinn der Arbeit, Catholic University of Leuven, Belgium, Licentiate thesis, 1964.

Facilities @ Management, Concept - Realization - Vision, A Global Perspective, First Edition. Edited by Edmond Rondeau and Michaela Hellerforth.

In Austria's capital Vienna just then a major public building project had generated huge publicity by a spectacular example of what nowadays is called investigative journalism. The at first sight financial and white-collar crime around the General Hospital in Vienna (Allgemeines Krankenhaus Wien, AKH), then and today still one of the biggest facilities of that kind in entire Europe, turned out at the end to be a very big political affaire, when in the mid-1970s financial irregularities of a large extent were revealed, the immense covering by all media nationally and internationally became a stumbling block not only for politicians of all couleur, including several Ministers and high-ranking public officials of the municipality, but also for many involved managers, planners, architects, engineering and construction companies.[2]

The planning for this huge and socially important building project had started in the mid-1950s, the building process began in the late 1960s, stretching on into the 1980s, with repairs, re-planning, and debugging aftermaths still going on today. But then also the lack of an effective control system became visible, as control of these processes was insufficiently performed by the political and financial authorities, relying on accounting control of local and international auditing firms, unfortunately entangled themselves, as completely overwhelmed by such "socio-dynamic" structural problems, as already experienced in other countries as well.

As these aspects of management and consulting control mechanism together with decision making theories were the main research focus of Hofmann's time as a visiting professor at Harvard, this "planning and building issue" inspired him to install within his chair in the early 1980s an interdisciplinary "Consultancy Study Group" (CSG) together with some of his post graduate students and a major architectural company he already coached in management. The aim was twofold: academically to detect potential conflicts of consulting in combination with auditing, as many consulting firms frequently were offshoots of only a few international auditing companies, and practically to conceptualize and develop decision making, control, and feedback models in complex professional and organizational structures and hierarchies.[3]

This unusual private–public initiative was meant as a platform to foster cooperation between academic research of universities canalized by Hofmann's Management Institute at Vienna's WU, and Sertl's Institute for SMB (Small and Middle-sized Businesses), to identify qualified partners in public institutions (e.g., in the health care sector) and private corporations/companies (e.g., the state-owned industries in Austria) alike, interested in supporting case study research projects of mutual benefit, and of course to eventually publish the results.

CSG soon also had received offers to co-operate with Austrian and German consulting companies, and through Hofmann's research network going back to his time in the USA, even from Arthur D. Little, Inc. of Cambridge, MA.[4] But when Prof. Hofmann was approached by ADL International's German office in Wiesbaden,[5] for a recommendation regarding staffing and establishing an ADL branch office in Vienna to develop the interesting Austrian market, the suitable candidate he proposed, Dr. Mack, his disciple, refused out of personal and academic consideration. He wished to rather stay an independent freelancer reseacher than an employee of a consulting company.

2 Rampant information on this issue on the Internet.

3 In November 1986 officials of the Government Accountability Office, GAO, an institution of the US Senate, Washington, DC, were invited by CSG to Frankfurt/Main, Germany, to be part of a "Fact Finding Meeting on Accounting and Auditing in other Countries," organized by CGS's German representative, Dr. Josef R. G. Mack, in reaction to the recent events in Austria.

4 The successor company of the world's first consulting company, founded in the year 1909 by the name giver Arthur Dehon Little, an American Chemist and chemical engineer, who could not complete his studies at MIT for financial reasons, but was rewarded later on for his scientific achievements with an honorary doctor of MIT.

5 In 1982 the maintained contact with Arthur D. Little, Cambridge, MS, had led to a personal search request by Dr. Tom Sommerlatte, Vice-President and CEO of ADL International's German branch office in Wiesbaden, to install ADL in Vienna for serving the Austrian market, so far handled from Germany. Executed through the agency of Dr. Mack, a fellow student of HfW-times, Dr. Manfred Kunze, and later colleague researcher in the reorganization project of the Austrian Health Care System directed by the Ministry of Finance and the Ministry of Health, was won over by him, to start a new career with ADL. But plans exchanged and discussed between CSG and him as CEO of ADL Austria for cooperation in Consulting and Strategic Leadership research were not to be realized in the envisioned extent (except for a minor involvement in conceptualizing a publication of ADL, published in 1985 with Gabler, Wiesbaden, on "Management im Zeitalter der strategischen Führung," Arthur D. Little International (ed.)). Austria, however, became a very successful and profitable business engagement for Dr. Kunze and ADL.

However, legal and financial considerations regarding the planned publications and pending contract research assignments in consultancy, required finally an adequate legal status, different from the university attached CSG. In the year 1985 an Austrian limited company detached but in co-operation with academia and CSG was founded in Vienna and named "MCI Management Consulting Institut."[6] With due lead-time MCI started two series of books, the first on the subject "Management Consulting" and in 1989 the second on "Mergers and Acquisitions." Until the middle of the 1990s five volumes of each of these series had been published by MCI as editor with the renowned German publishing house "Kohlhammer Verlag" in Stuttgart, the contents compiled of abstracts from CSG's research projects like theses, seminars, lectures, and consulting cases.

Then on November 9, 1989, an unexpected and for certainly the majority of people almost unbelievable event took place: the fall of the Berlin Wall, which promised completely new perspectives for unprecedented international activities after WWII and the era of the Cold War. The fundamental political change in the USSR, "Perestroika" introduced by Mikhail Gorbachev, brought first for East Germany, then for almost the entire Eastern Block a complete realignment. For Germany a division of more than 28 years between the two parts of Germany, alienated by their membership in two different political systems, hostile to each other for decades in the so-called "Cold War" with an "Iron Curtain," had ended without one shot and an exchange on all levels including visits to the German Democratic Republic (GDR) were again possible finally.

Austria as a neutral country had maintained contacts with the GDR especially in the (state-owned) heavy industry sector located in Linz as well as very good commercial and academic relationship was able to gain from existing links, the latter used by MCI to take up contact with the premier publishing house of the GDR for scientific and political publications, the Akademie Verlag in East Berlin, offering joint publication activities. Already at the end of November 1989 a visit to East-Berlin led to an immediate (under the circumstances necessarily sponsored by MCI) publication in early 1990, a Handbook on Management Techniques authored by Wolf D. Hartmann, a professor of the GDR, and also by him a *German-German Dictionary for Business Managers.*[7]

After the fall of the Berlin Wall, I had also looked forward to the diverse tasks and possibilities for management consulting beyond Germany in other countries of Central and Eastern Europe. Under the impression and hope the change in politics could impact us in Europe, I gave the fifth volume of our series, now under my sole editorship, the title "Middle- and Easteurope: A Challenge for Management Consultancy,"[8] with inspiring sentences on the book back, like "The establishment of democratic structures and market economy is a pan-European task, which also holds a special challenge for Management Consulting."

However, as a sociologist I had not forgotten that there has so far then been no concrete approach of management consulting from a sociological point of view for developing countries on other continents. Africa and Southern America certainly are of important geo-economic and geopolitical concern too, bearing still unsolved problems, for long already as urgent tasks for Europe's industrialized countries.

By this time, MCI (in Vienna and after the reunification of Germany also registered as a German Ltd. in Berlin) had already published in its first edition of the MC-series, which I was co-editing together with Walter Sertl, in part

---

6 MCI's founders and CEOs ("Geschäftsführende Gesellschafter") were Dr. Josef R. G. Mack and Dr. Peter Petersen (also a postgraduate of Hofmann), partnering in the company's activities, with a special focus on the support and assistance of CSG's academic research activities and the publication of the results. All of these efforts ever since starting in the Health Care field were international, yet also under consideration of the development in the EU. As Austria's joining was to be expected (which took place in 1995), a close research cooperation with a researcher in the UK, Dr. John H. Smalley of Leicester, England was strategically formed into a legal partnership between him and Dr. Mack, as a German partner, by forming the IMI Innovation Management Institute EEIG, as a not-for-profit-company under EU-law in Cardiff, Wales, acting as default guarantor for publication's.

7 Wolf D. Hartmann, Edition Management Know How, Handbuch der Managementtechniken, Akademie-Verlag, Berlin/Ost, 1990, and, also by Hartmann, Deutsch-deutsches Wörterbuch für Wirtschaftsmanager, Heinrich Bauer Verlag, Hamburg,1990.

8 Klaus Zapotoczky (Ed.): Mittel- und Osteuropa: Eine Herausforderung für die Unternehmensberatung, Management Consulting 5, Kohlhammer, Stuttgart, Berlin, Cologne, 1993.

IV problems with state trading countries as a topic, not only the East-West-Relationship, but also China.[9] Whereas in the second volume, Facility Management had been identified as an interesting consulting subject with a great potential for professions involved in planning and engineering of – in today's terminology – the built environment.[10]

The relationship between Sociology and Consulting and different approaches thereto were later expanded in an article in volume 3 of the MC-series, edited by Urs M. Rickenbacher, titled the "Future-oriented Training of Management Consultants,"[11] and deepened in MC 5 together with Michael Hofmann and Lutz von Rosenstiel, with the title "Socio-cultural Basic Condition for Management Consultants."[12]

The digression to the beginning of my interest in Facility Management after having been involved in Management Consulting issues I regarded as necessary to explain, as it seems for me important for the understanding of my approach as a sociologist: MC just like FM are beside all technical support nowadays available and useful by IT and AI maybe even more an issue of and for human beings in their social inclusion and context. Not only the people in Facility Management's People, Place, Process concept are to remain the center of all human effort.

Back to the historic development of my dealing with the subject FM, already before Autumn 1989, when the 1st Symposium Facility Management took place in Germany in Frankfurt/Main on September 28 and 29, organized in cooperation with the publisher "mi modern industry" of Landsberg, Bavaria, and a then just founded first FMI Facility Management Institut(e) in Europe, as a Swiss AG (stock company), an initiative of MCI's Dr. Mack and a Stuttgart-based design company by the name of "Partner."

This venture was a result of Hofmann's seminar activities in Vienna with renowned architects of Austria and Germany, started in the mid-1980s with research support of a handful of post graduate students and their research in the field of innovation in architectural IT-application (then CAD and CADD), all of them supervised mainly by my colleagues Hofmann, Sertl, and me, in close academic exchange and cooperation with experts in practice and academia of the USA, the Netherlands, England, and Scotland.

Specific aspects of management in planning, architecture, and construction interested one of Hofmann's seminar attendees especially, Hans Kahlen, an architect from Aachen in Germany. He had started using CAD in a large-scale project of his architectural company in Munich, Bavaria, and became interested in this tool's development from the point of view of an architect and manager of an atelier with more than 50 employees, not only designers and planners, but also managerial staff to manage projects all over Germany. Intrigued by FM's concept of "People, Place, Process," he approached the integration of IT (then called "EDP Electronic Data Processing" – in German "EDV Elektronische Daten Verarbeitung") from a very practical side, as research in this field was in its beginning, and so was FM's relationship with IT too.

As described before, after the experience with CSG and MCI as path-makers, FMI had a more targeted approach toward FM as the intellectual basis and technical resources to be used and integrated, seemed to allow genuine interdisciplinary research, as the concept's basic idea from the beginning was meant to overcome professional boundaries as existing in praxis and in the thinking of specialists because of their self-understanding. Therefore FMI started a series with Kohlhammer from the theoretic side, with FM 1 in 1989 on "The Power of Planners and Architects," FM 2 in the same year about "CAD-Use in Architecture," and FM 3, an anthology on "Planning, Building, Utilization, and Maintenance of Facilities," in mid-1993, overtaken by FM 4, which was already released in early 1991, also an anthology paying attention to "Building in the New Federal States" (of Germany), and also by FM 5 in 1992, a posthumous honor of Horst W. J. Rittel, "Selected Writings on Theory and Methodology," compiled and edited by his disciple Wolf D. Reuter, who was at the series' beginning and at its end.

9 Michael Hofmann, Walter Sertl (Eds.), Selected Problems and Tendencies in Development in Management Consultancy, Management Consulting 1, Kohlhammer, Stuttgart, Berlin, Cologne, 1989 (2nd edition).

10 Walter Sertl, Klaus Zapotoczky (Eds.), Neue Leistungsinhalte und internationale Entwicklung der Unternehmensberatung, Management Consulting 2, Kohlhammer, Stuttgart, Berlin, Cologne, 1989.

11 Klaus Zapotoczky, „Soziologie und Beratung", in: Urs M. Rickenbacher (Ed.), Zukunftsorientierte Ausbildung von Unternehmensberatern, Kohlhammer, Stuttgart, Berlin, Cologne, 1991, p. 61 ff.

12 Michael Hofmann, Lutz von Rosenstiel, Klaus Zapotoczky (Eds.), Die sozio-kulturellen Rahmenbedingungen für Unternehmensberater, Management Consulting 4, Kohlhammer, Stuttgart, Berlin, Cologne, 1991.

All of this had happened around the fall of the Berlin Wall. The first visit in the last week of November 1989 just when it was made provisionally possible was, as mentioned before, with the new contact at Akademie Verlag in East Berlin. The Chief Editor proposed to the visitors from "the West" two further contacts. One being the already mentioned Prof. Dr. Wolf D. Hartmann of "HTW Hochschule für Technik und Wirtschaft (Ost-) Berlin" (a University of Applied Sciences) in the field of economy, the other to the head of an institution with the name "Bauakademie der DDR" (Building Academy of the GDR).

Both contacts have been very special, and in several meetings, much exchange, and mutual visits that followed, possibilities for cooperation were negotiated with focus on the differences in management aspects in the building industry, heavily concentrated on industrial prefabrication. As the building sector was of great importance for the government of the GDR, and the Bauakademie as the only institution entitled to plan, coordinate, build, and maintain the entire building process in the planned economy system of the GDR, a very interesting field for research seemed to be identified.

However, the unification took a faster pace than any academic research project, and subsequently the Bauakademie was dissolved – there was no interest from the political side (West) for any integration of even worthwhile experience and know-how as originating from a planned economy (of the East).

None the less the academic aspects of this unique chance for system comparison found quite fast some expression in the first book series on FM in German, published by a now Berlin-based FMI GmbH (a German Ltd.) founded by Peter Zerahn and Dr. Mack in Berlin as then "FMI Facility Management Institut Forschungsgesellschaft GmbH" (later, as Affiliated Institut of CTU, FMI Facility Management Institut GmbH), where Zerahn at that time was a doctoral candidate of my colleague Sertl and me (thesis approved in 1991). FMI became in 1993 not only the status of an Affiliated Institute ("An-Institut") of the Cottbus Technical University but was also honored by a Certificate of Appreciation dated 6/27/93 sealed and signed by President Diane H. MacKight of IFMA, International Facility Management Association, which was handed over within an inauguration celebration by IFMA Executive Director Dennis Longworth together with Donald Young also with IFMA.

In autumn 1994 the very first Facility Management Forum at the Hannover Industrial Fair was made possible with the help and contacts of Prof. Dr. Volker H. Hartkopf, Center for Building Performance and Diagnostics, Carnegie-Mellon University, Pittsburgh, PA. At the fair a Facility Management Forum was hosted by FMI, with international assistance, support, and partially even personal attendance of guests and visitors. The success there achieved would certainly not have been so great without the help and backing of many supporters: Especially to be named from Austria, Prof. Degenhard Sommer, Technical University of Vienna; the USA Prof. Franklin D. Becker, Cornell University, Ithaca, NY; Kreon Cyros, INSITE, MIT, Cambridge, MA; Deane M. Evans, AIA/ACSA, Council on Architectural Research, Washington, DC; Layne S. Evans, CCB Information Systems, National Institute of Building Sciences, Washington, DC; Jeffrey M. Hamer, The Computer Aided Design Group, UCLA, Los Angeles, CA; the United Kingdom (Prof. Francis Duffy, DEGW, RIBA, University College, London, England; Prof. Keith Alexander, University of Glasgow, Scotland; Gareth E. Mulligan, University of Reading, England), and from the Netherlands Philip Tidd, DEGW/Twynstra, Amsterdam.

The by-then established contacts allowed a lively exchange in research and teaching internationally and in 2001 FMI (Berlin) launched a new edition on FM at the scientific Springer Publishing House, Heidelberg, Berlin, one volume edited by Edmond P. Rondeau, IFMA Fellow, and Walter J. F. Moslener, IFMA Fellow (Founder of GEFMA, German Facility Management Association and IFMA Deutschland).

But the sudden death of Dr. Moslener at the age of only 66 on July 1, 2001, just shortly after the release of the twin-pack-edition ended the very close collaboration, also with MCI, FMI, and the professors involved, which had developed from scientific research into a personal friendship and cooperation. This affected unfortunately also planned activities in Europe with Edmond P. Rondeau in the field FM education, as well as Kreon Cyros and the promotion of the technology transfer mission by his Consortium's INSITE System, with a huge potential for developing university and health care facilities in Europe.

Despite these personal effects, the target set, fostering the promotion and education of FM in the German-speaking countries remained academically with FMI, but here too, the loss of the front marketing man and his personal contacts to the upspringing FM-industry was fatal, Dr. Walther J. F. Moslener was not to be replaced by anybody available of FMI's staff. The accent of research then shifted back to where it had started from: FCA, a German Limited, founded in 1990 by Norbert Gerhards as "Research Company for Computer Applications in Architecture," active in Aachen ("FCA" Forschungsgesellschaft für Computeranwendungen in der Architektur mbH), took over the further development of necessary tools and support. As co-founder of the first FMI AG together with Günter Neumann et al. and sometime before partner of Dr. Mack in computer-related issues continued FMI's research and development for architects and took over assignments within Corporate Real Estate Management (CREM) including concepts and problem-solving in the field of commercial real estate projects.

Unique though remains FM 4 "Building in the New Federal States," in which the editor and his 12 co-authors proceeded primarily from the following basic assumption: "Integrative facility management in the unity of planning, building, using and maintenance will develop into a promising new market in the CAD environment in Germany and Europe in the coming years." In addition, in this volume the editor made the[13,14] Dessau Declaration of June 24, 1990, by 36 German architects on the subject of "On the way to a democratic building culture" accessible to a wider public.

In the following 14 focal points, considerations are made as to what a European building culture of the future could look like:[15]

1) Urban renewal versus unemployment
2) Citizen activation
3) Municipal planning
4) Local cooperation
5) Quality of planning
6) Requirements for public planning
7) Supra-local planning and planning instruments
8) Planning content
9) Power of the market
10) Builders with a sense of responsibility
11) Architects as trustees
12) Construction industry as a partner
13) Separation of powers also in the European internal market
14) Building as a cultural task

These – fortunately – short considerations of principle were meant to be taken up and tried to be implemented not only in the construction industry, but – in an appropriately adapted form – also in other areas of life, e.g., education and health, in Europe and worldwide. Together with Michael Hofmann and Lutz von Rosenstiel, I have tried to present different levels of sociocultural conditions for management consultants with the help of concrete consulting examples. Unfortunately, perhaps because of the physical distances and the lack of concrete opportunities, this hopeful approach did not result in a long-term collaboration.[16]

However, Facility Management continued to evolve and also focused on the different forms of company takeovers and their preparation and execution, both theoretical considerations and concrete practical experience and their interaction in the redesign of company takeovers. Other priorities as well as the retirement of professors from active service may have been decisive. An interesting stimulus for open, holistic scientific research did not revive the Austrian universities involved so far and possible collaborations lost their challenges. Also, at the

13 Hans Kahlen, CAD-Einsatz in der Architektur, Verlag W. Kohlhammer, Stuttgart, Berlin, Cologne, 1989.

14 Hans Kahlen (Ed.), Bauen in den neuen Bundesländern, Verlag W. Kohlhammer, Stuttgart, Berlin, Cologne, 1991, p. V.

15 Dessau Declaration of June 24, 1990, in: Hans Kahlen (Ed.), Bauen in den neuen Bundesländern, Verlag W. Kohlhammer, Stuttgart, Berlin, Cologne, 1991, p. 295 ff.

16 Michael Hofmann, Lutz von Rosenstiel, Klaus Zapotoczky (Eds.), Die sozio-kulturellen Rahmenbedingungen für Unternehmensberater, Verlag W. Kohlhammer: Stuttgart, Berlin, Cologne, 1991.

meta-level of Management Consulting and Facility Management, some things were lost or were no longer brought out. Suggestions for philosophical considerations and challenges for new theoretical approaches, which – in the long term – (can) also dynamize practice, no longer took place.[17]

Every scientific finding changes our possibilities. Applied to construction and its support options, this means concretizing the respective understanding of "work" for building. Like the three main meanings of work, to which Max Scheler (1874–1928) already pointed out, namely plan, concrete activity and product, building is also to be seen in a differentiated way, as the German language suggests to us: construction, dismantling, expanding, shoring, rebuilding. Early on, Wolf D. Reuter drew attention to the power of planners, but we must also pay attention to the power of those who can dispose of the product, the building and its possibilities of exploitation.[18,19]

We can not only build plant and buildings and accordingly employ architects and builders and their cooperation partners, but we can also plan, reproduce, and perhaps recreate fabrics, plants, animals, human limbs, human bodies, perhaps even human beings – but with human intelligence or AI?. Do we live in a world of unlimited possibilities, or do we want to (consciously) limit ourselves? Who can want to be a "limited" person?

What does this mean for the future of FM in a broader sense? Is a code of ethics necessary for FM in the broader sense, or should everything that is possible be allowed? Should we take the principle of the brain researcher Gerald Hüther, "Doesn't everyone who violates the dignity of another human being actually violate his own dignity" seriously as individuals and society and ask ourselves:[20]

- What is our dignity as human beings (as individuals and as a society)?
- How can we become aware of this dignity (as individuals and as a society)?
- How can we live this dignity (as individuals and as a society) before we die?

Hüther has given us suggestions for answers to these questions that we can and should consider. We need to go beyond this approach and develop mindfulness towards everyone, things, plants, animals and people, perhaps even including ourselves and God, the Other. But what does it mean for all our actions as individuals, as a society and as a nascent world society, to be mindful? Does comprehensive carelessness, ignorance of things, plants, animals and humans perhaps characterize our behavior and does misunderstood individuality today mean a kind of self-destruction as individuals and as a society?[21,22,23]

As a sociologist who feels jointly responsible for the shaping of a world society, the further development of the octagon of a democratic social design by Stein Rokkan seems to me to be a way to avoid the self-destruction of humanity. In my opinion, a continuation of the spiral of violence leads to the "common path into the abyss," as described by Friedrich Glasl for individuals and organizations.[24,25]

In the following, an attempt is made to further develop the octagon of a democratic social design for a sustainable understanding of facility management in a broader sense (see Figure 1):

---

17 Claus Bressmer, Anton C. Moser, Walter Sertl, (Eds.), Vorbereitung und Abwicklung der Übernahme von Unternehmen, Verlag W. Kohlhammer, Stuttgart, Berlin, Cologne, 1989.

18 Max Scheler, Erkenntnis und Arbeit – Eine Studie über Wert und Grenzen des pragmatischen Motivs in der Erkenntnis der Welt, Francke Verlag, Bern, 1960.

19 Wolf D. Reuter, Die Macht der Planer und Architekten, Verlag W. Kohlhammer, Stuttgart, Berlin, Cologne, 1989.

20 Gerald Hüther, Würde – Was uns stark macht – als Einzelne und als Gesellschaft, Knaus Verlag, Munich, 2018.

21 Klaus Zapotoczky, Wer bin ich eigentlich? Versuch einer Reflexion der Lebensgestaltung, Trauner Verlag, Linz, 2021.

22 Max Horkheimer, Die Sehnsucht nach dem ganz Anderen – Ein Interview mit Kommentar von Helmut Gumnior, Furche Verlag, Hamburg, 1970.

23 Rüdiger Safranski, Einzeln Sein – Eine philosophische Herausforderung, Carl Hanser Verlag, Munich, 2021.

24 Rokkan Stein, „Die vergleichende Analyse der Staaten- und Nationenbildung: Modelle und Methoden", in: Wolfgang Zapf (Ed.), Theorien des sozialen Wandels, Verlag Kiepenheuer & Witsch, Cologne, Berlin, 1969.

25 Friedrich Glasl, Konflikt Management – Ein Handbuch zur Diagnose und Behandlung von Konflikten für Organisationen und ihre Berater, Verlag Paul Haupt, Bern, 1994.

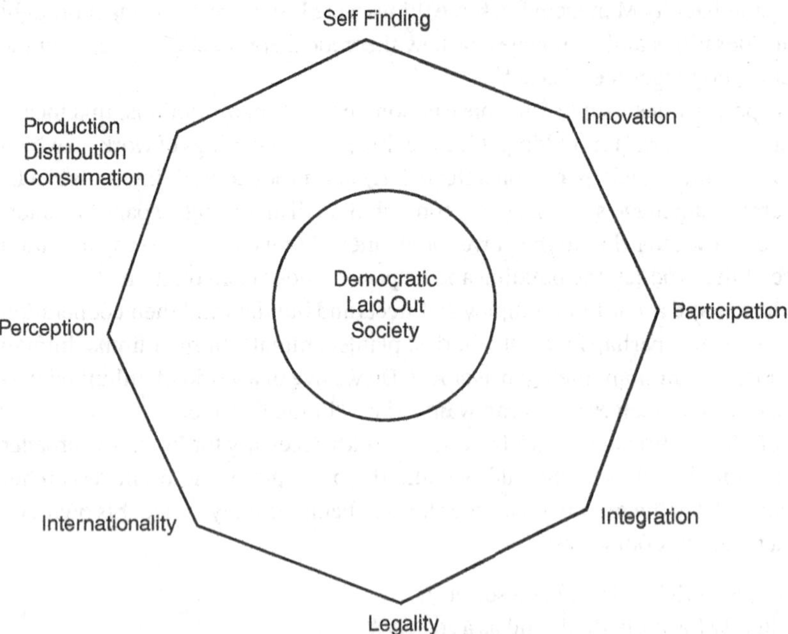

Figure 1 Tasks of a democratic society. (*Source:* Klaus Zapotoczky).

FM of the future will have to go beyond the facilitation of planning, construction, conversion and expansion as well as the reduction and elimination of damage to the environment of sites used, including possible water pollution or impairment, and must include questions of regeneration for all parties involved.

In a globalized world, people living far away can also be affected, which may require appropriate information. At the same time, inventions and patents can be important and the handling of them, which can take very different concrete forms, which may be legal, but can also have criminal traits. How can we ensure that the best for all, and not the benefits for a few, becomes the principle of realization? According to experts, the constitutional state will not be able to achieve this with its own resources but will have to build on conditions that it cannot create itself. But how do these facilities emerge in a broader sense, how do groups, societies and states become humane? The French atheist André Comte-Sponville has proposed an "encouragement to untimely living" and the Swiss theologian Hans Küng (1928–2021) has developed a project Global Ethic. Others are pinning their hopes on digitization, which may be able to cope with the huge quantity of data but is struggling to capture quality. People will have to learn (and for this important task appropriate help, facilities in the broader sense, will be necessary) to understand themselves as groups, societies and as a world society in mutual dependence and dependence on each other and[26,27,28,29,30] to develop a corresponding self-image.

At the same time, a legal mentality will have to be developed – supported by as many people as possible – which should consider the old insights of Max Weber and intercultural developments of this European approach. Perhaps

26 Robert Laughlin, Das Verbrechen der Vernunft – Betrug an der Wissenschaft, Suhrkamp Verlag: Frankfurt/Main, 2001.

27 Ernst Wolfgang Böckenförde, Staat, Nation, Europa – Studien zur Staatslehre, Verfassungstheorie, Suhrkamp Verlag, Frankfurt/Main, 2011 (4th edition).

28 André Comte-Sponville: Ermutigung zum unzeitgemäßen Leben – Ein kleines Brevier der Tugenden und Werte, Rowohlt Verlag, Reinbek, 2001 (2nd edition).

29 Hans Küng, Karl-Josef Kuschel, Wissenschaft und Weltethos, Piper Verlag, Munich, 2001.

30 Armin Nassehi, Muster. Theorie der digitalen Gesellschaft, Verlag C. H. Beck, Munich, 2021.

the wisdom of Confucius can also show how sustainable attitudes can be built in the world. The knowledge would be there, the interest is low for incomprehensible reasons.[31,32]

Following Stein Rokkan, it seems necessary to really involve as many people as possible in social activities in an appropriate way, to facilitate a prosperous cooperation of the many and to help those responsible at the various levels and areas to better unite with the people. Many things are possible, but too little is actually perceived at the moment. The reasons for this social deficit must be investigated and ways of better participation, mutual understanding and greater interest in the well-being of fellow human beings must be found and permanently followed. In this way, participation and perception can be filled with life.

For a long time, sociologists have believed that integration can be achieved through egalitarianism and uniformity. Today, integrated projects of all kinds show that good management can motivate and move very different people to work together – often over large spatial and cultural distances. Worldwide teams of various kinds are created and function. Integration must be desired by all parties involved and cannot be decreed.

Innovative ideas must be developed, and ways of realization can and must be open to many. Small experiments must be possible, from the success or failure of which it is possible to learn. Prohibiting small experiments and preventing the failure of large institutions, while at the same time upholding the principles of competition and personal development, contributes to the decline in the credibility of existing systems among the population. This is also regrettable because the complexity of modern societies and all their areas is increasing and, according to social science findings, can be reduced primarily through trust and confidence-building measures. But what does it mean when complexity and mistrust increase in one area and in society as a whole? What are appropriate socio-political strategies in such a situation? Is there a socio-political emergency? But "where there is danger, the rescuing also grows," says Friedrich Hölderlin. a German poet ("Wo aber Gefahr ist, wächst das Rettende auch."). If enough people persistently search, it will be possible to find ways out of this impasse.[33]

In all areas of society, people plan, produce the most diverse values (material and intangible, valuable and less valuable, expensive and cheap), distribute these values (not always quite fairly) and consume these values (with more or less respect for the values and their producers and distributors).

Increasing globalization makes it difficult to implement the other seven principles (self-discovery, legality, participation, perception, innovation, and integration, as well as dealing with the values of planning, production, distribution and consumption) because it is not possible to implement them concretely for everyone worldwide, but a subsidiary approach is required. In smaller units and where concrete needs are met directly, everyone should take responsibility and ask the question: Where is your brother, your sister? Not with the excuse: Am I the guardian of my brother, my sister? but answer with the servant willingness: Here I am, and I want with all my strength that everything is done for those in need to shape their situation in the best possible way. To do this, it will be necessary for everyone (as many as possible) to learn to take responsibility for their environment, their area. The UNESCO Commission has pointed out this task in its *Education Report for the 21st Century* and all member states are obliged to develop appropriate concrete steps.[34]

The prerequisite for this to actually become possible is that many people learn to take responsibility in all areas and at all levels. Above all, this requires a willingness and ability to take responsibility. Neither is it sufficiently developed in today's Europe and today's world. We can learn from role models such as Martin Luther King, Nelson Mandela, Vaclav Havel, or even Angela Merkel (and others) or study texts and try to turn them into reality.

---

31 Max Weber, Wirtschaft und Gesellschaft – Grundriss der verstehenden Soziologie, Verlag C. J. B. Mohr (Paul Siebeck), Tübingen, 1980 (5th edition).

32 Hans van Ess, Konfuzianismus, Verlag C. H. Beck: Munich, 2022 (3rd edition).

33 Hans Küng, Vertrauen, das trägt – Eine Spiritualität für heute, Verlag Herder, Freiburg, Basel, Vienna, 2008 (3rd edition).

34 German UNESCO Commission (Ed.), Learning Ability: Our Hidden Wealth, UNESCO Report on Education for the 21st Century, Luchterhand Verlag: Neuwied, Kriftel, Berlin, 1997.

Taking responsibility is at the same time a question of knowledge, ethics and willingness, in short, it requires the whole person. All facilities should serve this life's work. The power of a few can prevent many things for a long time, but in the end the power of the (many) powerless will prevail.

As the "Three Ps" "People, Place, Process" concept led to Facility Management:, the concept ought to be now expanded toward "Society, Built Environment, Technology,". The adaption the concept globally beyond workplaces, offices, and even private dwellings, As the pandemic showed by requiring work from home, which included a not only critical but constructive discussion of the use of infrastructure of all kind. Energy, logistics, including their resource-saving treatment from the standpoint of sustainability have priority. Questions which entail economic and ecological, social and political dimensions with great impact on all systems and societies, making radical changes unavoidable, otherwise future research sees manhood doomed to disappear.

Prof. Dr. Klaus Zapotoczky
formerly with Johannes Kepler University
Linz, Austria

After receiving his doctorate in law from the University of Vienna in 1961, he was a lecturer in sociology and development policy at the German Youth Academy in Klausenhofen from 1964 to 1966. After studying social sciences at the KU Leuven, he was an assistant at the Institute of Sociology at the University of Social and Economic Sciences in Linz with Erich Bodzenta from 1966 to 1971. After his habilitation in sociology at the University of Vienna in 1976, he was Professor of Politics and Development Research at the Institute of Sociology at the University of Linz from 1976 to 2006.

Prof. Zapotoczky is (co-) editor and (co-) author of more than 60 books and more than 200 journal articles.[35] In the last years two more books have been published, one of them about a necessary shift in our society, where the members are more and more detached from each other member, a publication in the series on "Health – People – Society".[36]

35 Further details can be found in the commemorative publication by Christian Pracher, Herbert Strunz (Eds.), Wissenschaft um der Menschen Willen, Duncker & Humblot, Berlin, 2003, and the commemorative publication by Bernhard Hofer, Claudia Pass, Christian Pracher, Herbert Strunz (Eds.), Religion und Gesellschaft, Mercur Verlag, Wien und Berlin, 2018.

36 Klaus Zapotoczky, Die Losigkeitsgesellschaft verwandeln, in: Klaus Zapotoczky (Ed.): Schriftenreihe Gesundheit – Mensch – Gesellschaft Band 19, Trauner Verlag, Linz, 2022.

3.5

We Are All Jugglers. Still.

Fred Kloet

Remember the video made by Steelcase in 1992? Yes, it's old. You can still find "The Juggler" on YouTube. Show it to your students. It will make them feel what it is to be the jack of all trades in an organization. To be the oil that connects the dots. The one that best understands the culture of an organization. Juggler Michael Davis explained what a facility manager does. He is a magician. Someone who can handle all kinds of tools, processes, cultures, demands and services. I have an original copy of the video in my small library at home. In my 33 years in FM, I have never experienced anything like it. The video should be part of the online anthology of FM.

Many years later integration of processes became one of our core competencies. And today data driven FM is doing the exact same thing. In a way nothing has changed since The Juggler. However, everything has become a lot smarter and less corrective. Clients, buildings, providers, systems and workplaces all "evolutionized." Technological innovation has always been driving FM products and allowed for smarter and more customer focused services. But one thing never changed. We were still working in silos. We still needed a magician to connect the dots. Yes, service providers could be hired to offer IFM or other types of high-level outsourcing solutions. But with outsourcing the silos often were just relabeled or started collaborating only. The professional specialists and their specialized markets still managed to sell their high-level output. Organizing shared service centers was the last attempt to bridge islands and bring together the silo's forming FM.

With the possibility of creating a Digital Twin and Metaverse of the workplace the coin flipped. The mindset changed. The real individual user perspective became leading for FM services and workplace concepts. Where co-working was still a combination of services outsourced to another organization offering a third location, nomadic working was all about the individual demands and wishes for a place to work and play. In a way the Active Digital Twin and Metaverse are the final step toward freedom from the location-based facilities. And it sure makes us in FM thinking about how to be the juggler in an online workplace environment.

I founded four FM-related companies and IFMA Netherlands, acted as EuroFM vice-president, and served on the CEN EN 15221 technical committee for many years. I really enjoyed working with many of the experts now also represented in this book. We all collaborated in professionalizing FM. And we succeeded. There is only one thing I hope we can still achieve: benchmarking based on the same space and area measurement standard. An ISO-standard based on the praised by many EN 15221-6. Not a commercial standard such as the IPMS. With the current architectural designs based on Building Information Modeling technology the opportunities to do cross-border FM-benchmarking becomes a piece of cake. No juggler needed. We already knew measuring a building with different national standards generates different outcomes. Jones Lang LaSalle discovered this many years ago. But this time there was no place to hide. Just recently FM officials at the NATO HQ in Brussels showed me their real-life use-case confirming that the use of FMBIM saves them 20% on cleaning costs. 20%!!

Facilities @ Management, Concept - Realization - Vision, A Global Perspective, First Edition. Edited by Edmond Rondeau and Michaela Hellerforth.

So, what is the learning curve here? We started with professionalizing FM. Many professors helped us. Some of them, like Keith Alexander even chaired our associations. When finished we focused on the customer needs. We tried to understand how, when and where groups of people want to work and how individuals within those groups could best be served. Working from home was made possible, but without FM support still. Workers started making their own choices. Forcing us to become hospitality managers and more flexible than we were ever before. And today we ended up in a world of location-free working. And we need to be even more of a magician and juggler all over again. We try to label what's happening. Calling it hybrid working. We try to get them back to the office. If possible, for at least 50% of their time. Like everybody we are trapped in the Maslow-pyramid. We need a roof, safe space, and a sense of belonging.

But this time it's different. We all have become aware of the limitation of available natural resources such as healthy air and drinkable water. We all know the demographic shift towards a grey population needs more radical changes. Combining Digital Twins and the Metaverse with AI and ChatGPT could potentially make workplaces service themselves and save resources. And because younger generations embrace technology and working online anywhere, the investment value of online real estate used for FM services increases. No cleaning and catering needed. In the Netherlands we already see that the occupancy of many offices on a Wednesday and Friday is no more than 10%. We are heading towards a four-day week at the office. No fulltime rental contract needed.

And still. The facility manager acts like a juggler. Searching for effective and efficient combinations. Finding creative and smart solutions. Integrating processes. The biggest difference with 33 years ago is we no longer juggle with tools and services but with People, Place, and Processes. We use data to help us minimize the use of Place and need for Processes. More and more we facilitate People in making better choices instead of choosing for them. And maybe this brings what we told ourselves a real facility manager should be: not necessary because all facilities and services are always as requested.

Fred Kloet
Arnhem, the Netherlands

In 1988 Fred graduated in Facility Management from Akademie Diedenoort. The first university of applied sciences (BSc) in The Netherlands teaching FM. He started working as staff member #2 at the consulting and training company founded by Henk Klee. The former European Facility Manager for Digital Equipment. After ten years Fred left the company as staff member #121 and member of the management team. As senior consultant and Business Development Manager he led projects ranging from seventy workplaces to 2.000.

When working for the European Council of Ministers in Brussels Fred discovered his intercultural side. He founded IFMA Holland and became active in various international non-profit workgroups focused on professionalizing FM. His curiosity led to his role as Editor-in-Chief of the magazine The European Facility Manager. After having worked as senior consultant with engineering firm Royal Haskoning DHV Fred founded Villa FM, the European Center for FM near Rome, Italy in 2003. He also joined the CEN/TC 348 defining the European standards for FM (EN15221) and representing The Netherlands in the workgroup creating the European space and area measurement standard (EN15221-6). He joined the EuroFM board as vice-president. He was responsible for renewing the governance, organizing a European conference and writing the text explaining the global history of FM that is still on www.eurofm.org

After returning to The Netherlands in 2012 Fred founded two FM-related companies. With PROCOS Netherlands he brought back Archibus to The Netherlands. With Smart WorkPlace he created a community around FM, real estate, workplace management and HR. Both companies are still successful. Fred moved on to BIM-Connected joining as a staff member focusing on FMBIM. He now acts as a senior consultant with FMHaaglanden. This organization is part of the Dutch Ministry of Internal Affairs and responsible for 20.000 workplaces. Fred also acts as the chair of the building SMART International project "FM and open BIM".

3.6

On the Brink of Change

Mark P. Mobach

Introduction

These are very exciting times for the Facility Management (FM) profession. Contemporary developments create unprecedented opportunities for FM, allow it to have positive impact on organizations and their workforce globally, and to contribute to global challenges. These opportunities also require new perspectives on FM and the built environment, different value propositions and investments, and the advancement of evidence-based design in FM. In short, the time for action is now. However, before reflecting on these opportunities we first need to better understand the nature of FM, also in its historical and contemporary contexts.

The Nature of FM

In 2014, together with Keith Alexander and others we argued that FM as a management discipline serves the primary process of organizations and is being associated with a broad spectrum of supportive activities varying from the design and management of buildings and technical systems to services as cleaning, security, and catering (Mobach et al. 2014). At the times, within the European Facility Management Network (EuroFM) it was generally accepted that the core value of FM was the "integration of people, process, and place." In 2002, FM had been redefined into space, infrastructure, people, and organization. It is pre-supposed here that infrastructure refers to the physical environment of organizations, such as buildings, interior, and technological devices. Infrastructure encapsulates space, which forms emptiness as well as air, light, scent, and sound. As such, space appears from within that infrastructure, space as an inversion of the physical world of organizations. In addition, people also organize themselves, mostly to reach specific objectives. In these organizations strategies appear as well as financial and hierarchical structures, HR and marketing policies, and cultures; to name just a few. Moreover, people are subservient to natural laws because they are part of a natural system. For instance, people need food, daylight, fresh air, and hygiene to stay healthy. In the provision of these important human needs FM can play a major role. In 2006 the network developed an agreed definition arguing that FM must be seen as "an integration of processes within an organization to keep and develop the agreed services which support and improve the effectiveness of its primary activities" (EN 15221-1, 2006), according to Mobach et al. (2014).

But there is still a high need to work on these topics. The literature on FM remains relatively scarce until today. This can be exemplified with a 2013 study in which we showed that insufficient research has been conducted on the relationship between the behavior of detainees and FM. From the literature research, the observations and

Facilities @ Management, Concept - Realization - Vision, A Global Perspective, First Edition. Edited by Edmond Rondeau and Michaela Hellerforth.

exploratory interviews, the conclusion was that FM does influence the behavior of detainees but that it hardly reaches practitioners. For instance, there is a great diversity between food and beverage supply systems in prisons. Why is this the case? If we have norms, why can we not deliver standardized services and still earn money? Take the fact that self-cooking for detainees contributes to better rehabilitation, but at the same time it creates problems about food safety. How to deal with such topics as a profession? What are our best practices when we also know that the cost of self-cooking is higher than that of meals that are cooked elsewhere in the prison and delivered into prison cells (Kuijlenburg et al. 2013). What do we learn and how does the profession use the knowledge? For instance, if we know that in comparison with waiting areas that are almost empty, customers in a waiting area with shopping facilities have more interaction with the physical environment, experience a shorter wait, were more satisfied with the prompt taking of orders, and spent more money (Mobach 2013). I believe that scholars and practitioners may have to be modest with expectations on progress and uptake. We must keep in mind that the built environment and related services of organizations are for most people like a façade: they do not know what is hidden inside. According to the ISO 41011 (2017), FM is the organizational function which integrates people, place, and process within the built environment with the purpose of improving the quality of life of people and the productivity of the core business. This seems a rather abstract definition that can benefit from examples to further clarify its meaning and potential impact on society.

In this context, in an interview with Facility Management Magazine, we described that the main challenge for the facility manager of the future is to create a connection of space and infrastructure with people and organization; always involving four different areas of interest: business, architecture, technology, and nature. The facility manager had to truly ease the primary process in his system and, above all, build in flexibility (Bekkering 2012). In the same year, FM had also to be about how to use spatial solutions to achieve business goals, such as promoting good collaboration or staff health (Visser 2012). In 2016, we found outcomes that suggested that a combination of the five characteristics defines an excellent facility manager: she or he has communication skills; acts results-oriented; is entrepreneurial; is sensitive to the needs of the organization; and demonstrates personal leadership (Roos-Mink et al. 2016).

In contrast, scientifically, the exact influence of the spatial environment on the function of humans in and around organizations has been surprisingly scarcely described in the context of FM. The scientific research proves to be fragmented and has revealed itself in very different scientific fields of study to decision makers in organizations. The main focus here will be on the coherence of the organization and the built environment – obviously through an interrelation of people, place, and process – and its supposedly related positive impact on the quality of life and the primary process. In this contribution I look to exemplify and elucidate the existing insights of FM and related fields and to inspire facility managers in practice. This approach should stimulate practitioners to better understand the workings of organization and space, and to have more positive impact on people in organizations.

Historical Context

FM does not only support organizations with building operations and related services that allow the workforce to work dry and comfortably, but in many cases, the profession actively looks to have positive impact on organizations and employees. For instance, to structure the work, to improve worker productivity, or to advance their health and wellbeing. The discipline may be very young, but related examples can be found throughout history. Buildings are more than only bricks, glass, and steel. The built environment has always been used to convey messages to others and to facilitate the primary process of organizations. History provides us with good examples of the coherent design of organizations and the built environment and its impact on humans. Pharaoh Kufu's Great Pyramid at Giza, the Flavian amphitheater by Vespasian in Rome, and the palace of Versailles by the sun king Louis XIV are just a few examples of how the built environment was deliberately used to communicate the power of a dominion. The FM industry may be relatively young, but it relates to many of these classical examples, and, by doing so, the field can rely on a rich history. Following some of my earlier work (Mobach 2009), I will describe three examples, the British Panopticon, the Larkin building, and the German AEG.

A built environment that serves the needs of organizations is exemplified in the Panopticon writings of the British Jeremy Bentham in the 18th century. Bentham looked to "create" better people during imprisonment and proposed an organizational-spatial design to do so. He envisaged a circular building with an inspector's lodge at the center and prisoner cells in the circumference. The cells had a window for lighting, during the night secured by small candles outside the window, and light iron grating. This allowed the inspector to view every single movement of the prisoner. The essence was that the prisoner never knew when he was being watched but was certain that it could be any time of day. As a result, there would literally be no space for undesired behaviors of the imprisoned and they would automatically turn into good people. Although the Panopticon was never built, Bentham's Panopticon remains influential until today.

The Larkin soap company manufactured toilet soap and detergents in Buffalo, US. Due to a mix of events Larkin grew substantially in the mail order business and the company was in urgent need of a new building. Frank Lloyd Wright was contracted and designed the Larkin Administration Building in 1903. The building has a strong coherence with the primary process. The key idea was that a clean, safe, and attractive office building would increase productivity and attract new highly qualified staff. This strong vision stimulated innovations and early adopting such as elevators, space for personal care and relaxation, climate control, air filtration, daylight, modern artificial light in darker areas, sound absorption, fire protection, and its particularly modern appearance. The building was also used to market Larkin as a company: tours were held, and much was written about the building by the company. As such, the building had great impact on staff and market.

In 1907, AEG a German electro-technical company invited Peter Behrens to Berlin for artistic advice on products and advertisements. Over the years, Behrens became directly responsible for the design of graphical work, products, and buildings, and, by doing so, he designed the entire corporate identity of AEG. For the first time ever, a designer influenced all industrial buildings and projects of a large corporation. From large buildings to tiny objects like ashtrays, he designed all. As such he is the first-known industrial designer. Famous are his turbine hall in steel and glass in Berlin from 1908 and the factory for light engines and the assembly hall at Wedding from 1913. With this initiative AEG stimulated innovation, not only within its company, but also in architecture. Behrens studio attracted and inspired many talented young people, among which were Mies van der Rohe, Le Corbusier, and Walter Gropius.

It seems that organizations and facilities have a lot in common. Construction or re-construction is therefore a wonderful opportunity for any organization. It can be used to support organizational design and change or to improve current organizational processes and outcome. For instance, to support the workplace, to reduce patient stress in a hospital, or to stimulate chance interaction on university campuses.

Contemporary Developments

Since COVID-19, a large proportion of the global workforce seems on drift with major implications for the value of the corporate real estate portfolio in various sectors. This development also blurs the boundaries between the work and the home, and which – let us be particularly aware of this, regardless of what has been argued in various disciplines over the last decades – has been relatively stable since the industrial revolution. Where the industrial revolution had created a strict and walled division between the worlds of home and work, COVID-19 has exposed the global workforce to new ways of working. These blurred lines between both worlds created unexpected possibilities, which were also reported to have worked surprisingly well. Consequently, hitherto strict dividing lines have become more fluid than ever before and, in some cases, even became obsolete.

For centuries, each morning many people left home and went to work. Both spaces were confined and pure. Employees commuted, worked, and lived mostly at their workplace. The factory that employed them, the office, or the shop. After working hours, they commuted again and went back home. Even though the digital transformation enabled working from home, the reality was that in many cases employees remained to work on-site for a large proportion of their working week. Since the pandemic, a dramatic shift in people's work routines and places had become an overnight reality. This has had an impact on their workplaces and routines until today. For many there is no clear identifiable "workplace" anymore, let say supposed routines with constituents like "each morning," "commuting," "working," or "pause." These days it is hard to imagine that we will go back to "business as usual." Corporates and many other organizations must learn to adapt to the fluidity of the current transitional state of the system and anticipate related insecurity and challenges.

For the FM profession this means that employees need support in other places than the places that were formerly known as the workplace at the office. A perspective of the worker journey may be key to better understand the wants and needs of end users and its relevance for performing a task. Especially how to facilitate people working best, also from home, with furniture, advice, and information and communications technology (ICT), and related implications and changes at the office.

Moreover, the FM industries are faced with global challenges. On behalf of the Research Advisory Committee of the International Facility Management Association (IFMA), Mobach et al. (2022) argue that buildings' environmental footprints are massive and need to be reduced. They contend that the construction and building sectors are responsible for one-third of global resource consumption and almost half of global waste. According to the authors some questions needs to be addressed soon, for example, what can facility managers do to decrease buildings' greenhouse gas impacts? How can facility managers reduce, narrow and close materials, energy, and water flow into and out of the built environment? How can the built environment and surrounding areas restore local biodiversity? (Mobach et al. 2022).

In the summer of 2022, IFMA Foundation was granted a special consultative status with United Nations Economic and Social Council (UN ECOSOC) as nongovernmental organization (NGO). In this context, it is argued that FM has the capacity to positively influence all 17 Sustainable Development Goals (SDG) of the UN. Matthews and Mobach (2023) contended that the logistic, event, and spatial ability of FM is highly relevant not only to avoid waste and promote responsible consumption and production, but also to help others with the redistribution of food and housing. FM may also contribute to the SDG with buildings that promote and advance health and well-being of users. As a network of researchers and educators, it can demonstrate its added value, and by doing so, the profession can contribute to a resilient built infrastructure, promote inclusive and sustainable industrialization, and foster innovation. FM can give these innovations among young people, students, and scholars; this allows the sector to learn and improve. With its workforces they can contribute to gender equality, stimulating fair and equal payment as a start, decent work and economic growth as a next step. Within its community, they can share their expertise to provide clean water and sanitation as well as affordable and clean energy and safe, peaceful, and resilient communities to allow this to reduce inequality within and among countries. FM can highlight how to act; with a built environment that is ready for excessive temperatures, water levels, winds, fires, and even pandemics. And with logistic systems with which it buys responsible non-pollutive and circular goods and services. Overall, according to Matthews and Mobach (2023), FM demonstrates that the strong ties within its community allow them to implement all SDG and to revitalize the global partnership for sustainable development.

So, our profession can be at the heart of positive change. The unique contribution of FM to the SDG is its role as a change agent in global society. There are four reasons why FM provides the UN with unprecedented opportunities to implement all SDG in practice. Firstly, because it has outstanding expertise and wide experience of building operations. This seems crucial as UN reports that, for instance, decarbonizing the buildings sector by 2050 is critical for decarbonization and requires immediate action (UNEP 2021). Moreover, it is argued that workers need effective support and action to better stimulate health at work (Yonah 2022). As such, the discipline can stimulate global change on a large variety of topics. Secondly, in this context, it is also crucial to know that FM has a focus on practical implementation. As a discipline it is used to make things happen, and to do so very precisely and in compliance with existing standards, norms, and procedures (e.g., ISO 41001; EN 15221). Thirdly, it has a key position in the decision making of buying and procurement processes in organizations. From decisions about pencils, desks, chairs, computers, buildings, food, cleaning, safety, and energy; discipline is involved. Fourthly, FM is omnipresent in the world. For the UN, this supplies a new and vibrant entrance for SDG-implementation, in an area that is surprisingly enough currently enormously underexplored.

Exemplification of Organization and Space

The research group Facility Management at Hanze University of Applied Sciences started its research in FM with its knowledge and practice partners in Groningen, the Netherlands, in 2012. Until today, the group looks to better integrate organization and space, with a special attention for the health and wellbeing of end users in the built environment. The organizational focus comprises general management and organization topics as well as a special focus on facility services. Over the years, the group has worked on a variety of topics, especially in healthcare, offices, education, and on cleaning. I will describe some of our group's main findings here.

Healthcare

In 2013, we started the living lab Health Space Design, in which scholars, students, and practitioners started to explore how space, infrastructure, and services can contribute to quality of care, more satisfied patients, residents, and staff, and greater efficiency of the work. In the same year, in an explorative study in the context of discovery a facility management perspective on organization and space was applied to explore its relationship with reported incidents of intellectually disabled residents in an institute. It was shown that reduction of stimuli can improve the quality of life of residents. Stimuli can occur both in organization and in space. Such stimuli, for instance, may be team changes and group accommodation. The study showed that in periods where there were an individual and stable team, a standardized approach and individual accommodations less incidents were reported than in periods with a group team, team changes, no standardized approach and accommodation of residents in group accommodations. Moreover, better adjustment of the individual care program to the characteristics of the residents supported the residents' control over spatial stimuli. Finally, it was very surprising to see that in a room for lunch and coffee and tea breaks a large number of stimuli were observed even though it is a well-known fact that these residents have serious problems coping with stimuli. These preliminary results suggested that a fit between the needs of residents with organizational and spatial designs can improve the interaction between residents and staff, and by doing so, positively influence the well-being of residents (Daatselaar et al. 2013).

In 2013 we also decided that relocation influences the wellbeing of nursing home residents. Positive or negative influences are determined by the organization of the relocation. Positive influences can be expected, for instance, if residents are mentally well prepared prior to the move, if extra staff is deployed to stay in close contact with residents during the move, and if facility managers develop after-care programs until residents are completely accustomed to the new situation (Tjeerdsma et al. 2013).

As of 2014, we explored if facilities planning at a diagnostic outpatient clinic can increase the level of space use and the speed of care delivery. This was confirmed. The actual space utilization level deviated from the planned utilization level. In this case the actual fluorodeoxyglucose (FDG) whole-body examinations on a positron emission tomography-computed tomography (PET-CT) scanner took less time than planned and, in addition, the weight of patients significantly influenced the actual examination times. Patients with a heavy body weight took more time than patients with low weight. Information on the properties of end-users, in this case the weight of the scanned patients, allows for better planning. Such information not only allows for better planning, but also reduces non-care-related activities that better meet the needs for increasing demand for care delivery; all promoting efficient space use at hospital buildings (Zijlstra et al. 2014).

In 2016, we started focusing on the patient journey in hospitals. Firstly, we performed wayfinding experiments with students wearing gerontological suits in a real-life hospital. We showed that people on more complex routes in hospitals (i.e., more floor and building changes) walked less efficiently than people on less complex routes. In addition, simulated elderly participants perform worse in wayfinding than young participants in terms of speed. Moreover, it was shown that simulated elderly persons had higher heart rates and respiratory rates compared to young people during a wayfinding task, suggesting that simulated elderly consumed more energy during this task (Zijlstra et al. 2016). In 2017, we confirmed that natural environments – in this case even digital images of nature – can positively influence people. In a study we investigated whether the use of motion nature projection (vs. no intervention) in computed tomography (CT) imaging rooms was effective in mitigating psycho-physiological anxiety using a quasi-randomized experiment. Motion nature projection had a negative indirect effect on perceived anxiety through a higher level of perceived pleasantness of the room. Moreover, heart rate and diastolic blood pressure were lower when motion nature was projected (Zijlstra et al. 2017).

In a series of studies starting in 2016, we determined that noise levels in hospitals were reported to be above WHO guidelines. Two main sources of noise were found: human-related and technical-related noises. The first being, for instance, socializing at the nurses' station, talking in hallway, television, noises of fellow patients (coughing, snoring, gagging, and moaning), closing rubbish bins, moving chairs and tables, doors (opening, closing, slamming). The latter being, for example, alarm of equipment, squeaking parts on beds or equipment as well as flushing toilet, shower use, rolling walker, telephone, vacuum cleaner, and air conditioning. We found that noise can have a negative influence on the wellbeing of patients and staff. We concluded that FM could stimulate noise-reducing measures through smarter procurement and better facility design (Roos-Mink and Mobach 2016). In 2019, we elaborated on these findings by employing a naturally occurring field experiment to assess the influence of a non-talking rule on the actual sound level and perception of patients in an outpatient infusion center. In the control condition, patients ($n = 137$) were allowed to talk to fellow patients and visitors during the treatment. In the intervention condition patients ($n = 126$) were requested not to talk to fellow patients and visitors during their treatment. This study measured the actual sound levels in dB(A) as well as patients' preferences regarding sound and their perceptions of the physical environment, anxiety, and quality of health care. We demonstrated a significant, but rather small reduction of the non-talking rule on the actual sound level with an average of 1.1 dB(A). Half of the patients preferred a talking condition (57%), around one-third of the patients had no preference (36%), and 7% of the patients preferred a non-talking condition.

Our results suggest that patients who preferred non-talking, perceived the environment more negatively compared to the majority of patients and perceived higher levels of anxiety. In conclusion, a non-talking rule of conduct only minimally reduced the actual sound level and did not influence the perception of patients (Zijlstra et al. 2019). In 2020, we did another study on sound in healthcare. We sought to gain a better understanding of the experiences of patients in an outpatient infusion center. Findings showed that patients perceived a lack of acoustic privacy and therefore tried to emotionally isolate themselves or withheld information from staff. In addition, patients complained about the sounds of infusion pumps, but they were neutral about the interior features. Patients who preferred non-talking desired enclosed private rooms and perceived negative distraction because of spatial crowding. In contrast, patients who preferred talking, or had no preference, desired shared rooms, and perceived positive distraction because of spatial crowding. Including these preferences in space-related treatment planning, for instance, asking patients in which room they would like to receive a treatment, can potentially improve patient outcomes (Zijlstra et al. 2020).

In 2016, we also explored the impact of FM on palliative care. Patients in the final stages of life need a last refuge that requires a higher standard when compared to regular healthcare environments. The spaces and service delivery processes at hospices seem to be optimal while in other (hospice) care settings users miss adequate spaces and services. In addition, management in care systems needs to reconfigure accordingly to offer flexible customization, e.g., allocation of staff. Several space and service requirements were identified, like domesticity, layout, style of décor, space for loved ones, quiet, and personal artefacts (Martens et al. 2016).

In 2022, we asked ourselves how architecture contributes to reducing behaviors that challenge? Behaviors that challenge might prevent intellectually impaired individuals from experiencing a good quality of life (QoL). These behaviors arise in interaction with the environment and can be positively or negatively affected by architecture. In a scoping review we explored how architecture contributes to the QoL of individuals engaging in such behaviors. The review showed that architecture, QoL, intellectual impairment, and behaviors that challenge have not yet been studied jointly (Roos et al. 2022). We decided to fill that gap with a new study (Roos et al. 2023).

Workplace

In 2012, we reported that in the industrial era, work and home were strongly separated. At work, there was really no room for homeliness. The general perception was that industry was destructive to human relationships. With the emergence of the human relations approach, even in factories there began to be increasing attention to domestic features to make the worker comfortable. At this time, hedonism has fully penetrated the office, a homely atmosphere complete with its own music and pets is an important part of this. With that, homeliness and work are currently very different from the last two centuries, no longer necessarily by definition to be understood as an antagonism. If used intelligently and deployed in moderation, the two seem to be mutually reinforcing at this moment in time. In this context, a very old love between the sphere of work and home seems to be reborn (Mobach 2012a).

In 2016, our research group FM started the living lab Healthy Workplace, in which scholars, students, and practitioners studied how office environments can contribute to healthy behaviors, and by doing so, healthy staff and organization. In 2016, we studied indoor environmental conditions in offices. We determined that office workers who experienced an indoor temperature of 20 °C graded this temperature the highest (6.7 on a scale of 1–10). At 20 °C the percentage of workers that was dissatisfied was the lowest (30%). The study also showed that female workers were more likely to have the sensation that it was too cold than male workers. European and Dutch standards prescribe that an indoor temperature between 21 °C and 23 °C should be the most ideal temperature during wintertime. This study indicates that an indoor temperature higher than 22 °C might be too warm for office workers in the Netherlands during wintertime and that application might influence office workers' satisfaction negatively (Brink et al. 2016).

As of 2015, we started to report on a series of studies in Activity-Based Working (ABW) environments. We first developed, tested, and validated a tool for measuring and optimizing workplace utilization, using work sampling and mobile technology. This new tool, "MyPlace2Work," uses a mobile application, to be installed on any smartphone or tablet, enabling on-the-spot collection of self-report data. Participants provided data about the activities they performed, the workplaces they used and associated levels of satisfaction, several times a day during a measuring period. This data were combined with the answers to a questionnaire covering several psychological and job characteristics. The tool had been tested extensively, resulting in some improvements of the software and a protocol for the application in field studies. Preliminary results confirmed the relevance and usability of the data for further research regarding the utilization of workplaces in ABW environments (Hoendervanger et al. 2015). In 2016, we found satisfaction ratings of 4% of the respondents who switched in offices several times a day appeared to be significantly above average. Switching frequency was found to be positively related to heterogeneity of the activity profile, share of communication work and external mobility (Hoendervanger et al. 2016). In 2018, we found significant correlates of satisfaction with ABW environments were found: need for relatedness (positive), need for privacy (negative), job autonomy (positive), social interaction (positive), internal mobility (positive), and age (negative). Need for privacy appeared to be a powerful predictor of individual differences in satisfaction with ABW environments. These findings underline the importance of providing work environments that allow for different work styles, in alignment with different psychological need strengths, job characteristics, and demographic variables. Improving privacy, especially for older workers and for workers high in need of privacy, seems key to optimizing the satisfaction with ABW environments (Hoendervanger et al. 2018).

In 2019 we found that perceived fit of person and environment is a function of activity, work setting, and personal need for privacy, with indirect effects on satisfaction with the work environment, and task performance. A misfit was perceived particularly among workers high in personal need for privacy when performing high complexity tasks in an open office work setting. Hence, we recommend that organizations facilitate and stimulate their workers to create better fits between activities, work settings, and personal characteristics (Hoendervanger et al. 2019). In 2022, as a follow-up we found that workers' perceived fitness was higher when they used to close rather than open work settings for individual high-concentration work. Furthermore, more frequent setting-switching was related to higher perceived fit. Unexpectedly, however, this relation was observed only among workers low in switching between activities. The findings indicated that user behavior may indeed be relevant to creating fit in activity-based work environments. To optimize workers perceived fit, it seemed to be particularly important to facilitate and stimulate the use of closed work settings for individual high-concentration work (Hoendervanger et al. 2022).

In 2021, we reported on the challenge of combining professional work and breastfeeding and why women choose not to breastfeed or to stop breastfeeding early in relation with FM. We posited that having access to a high-quality lactation room at the workplace could influence working mothers' satisfaction and perceptions related to expressing breast milk at work, which could have important longer-term consequences for the duration of breastfeeding. Specifically, we developed and validated a checklist for assessing the quality of lactation rooms and we found that objectively assessed higher-quality lactation rooms were associated with increased levels of satisfaction with the lactation rooms, perceived ease of milk expression at work, and perceived support from supervisors and co-workers for expressing milk in the workplace (van Dellen et al. 2021). As a follow-up, we reported that mothers exposed to the high-quality lactation room anticipated less stress, more positive cognitions about milk expression at work, more perceived organizational support, and more subjective well-being than mothers exposed to the low-quality lactation room in an online study in 2022. The effect of lactation room quality on perceived organizational support was especially pronounced for mothers who were higher in environmental sensitivity. In a field experiment we showed that use of the high-quality room led to less reported stress than use of the low-quality room. We also found that mothers who were higher in environmental sensitivity perceived more control over milk expression at work and experienced more subjective well-being in the high-quality condition than in the low-quality condition (van Dellen et al. 2022).

In 2023, we performed a COVID-19 study and reported that home workers had less favorable scores for concerns about and facilities of on-site buildings and workplaces upon return to work, but better scores for work quality and health than non-home workers. However, additional analyses also suggested that buildings, workplace, and related facilities may have had the capacity to positively influence employees' affective responses and work quality, but not always their health (Mobach and Onyeaka 2023).

Education

In 2013, we found indications that the perceived quality of facility services that are education-related and provide personal comfort to teachers have a positive relationship with study success. Layout, fitting out, and general facility services show no statistically significant relationship with study success, whereas (traditional) workplaces have a negative relationship. Also, we found that the size of the education institution strongly negatively relates to study success, and institutions with a Christian identity outperform non-Christian institutions (Kok et al. 2013). In 2015, we demonstrated a statistically significant positive relationship between the perceived quality of cleanliness, classroom conditions, front office, and ICT with study success. Closed environments like offices and meeting rooms, but foremost the size of the education institution, relate negatively to study success (Kok et al. 2015a). In the same year, we established that employees with management perspectives in higher education (i.e., top managers and facility managers) were significantly more positive about the facility design than frontline employees and their supervisors with providers' perspectives. Also, these providers attributed a more important role to facility design with respect to delivering interpersonal services than management did (Kok et al. 2015b).

In 2018, the research group FM started the living lab Campus Design, in which scholars, students, and practitioners sought to improve the impact of educational environments on different scale levels, for instance, from the design of university classrooms to campuses, on organizational and user outcomes. The latter dealt with the on-campus interactions between university and business employees that could be traced back to the design of spaces and services. In 2019, we started to publish a series of campus-related studies with a systematic review. We found evidence that spaces and services were mostly studied separately. Most studies were based on perceptions (surveys or interviews). We identified eight critical success factors: geographic proximity, cognitive proximity, scale, transitional spaces, comfort and experience, shared facilities and events, local buzz, and networks. These factors are interrelated. We presented a new relational model, from spaces and services, through interaction to innovation, visualizing how the identified studies were related (Selman et al. 2019). In 2021, as an empirical follow-up, we defined six categories of critical success factors that were reported to influence interaction on campuses – as identified by the facility directors of Dutch university campuses: constraints, motivators, designing spaces, designing services, building community, and creating coherence. The campus can be seen as a system containing subsystems and is itself part of a wider system (environment), forming a layered structure. Campus constraints and motivators are part of the environment, but cannot be separated from the other four categories, as they influence their applicability (Jansz et al. 2021). In 2022, we found four principal components for services (Relax, Network, Proximity and Availability) and three for locations (Aesthetics, Cleaned and Indoor Environment).

Personal characteristics as explanatory variables were not significant or only had very small effect sizes, indicating that a campus' design does not need to be tailored to certain user groups but can be effective for all. The pattern of successful locations was discussed, including the variables in each principal component. These principal components provide a framework for practitioners who want to improve their campus' design and to stimulate unplanned meetings, thus contributing to cooperation between different campus users, hopefully leading to further innovation (Jansz et al. 2022). In 2023, as a final step in these series of studies, we aimed to identify why some campus locations were successful in fostering unplanned meetings while others were not. We found three main themes: function (food, drinks, events, work, facilities), space (distance, experience, accessibility, characteristics), and organization (coherence, culture, organization). Time was an overarching constraint, influencing all other themes. There were three natural moments for unplanned meetings: during short breaks, lunch breaks, and events. The outcomes were used to propose a 5-minute campus as the environment of interaction; a campus where natural moments, locations, and travel time for unplanned meetings are designed and aligned: under 5 minutes walking for short workplace breaks, approximately 5 minutes travel time for lunch breaks, and over 5 minutes travel time for events, depending on the event length and anticipated knowledge gain (Jansz et al. 2023).

In 2020, we asked ourselves whether phone pods and office booths, hereafter referred to as pods, had proven their added value and popularity in open-plan offices, but how would that work in another context, such as in higher education? We found that students use these pods mainly for seven activities: meeting, project work, noise-free work, study, phone call, relaxation, or hang out. Students report a positive general experience of the pod, a very positive experience when entering the pod, and hardly any negative experiences. They feel at ease and the pods ensure better concentration. Finally, pod users reported to be a little less nervous than other atrium users (Offringa et al. 2020)

In 2021, we started publishing a series of studies on classrooms. We performed a systematic review to determine the influence of four indoor environmental parameters – indoor air, thermal, acoustic, and lighting conditions – on the quality of teaching and learning and on students' academic achievement in schools for higher education, defined as education at a college or university. The collected evidence showed that the indoor environmental quality (IEQ) can contribute positively to the quality of learning and short-term academic performance of students. However, the influence of all parameters on the quality of teaching and the long-term academic performance could not be determined yet. Students perform at their best in different IEQ conditions, and these conditions are task-dependent, suggesting that classrooms which provide multiple IEQ classroom conditions facilitate different learning tasks optimally (Brink et al. 2021). In 2022, in a follow-up, we developed and validated a systematic approach to measure the perceptions, internal responses, and short-term academic performance of participating students in higher education.

During the pilot study, the IEQ of the classrooms varied slightly. Significant associations ($p < 0.05$) were observed between these natural variations and students' perceptions of the thermal environment and indoor air quality. These perceptions were significantly associated with their physiological and cognitive responses. Furthermore, students perceived cognitive responses were associated with their short-term academic performance (Brink et al. 2022). In 2023, we showed that reduction of the reverberation time positively influenced students' perceived cognitive performance. A reduced reverberation time in combination with raised horizontal illuminance improved students' perceptions of the lighting environment, internal responses, and quality of learning. However, this experimental condition negatively influenced students' ability to solve problems, while students' content-related test scores were not influenced. This shows that although quality class A conditions for reverberation time and horizontal illuminance improved students' perceptions, it did not influence their short-term academic performance (Brink et al. 2023)

Cleaning

In 2018, we started a series of studies on cleanliness. We identified actual cleanliness, staff behavior, condition of the environment, scent, and the appearance of the physical environment as variables influencing perceived cleanliness and service quality. Moreover, the presence of litter, behavior and presence of others, scent, disorder, availability of trash cans and informational strategies were identified as stimuli affecting littering and other kinds of unethical behavior. Finally, the effect of perceived cleanliness on satisfaction, approach behaviors, physical activity, and pro-social behavior was registered (Vos et al. 2018a). Moreover, in another study we found that for actual cleanliness, cleaning staff behavior and the appearance of the environment were identified as the three main antecedents of perceived cleanliness. Client organizations tend to have a stronger focus on antecedents that are not related to the cleaning process compared to facility service providers (Vos et al. 2018b). In 2019, we also found that the presence of cleaning staff positively influenced train passengers' perceptions and satisfaction. Effects were stronger in the second study, after the second consecutive intervention (i.e., hospitality training, corporate uniforms). In both studies, the presence of cleaners positively influenced passengers' perceptions of staff, cleanliness, and comfort. The perception of atmosphere was only significant after the intervention (Vos et al. 2019a). In 2019, we also validated a new instrument to measure cleanliness perceptions.

The cleanliness perceptions scale (CPS) is more comprehensive than existing scales, which focus on how customers perceive the cleanliness of specific interior elements, thereby ignoring other dimensions, including the aesthetic quality of a service environment. Whereas existing scales are intended for restaurants and hotels, the CPS was specifically developed for the facilities management industry, which is responsible for cleanliness in a wide variety of service environments. A qualitative study followed by two quantitative studies resulted in the twelve-item CPS, which covers the three dimensions of perceived cleanliness: cleaned, fresh, and uncluttered (Vos et al. 2019b). In 2019, from another study of our team we learned that cleaning in a special context – in detention – had a positive impact on meaningful daytime activities and the autonomy of detainees. The work gives satisfaction, and detainees reported appreciation from others and a better day and night rhythm (Kuijlenburg et al. 2019). In 2020, we used the CPS and found that smooth seating materials and uncluttered architecture positively influenced perceived cleanliness (Vos et al. 2020).

Virtual Reality

In 2008, I asked myself: "Do virtual worlds create better real worlds?" (Mobach 2008). At the time the answer was "yes," virtual reality can help predicting the interactions between space and organization, and by doing so, improving the functionality of spaces for organizations. What happened since 2008?

In 2010, we expected demands for spatial visualizations to grow. In many areas of organized human activities organizations were expected to turn away from textual and numerical flatlands, and to rely on the convenient and multidimensional digital worlds. Virtual worlds for facility management, design, and planning were said to be no exception, it has an enormous potential to help organizations find the right spaces that fit the human activities they perform. However, a major take-up of virtual worlds in this context allowing a comparison between present and future, was still to come. Perhaps such applications, interweaving virtual and real worlds to design better facilities were at their beginning stages. One thing was clear: sophisticated applications may have remained absent until 2010, but it will come to us. Digital worlds will start to normalize, and the design of organizational spaces can benefit from that development. It was concluded that data on organizational performance serve as a linking pin between facility management and virtual worlds. Interaction can thus be improved by using organizational data as "subtitles" which stimulate a more active use of visualization (Mobach 2010). In an interview with Business Haaglanden, I reported that visualizations could encourage new ways of working, promote communication and collaboration in a design group, and make a design more flexible and sustainable. We demonstrated the importance to organizations of visualizations in the early stages of a design process (Visser 2012).

Ever since and nuances notwithstanding (Mobach 2012b), our studies have consistently confirmed the relevance of virtual reality for research and practice purposes (Vos et al. 2015; Hoendervanger et al. 2019; Siegelaar et al. 2020; Zijlstra et al. 2023). I can only advise scholars and practitioners to use virtual reality as a tool to predict the impact of design on organizational performance. And always to use and combine data of the primary process of the organization, the built environment (current building and new building), and related perceptions of end-users, to discuss the expected functionality of the design and change whenever necessary. Especially, I advise to focus on combining and comparing objective data with perceptions. In most cases it really helps to build better buildings for organizations.

Conclusion

As we have learned, the relevance of FM for organizations is often associated with better performance, of which productivity, cost reduction, health, and wellbeing are prominent ones. However, we must also be aware that it has particular relevance for us as experts. In contrast, for many people, the built environment plays no role at all; they do not notice it. They walk through spaces or work in it, nothing more. Some may argue that the influence of a physical layout, for example, through its size and shape, would be limited to the positioning of desks and machines, the flow of goods, and the sequence of tasks. Some may even contend that it would only be used to control and manage workers and not to stimulate or motivate them. In this line of thought a wall cannot motivate people as space is incapable to do so. In this perspective, only a very limited number of physical factors, such as temperature and light would directly affect employees. For some decision-makers in organizations, the built environment and related services do not make the workforce any better. It just represents cost.

Not everyone is therefore convinced in advance of possible positive impact, indeed many focus on costs and risks. And yes, human behavior is always of decisive relevance for organizations; and always will be. Some even coined the building as merely stone, mortar, and sophisticated technology; life is breathed into it by the people. This may be true for many organizations where the building is nothing more than a box to work dry, such as a hangar of airline; indeed, the fact is that an aircraft must fit in it and the mechanics must be able to work in and around the aircraft. Protected from wind and weather, no more and no less. But it is also true that an individual could function independently, without influence of such environmental conditions is highly unrealistic. In many cases, spaces in itself may hardly bring positive change to the organization, but without spatial changes positive change seems also unlikely.

Thus, the relevance of the built environment and related facility services for organizations should also not be overestimated. One could compare the relationships between buildings and organizations with marriage. You can live in a bad house, but if you later live in a villa and the relationship has become broken, you may look back with nostalgia to the time with the poor housing conditions. That may sound a bit strange in a book about FM, but it is always important to remain critical and nuanced. Only with the right realism and rationalism, which are so common in organizations, the relevance of FM will become clearer and better understood. The profession has to go to the heart of the organization. When decision makers are keen to invest more, we are on the right track. Despite these nuances, there is increasing attention at decision makers for this hidden but omnipresent factor of FM, with the potential to have a very positive impact on organizations globally. It is evidenced, we need to act now.

References

Bekkering, P. (2012). Facility manager moet bovenal flexibel zijn. *Facility Management Magazine* 25 (202): 11–15.

Brink, H.W., Mobach, M.P., Balslev Nielsen, S., and Jensen, P.A. (2016). Quality and satisfaction of thermal comfort in Dutch offices [Paper presentation]. *15th European Facility Management Conference*, Milan, Italy.

Brink, H.W., Loomans, M.G.L.C., Mobach, M.P., and Kort, H.S.M. (2021). Classrooms' indoor environmental conditions affecting the academic achievement of students and teachers in higher education: a systematic literature review. *Indoor Air* 31 (2): 405–425.

Brink, H.W., Loomans, M.G.L.C., Mobach, M.P., and Kort, H.S.M. (2022). A systematic approach to quantify the influence of indoor environmental parameters on students' perceptions, responses, and short-term academic performance. *Indoor Air* 32 (10).

Brink, H.W., Krijnen, W.P., Loomans, M.G.L.C. et al. (2023). Positive effects of indoor environmental conditions on students and their performance in higher education classrooms: a between-groups experiment. *Science of the Total Environment* 869.

Daatselaar, R., Schaap, M., Mobach, M.P., and Alexander, K. (2013). Added value of facility management in institutes for intellectually disabled residents (with a severe behavioural disorder) [Paper presentation]. *12th European Facility Management Conference*, Prague, Czech Republic.

EN 15221-1:2006 (2006 October 24). *Facility Management - Part 1: Terms and definitions.* British Standards Institution.

Hoendervanger, J.G., le Noble, V., Mobach, M.P. et al. (2015). Tool development for measuring and optimizing workplace utilization in activity-based work environments [Paper presentation]. *14th European Facility Management Conference*, Glasgow, UK.

Hoendervanger, J.G., Mobach, M.P., De, B.I. et al. (2016). Flexibility in use: switching behaviour and satisfaction in activity-based work environments. *Journal of Corporate Real Estate* 18 (1): 48–62.

Hoendervanger, J.G., Ernst, A.F., Albers, C.J. et al. (2018). Individual differences in satisfaction with activity-based work environments. *Plos One* 13 (3): 0193878.

Hoendervanger, J.G., Van Yperen, N.W., Mobach, M.P., and Albers, C.J. (2019). Perceived fit in activity-based work environments and its impact on satisfaction and performance. *Journal of Environmental Psychology* 65.

Hoendervanger, J.G., Van Yperen, N.W., Mobach, M.P., and Albers, C.J. (2022). Perceived fit and user behavior in activity-based work environments. *Environment and Behavior* 54 (1): 143–169.

ISO 41011 (2017). *Facility management — Vocabulary, ISO 41011:2017 defines terms used in facility management standards*. Publication date 2017-04, Edition 1. 11, Technical Committee, ISO/TC 267 Facility management.

Jansz, S.N., van Dijk, T., and Mobach, M.P. (2021). Facilitating campus interactions – critical success factors according to university facility directors. *Facilities* 39 (9–10): 585–600.

Jansz, S.N., Mobach, M., van Dijk, T. et al. (2022). On serendipitous campus meetings: a user survey. *International Journal of Environmental Research and Public Health* 19 (21).

Jansz, S.N., Mobach, M., van Dijk, T., and Tchounwou, P.B. (2023). The 5-minute campus. *International Journal of Environmental Research and Public Health* 20 (2).

Kok, H.B., Mobach, M.P., Omta, S.W.F., and Alexander, K. (2013). Can facility management contribute to study success? [Paper presentation]. *12th European Facility Management Conference*, Prague, Czech Republic.

Kok, H., Mobach, M., and Omta, O. (2015a). Predictors of study success from a teacher's perspective of the quality of the built environment. *Management in Education* 29 (2): 53–62.

Kok, H., Mobach, M., and Omta, O. (2015b). Facility design consequences of different employees' quality perceptions. *The Service Industries Journal* 35 (3): 152–178.

Kuijlenburg, R., Mobach, M.P., and Alexander, K. (2013). The influence of facility management on detainees [Paper presentation]. *12th European Facility Management Conference*, Prague, Czech Republic.

Kuijlenburg, R., Offringa, J., Brouwer, V., and Mobach, M.P. (2019). *De toegevoegde waarde van facility management voor detentie*. Service management, 6 februari.

Martens, R.M.G., Mobach, M.P., Balslev Nielsen, S., and Jensen, P.A. (2016). Spaces and services in Dutch hospice care [Paper presentation]. *15th European Facility Management Conference*, Milan, Italy.

Matthews, A. and Mobach, M.P. (2023). FM's impact on U.N.'s Global Challenges. *FMJ (Facility Management Journal)* 33 (2): 90–93.

Mobach, M.P. (2008). Do virtual worlds create better real worlds? *Virtual Reality* 12 (3): 163–179.

Mobach, M.P. (2009). *Een organisatie van vlees en steen*. Van Gorcum.

Mobach, M.P. (2010). Virtual prototyping to design better corporate buildings. *Virtual and Physical Prototyping* 5 (3): 163–170.

Mobach, M.P. (2012a). Huiselijke faciliteiten in organisaties. *Holland, Historisch Tijdschrift* 44 (3): 179–184.

Mobach, M.P. (2012b). Interactive facility management, design, and planning. *International Journal on Interactive Design and Manufacturing* 6 (4): 241–250.

Mobach, M.P. (2013). The impact of physical changes on customer behavior. *Management Research Review* 36 (3): 278–295.

Mobach, M.P., Nardelli, G., Kok, H. et al. (2014). Facility management innovation. *13th European Facility Management Conference*, Berlin, Germany.

Mobach, M.P., Sanquist, N., and Saunders, J. (2022). *Seeking higher ground: navigating the FM industry's transformation*. Houston: International Facility Management Association (IFMA).

Mobach, M.P. and Onyeaka, H. (2023). Workplace impact on employees: a lifelines corona research initiative on the return to work. *Plos One* 18 (1).

Offringa, J., Roos-Mink, A., Roosjen, M.A., and Mobach, M.P. (2020). Living in a pod: the impact of tiny spaces on a Dutch university campus [Paper presentation]. *European Facility Management International Conference (EFMIC) 2020*, online.

Roos, B.A., Mobach, M., and Heylighen, A. (2022). How does architecture contribute to reducing behaviours that challenge? A scoping review. *Research in Developmental Disabilities* 127.

Roos, B.A., Mobach, M., and Heylighen, A. (2023). Challenging behavior in context: a case study on how people, space and activities interact. *Health Environments Research & Design Journal* 16. Accepted, forthcoming.

Roos-Mink, A. and Mobach, M.P. (2016). Hospital noise. In: *Hospital Healthcare Europe*, 125–127. Hospital Healthcare Europe publisher.

Roos-Mink, A., Offringa, J., de Boer, E. et al. (2016). A research-based profile of a Dutch excellent facility manager [Paper presentation]. *15th European Facility Management Conference*, Milan, Italy.

Selman, J., Spickett, J., Jansz, J., and Mullins, B. (2019 March). *Confined space rescue: A proposed procedure to reduce the risks*. Safety Science, Elsevier. Crown Copyright © 2018 Published by Elsevier Ltd. All rights reserved.

Siegelaar, A., Boersma, F., Zuidema, S.U., and Mobach, M.P. (2020). Controlling the stimulation of senses in design for dementia [Paper presentation]. *European Facility Management International Conference (EFMIC) 2020*, online.

Tjeerdsma, A., Mobach, M.P., and Alexander, K. (2013). Relocating a nursing home [Paper presentation]. *12th European Facility Management Conference*, Prague, Czech Republic.

UNEP, United Nations Environment Programme (2021). *2021 Global status report for buildings and construction – towards a zero-emissions, efficient and resilient buildings and construction sector*. Nairobi: United Nations Environment Programme.

van Dellen, S.A., Wisse, B., Mobach, M.P. et al. (2021). A cross-sectional study of lactation room quality and dutch working mothers' satisfaction, perceived ease of, and perceived support for breast milk expression at work. *International Breastfeeding Journal* 16 (1).

van Dellen, S.A., Wisse, B., and Mobach, M.P. (2022). Effects of lactation room quality on working mothers' feelings and thoughts related to breastfeeding and work: a randomized controlled trial and a field experiment. *International Breastfeeding Journal* 17 (1).

Visser, M. (2012). Facility management: De link tussen ruimte en organisatie. *Business Haaglanden* 15 (6): 43.

Vos, M., Mobach, M.P., and Kok, H.B. (2015). *Light and the perception of safety, cleanliness and ambience at train stations*. Master thesis. Wageningen University.

Vos, M.C., Galetzka, M., Pruyn, A.T.H. et al. (2018a). Cleanliness unravelled: a review and integration of literature. *Journal of Facilities Management* 16 (4): 429–451.

Vos, M.C., Galetzka, M., Mobach, M.P. et al. (2018b). Exploring cleanliness in the dutch facilities management industry: a delphi approach. *Facilities* 36 (9–10): 510–524.

Vos, M.C., Sauren, J., Knoop, O. et al. (2019a). Into the light: effects of the presence of cleaning staff on customer experience. *Facilities* 37 (1–2): 91–102.

Vos, M.C., Galetzka, M., Mobach, M.P. et al. (2019b). Measuring perceived cleanliness in service environments: scale development and validation. *International Journal of Hospitality Management* 83: 11–18.

Vos, M., Galetzka, M., Mobach, M.P. et al. (2020). Cleaning with services and spaces: effects of seating materials and architectural clutter on perceived cleanliness [Paper presentation]. *European Facility Management International Conference (EFMIC) 2020*, online.

Yonah, A. (2022 February 2). Healthy and safe telework: a WHO/ILO technical brief. *World Health Organization* 13: 304.

Zijlstra, E., Mobach, M., van der Schans, C. et al. (2014). Facilities planning promoting efficient space use at hospital buildings [Paper presentation]. *13th European Facility Management Conference*, Berlin, Germany.

Zijlstra, E., Hagedoorn, M., Krijnen, W.P. et al (2016). Route complexity and simulated physical ageing negatively influence wayfinding. *Applied Ergonomics* 56: 62–67.

Zijlstra, E., Hagedoorn, M., Krijnen, W.P. et al. (2017). Motion nature projection reduces patient's psycho-physiological anxiety during ct imaging. *Journal of Environmental Psychology* 53: 168–176.

Zijlstra, E., Hagedoorn, M., Krijnen, W.P. et al. (2019). The effect of a non-talking rule on the sound level and perception of patients in an outpatient infusion center. *Plos One* 14 (2).

Zijlstra, E., Hagedoorn, M., Lechner, S.C.M. et al. (2020). The experience of patients in an outpatient infusion facility: a qualitative study. *Facilities* 39 (7–8): 553–567.

Zijlstra, E., van der Zwaag, B., Kullak, S. et al. (2023). *The impact of design principles on patients*. Under review.

Mark P. Mobach, PhD
Professor
Hanze University of Applied Sciences
Groningen, Netherlands

Mark Mobach, PhD, is a professor in Facility Management at Hanze University of Applied Sciences Groningen, a leading professor of the Hanze Research Centre for Built Environment, and professor of Spatial Environment and the User at The Hague University of Applied Sciences, the Netherlands. He has studied and worked at the University of Groningen (the Netherlands), where he took his PhD cum laude, the University of Stockholm (Sweden), and Wageningen University (the Netherlands).

With his research team in Facility Management, he seeks to create better buildings for people. They focus on positive change: better buildings for people and better coherence between spaces and services. There is particular interest in the advancement of Facility Management and of the health and well-being of users in the built environment. With his research team Mark is involved in research projects of, for instance, healthcare, offices, education, and cleaning. Projects are always multidisciplinary – across a wide range of academic disciplines – and preferably on the crossroads of research, practice, and education. Knowledge development in facility management is combined with disciplines such as (interior) architecture, building physics, medicine, nursing, human movement sciences, organization science, real estate, psychology, and spatial sciences.

Mark Mobach is program leader of the research program Healthy Buildings, with a focus on education, healthcare, and offices. In 2019 Mark received the prestigious Dutch Deltapremie for his research in FM; the same year the research center he affiliates was rated excellent. He is actively involved in various facility management projects and is also an invited speaker to conferences and master classes relevant to the field. By doing so, with his team he seeks to advance and disseminate knowledge and innovations in facility management regionally, nationally, and internationally. He has held and holds several positions relevant to the field, e.g., at FMN, EuroFM, and IFMA. He is currently member of the IFMA Research Advisory Committee, member of the EuroFM Scientific Advisory Committee, and member of the United Nations Task Force of IFMA Foundation. In 2023, Mark Mobach was elected a full member of Sigma Xi.

3.7

Workplace Evolution in Ireland and Beyond

Bernie Gorman

World events of the past few years have seen seismic shifts in how and where people work. We are in the midst of a workplace evolution with both challenges and opportunities for the built environment and in order to examine where we are headed we must first look to identify the drivers of change and how we use this to re-imagine the customer journey.

The economy, technology, society and the environment are all key drivers for change. The large-scale disruption and crisis that was the COVID-19 pandemic affected multiple drivers all at once and triggered a colossal change in facility management (FM). The pandemic response resulted in an overnight shutdown of offices and workplaces globally. Workers set up temporary workspaces in the home and businesses scrambled to get them the supplies needed. Demand for IT equipment and home office furniture outstripped supply. Access to global supply chains for goods and services was severely impacted.

Over time, workers advanced from the initial juggling of the multiple demands of working from home, childcare, home-schooling and family life to become organized and structured in a new way to support their new normal. Temporary home office setups evolved into more permanent, personal workspaces and with the daily commute a thing of the past workers eased into a new rhythm. Many offices, however, remained near empty or closed.

In terms of FM, this represents a challenge for companies, as office space is being underutilized. Spaces intended for 100+ people are being used by 5 people, resulting in a lot of dollars being spent on empty space. Companies are working on analyzing space utilization to optimize for employee usage, while also examining the overall property's functionality. Focus for companies will be on right-sizing the properties that meets the future needs of the workforce and boosts productivity and employee satisfaction.

Today, in addition to on-site working, remote working and hybrid working models are the new normal. "According to *Forbes*, 32% of employees want to work fully remote, compared to 59% who choose hybrid, and just 9% who want to return to the all-office schedule in the future." Although there are some geographical variances in statistics, broadly speaking the message remains the same. Some people want to work remotely all of the time and some people want to return to the office some of the time. What is not explicit from any of the data is what impact decisions being taken now will have on the workforce in the future.

What is clear, however, is that in order to align the employee experience with an organization's operational investments, a precise and effective return-to-office and long-term real estate strategy is required. Creating a framework replaces guesswork and provides an opportunity for people-centric experiences that inspires, retains, and attracts talent and drives company performance and growth. Along with workplace and culture, company policy is a key driver for how employees engage and interact. Employers must focus on the attributes that constitute a positive employee experience and do more of that.

Facilities @ Management, Concept - Realization - Vision, A Global Perspective, First Edition. Edited by Edmond Rondeau and Michaela Hellerforth.

What is also certain is that it does not work to mandate employees return to the office. Companies that offer autonomy and choice of where and when to work have more productive and engaged employees with higher retention rates. Teams should be encouraged to come together to develop their own schedules and plans for office attendance that works for all.

Recently we have seen companies such as Amazon, Apple, and Disney issue return to work mandates to employees instructing them to be in the office 3–4 days in the week. It is reported that there have been threats to employees to terminate their employment for noncompliance. Employees, however, are pushing back on plans, organizing and filing petitions against the change. This type of arbitrary action does not seem to serve the employee or the company.

How we work, where we work, and what we do at work have changed, and so too must the office. We are seeing a complete paradigm shift and with this our thinking must also evolve. Offices are now essentially an "on demand" resource. In mapping the office of the future we must recognize that one size does not fit all. We must embrace this opportunity to re-imagine our customer journey and re-define the purpose of the office.

Hybrid working will not only impact offices but will have a ripple effect on cities too. With a reduced appetite for commuting into city centers, we are seeing the adoption of the Hub and Spoke model in some areas. With the headquarters as a hub and smaller offices or co-working spaces near where people live. This change in concept, design and location will impact how our cities look in the future.

For employees, flexibility is a top priority, with focus work being predominantly carried out at home and the office being used as a space to reconnect. The office of the future will be a social destination, a place to build social capital for the organization, a space to collaborate and connect. It will be a space for communities to come together.

Companies are seeing difficulties around connectivity and striking the balance between digital connection versus the one to one in person connection. Leadership and mentorship is presenting an increasing challenge of how younger workers can learn if they are not benefiting from learning by showing or doing. The key reason for any return to the office is connection. Connection to colleagues for conversation, collaboration, innovation, and socializing. Connection to company culture and connection to learning.

We have seen in the past that technology has been a key driver for our way of working. Where once large desk bound computers dominated the landscape and offices were built to support this technology, technology now too provides new and exciting ways to leverage human interaction. The growth of digital infrastructures and digital technology integrations with traditional technologies are changing the way people live and work all around the world.

Disruptive technologies such as AI (Artificial Intelligence), robotics, apps, blockchain, and cloud software will be part of our everyday workplace lives and will revolutionize the way we work, bringing advancements to increase efficiency and productivity, promote innovation, create competitive advantage and improve value chain and agility. We can expect to see an acceleration of digital transformation with AI augmenting all of our business processes and jobs.

Workplace design will also be influenced by these new technologies. Smart buildings are using interconnected technologies to make buildings more responsive and intelligent, ultimately improving performance and facilitating better decision making. IoT (Internet of Things) sensors for example can be used to support office design optimization through learning employee work habits.

Wellness, mental health, and work–life balance have become hot topics in the workplace. Intentional workplace design supports feelings of wellness, calm and comfort for the people using the space. It promotes physical and mental health and well-being, ultimately supporting positivity and productivity for the users. Workplace designers are integrating biophilic design which is a design approach that seeks to connect the occupants of the building more directly to the natural environment by incorporating natural lighting, ventilation, and landscape features, underpinning the positive effects that human–nature connectedness can bring to the occupants and the environment.

Similarly, considering neurodiversity in workplace design will help organizations be more inclusive to an increasingly neurodiverse workforce. Post pandemic we are also seeing workers without typical neurodiverse conditions, but who are hypersensitive to their surroundings. For example, having low-stimulation environments for hyper focus, social spaces for stimulation breaks and low traffic areas to ease social anxiety can help neurodivergent workers to thrive.

As the reasons for employees coming into the office changes, we will see a shift in office design and services. Offices as an "on demand" resource may operate more as a specialist hospitality/concierge service. Employees may not have an assigned desk but instead come together to interact, collaborate, and brainstorm. This would see a reduction in dedicated desk space and an increase in collaboration spaces and meeting rooms with large floor to ceiling video walls for enhanced visibility and communication.

For now, we find ourselves with a foot in two worlds. The world that we knew pre-pandemic, where we predominantly all worked in the office and where the workplace, its design, its services and how we culturally interacted was contained in a neat Monday to Friday, 9 am–5 pm structure. Where we commuted to and from work in rush hour traffic, rushed to care for children, cook dinner, prepare for the next day before falling into bed and doing it all over again.

Post-pandemic where schedules and working days vary, with some people working from home some of the time and some people working at home all of the time. The commute to the office is a quick walk downstairs and the time previously spent in traffic is invested in other areas of life. There is no neat one size fits all structure in which to operate. Everyone has different requirements. Therein lies the challenge for FM and organizations.

As we move between worlds, while the structure of one breaks down and the other is not yet fully set up, change management will be vital as we realign the workplace and the workplace experience. Companies may transition from single to multiple locations and new technology may open up ways for us to work in new ways. Focused change programs will be required to ensure successful outcomes.

Employees, broadly speaking, benefit from remote working. They have more time because they have less of a commute, are better off financially, have flexibility to live where they choose, have flexibility in how they spend their time, have flexibility to choose their work location, and have flexibility on their career options.

Still there are challenges to working remotely. Feelings of loneliness and isolation, difficulty staying motivated, difficulty with collaboration and communication, working more, and not being able to unplug with no separation between work and home life. Understanding the complexity of challenges faced by employees is key for FM and organizations when making decisions for the future of work and the future of offices.

Aside from managing the complexities of change around the built environment, sustainability will move to the top of the agenda this year. We are likely to see legislation for business around environmental, social, and corporate governance (ESG). Conscious consumers and conscious investors will assess organizations on their environmental and social impact.

Companies will need to weave sustainability into the fabric of their business, examining their business processes and supply chain to remove or minimize any negative effects of their business. Additionally, companies will demand that future workplaces expand their operational efficiency, energy management, waste reduction and improve procurement practices and strategies in line with sustainability goals.

As economic instability continues, with the continuing fallout from the pandemic, rising inflation, rising energy prices, recession and war, companies are working on strategies to reduce exposure and improve resilience. They are examining the way that they work, shifting focus to skills not roles and examining what competencies are needed for future competitive advantage.

We will likely see less hierarchical structures in favor of more fluid flow. We will also likely see a shift to workers working for multiple organizations, offering a particular set of skills to fulfill a need at a particular point in time.

History informs the future. We can see the evolutionary pattern of work, workplaces, and the built environment over time. Looking at the cycles of change brought about by the First, Second, & Third Industrial Revolutions. We see a common theme appear across all of the cycles, themes that we can also see in our current evolution.

- New technologies are discovered.
- Increased demand for labor
- Increase in production and consumption.
- Changes to human settlement, work and family life
- Workers organize to defend their interests.
- New civil and political rights
- Workplace evolution

The First Industrial Revolution in the late 1700s saw the advent of the use of steam power and mechanization of production. The shift to larger scale mechanized agriculture overtook the subsistence farming previously practiced.

New technologies in the iron, steel, transport and cotton textiles industries, created a growing demand for factory labor. The introduction of new technology also triggered changes to patterns of human settlement, work, and family life. As factories expanded, people were needed to manage the workers and the administration around the factory. It is here that office work and offices became an extension of the factory.

The Second Industrial Revolution (Technological Revolution) followed from the late 1800s. This was a time of huge scientific discovery, mass production, and standardization. Advances in steel production, the automobile, and electricity fueled the production of mass produced goods which could now be transported to expanded territories. The need for offices grew with the expansion of factories and rail networks. Planes, trains, and automobiles made it easier to get around and communication was also made easier through newspaper, radio, telegraph, and telephone.

From the late 1900s the Third Industrial Revolution (Digital Revolution) saw the shift from mechanical and analogue electronic technology to digital electronics. Here we see the rise of digital technology, the invention of the internet and home computing, both of which transformed how we produce and deliver goods and services and how we do business and interact in the world. For offices, large desktop computers, fax machines, printers, scanners, etc. dictated how they were designed and utilized.

Today our computers and devices are mobile. Workers do not need to go into the office to do their work. Technology again is a driver for change. As we move through the latest evolutionary changes, FM leaders will need to develop a workplace strategy which aligns key business interests across, Finance, HR, IT, and other key business stakeholders which serves both the immediate and future needs of the business. The strategy to measure the employee workplace experience should be evidence-based and data-driven and should look to align operational investment. What an amazing opportunity to reimagine the customer journey and the future for the built environment!

Bernie Gorman
Sr. Facilities Manager
International Real Estate Operations
WPEngine, Inc.
Limerick, Ireland

Bernie Gorman brings 15 years of Project Management experience, in senior positions, collaborating with private clients and businesses ranging from small startups to large corporations across multiple industries.

With a proven track record in managing workplace projects and programs across multiple geographies and time zones through all aspects of the process from planning to delivery and handover.

Bernie is passionate about delivering creative, vibrant workplace environments that reflect individual and team needs, supports company culture, supports community, inclusion, and connectedness, and in turn brings the company core values to life in the physical workplace.

Experienced at working effectively cross functionally with multiple stakeholders to ensure projects are delivered on schedule and within budget, Bernie optimizes outcomes for all and positively contributes to the bottom line and company growth and expansion.

Bernie holds a Bachelor of Commerce (Hons) degree from the University of Galway (formerly, NUIG), Ireland.

3.8

The Role of Facility Management in the Wake of Global Resource Scarcity

Joint Contribution By Christin Kuchenbecker and Andreas Kühne

The growing world population and increasing prosperity are leading to an ever-increasing demand for consumer goods. Since the production of consumer goods takes place using natural resources, the increasing demand leads to a worldwide shortage of resources and thus to an increase in the price in the market.

This increase in prices, triggered by the scarcity of resources, is also felt in the real estate and FM sector. According to the Federal Statistical Office, construction prices for residential real estate in Germany rose by 17.6% in May 2022 compared to the same month last year. This is the highest increase in 52 years. The increase in construction prices for office properties in May 2022 was 19% and for commercial properties 19.4% compared to the same month last year.[1] The increase in operating costs can be summed up even more clearly: they have almost doubled since 2018.[2]

Under the impact of the rise in construction prices, a decline in new construction projects with a focus on existing properties is to be expected. But not only from an economic, but above all from an ecological point of view, the turn to existing properties, including the reduction of CO_2 consumption, is an urgent requirement. In 2019, the European Commission set itself the goal of reducing net greenhouse gas emissions in the European Union to zero by 2050 as a roadmap for a sustainable, finance-oriented EU economy and thus becoming the first continent to become climate-neutral (so-called European Green Deal). This means that by 2050, the entire building stock in Europe must be climate-neutral.[3]

Current discussions often focus on operational CO_2 emissions. Operational carbon describes the emissions caused by the use of the building, e.g., by heating and cooling.[4] However, in the context of the CO_2 discussion, the property must be considered in its entire life cycle, i.e. from its construction to exploitation. Especially in the phase of construction of the property, a not insignificant part of the CO_2 emissions resulting from the process of manufacturing the building materials and components as well as the construction of the entire end product building (so-called embodied carbon) is emitted. While emissions from the operation of a building are proportional to the use and duration of its life cycle, the embodied carbon is released into the environment within short periods of time.[5]

1 Development of construction prices in Germany from August 2019 to August 2022, Federal Statistical Office (Statista), 2023.
2 Bauakademie (Ed.), NEO Office Impact Bench Market Report 2022.
3 European Commission, European Green Deal, https://commission.europa.eu/strategy-and-policy/priorities-2019-2024/european-green-deal_de, as of 07.01.2023.
4 Bauakademie (Ed.), NEO Office Impact Bench Market Report 2022.
5 Bauakademie (Ed.), NEO Office Impact Bench Market Report 2022.

Facilities @ Management, Concept - Realization - Vision, A Global Perspective, First Edition. Edited by Edmond Rondeau and Michaela Hellerforth.

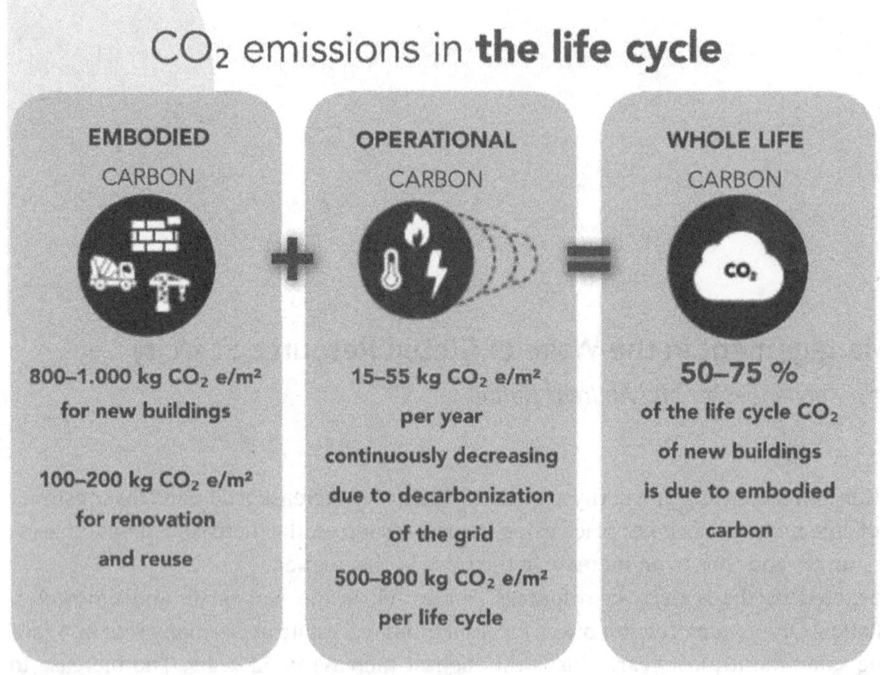

Figure 1 Distribution of emissions in the building life cycle based on Alstria Office REIT. (*Source:* Adopted from Alstria Office REIT).

The NEO Office Impact Report 2022, newly published in Germany, has shown that operational CO_2 emissions account for 25–50% and the grey emissions of embodied carbon are 50–75% (see Figure 1).

Due to the increased construction costs and the high CO_2 emissions during the construction phase, a move away from new construction properties and a shift towards existing properties is therefore to be expected. This means that existing properties and their management are becoming more important and bringing new perspectives for Corporate Real Estate Management (CREM) and Facility Management (FM). Together with their company, the BAUAKADEMIE Group based in Berlin, affiliated institute of the Berlin University of Applied Sciences, the authors strive to develop solutions that focus on both the optimization of operating costs by means of benchmarking and long-term maintenance costs by means of condition-based cost forecasts.

Original Benchmarking Principles

The term "benchmarking" has been coined by established initiatives in German real estate and facility management for many years. A closer look at the results achieved should be thought-provoking. The starting point should be the original principles of benchmarking, which can be vividly described with two historical milestones that have made benchmarking one of the most popular "evergreens" in the management toolbox in the first place:

In 1914, Henry Ford revolutionized the automobile market with the legendary Model T by being the first manufacturer to successfully produce an industrial series production. The decisive idea for converting production to assembly line technology is an adaptation from meat processing. Ford engineers took over and mechanized the idea and constructed the first "moving assembly line" and, on this basis, the first factory for mass production in

Detroit. As a result, the manufacturing costs per vehicle were reduced by 70% and the production volume increased eightfold.[6]

In 1979, the American company Xerox, which until then had been the world market leader and patent holder for fully automated office copiers, was in a completely different situation. However, in recent years, Japanese manufacturers have increasingly launched cheaper models on the market and subsequently pushed Xerox to the brink of insolvency. In a final rescue act, a team of consultants had dismantled and analyzed the competitors' products into their individual parts. Subsequently, a new device was constructed, which was then successfully positioned on the market at competitive prices and ultimately averted the demise of the former inventor.[7] Today, this success story can be seen as the initial spark for the firm establishment of benchmarking as a management method.

In summary, benchmarking can be defined as a "process of systematic comparison with successful market participants with the aim of identifying and adapting best performance in order to improve one's position."[8]

Current Maturity Level

In order to be able to assess as objectively as possible the degree of maturity of benchmarking in real estate and facility management in Germany today, it is helpful to start with an overview of the different types of benchmarking. With regard to the benchmarking object, a distinction is made between product, process, and organizational benchmarking. Product benchmarking is the aforementioned example of Xerox as well as most of the forms of real estate benchmarks known today in which, for example, operating costs are compared.

With regard to the reference class, i.e. the target group of the intended comparison, a fundamental distinction is made between internal and external benchmarking. External benchmarking differentiates between "market," i.e. comparison with direct competitors, and "functional" (usually targeted cross-comparisons with other industries) and "generic" at the highest level. The latter is the global, cross-industry search for best practices. Depending on the reference class, the identified best performance is described differently (see Figure 2).

Characteristic of the current level of maturity in Germany is the organizational form of a benchmarking project, i.e. the way in which benchmarking is carried out, of importance. There are basically three forms here: In a star organization, there is a central coordinator who designs and controls the benchmarking, receives and evaluates all data. As a rule, there is no contact between the companies participating in the benchmarking. In a circular organization, on the other hand, the participants exchange all information directly with each other, completely without a coordinator. For comparisons between competitors, this form is generally to be regarded as questionable under antitrust law, because, among other things, the necessary anonymization of the exchanged data would not be guaranteed. Therefore, there is the organizational form as a wheel. As with star, a central coordinator controls the entire benchmarking process, but which also includes targeted interaction between the participants.

From Benchmarking to Benchlearning

How important are these interactions? Experience has shown that participants in benchmarking need a common understanding of the objectives at the beginning and derived from them the data to be collected. Further coordination is conducive to agreeing on criteria to be used to determine the benchmark. This is particularly important when benchmarking maintenance and personnel costs. In a classic cost benchmarking, the cheapest

6 Ford: Success in Life – My Life and Work, 1952.

7 Camp: Benchmarking: The Search for Industry Best Practices that Lead to Superior Performance, 1989.

8 Kühne: Benchmarking 2.0, 2012.

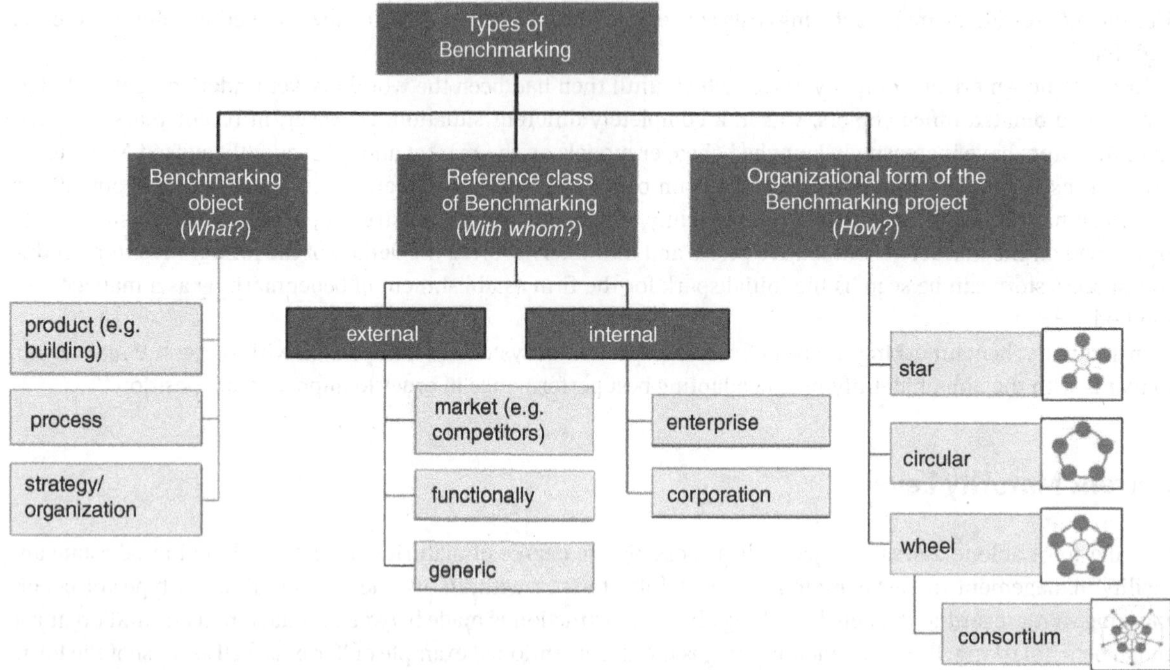

Figure 2 Overview of benchmarking types. (*Source:* Adopted from BAUAKADEMIE).

participant would be regarded as a benchmark and the delta to the other participants as optimization potential. The last and perhaps most important interaction for the participants is to derive the appropriate measures for their own company from the benchmark results and to adopt lessons learned from other participants. In compliance with the principles of competition law, which are largely regulated in EU law and the German Act against Restraint of Competition, participants in benchmarking projects can exchange structured information in order to learn from each other. This form of interaction, i.e., learning from each other on the basis of benchmark results, is now referred to as benchlearning (see Figure 3).

As part of a scientific study,[9] ten benchmarking initiatives for real estate and facility management offered in Germany were examined. With regard to the benchmarking subject, 90%, i.e., nine out of ten initiatives, can be classified as product benchmarking. This result is an expression of the prevailing desire to always find as many similar properties as possible for comparison. With regard to the form of organization, 70% are organized as a star. Consequently, only three initiatives provide for the possibility of interaction. Finally, qualitative aspects that are crucial for successful benchmarking with the aim of generating reliable results were examined. This includes, for example, the application of standardized structures and definitions in the context of data collection. In this way, a uniform understanding of the benchmarking object is usually established. This is particularly necessary if the benchmarking participants are not involved in the definition, but only from the time of data collection. As a result of the study, it was found that only 60% of the initiatives use standardized structures and definitions.

Finally, the generated key figures, which are used in the presentation of results, were analyzed. Here it is striking that 60% only show average values and ranges. In the sense of the definition of benchmarking shown above, the minimum or maximum values would actually have to be shown in order to actually compare with best performances, but not only with the "mediocrity." In summary, it can be stated that the benchmarking initiatives established in Germany for real estate and facility management:

9 Kühne: Benchmarking 2.0, 2012.

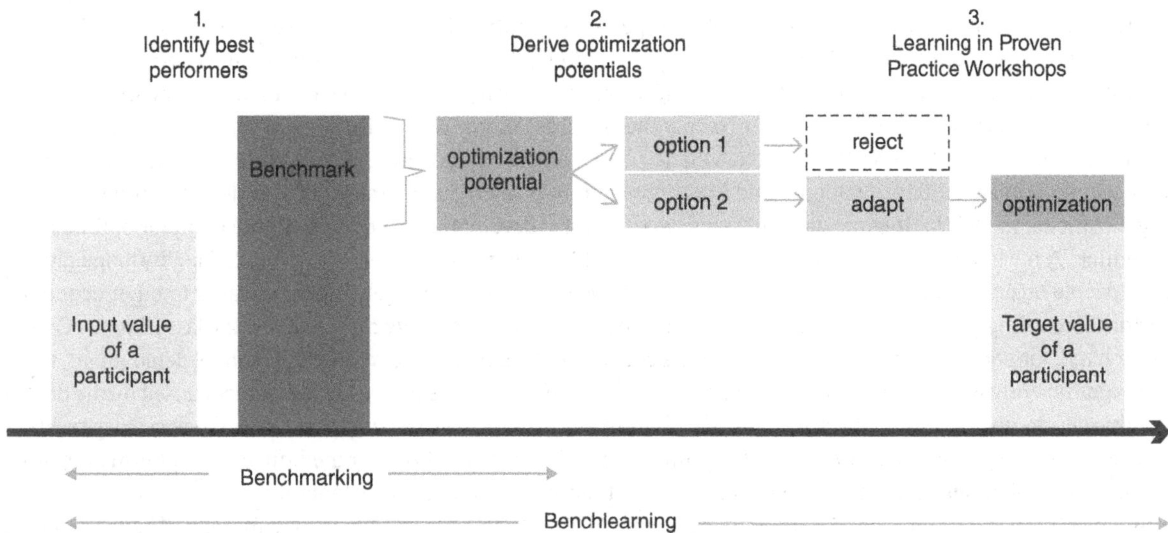

Figure 3 Benchlearning process. (*Source:* Adopted from BAUAKADEMIE).

- are predominantly to be described as a "black box" because, as a "star," they do not allow interaction between participants;
- publish almost exclusively cost-oriented key figures, the consistent application of which in a target system implies an "infinite" cost reduction;
- the statistical methods used predominantly reflect "mediocrity" instead of identifying best performances.

Taking into account some positive exceptions, it must be stated that the majority of the initiatives examined do not correspond to the original principles of benchmarking. A benchmark is more than a market comparison – it is the result of a targeted search for optimization potentials generated from a systematic comparison with successful market participants. The authors therefore argued that, in order to further professionalize the market, it would be necessary for initiatives which do not fulfil the above principles not to be referred to as benchmarking. More suitable would be terms such as market reports or indices such as those used for rental prices or construction costs. The good work of the benchmarking pioneers at the turn of the millennium must be continued and further developed with new creativity. The most urgent needs for action include:

- organizing new formats and initiatives aimed at real benchmarking.
- the application of qualitative assessment methods to broaden the thematic spectrum;
- Development of output-oriented key figures that enable performance and success measurements.

As a result, new confidence-building and reliable comparative values can be provided, which will make a decisive contribution to solving the challenges of the real estate industry described above.

Example of New Formats: Benchlearning Community and COMPAS Roundtable

The BenchLearning Community is an exclusive community of over 70 large companies that systematically and regularly practice benchmarking on all topics relating to the operation of real estate and industrial sites in compliance with competition law principles. This community is organized in specialist circles, in which 10 to 15 companies participate, which are designed as a "round table" with a neutral benchmark coordinator. BAUAKADEMIE Performance Management GmbH, a subsidiary of the BAUAKADEMIE Group, assumes this function and has systematically built up the benchlearning community since 2004. It works to ensure that reliable benchmarks are created in compliance with the principles of European competition law and that sustainable optimization successes are derived from them. The benchlearning methodology used is certified annually in accordance with EU competition law, ISO 9001 and ISO 27001 (information security). Once a year, all members meet for the Benchlearning Summit in Berlin, which presents the Benchlearning Awards for proven top performance.

An example is the COMPAS Roundtable, which focuses on the organization of corporate real estate and facility management organizations of national and international non-property companies. Its[10] participants aim to analyze organizational models taking into account trends and industry developments and to examine them for optimization potentials by means of benchmarking. A central component is regularly personnel requirements. Facility management is a "people business," i.e., the share of personnel in the value added is relatively high compared to other economic sectors. Accordingly, education and training in Germany has been intensively promoted over the last 15 years. On the other hand, for personnel measurement, i.e. for the concrete calculation of the required personnel deployment, there are no sufficiently well-founded studies available for clients. In view of the increasing dynamics caused by the demographic development in companies, the lack of well-trained and available specialists as well as the necessary reaction speeds to changing economic developments, companies need reliable methods for medium and long-term personnel requirements planning and the associated allocation of their management competencies.

Against this background, the companies participating in COMPAS have set themselves the goal of benchmarking the deployment of personnel for the function of facility management and developing a practice-oriented recommendation for a procedure for personnel measurement from the lessons learned and underpinning it with empirically proven empirical values. For this purpose, the job descriptions used in the companies were first compared and a harmonized standard, based on the international standard for facility management (DIN EN ISO 41001 ff.), was derived. Subsequently, capacity drivers were described, i.e. parameters that have a causal effect on the question: How much space can an object manager optimally manage? Of course, the demands of the participating companies on the care of the employees employed and the quality of the work results were taken as boundary conditions.

A total of 15 capacity drivers were examined, starting from the technical characteristics of the object to be supervised to the qualification of the personnel deployed, and differentiated relevant characteristics for each driver in practice. On this basis, all participants surveyed their actual personnel capacities – in each case for a self-selected reference location, i.e. a location that is regarded as an internal model solution, and for the entire real estate portfolio managed in Germany. Over a period of 26 months, a total of almost 450 facility managers and 35 million m^2 were empirically surveyed and their specific distribution to all capacity drivers analyzed. As expected, the concrete measurement results have sometimes fluctuated greatly between the companies – but as a result, causal correlations could be proven for ten capacity drivers. Specific surcharge and discount factors were determined for all characteristics and the capacity drivers were weighted among themselves. On average, a facility manager manages 40,000 m^2 GFA, which may fluctuate between 25,000 m^2 and 70,000 m^2 under concrete application of all surcharge and discount factors (see Figure 4).[11]

Example of New Methods: Hedonic Evaluation

The theory of hedonic models assumes that information about the valuation of the individual value-determining properties is "hidden" in the market price of an asset. This theory has been successfully applied for several years to determine purchase and rental prices, where sufficiently large amounts of data are available in publicly accessible sources. Applied to the benchmarking of real estate costs, the influencing variables on the level of operating costs can be divided into four groups:

- Macroeconomic factors such as the inflation rate, taxes or the development of industry minimum wages, which have an influence on maintenance and surveillance costs, for example.

10 BMW GROUP, BOSCH, COVESTRO, DEUTSCHE BAHN, EnBW, EUROPEAN CENTRAL BANK, EVONIK, PANASONIC INDUSTRIES, QVC, THALES.

11 Study and tool are published under www.compas.benchlearning.de.

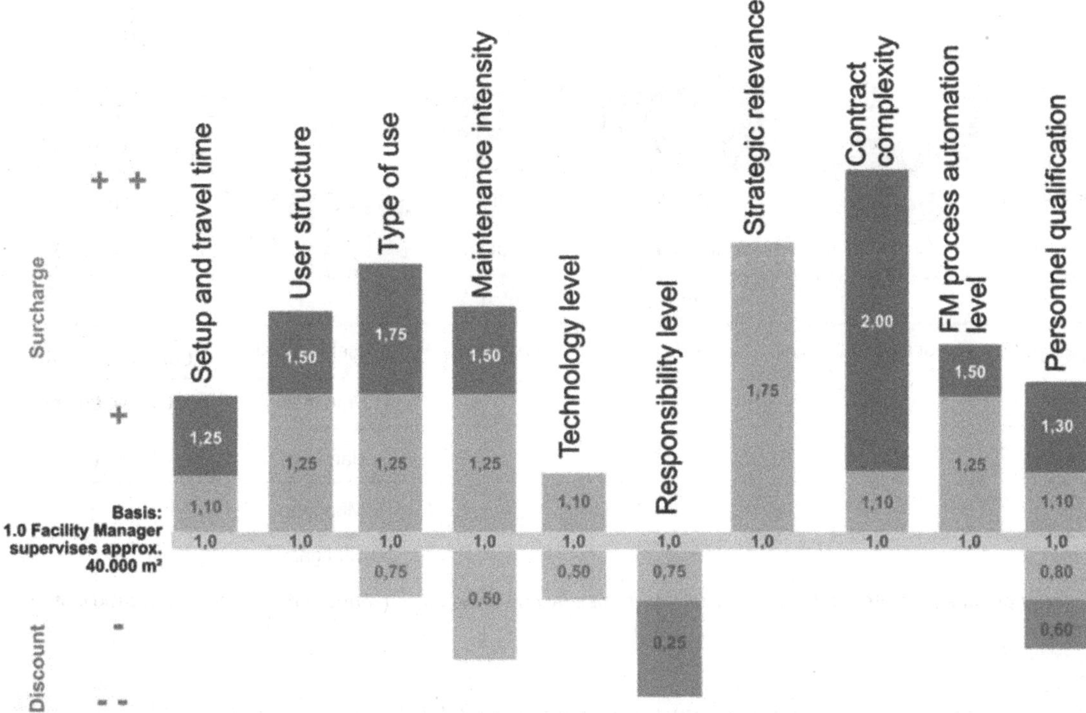

Figure 4 Capacity driver model of personnel requirements in object management. (*Source*: Adopted from BAUAKADEMIE).

- The location that influences, for example, the amount of municipal taxes for property tax and street cleaning.
- The quality of the building, e.g., the age and condition of the structure and the technical facilities, which influence the level of maintenance and inspection costs.
- User requirements, e.g., the frequency of hygiene and cleaning services and the staffing times of reception services.

Macroeconomic influencing variables are sufficiently determined by public statistics, accessible as a public data source and thus generally available for application to specific real estate. A challenge is posed by the not publicly available data for the object-specific analysis and calculation of operating costs. While the location of a property is naturally one part of the established master data, experience has shown that the data availability for building type, size, age and type of air conditioning is severely limited. The same applies to user requirements. These are not standardized anyway but can be described approximately in service level standards. Because this data is not relevant for the rental settlement of operating costs, it is not available as standard in property management. For their collection, data sources from other participants must therefore be regularly integrated.

Although there have been various benchmarking initiatives specializing in operating costs of office properties for many years and combining these different data, the data collected does not cover all combinations of location, building quality and user requirements, etc. (see Figure 5).

With the extensive database of the NEO Office Impact Initiative and its predecessor, the OSCAR (which was published jointly by JLL and BAUAKADEMIE until 2019), a corresponding pilot process was developed by BAUAKADEMIE Performance Management GmbH and successfully tested in large portfolios.[12] In this way, operating costs for office buildings for which no comparison objects are actually available can be predicted for the

12 Kühne, Fuhr, Emmrich: Effective Cost Handling. Published in Data Intelligence on real estate data-based decision support of the Gesellschaft für immobilienwirtschaftliche Forschung e.V. 2022.

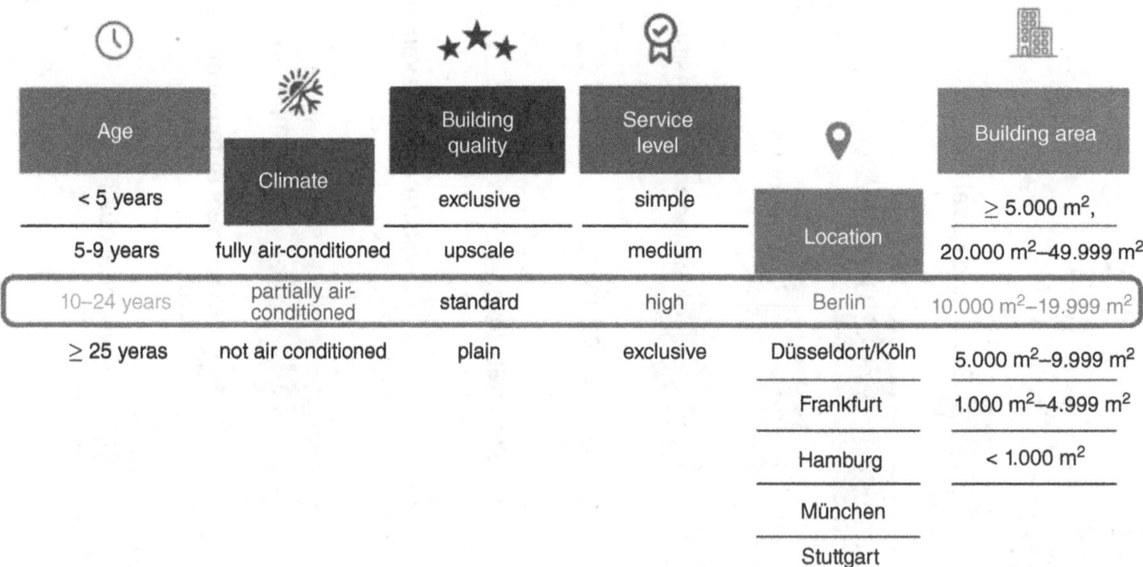

Figure 5 Example of a possible combination to calculate specific operating costs. (*Source:* Adopted from BAUAKADEMIE).

first time without great effort. This method is currently being further developed using modern AI-based methods and is also being transferred to other asset classes.

Importance of Condition-based Cost Forecasting

Due to the increasing importance of existing properties, strategic management of existing properties is becoming increasingly important. Real estate accounts for about 5–20% of a company's fixed assets. Depending on the core business and company size, the real estate costs are 10–20% of a[13] company's total costs. Thus, in addition to a pure cost analysis (see benchmarking), the value retention of the real estate is an important lever to ensure optimal provision and control of existing resources.

Under the term "Asset Intelligence," the BAUAKADEMIE Group has developed various modules that enable companies to first systematically examine their portfolio properties on the basis of data-supported analyses in order to then make strategic deductions in the sense of a portfolio, asset, or FM strategy (see Figure 6).

Module Asset Intelligence

The modules "Condition assessment" and "Cost forecast" form the starting point for a systematic examination of existing properties. As part of the condition assessment, the degree of wear of buildings and components is determined by means of uniform evaluation criteria. In addition, tried-and-tested help texts are used to create a comprehensible and objective documentation of the actual state of buildings (including the same components and systems). Specific acquisition algorithms are available for office and administrative, research and production buildings as well as for warehouses, workshops, and outdoor facilities. The condition levels are plant-specific, as the following graphic shows by way of example using load-bearing outer walls (see Figure 7):

13 Glatte: Corporate Real Estate Management – Quick Start for Architects and Civil Engineers, 2019.

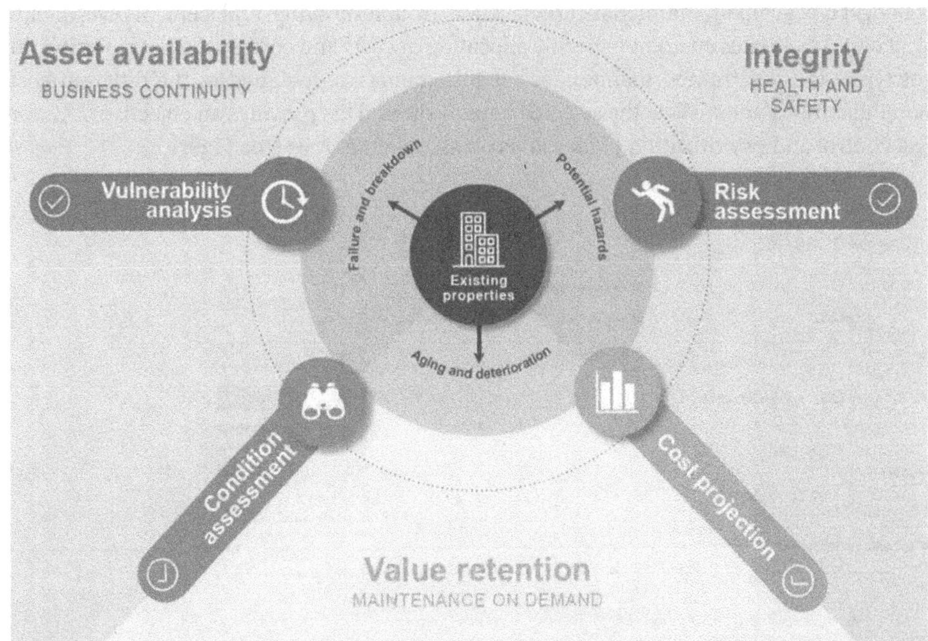

Figure 6 Categories for the systematic consideration of existing properties. (*Source:* Adopted from BAUAKADEMIE).

Cost group 331 according to DIN (german industrial standard) 276: Load-bearing exterior walls

Condition stage 1	Condition stage 2	Condition stage 3	Condition stage 4	Condition stage 5
No damage visible. No cracks. Surfaces as good as new.	First signs of use are visible, e.g. mechanical abrasion.	Spalling, cracks. Insufficient concrete cover.	Large spalling and cracks. Exposed reinforcing steel. Corrosion damage to steel.	Advanced carbonation. Advanced corrosion of the reinforcing steel. Reduction in cross-section due to extensive spalling.

Figure 7 Practical help texts for condition assessment using the example of load-bearing exterior walls. (*Source:* BAUAKADEMIE / https://bauakademie.de/wp-content/uploads/2021/03/Asset-Intelligence_Flyer.pdf).

The degree of wear and tear of the existing property is determined by depositing the years of construction or modernization as well as the assessment of the condition per component from condition level 1 to 5. This serves as the basis for rolling planning and prioritization of maintenance measures. The result of the condition assessment is an overall valuation for the existing property under consideration or, in the case of several properties, an evaluation of the entire real estate portfolio, which is determined with the help of a digital tool from the states of all existing components.

Based on this condition assessment, the "Cost forecast" module systematically determines the maintenance requirements of the properties and their technical equipment for a freely selectable period of time (e.g., 10 years). For this purpose, the result of the condition assessment is imported into the cost forecasting tool. Dynamic algorithms ensure a condition-modified and manufacturer-/user-defined consideration of the remaining useful life in order to flexibly map the aging behavior. On the basis of stored benchmarks and the determined wear and tear behavior, the maintenance requirements of the property and its technical equipment for the selected period are calculated.

The consideration of six budget types (repair, maintenance/inspection, troubleshooting, replacement investment, demolition, ancillary project costs) facilitates customer-specific allocation of costs and corresponding reporting. In addition to the six budget types, the determined maintenance requirements are assigned to the different cost groups of DIN 276 (German industrial standard) via the selected annual slices. This provides an objective basis for decision-making for budget control and prioritization of measures on an ongoing basis (see Figure 8).

Compilation of the projected values for the object "XYZ 123"

Base year:	2023	**Prognosezeitraum:**	2022-2031
Replacement value:	18.884.000	**Building type:**	G = Functional technical building
Utility area	Prduction		

Costs for the production of a defect-free condition (planned maintenance):	8.117.849
Costs for maintenance / inspection:	2.152.776
Costs for troubleshooting (unplanned repair):	717.592
Maintenance costs:	10.988.217
Correction factor due to object-specific conditions (for repair)	1,0
Justification of the object-specific conditions:	
Total cost of maintenance:	10.988.217
Ancillary costs:	188.840
Costs for the production of documentation suitable for the operator	188.840
Ancillary building costs	0
Project cost:	1.500.000
Total costs for the complex revitalization of the property in the projection period.	12.677.057

All prices are exclusive of the statutory value-added tax.

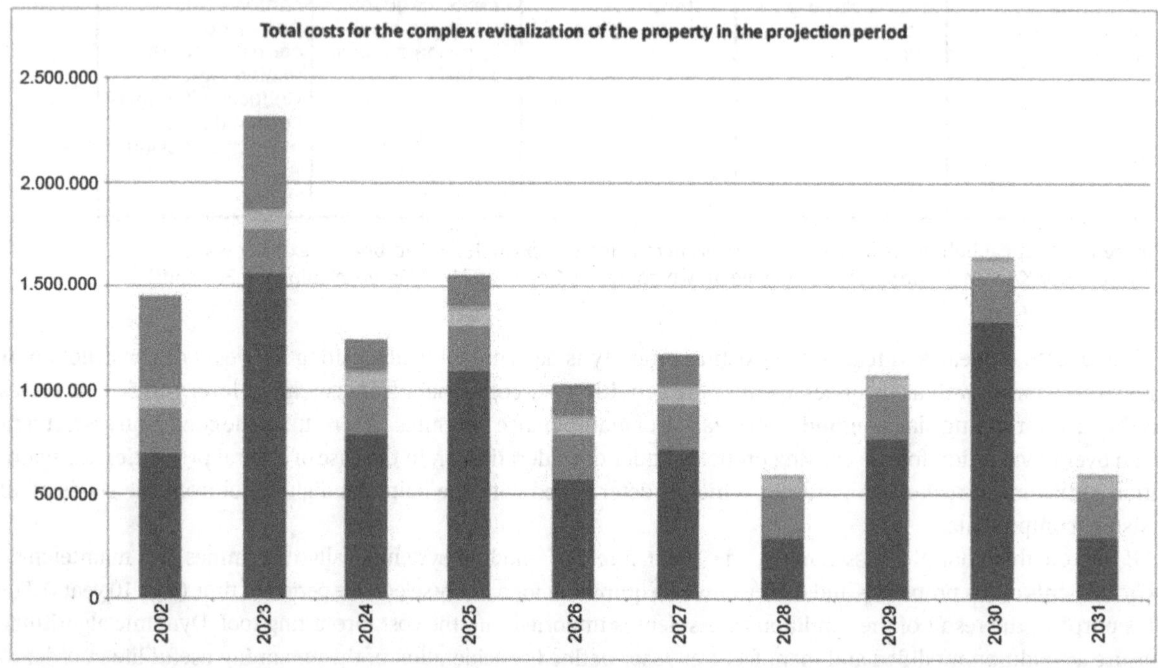

Figure 8 Aggregated summary of cost forecast at portfolio level. (*Source:* Adopted from BAUAKADEMIE).

These two modules can be supplemented by the modules "Risk assessment" and "Functional safety." The aim of the risk assessment is a systematic recording and evaluation of risks from building operation for health and the environment. The content of the system is based on the Industrial Safety Ordinance and is methodically structured like a risk analysis. In 12 predefined hazard groups, potential risks are examined and documented on a property-specific basis and evaluated with regard to their event potential and probability of occurrence. In addition, the costs for suitable countermeasures are recorded.

As part of the "Functional Safety" module, the systematic assessment of the failure risk of components is carried out from the perspective of business continuity management. The result is a measure-related, prioritized target/actual comparison to ensure the availability of the relevant components for the intended use.

Based on the information obtained from all modules, the costs can be combined in a profitability calculation. This is based on the central corporate specifications (e.g., targeted interest rate) at building level and compares the expected expenses and income. The results obtained in this way form the input for the strategic derivations in the sense of a portfolio, asset, and maintenance strategy (see Figure 9).

BAUAKADEMIE Professional Development GmbH, a subsidiary of the BAUAKADEMIE Group, uses asset intelligence to support its customers in deriving demand-oriented portfolio, asset, and maintenance strategies as a prerequisite for optimal provision and management of existing properties.

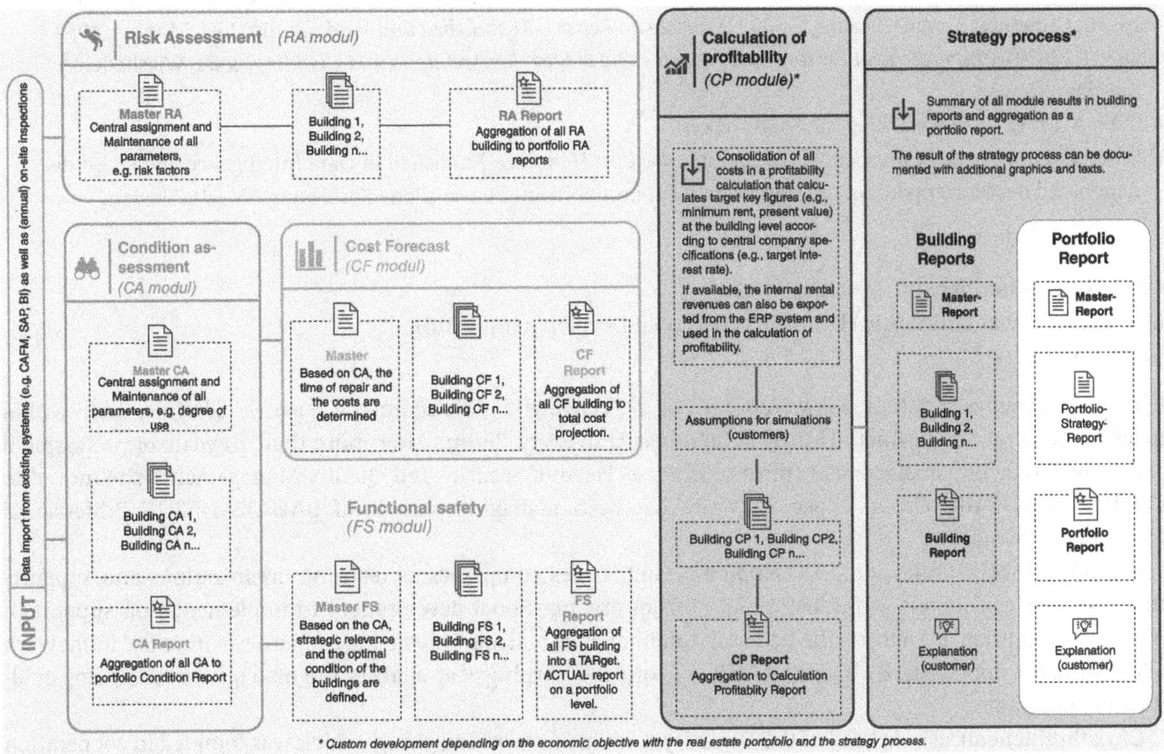

Figure 9 Interaction of all modules for data-supported strategic management of real estate. (*Source:* Adopted from BAUAKADEMIE).

Summary

In conclusion, it can be stated that existing properties and their management will gain in importance under the growing pressure of scarcity of resources and reduction of CO_2 emissions. The role of facility management is thus also coming into focus. A structured exchange of information (benchmarking) and mutual learning (benchlearning) has the potential to further improve management. Additional possibilities for optimization lie in strategic management of existing properties (asset intelligence).

Bibliography

BAUAKADEMIE (ed.) (2022). *NEO office impact bench market report*. Berlin.

Camp, R.C. (1989). *Benchmarking: The Search for Industry Best Practices that Lead to Superior Performance*. New York: Productivity Press.

European Commission, European Green Deal. https://commission.europa.eu/strategy-and-policy/priorities-2019-2024/european-green-deal_de, as of 07.01.2023.

Federal Statistical Office. *Development of construction prices in Germany from August 2019 to August 2022*. https://de.statista.com/statistik/daten/studie/164936/umfrage/entwicklung-der-baupreise-in-deutschland, as of 07.01.2023.

Ford, H., Crowther, S., and Thesing, C. (1952). *Erfolg im Leben – Mein Leben und Werk*. Berlin: List, Munich, 1952.

Glatte, T. (2019). *Corporate Real Estate Management – Quick Start for Architects and Civil Engineers*. Wiesbaden: Springer Vieweg.

Kühne, A. (2012). *Benchmarking 2.0*. Saarbrücken.

Kühne, A., Fuhr, S., and Emmrich, F. (2022). *Effective Cost Handling*, Published in Data Intelligence on real estate data-based decision support of the Gesellschaft für immobilienwirtschaftliche Forschung e.V., Wiesbaden.

Christin Kuchenbecker, Ass. Jur.; MBA
Managing Partner of BAUAKADEMIE Professional Development GmbH.
Berlin, Germany

Christin Kuchenbecker studied law (Ass. jur.) at the University of Potsdam and business administration with a focus on General Management (MBA) at Steinbeis University Berlin. After more than 10 years of professional experience in facility management (most recently as Head of Security and Quality Management), she moved to the BAUAKADEMIE Group. Since 2021 she has been managing partner of BAUAKADEMIE Professional Development GmbH.

BAUAKADEMIE Professional Development GmbH sees companies as learning organizations and supports them in their continuous self-transformation from organizational development to implementation support in practice. In addition, it forecasts the necessary maintenance budgets for existing properties within the framework of construction appraisals. BAUAKADEMIE supports the learning of specialists and managers by designing qualification programs.

Christin Kuchenbecker is head of the Diversity & Inclusion working group, which was founded in cooperation with RealFM e.V. Association for Real Estate and Facility Managers and GEFMA German Association for Facility Management e.V. She is also a lecturer at the Berlin University of Applied Sciences.

Andreas Kühne, M.A., Dipl. Bw. (BA), CREM (ebs)
Managing Partner of BAUAKADEMIE Performance Management GmbH
Berlin, Germany

Andreas Kühne studied business administration and real estate economics in Leipzig, Oestrich-Winkel and Berlin and has been managing partner of BAUAKADEMIE Performance Management GmbH since 2014. This company initially specialized in syndicate benchmarking and has commercial real estate worth 180 billion Euros from over 70 corporates "under benchmark."

With the NEO Impact Bench, another benchmarking project exists since 2020 that produces an annual market report on the development of operating costs, energy and space consumption as well as CO_2 emissions. Andreas Kühne is author of several books and lecturer at the International Real Estate Business School of the University of Regensburg, the German Real Estate Academy in Freiburg/Breisgau, and the Berlin University of Applied Sciences and Technology.

3.9

Paying It Forward Through the IFMA Foundation

Diane Levine

My dream as a youth was to become a classical violinist and play the Bruch Violin Concerto on stage with a symphony orchestra and the audience wrapped with me as one in a beautiful state of entrainment. A funny thing happened on the way to the concert hall. Life experiences took me in exciting directions, and I wouldn't change any of the detours I've taken from a full-time music career to my current job as Executive Director of the IFMA Foundation. Why? Because along my journey, I accomplished more good than I ever imagined and, at the same time, discovered talents that I never knew existed. Had my career only focused on music, I would not have been exposed to opportunities to change lives, mentor staff, learn about the built environment and FM, experience managing people, create innovative workplaces, make global friends, write publications, speak in public, and travel the world. Incidentally, as you will learn later, my music and orchestral training was instrumental (no pun intended) in my success within IFMA and my career.

In college, I always pictured myself running a nonprofit, exposing disadvantaged children to the world of music and the arts, taking them to orchestra concerts, art museums, theaters, and training students in the various arts disciplines. I was afforded this opportunity growing up in a single parent home, given a violin in a school program and was able to obtain music scholarships to St. Xavier's Academy, a New England all girl prep school and Rhode Island College. I performed in Kennedy Center with Young Peoples Symphony of Rhode Island and again in college, where our ensemble won a bronze medal from President Jimmy Carter.

As a music major, I knew it was my destiny at a point in my career to give back and either create or lead a nonprofit. I even obtained a Non-Profit Management in the Arts certificate while in college. I never imagined that the nonprofit I would finally lead would be in the field of facility management. When Jeff Tafel resigned from the Foundation as Executive Director, I never wanted a job as much as I wanted this one. The potential for growth and the opportunity to change lives and expose youth to a career that afforded me so many opportunities, not to mention a great salary, was stimulating. I knew through the IFMA Foundation's Global Workforce Initiative (GWI), I could expose people like me, an underprivileged youth, to an exciting field providing them education and economic mobility. When Jeff left the Foundation, GWI was in its initial stages and there was so much opportunity for funding through grants and corporate interest in supporting diversity, Environmental, Social, and Governance (ESG), and the UN2030 sustainable development goals, which is all connected to FM workforce development and the mission of the Foundation.

Facilities @ Management, Concept - Realization - Vision, A Global Perspective, First Edition. Edited by Edmond Rondeau and Michaela Hellerforth.

From Librarian to FM

Growing up poor, my siblings and I qualified for a youth workforce development program where I began working full-time in the summer beginning at age 14 and part-time when high school was in session. By the time I graduated college, I had a head start on my peers, with 8 years of work experience in libraries working my way up to a supervisory position. At the time of college graduation, Rhode Island was economically depressed and upward mobility in a career was scarce. My older brother invited me to join him in California, so I bought a one-way ticket, sold everything I owned, and left for Los Angeles with a suitcase full of clothes, a suitcase full of sheet music, my violin, a cast iron frying pan (an heirloom from my grandmother) and $1,000.

Finding a job in California was quite easy because, at the time, unemployment was 1.5% and I quickly found a management job at Chapman College Library. All I needed to do was speak in my Rhode Island accent because New Englanders were thought of in California as hard workers. From there, I moved on to records and information management consulting and pursued a business degree as I had a great desire for a different career in addition to my part-time music gigs. I was offered a position as a Records Management Supervisor at Orange County Transportation Authority (OCTA), which paid more than I could have made as a full-time musician.

When my boss left the organization, I was made Acting Manager of General Services. I knew this would be my ticket out of libraries so I soaked up as much as I could from vendors and joined IFMA. My first week on the job, we had a move of 200 people. The architect, contractors, movers, furniture vendor all helped in my FM education. I was an eager student ready to soak up all that I could learn from them. Eventually, at age 28, I was offered my boss's position and a higher-than-expected salary moving me further away from a full-time music career.

As the Manager of General Services, I was responsible for facilities, real estate, office services, telecommunications, records management, and in-plant printing. With both owned and leased facilities and over 20 miles of railroad right of way land leases to manage, I had a full plate. During the merger of OCTA, working with Urquiza Group Inc. architecture firm and CBRE, I had the privilege to lead the consolidation of OCTA's real estate and facilities, and create and execute a master plan which was one of the largest real estate deals in Orange County California at the time.

The project was so successful, I was asked to help a new startup public agency, Orange County Health Authority dba CalOPTIMA, establish their real estate and FM master plan. OCTA paid my salary and loaned me out to CalOPTIMA. What an exciting opportunity to be at the ground floor of a startup creating everything from scratch including strategic plans, policies, procedures, and hiring staff. After six months at CalOPTIMA, they made me an offer I couldn't refuse and became the Director of Facilities and Procurement. I continued to play part time gigs in string quartets and the Crystal Cathedral Symphony Orchestra, occasionally appearing on TV.

Musicians Make Great FMs

As a musician, FM was an extremely attractive career to me. It's no accident that musicians are in the profession as FM, like music, requires creative thinking, discipline, leadership, collaboration, and problem solving.

- *Problem Solving* – Musicians can easily improvise and solve problems. Music is math and musicians have good pattern-recognition skills, insight into symmetry, strong systematic and categorical thinking abilities. Music theorists, like experts in other disciplines, use mathematics to develop, express and communicate their ideas. Notations of composers and sounds made by musicians are connected to mathematics including counting, rhythm, scales, intervals, patterns, symbols, harmonies, time signatures, overtones, tone, and pitch.
- *Collaboration and Teamwork* – Musicians must be highly attuned to their fellow performers, constantly listening and adjusting. It's about giving and taking continually throughout a performance. In an orchestra, instrumentalists watch and follow the conductor, read the sheet music, while keeping an eye on their section lead and

actively listen to fellow players. In rehearsals, they constantly receive feedback and get a well-rounded sense of what's working and what isn't. Being a musician improves the ability to cooperate with others and promotes camaraderie. Great orchestras and bands have musicians that seem to blend into each other, playing as if they were one. When the audience feels this oneness, it is called "entrainment."

- *Project Management* – Managing a construction project is like conducting an orchestra or a band. Everyone has their part and needs to play their music at the right moment it appears in the score. Musicians must understand not only their part but the entire score to know when to come in and when to exit. Having keen pattern recognition and arranging skills allows musicians to lay out the entire project in their head from beginning to end before writing it down. Musicians can easily visualize the score and arrange all the pieces and resources for maximum productivity. They are conductors, aligning and realigning the score in the most productive configuration possible. They are used to juggling a lot of balls at once and flexing quickly when well laid plans get derailed.
- *Leadership and Strategy* – Musicians are visionary and strategic. They're always looking ahead at the next measure while playing the current notes. Leadership in an organization is like leadership in a music ensemble as they both help set a common vision, generate enthusiasm, empower, and develop people, find solutions to roadblocks, and help their team be the best that they can be. Mutual learning and development are a priority. Musicians, like leaders are self-disciplined. Musicians will practice and play a measure of music over and over until they get it right. Good musicians are used to making smart decisions even under intense pressure. They can do this when they are adept on everything around them to accurately anticipate the result of their decision. This is called informed improvisation and is a result of countless hours of practice.

Getting Involved in IFMA

As you can see, my music training helped me as an FM and, I realized in 2003, that I needed to use principles applied to mastering a musical instrument to my FM career. If this was going to be my career, I wanted to be the best FM I could be and finally realized that to do this, I needed IFMA to be my teacher and conductor. During this revelation, I was offered a job as Director of Facilities and Real Estate at SCAN (Senior Care Action Network) Health Plan, and, lucky me, they were willing to support IFMA activities. While I had been a member of IFMA since 1990, my real involvement began in 2004 in Salt Lake City at my first World Workplace conference. That's where I met IFMA Fellow, Jennifer Corbett-Shramo who took me under her wings, ensured that all the important conference events were on my schedule and introduced me to IFMA VIPs and import local southern California members like IFMA Fellow Phyllis Meng, Tony Soriano, IFMA Fellow Bill Conley, Oren Gray, and others.

Almost immediately, I joined Tony's golf committee and Jennifer's Foundation Fund-raising Committee and before I knew it, Bill Conley and Phyllis Meng began mentoring me as Vice President and eventually President of the IFMA Orange County (OC) chapter. It all happened so fast! The support and encouragement from IFMA leaders were remarkable. As a chapter leader, I was able to hone my FM, strategic planning, leadership, and public speaking skills. Together with the board, we grew chapter membership and created outstanding educational programs. IFMA staff supported us with on-site balanced scorecard training and helped us create our strategic plan.

The following year at World Workplace in Philadelphia, I met Janice Cimbalo, attorney, and real estate broker. As the Vice President of the IFMA Corporate Real Estate Council, Janice invited me to join the council and attend an event in the spring in San Antonio. This event became the pre-cursor to Facility Fusion. This was an opportunity to meet and bond with other FMs who managed real estate, many of whom are now dear friends. I implemented ideas learned and gained a network of fellow Real Estate Council members that I could call on for advice. Before long, I became the Treasurer of the Real Estate Council and started helping Janice develop real estate program sessions for world workplace and CREC events.

The AWESOME Project, Work on the Move, and the Birth of the Workplace Evolutionaries

It was during this multi-IFMA council conference in 2006 that I met Dr. Jim Ware and Dr. Charlie Grantham, authors of respected publications including *Corporate Agility: A Revolutionary Model for Competing in a Flat World*, who were speaking about their ROI calculator for the hybrid workplace. Little did I know that this meeting would be life changing and these two intellectuals would become mentors, friends, and co-authors of three books.

My boss, Dennis Eder, CFO, and I were exploring ways to save real estate costs and interested in creating a hybrid workplace which was innovative at this time. Jim and Charlie had the information we needed to convince the C-Suite to move to agile working. SCAN Health Plan was having challenges due to Rapid growth in customer base and staff, while at the same time experiencing a decrease in health care revenue from the government. The corporate facility was at capacity and the workplace didn't meet the company's 2012 strategy. Workforce support costs were high and growing.

After 18 months of planning, in 2007, we launched an innovative workplace strategy which won the IFMA George Graves Award and a CoreNet Remmy award. The facilities team took leadership in aligning itself with human resources (HR) and information technology (IT) departments to spearhead a workplace strategy. This strategy was referred to as "The AWESOME Project," which stands for alternate workplaces engaging staff and office management efficiencies (AWESOME). The project involved three major changes made at SCAN:

1) A *flexible work program* that enabled 30% of employees to work from home, at other locations or anywhere in the building.
2) An *aggressive redesign and repurposing* of the workplaces within the company's corporate headquarters facility; and
3) A *fundamental redesign of the process* by which individual and team workplaces were configured and provisioned.

The AWESOME team (Jim Ware, Charlie Grantham, Manuel Urquiza and the Urquiza Group, Janice Cimbalo, Steelcase, Knoll, and Innerspace Constructors) realigned the workplace strategy with SCAN's 2012 vision and created a new flexible, agile, and sustainable work environment. The AWESOME project was written up in magazines including Business Week, Return on Investment Magazine, and in several books.

After receiving the George Graves award for the AWESOME project at the 2009 IFMA World Workplace, I met Nancy Sanquist, who was interested in learning more about the project. Little did I know that this would be the start of our years-long creative relationship where we would co-edit and co-author books, co-create the IFMA Foundation Global Workforce Initiative, co-present at conferences, and write the Foundation's successful UN NGO application. That evening, at the awards banquet, Nancy introduced me to George Grave, founder of IFMA, and the three us danced the night away. During the excitement of the success of the AWESOME project, Pat Turnbull and Jennifer Corbett-Shramo encouraged me to apply to become an IFMA Foundation Trustee. After my second Board of Trustees (BOT) meeting, fellow Trustee Sandra Oliver recommended that I author a book about the AWESOME project. Jim, Charlie, Janice, and I had been speaking at conferences and judging from the high number of people at our presentations, we saw that there was a great need at IFMA for workplace strategy education. Jennifer created a model for the IFMA Foundation to produce books and one publication she led in 2009, *Cut it Out: Save For Today, Build for Tomorrow* included Jim, Charlie, and I as co-authors along with others. This was the beginning of my experience in producing books.

As a member of Jim and Charlie's Future of Work Consortium, I met the leading innovators at that time in workplace strategy including Chris Hood, Kate North, Glenn Dirks, Joel Ratekin, and others. They were all implementing innovative hybrid workplace projects. This gave me the idea that the book needed to be less about SCAN and more about what was happening globally in this area and what the future held for workplace. Working with Sandra, Jim, and Charlie, we developed the crazy idea to create a collaborative book on workplace strategy, pairing 19 authors all over the world, who had never met and requiring them to work in a virtual environment to create the first IFMA Foundation "Work on the Move" book. It was a great experiment. The authors met for the first time in Phoenix at the 2012 World Workplace conference.

I was in over my head during the production of the book and had the good fortune to bump into Nancy Sanquist at a Facility Fusion conference in Boston. I learned that she had editing experience and, after several glasses of wine at a CREC, LLLC event at Legal Seafoods, I convinced Nancy to co-edit Work on the Move. We met in spectacular hotel lobbies along the coast of Southern California, sipped tea, and, depending on the time of the day, sometimes wine, and discussed the book. When we came to an impasse, Nancy would say emphatically "go play your violin." Practicing violin would remove the cobwebs in my brain and spark creativity. Thankfully, Nancy saved the book and so did Michael Schley and FM:Systems as they assisted with production to meet our deadline for World Workplace.

After the book launch in 2012, a group of authors and sponsors including Nancy, Michael Schley, and Erik Jaspers wanted to continue the workplace strategy conversation and create ways to continue the collaboration. We collectively developed the idea of an IFMA Foundation Workplace Strategy Research Summit. Michael Schley led our team, including Nancy, Alexi Marmot, and I, in developing the first Summit in conjunction with Cornell University with luminaries in the field speaking including Frank Becker, Michael Joroff, and Sir Francis Duffy. Members of both IFMA and CoreNet Global's workplace community attended.

Book Author of the Year Award with Nancy Sanquist for Work on the Move (see Figure 1). The book resulted in global speaking opportunities for the authors and I in Europe, Asia, and the Americas. For many years, Pat Turnbull and I were guest lecturers together on workplace strategy at Vienna University of Technology in Austria. It was apparent from these presentations and the Summit that IFMA members were craving workplace strategy education, and, through the book and the summit, a small community was started at IFMA. When CoreNet disbanded their communities of practice, Kate North approached Nancy and I and together we asked IFMA if we could form the Workplace Evolutionaries (WE). We had two months to develop a program for Facility Fusion in Los Angeles. With our red berets and under Kate's leadership, collaborating with members of IFMA and the CoreNet Workplace Community, WE was launched.

In 2014, the same team launched a second IFMA Foundation Workplace Strategy summit in England in conjunction with University College London. The Foundation published a synopsis of each summit edited by Michael. Nancy and I co-edited Work on the Move 2 in 2016. Five years later, in 2021 Michael Schley and Alexi Marmot co-edited Work on the Move 3 (see Figure 2). With the authors donating their time, all three books won IFMA book author of the year awards and raised over $180,000 for the Foundation. These books included case

Figure 1 Award Ceremony at IFMA World Workplace. (*Source*: IFMA).

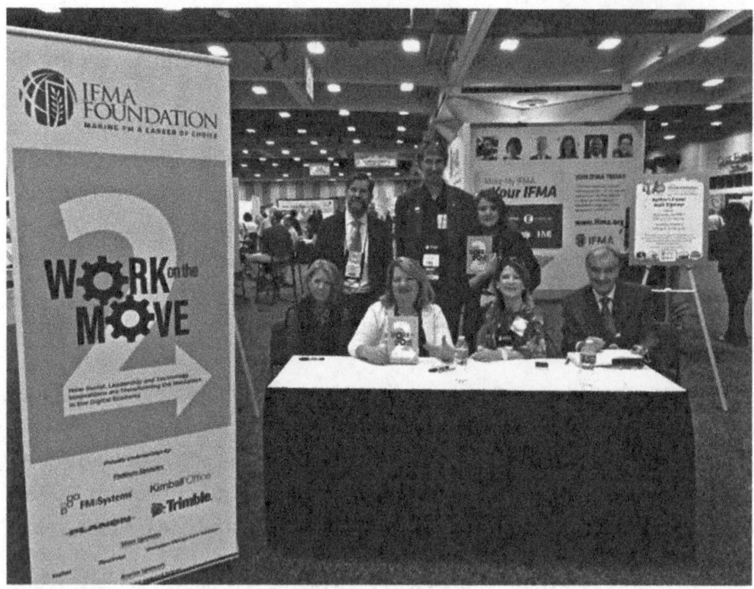

Figure 2 Work on the Move 2 Launch with Co-Authors at IFMA World Workplace 2016. (*Source:* IFMA).

studies from around the world, thanks to Kate North, and became textbooks for facility management accredited degree programs. They are among the most popular books published by the Foundation.

The Global Workforce Initiative (GWI)

In 2015, under Jennifer Corbett Shramo's leadership as Chair of the Foundation, she invited Patrick McGee, an economic development expert and board member of the Industrial Asset Management Council (IAMC), to attend our Trustees offsite retreat. As an outsider, Patrick gave us insights that led to the creation of the Global Workforce Initiative. He helped us see the connection between the Foundation, Economic and Workforce Development. As Trustees, Nancy and I spent an entire summer researching the future of education and how the Foundation might connect with government agencies, economic and workforce development to bring more people into the profession using outside funding.

Nancy's research uncovered enlightening information about the future of academia, P-Tech schools and the growing trend of industry certificates and credentials in community colleges. With increasing costs of higher education, particularly in the USA, we knew a different approach was needed for those who couldn't afford college. Patrick reminded Nancy and I of a common friend among the three of us who could help us think through GWI. Mary Jane Olhasso, CFO of the County of San Bernardino, California, and former head of Economic and Workforce Development had a global reputation in the field as an innovator. Mary Jane helped us strategize and create the GWI model.

We learned from Mary Jane that, to be successful, the Foundation must work closely with regional and local governments, multi-national corporations, national companies, and small- to medium-sized businesses to identify, design and implement alliances that leverage the strengths of the corporate and government sectors on behalf of facility management for social good. She introduced us to the President of Chaffey College and, working with Jim Caldwell from the California Community College Chancellor's Office, our first GWI program was launched in 2017.

Figure 3 Chaffey College Student Chapter 2018 with Phyllis Meng, Diane Levine, Tony Soriano and County of San Bernadino Economic & Workforce Development department staff. (*Source:* IFMA / Chaffey College Student).

Jim Caldwell secured a grant through California Community Colleges to fund the Essentials of Facility Management (EoFM). Sixty percent of the students at Chaffey are economically underserved and 70% require financial aid. The FM program connects to the Associates Degree in Business and includes EoFM, a project management course, and a mandatory internship. Phyllis Meng became the instructor and, in the first three months, helped to create a student chapter. In 2019, the Chaffey College student chapter won chapter of the year (see Figure 3).

Chaffey students have received IFMA Foundation scholarships and obtained FM jobs with our GWI Advisors JLL, Sodexo, and ABM Industries. The County of San Bernardino supported the program and collaborated with local companies to provide jobs for students. Their workforce and economic development team attended Chaffey IFMA student chapter meetings and worked to break down any barriers to achieving a degree. Mary Jane Olhasso said that "the Chaffey FM program was literally taking families out of poverty." Students at Chaffey are now FMs and leaders of the Inland Empire Chapter. The "Chaffey model" expanded to other community colleges in northern and southern California and Missouri and became the catalyst for Foundation grant programs.

My husband, Dr. Art Levine, became part of the IFMA family. He advocated for the profession and promoted the Foundation and IFMA on his southern California based television show "Straight Talk TV." Jeff Tafel, Tony Keane, and many other IFMA VIPs were frequent guests discussing workforce development. Art even hosted a mock TV show as a World Workplace Keynote speaker in 2013 interviewing CEOs from major FM outsourcing firms.

In 2018 Jeff Tafel announced his departure as the Foundation's Executive Director. Concerned about the future of GWI, after all those years working to help it grow, I decided to apply for Jeff's position and became the new Executive Director. My dream to lead a nonprofit, while not a musical one, finally came true.

That same year, Dean Stanberry introduced me to the Denver Economic Development department. After learning about the Chaffey program, the City of Denver was intrigued and together we developed our first government contracted pilot program. Using the "Chaffey model," I wrote a proposal to Denver for FM workforce training and development. Luckily, it was approved. This was the beginning of the Foundation's GWI FM Talent Development Pipeline (TDP) contracted programs. The pilot included 10 students, 50% were women, and 50% men with half being veterans.

The success of the Pilot led to the City of Denver recommending to the State of Colorado that the Foundation create a US Department of Labor (DOL) FM Registered Apprenticeship program. The staff at the city and state generously walked us through the entire process. Irene Thomas Johnson, Trustee and JLL Executive Director, and others helped to create the work competencies, as JLL had experience creating an apprenticeship program. In 2021, the Foundation became the US DOL official sponsor for the "Facility Manager" registered apprenticeship program.

During COVID, the Facility Manager at Hot Bread Kitchen (HBK) in New York City contacted me to learn more about FM Education. Hearing that HBK was a nonprofit involved in educating immigrant women and women of color in the food service industry, and restaurant jobs were scarce due to the pandemic, I developed the idea to ask HBK to expand into FM training. After speaking to the president and staff, we started another FM Talent Pipeline program at HBK. The IFMA NYC Chapter was instrumental in providing instructors, mentors and working with the IFMA community to find employment for the graduates. They set the chapter GWI model and won an IFMA Foundation Award of Excellence for their efforts.

Now, with experience in contracted programs in Denver and New York City, along with the US DOL approved apprenticeship program, the Foundation was ready to seek grant funding to expand GWI. In February 2021, we created a Grants committee and five months later Jim Caldwell, a grant expert, became a Trustee and leader of the committee. Jim orchestrated our grant applications strategy, and, within 4 months of his arrival, we started applying and winning grants. The Foundation was awarded two grants totaling $650,000 from Texas Workforce Commission for an FM Pre-apprenticeship program and Denver Climate Action, Sustainability & Resiliency Office (CASR) for a Good Green Jobs program. Partnering with Goodwill of Houston, Goodwill of Colorado and the IFMA Houston and IFMA Denver chapters, together we are providing green jobs and economic mobility for students enrolled in these grant programs.

We won two additional grants totaling of $462,000 from Arapahoe County Colorado and Minuteman Technical Institute Senator Donnelly Grant in Massachusetts. The Grants Committee in their first year achieved $1,112,000 in FM workforce development funding and we hired James Morante as our contracted Workforce Development Advisor to manage these programs. These funds are providing FM workforce development and education to economically underserved communities, women, veterans, and workers displaced by COVID.

The success of GWI and grant programs is accomplished with support from IFMA Foundation GWI advisors JLL, Sodexo, and ABM Industries, who not only fund the Foundation but also provide employment and career pathways for students graduating from accredited degree and grant programs. Irene Thomas Johnson champions the GWI Advisor program within JLL and her efforts enhance the GWI Advisor model for others to achieve maximum benefits of the partnership. All the Advisors meet on a regular basis to provide strategic direction in Foundation workforce development programs and share best practices in talent acquisitions, internship, and apprenticeship programs. The Foundation and GWI Advisors partner together on innovative ways to attract new talent to the profession and make a difference in society. Last year, 2021, JLL alone hired seventeen students through these programs.

The IFMA Foundation and the United Nations

Christina DeBono, CEO of Cleartech Media and IFMA SoCal WE Hub Leader, insisted that I meet Susan Angus, Executive Director of the Commission on Voluntary Service in Action (CVSA) who, as Christina said, was "doing amazing work on the UN 2030 SDGs." Susan was not only raising awareness about the Sustainable Development Goals, but she was also involved in their development, as CVSA has been a UN Non-Governmental Organization (NGO) since 2003. Before long, I was asked to join CVSA's advisory board and learned more about UN NGOs and theualifycations. Susan encouraged the IFMA Foundation to apply as an NGO and advised us on the application process. Nancy and I spent two weeks day and night writing the application submitted in April 2020. Two years later, after UN committee votes, the IFMA Foundation became a UN NGO.

The Foundation is in a unique position to contribute to the work of UN Economic and social Council (ECOSOC) because our mission with GWI addresses the three pillars of sustainable development: economic, social, and environmental. Meeting this demand is critical to the energy efficiency of the built environment, reducing greenhouse gas emissions, increasing economic performance of these buildings, and providing education and employment opportunities in the profession.

FM is one of the few professions than can impact all the seventeen UN 2030 SDGs and, this fact is used to attract students to facility management degree programs, particularly in Europe. Sustainability skills will remain essential for FMs in the future and will increase the leadership potential for those with demonstrated expertise in maximizing efficiencies, streamlining building operations, and implementing projects that save money, increase worker well-being and performance, operate healthy buildings, and positively impact the community and the planet.

As a UN NGO with Special Consultative Status, the Foundation will raise awareness of the profession and the work of ECOSOC and the UN 2030 Sustainable Development Goals, as well as work with UNESCO NGOs in the built environment. We intend to share ideas, learn about international developments, collaborate, and form partnerships in government, the private sector, and civil society to further our mission, share innovations, and participate in the latest progress to advance and achieve the UN 2030 agenda. As the accrediting body, together with ABET for Facility Management degree programs, we have a constituency of global experts in facility management and the built environment, through the FM Accredited Degree programs to call upon to assist in supporting the work of UN ECOSOC.

Conclusion

I am living proof that education and workforce development programs improve lives. Lacking financial resources to reach my goals, I faced the same barriers that students in lower socioeconomic status face today. Luckily, I found my pathway to success through a myriad of resources included a local government workforce program, scholarships, and teachers and mentors both at work and at IFMA who guided me in the right direction. I am fortunate in my wide-ranging, global experiences in the FM profession, IFMA, and my current position at the Foundation and am grateful to use this knowledge and pay it forward. And, of course, I continue to play my violin as a part-time musician, which sparks creativity in my work at the Foundation (see Figure 4).

Figure 4 Diane Fiddling for the Foundation – Raising funds at the IFMA LA St. Patrick's Day Golf Tournament 2021. (*Source:* IFMA).

Diane Coles Levine, IFMA Fellow, MCR
Executive Director, IFMA Foundation
Houston, TX USA

As the Executive Director of the IFMA Foundation, Diane is passionate about igniting the future of Facility Management. She is one of the founders of the IFMA Foundation Global Workforce initiative and Past Chair of the IFMA Foundation. With over 25 years of experience in facility management, when practicing, Diane specialized in corporate real estate, facility management, business resilience, and workplace strategy.

Diane's work has been written up in *Business Week, Return on Performance Magazine*, and other publications. She's won numerous awards, including the IFMA George Graves Facility Management Achievement Award, the CoreNet Southern California REmmy for "Workplace Innovation," and was named a Southern California Real Estate Journal "Woman of Influence." Diane is the co-founder of the IFMA Foundation Workplace Strategy Research Summits held at Cornell University and Wokefield Park, London. She is a Guest Lecturer at MIT and Vienna University of Technology.

Diane is the editor with Nancy Johnson-Sanquist and co-author of the "Work on the Move" book series and co-author of *Cut It Out: Save for Today, Build for Tomorrow*. She is a contributing author to *Facility Management Journal* (FMJ) and other industry publications with over 30 articles published. Diane is a past member of the IFMA Board of Directors and was recently named an IFMA Fellow in 2020. She holds a Master's in Real Estate (MCR) designation from CoreNet, a Bachelor of Science in Business and Management, a Certificate in Facility Management from University of California, Irvine, along with a Bachelor of Arts in Music, Instrument – Violin.

3.10

Facility Management

Whence It Cometh, Whither It Goeth?

John Adams

In the Beginning...

The concept of Facility Management evolved in the 1970s and early 1980s as the result of a confluence of several forces. At a time of unparalleled growth in what Peter Drucker termed "knowledge workers," the office emerged as a workplace of vital concern to the health and growth of businesses around the world. But unlike manufacturing environments and agricultural environments, the world of office workers and office work was largely an overlooked domain as well as an unmanaged resource.

One of the key insights informing this new perspective was the recognition that change is the only constant in a dynamic world. The first to recognize these changes and their implications was Robert Propst, President of Herman Miller Research Corporation. His seminal book, *The Office: A Facility Based on Change*, was first published in 1968, and led to the development of flexible, adaptable office furniture systems such as Action Office, that enabled corporations to manage their work environments in an active and dynamic way, rather than as static and fixed places. The office furniture industry was forever changed as many companies followed suit and developed similar systems furniture products.

But Propst, working with David Armstrong, a university educator at Michigan State University, quickly realized that innovative products themselves were not sufficient to realize the potential of these new ideas. Propst and Armstrong recognized that large organizations did not view their environments as management tools to advance their aims and business objectives. Rather the environment was largely a "backwater" of management attention. Being in charge of corporate facilities was perceived largely as a caretaker position and certainly not the fast track to the executive ranks. At the same time there were a few agile and visionary practitioners out in the corporate world, whom Propst and Armstrong characterized as professionals without a profession. The tools, research, support systems, and professional infrastructure to support their efforts simply did not exist.

Propst and Armstrong, working in concert with those visionary practitioners, concluded that a concerted effort and investment needed to be made to professionalize the nascent field of facility management, and in 1979 with the support of Herman Miller Inc., the Facility Management Institute was founded in Ann Arbor, Michigan.

At the root of these investigations, insights, and conclusions lay the underlying belief that environments matter, to their occupants, to their owners, and to the organizations with the responsibility to manage valuable corporate assets wisely.

Facilities @ Management, Concept - Realization - Vision, A Global Perspective, First Edition. Edited by Edmond Rondeau and Michaela Hellerforth.

The Rest Is History...

FMI developed a program consisting of education, research, and consulting initiatives focused entirely on managing workplaces more effectively. The model utilized had its roots in the philosophical approach of the university model of agricultural education from which Armstrong came. The Research program aimed at describing, measuring, and exploring the dimensions of facility management problems and their possible solutions. The education program had a strong emphasis on personal development and commitment, challenging participants to re-imagine the role of FM in their organization and their personal contribution to it. These two programs were coupled with the equivalent of agricultural extension whereby newfound knowledge and know-how is delivered to the practitioners, i.e., the farmers. FMI developed a robust consulting practice to disseminate know-how to organizations in need.

In addition, the FMI seeded the development of a professional guild for facility managers, which until that time did not exist. The National Facility Management Association was organized in 1980 initially with 16 founding members, all practicing facility managers who shared the vision of professionalizing the field. At the end of its first year, the organization had grown to several hundred. In 1982 a chapter was organized in Toronto and the association was renamed the International Facility Management Association. Today the association has over 22,000 members in 100 countries. In addition, the association offers a variety of programs and services to its membership including educational seminars, an annual conference & exposition, a professional journal, a certification program, and a scholarship program. Clearly, a significant need had been detected, and FMI was instrumental in responding to that need for professionalizing the field.

As the IFMA grew and matured, and its programs evolved and developed, it became logical to transfer the research, education, and consulting programs of the FMI to IFMA. In the late 1980s those programs came under the auspices of IFMA, and the operations of the FMI ceased.

Flexible Office Systems and Facility Management: A Mixed Blessing?

Armstrong often called for "facility audits." That is, an evaluation of how well the work environment supported the people and work processes of the organization. We should now reflect on the last 40 years and audit the impact of FM and flexible office systems on the quality, productivity, and satisfaction of the office workforce. Is the world a better place for FM to have been recognized as an essential, proactive corporate activity? How well has the field of FM performed in accomplishing its mission? What are the tangible results?

Many corporations, working through their FM departments, focused not on creating more supportive and invigorating workplaces, but on picking the low hanging fruit of space savings. The cubicle became ubiquitous. The stereotype of the office worker jammed into a small cubicle came to symbolize the drudgery of office work and the sterility of corporate office culture. Caricatures such as Dilbert satirized life in the office. It is ironic that furniture systems designed to enable more sophisticated and supportive environments were used by many organizations simply to reduce number of square feet per office worker. The second result of this trend was the use of office furniture systems to reduce cost of reconfiguration of the environment. The so called "churn rate" of organizations, was tracked and measured. The promise of flexible office furniture systems and of more sophisticated office planning and facility management was only partially fulfilled.

The professionalization of FM conferred legitimacy and status on facility managers, but did work environments get better? What were the results? Did office workers become more satisfied with their environments and productive in their work due to advances in FM? Not so much. A few fresh and innovative environments were created, but far too many dull cubicles diluted the workplace. The **profession** of FM has developed and advanced, but the quality of work environments has improved only marginally. The **purpose** of facility management has been only partially realized.

Fast Forward 40 Years and Lo and Behold...

If the singular conclusion that change was the only constant was central to the genesis of FM, then past is certainly prologue. In the 40+ years that have ensued, the world of business has seen unprecedented change, much of it only vaguely anticipated from the vantage point of the late 1970s and early 1980s.

From Peter Drucker's emphasis on the importance of managing the productivity of knowledge workers and viewing a well-designed and maintained work environment as a benefit of employment, the world of work has devolved into the gig economy characterized by chronic employment and income insecurity, and the social contract between workers and their employers has become a challenge for many. Some of today's organizations seem scarcely concerned with attracting and retaining human capital. The current attitude by some organizations appears to be that whatever skills are needed can be purchased transactionally, and workers may be seen to be simply a commodity to be acquired by the purchasing department. This implies that corporate settings, and by extension corporate facilities, may no longer be a primary source of identity, purpose, and community for the work force.

Aiding and abetting this commoditization of work and workers has been the globalization of the work force, further splintering what was once an integrated corporate whole into more disparate pieces through "outsourcing."

Work from home is now widely accepted and preferred by a plurality if not a majority of the workforce. Many office workers see themselves now as visitors to the office not occupants. The era of dedicated workstations has shifted radically, largely as the result of world's response to the COVID-19 epidemic. Along the way the cubicle approach to office planning has largely collapsed under its own weight and we now see more experimentation with new and different approaches to structuring the workplace.

In the early days of FM, office automation was a concept that promised many changes to the world of office work. Starting with the advent of the PC, word processing, voice mail, and a variety of business information systems, office planning experienced significant change. In the 40 years since, the technologies of laptops, wireless networks, smart phones, and the internet have changed the environment even more. Many of these changes were not widely anticipated by FM, but rather they reacted to and accommodated these changes as they occurred. It seems certain that profound changes await us over the next 40 years, and we can only dimly speculate about how they might change office work, office life, and office environments.

In a profound way, Propst got it right in *The Office: A Facility Based on Change*, and the premise is no less applicable today than it was in 1968 when the book was first published. It seems a safe bet that we can anticipate fundamental changes in office work that rival or exceed the changes we have experienced over the last 40 years.

Prophecy Is Very Difficult, Especially with Respect to the Future

So, what's next? What forces of change will bring FM a new set of challenges in the years to come?

1) The aging of the population, the end of retirement as we once knew it. People are living longer, more active lives. 10,000 people turn 65 everyday in the United States. Many of these age cohorts are choosing to work beyond traditional retirement age both by choice and by necessity. Many will participate in the gig economy in one way or another. They will be a part of the workforce which FM must support and facilitate. In the residential environment there is a strong move towards "aging in place," meaning adapting existing home environments to be more supportive of older people, rather than the traditional model of moving to retirement facilities. An older work force implies that the work environment must also become more supportive and accommodate the needs of an older population with a variety of mobility and health needs. The environment can embody prosthetic features, i.e., the environment can compensate for the loss of abilities as people age. The concept of Universal Design is gaining currency in residential settings, and office work environments will not be exempt from this need and trend.

2) We need to accomplish some serious downsizing, eliminate overconsumption, and make a serious commitment to sustainability as the world grapples with climate change. We must do more with less. The likelihood of massive climate migration, relocation, and revaluation of real estate driven by climate change and rising sea levels has been widely predicted. This will inevitably have a profound impact on both commercial and residential real estate markets and by extension the field of FM.
3) The decentralization of everything? The modern world is highly dependent on distribution systems. Our electricity comes from somewhere else. Our food comes from somewhere else. Our water comes from somewhere else. Our clothing comes from somewhere else. When distribution systems falter, there are profound effects on our ability to live our lives as we expect. Witness the impact of the chip shortage during COVID as well as other logistical backlogs. There is some indication that the pendulum will swing back toward local production of the goods and services needed for our daily lives, making us in turn less dependent on systems of distribution.
4) The continuing evolution of computing technology, the internet of things, AI, social media, etc. will without doubt continue to change the nature of work and therefore the environments in which work is done. There is every reason to believe that the impact of these new technologies will be very much as revolutionary as what we have experienced in the last 40 years.
5) New work forms, new work settings, and new work communities seem to be an inevitable consequence of the changes that are in process.

Implications for FM

The trends and impending changes outlined above strongly imply that there will be no shortage of challenges and opportunities for the field of FM and its practitioners. Is there a compelling reason to change the orientation and practice of FM?

If we are in fact entering an era characterized by the need to limit consumption and share scarce resources, then systems for sharing these resources will need to be devised and managed. Take for example the simple example of the "hot desk," a work environment shared by multiple workers at different times. This is not a new idea at all, and its re-emergence will likely change the paradigm for office environments. No longer will there be a dedicated piece of real estate and related furniture and equipment allocated and assigned to each individual worker. Instead, a smaller footprint of resources will be utilized more intensely and efficiently. Now think about the systems for sharing, scheduling, and maintaining these environments so they are worker-ready in real time. We are describing a much more dynamic environment which will require active, energetic management.

If multiple workers are utilizing these environments in rapid succession, how are the environments tailored to the needs of the individual worker and his or her unique work processes and needs? The notion of adjustability of the environment needs to be re-imagined in a faster, more responsive time scale, i.e., minutes and seconds, not days and hours. One can anticipate that the adjustment and tailoring of the environment to meet these needs will become a personal activity of the occupant, a part of quickly setting up shop for a limited period of time. These changes need to happen "at will" and with minimal support and intervention by FM personnel. The kitchen table at home can be used for homework by the kids, then quickly reorganized for a family dinner, and afterwards used for work at home by a spouse. This implies new levels of functionality in furniture, systems, and equipment. It also implies more personal involvement by workers in making these changes and adjustments. Environments will become more personal and less corporate.

The shift to work from home, accelerated by our response to the COVID virus, suggests that much of the "facility management" will occur in the home, not the office. If so, what is the role of the traditional corporate FM organization? Devising support systems and supporting workers indirectly rather than providing formal project management services in a corporate environment seems a likely trend. Word processing departments no longer exist in the corporate world because workers can use the computer-based tools and systems directly. Will the same be true for FM? We can envision FM becoming a more personal activity which is intrinsic to the work itself, not a request for service from an FM department. Whatever form these new patterns of managing work environments may take, we can safely predict that there will be an ongoing "need for nimble" by all concerned.

We can further anticipate, with the scope of changes mentioned above, that there will be a great deal of experimentation with new environments, new ways of using and managing them, and the results will be mixed. Both organizations and their workers will need to be engaged in trial and error, figuring out what works and what doesn't, extracting the lessons learned, and distributing the know-how throughout the organizations in the mode of best practices. FM can be a catalyst for capturing this emerging knowledge and know-how and can play a role more as facilitator for the work force and a provider tools, systems, and techniques.

All of this strongly implies that FM will need to transform itself, abandon obsolete roles, services, and practices, and become a true learning organization. Otherwise, it risks becoming irrelevant to the organization. There is indeed a compelling reason to change. FM needs to be reimagined in the same way that office work and office work environments themselves are being reimagined as we adapt to the changes that are underway in our world.

As the Chinese proverb states: May you live in interesting times.

John R. Adams
Consultant
Chicago, IL USA

For the past 22 years John Adams has led and managed two companies in the Chicago area: HomeCrafters LLC a residential design build construction company and Vesta LLC, a design, planning, and construction consultancy.

Prior to that he held a number of management and executive positions at Herman Miller Inc. He served as Director of Herman Miller's workplace consulting organization and served as Vice President of Milcare Inc., Herman Miller's health care subsidiary with responsibility for research, marketing and communications, product development and engineering.

From 1979 to 1985 he was a member of the management team of the Facility Management Institute in Ann Arbor, Michigan, an organization formed and funded to promote the professionalization of the field of facility management. He served as Assistant Director of Research, Associate Director of Consulting, and ultimately served as Director of the Institute with general management responsibility for its programs in research, education, and consulting.

From 1974 to 1979 Adams held the position of Research Associate at Herman Miller Research Corporation in Ann Arbor with responsibility for managing various research and development investigations and for managing several experimental facilities developed in concert with several Fortune 500 companies both in the US and abroad.

He holds a Bachelor of Architecture degree from North Carolina State University and a master's degree in environmental psychology from the Pennsylvania State University.

3.11

Integration of Business and User-Centered Corporate Real Estate and Facilities Management

Theo van der Voordt

Personal Involvement in FM

A red thread in my academic carrier of over 40 years is the development and sharing of knowledge about the experience and use of the built environment, including offices, health care facilities, childcare facilities, learning environments, housing, and public spaces. The latter is currently part of urban FM. My main drive to do this for such a long time is the willingness to contribute to an environment that enables people to conduct their activities in a satisfactory, comfortable, efficient, and effective way, and that fits with the preferences, needs, and values of organizations, customers, end users, and society as a whole. In other words: to contribute to user-centered and value-based briefing, design, management, evaluation, and further improvement of buildings, facilities, and services.

I started my carrier as an engineer. However, during my study in civil engineering I noticed that I am less interested in concrete, steel and building technology, and more in what drives human beings. I became more aware of this in my first job as a construction engineer. After thorough consideration I left my job and applied for a job as a research assistant at the Centre for Architectural Research at the Delft University of Technology. This small group included sociologists, psychologists, an expert in acoustics, an economist, and an urban planner. My first research regarded the image of a city, based on the ideas of Kevin Lynch (1960): what environmental characteristics can help people to build a clear internal image and mental map of a city, and as such support them in spatial orientation, wayfinding, and creating a sense of belonging and being connected? Quite soon our team started to study health care centers and childcare centers, in order to develop guidelines for briefing, design and management. In addition to Post-Occupancy Evaluations (POE), using observations, walk-throughs, questionnaires, interviews, and group discussions, we developed the method of comparative floor plan-analysis (CFA). Key in CFA is an assessment of a collection of floorplans on similarities and similarities regarding the access of the buildings, inner circulation patterns, spatial-functional layouts etc., and trying to understand the motives behind design and management choices by interviews with clients, architects and end users, and other POE methods. This makes it possible to develop so-called annotated typologies of buildings: design alternatives with comments on what works and what works not, why, and for whom.

Other main research topics in the eighties and nineties of the last century were crime prevention through environmental design (CPTED), Universal Access, i.e., Design for All, building adaptable housing (to make them more suitable for people with physical impairments), and assisted living facilities for the elderly, including people with dementia. The insights from these two decades were summarized in the book *Architecture in Use* (Van der Voordt and Van Wegen 2005), which is also available in Dutch and Portuguese. The methodological lessons learned regarding POE and CFA and different types of design research, design studies and typological research

Facilities @ Management, Concept - Realization - Vision, A Global Perspective, First Edition. Edited by Edmond Rondeau and Michaela Hellerforth.

have been incorporated in the book *Ways to Study and Research Architectural, Urban, and Technical Design*, edited by Taeke de Jong and myself (2002), with over 40 contributions of other staff members of our faculty.

As common in most large organizations, I had to cope with various reorganizations. In the mid-1980s our Centre for Architectural Research merged with the Centre for Urban Research. Later on, our staff was re-allocated to different departments. I moved to the department of planning, design, and management of buildings. Although I learned a lot from my new colleagues – mainly architects – I noticed that the end users were not key in this group. So, in the late 1990s I moved again, to the department of real estate and project management, later renamed as Management in the Built Environment (MBE). Here I got familiar with corporate real estate management (CREM) and facilities management (FM). My focus shifted toward workplace studies, in particular regarding the impact of activity-based work environments on employee satisfaction, perceived productivity and wellbeing, and performance management and measurement.

Other topics in the last decades are adaptive reuse of vacant office buildings, models and tools for the development, implementation, evaluation, and monitoring of successful CREM and FM strategies, and adding value through appropriate design and management of buildings, facilities, and services. A special POE in which I was personally involved regarded a user survey of another Faculty Building to which we had to move after our former faculty building burnt down in 2008 (Gorgievski et al. 2010). An overview of the teaching and research of the MBE department can be found in the book *Dear is Durable* (2016), edited by Arkesteijn et al. (2016). Overall, I was and still participate in a huge variety of FM and FM-related topics and co-authored and co-edited numerous publications with different people (see the bibliography at the end of this contribution). This shows the wide scope and richness of our field.

Another red thread in my carrier is lifelong learning. With a background in civil engineering, I had to learn new theories and tools from the fields of environmental psychology, sociology, statistics, architecture, and corporate real estate and facilities management. I got inspired by journals like the Journal of Environmental Psychology, Environment and Behavior, Building Research & Information, Facilities, the Journal of Facilities Management, the Journal of Corporate Real Estate, and many more. Furthermore I got inspired by many conferences, in the 1970s and 1980s mainly of the International Association of People and the their Physical Surroundings (IAPS) and the Environmental Design Research Association (EDRA), since the late 1990s in particular by the European Facility Management Conferences (EFMC) of EuroFM, the European Real Estate Society (ERES), the International Council for Research and Innovation in Building and Construction (CIB), and more recently also of the Transdisciplinary Workplace Research network (TWR).

Currently I find myself as a teacher and researcher of the built environment, not knowing whether I should call myself an environmental psychologist, design researcher, specialist in people-environment relationships or FM/CREM expert. But what's in a name? The most important is to contribute to a better fit between business and people and their physical surroundings.

An anthology does not provide the space to write extensively about all mentioned topics, even when it is a personal one. In the next sections I will discuss three topics: (1) similarities and dissimilarities between FM and CREM; (2) trends in CREM and FM strategies; and (3) adding value through FM and CREM.

Similarities and Dissimilarities between FM and CREM

ISO 41011:2017 – Facility management, Vocabulary, defines Facility Management (FM) as an organizational function which integrates people, place, and process within the built environment with the purpose of improving the quality of life of people and the productivity of the core business. A former definition defined FM as the integration of processes within an organization to maintain and develop the agreed services which support and improve the effectiveness of its primary activities. Corporate Real Estate Management regards the management of the real estate portfolio of a corporation by aligning the portfolio and services to the needs of the core business, in order to obtain maximum added value for the business and to contribute optimally to the overall performance of the organization. In the past, facilities management (FM) and corporate real estate management (CREM) used to be rather separated. Both fields have a different history, different key objectives, concepts, theories, data, and tools (Van der Voordt et al. 2016).

FM originates from professionalizing IT services and is traditionally linked to facilitating people and business processes in buildings-in-use, by appropriate furniture, plants, cleaning services, security services, and so on. CREM regards accommodating people and is usually linked to the whole life cycle of buildings and real-estate portfolios, from the first initiative and briefing and design phase till managing of buildings-in use, renovation, adaptive reuse, or demolishment and new building. FM and CREM have their own journals, too, e.g., *Facilities* and the *Journal of Facilities Management* versus the *Journal of Corporate Real Estate* and *Corporate Real Estate Journal*. However, conferences such as the European Facility Management Conferences and the CIB World Building Congresses discuss topics that are related to both FM and CREM, like workplace management, performance measurement, benchmarking, added value, maintenance, and sustainability. Conferences under the auspices of the International Real Estate Society's sister societies discuss CREM- and FM-related topics as well. Both FM and CREM are pretty young academic disciplines that are strongly based on practice. Both FM and CREM focus aim to provide well-designed and appropriately managed buildings, facilities and services that add value and fit with the needs of clients, customers, end users and society as a whole. Table 1 shows a number of similarities.

Figure 1 below visualizes the relationship between alignment (of demand and supply, and organization/end users and their physical environment) and adding value.

Trends in FM and CREM strategies

In 1993, Michael Joroff and his team[1] presented a five-stage real estate evolutionary model, from technical to strategic, from engineering buildings, minimizing costs, standardizing usage, and matching market options to convening the workforce. According to Joroff et al., the traditional role of maintenance by ad hoc interventions has

Table 1 Similarities between FM and CREM (adopted from Van der Voordt et al. 2016)

- management disciplines, including strategic management, procurement management, workplace management, information management, risk management, relationship management, financial management, project management, contract management, change management, quality management.
- connect people, processes, housing, facilities, services, and technology.
- facilitate the effectiveness and efficiency of the organization.
- shift from a one-sided focus on cost reduction to more attention for added value.
- facilitate new ways of working through workplace innovation and activity-based workplace concepts.
- search for the optimal balance between individual use and shared use of spaces and facilities, insourcing and outsourcing, centralization and decentralization, and shared service centers.
- increased attention to the entire lifespan of buildings and facilities.
- increased awareness that investment decisions have consequences for organizational performance and operating costs in the use and management phase.
- public-private partnership in major projects.
- involvement in an early phase of the plan development.
- increased attention to sustainability, adaptive reuse, smart technology, and wellbeing.
- increased attention for a variety of functions, including offices, healthcare, education, retail and leisure, and industry.
- evidence-based and data-driven decision-making.

1 Joroff, M, Louargand, M., Lambert, S., & Becker, F. (1993). *Strategic management of the fifth resource: Corporate Real Estate*. Industrial Development Research Foundation, United States of America. Report 49.

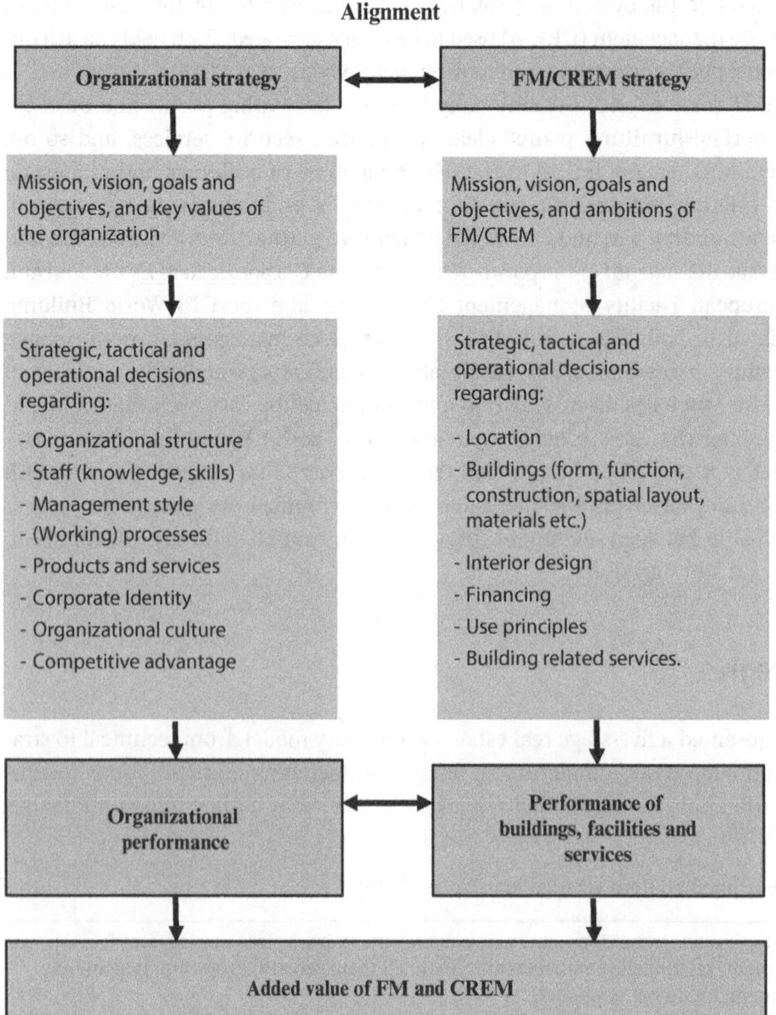

Figure 1 Relationship between alignment and added value of FM and CREM. (*Source:* T. van der Voordt).

shifted toward a more strategic role, with a cumulative integration of minimizing real estate costs and cost efficiency (controller), standardization of building usage (dealmaker), matching real estate with business plans and market options (intrapreneur) and management using performance indicators regarding costs and quality (business strategist); see Figure 1 and Table 2. The fifth stage includes acting in a planned and proactive manner in cooperation with other disciplines, strategically aligning the accommodation with the vision, mission, and goals of the organization and the external context and incorporating different stakeholders. The stages are cumulative: each subsequent stage builds on the preceding stages. Each more complicated stage adds a new role in the search for real estate value. The first three stages occur principally through project level work related to the internal needs of the corporation. Stage four addresses portfolio-wide needs, focusing on outward trends affecting the business units. Stage five tackles company-wide competitiveness, involving a myriad of stakeholders outside the corporation's more traditional bounds. Furthermore, as the organizational stages evolve from taskmaster to strategist, the benefits obtained by stakeholders evolve from short- to long-term, with a growing customer orientation and a need for continuous learning and change. Each successive level brings the real estate unit closer to the senior corporate management.

Table 2 Main characteristics of the five stages. Joroff et al. (1993) / Industrial Development Research Foundation

1	Taskmaster	Supplies the corporations' need for physical space as requested
2	Controller	Satisfies senior management's need to better understand and minimize real estate costs
3	Dealmaker	Solves real estate problems in ways that create financial value for the business units
4	Intrapreneur	Operates like an internal real estate company, proposing real estate alternatives to the business units that match those of the firm's competitors
5	Business strategist	Anticipates business trends, monitors, and measures their impacts, contribute to the values of the corporation as a whole by focusing on the company's mission rather than focusing only on real estate

Joroff et al. wanted to contribute to corporate real estate managers' awareness that "their business is not real estate, but the business of the business." The CREM maturity model provides a framework for analyzing, creating, and managing a strategy for change. It outlines a pathway for the evolution of how Corporate Real Estate can be managed as "a fifth resource of a firm," in addition to capital, people, technology and information.

Michael Bell, one of the team members, identified twelve shifts: (1) from real estate orientation to a business focus; (2) from a transactional to a process orientation; (3) from control-oriented to service-oriented; (4) from reactive to pro-active; (5) from decentralized to centralized; (6) from in-house expertise to collaboration; (7) from hiring experts to do a job to inviting service providers to become members of the team; (8) from automate to automation i.e. using information technology; (9) from relationships built on personal contact to interactions supported by information flows; (10) from big to small; (11) from standardization to customization; and (12) from real estate skills to general management capability. This was noticed already in the early 1990s! Issues such as activity-based working, teleworking, how to maintain a sense of community, cost savings, productivity, flexibility, satisfaction of the staff and senior management, and the added value of CRE were included in this report already as well. The report marked a paradigm shift in how corporation leaders understand the concept of "workplace" and perceive the "value" of the real estate that they own or lease.

Twenty years later, Joroff and Becker (2016[2]) argued that the evolution of corporate real estate reflects six primary shifts in how corporate real estate is viewed and how it can best be managed, with close connections to workplace management:

1) From financial to Business Asset i.e., a shift in the mindset that viewed corporate real estate as a passive financial asset with a high cost to one that perceives the real estate portfolio as an asset integral to the conduct of the business, with high use value and proactively promoting new ways of working that, along with more flexible, informal and open corporate management and culture, and transformative information technologies, enhances business performance.
2) Workplace as an integrated Ecological System, comprised of physical design and space, information technologies, workforce demographics, work processes, and organizational culture. The design and management of these interdependent factors aim to support different kinds of work, not only as a place that houses people to do assigned tasks, but also as a means of attracting and retaining the best and brightest employees and engaging and enabling their talent and energy. The "workplace" is more and more recognized as a system of loosely linked spaces inside and outside the "office" (the building) designed to support specific activities such as quiet work, informal communication, and client and group meetings, and relies on cyber as well as physical space.

2 Joroff, M., & Becker, F. (2016). Exploiting change and uncertainty to drive corporate value. In Arkesteijn, M., Van der Voordt, T., Remøy, H. & Chen, Y. (Eds.), *Dear is durable* (pp. 105–113). TU Delft Open.

3) Needs vs. Preferences. Where once the modus operandi of the corporate real estate function was simply to take orders from business units for property, and deliver it on time and within budget, now the role is to proactively work with business units to anticipate their needs and to sharpen their understanding of how to best meet these needs through real estate and workplace strategies.
4) Power and Opinion vs. Data. Decision-making about real estate and workplace investments is now more often underpinned by analytics and rigorous review of business context, data about real estate financial, individual, and team and department performance data, and how space is being used.
5) From stable/static to Agile. At the time when the corporate real estate paradigm began to shift, enterprises were largely perceived as relatively stable, with a known culture and known tasks and processes. Today, everything is subject to change. This requires facilities and arrangements for corporate tenancy that are flexible, in which space can be rapidly acquired and just as rapidly abandoned almost anywhere in the world.

Currently, IT-enabled time- and location-independent working has become a daily reality for knowledge workers. New work practices like blended working and activity-based working seem to become the new normal worldwide. Workers are more and more enabled and allowed or even encouraged to use different locations (e.g., corporate offices, client or partner offices, home office, co-working spaces, on the go) and work settings within the office (open and enclosed workstations, phone booths, lounge areas, project rooms). Along with the expanding range of choice, individual workers and teams are discovering and adopting their own preferred ways of working. Recent events like the COVID-19 lockdown may work as a catalyst in this process.

Organizational behavior has become more central in implementing corporate strategy. Particularly for knowledge-based organizations, desired outcomes are highly dependent upon behavioral patterns in the workplace (how workers collaborate, learn, concentrate, and recuperate). Hence, we see many corporate programs focusing on behavioral change, which is frequently linked to workplace change. "Nudging" desired behavior through workspace design is gaining attention in both practice and research. An example is the promotion of healthy behaviors in the workplace (physical movement, relaxation, social contact, nutrition), which receives growing attention in relation to reduction of sick leave, burnout, and sustainable employability.

Due to the global "war for talent," employers can no longer force employees to work in unattractive environments or at unattractive locations. Convening the workforce has become a necessity rather than an ambition. Optimizing the "workplace experience" has become a key topic. At the same time, the user-centered approach seems to be shifting from a focus on optimizing user satisfaction toward a more goal-oriented focus on specific user needs and behaviors that are important for organizational effectiveness.

Given these developments, a sixth stage has been added to the CREM maturity model (Wijnja et al. 2021; Hoendervanger et al. 2022), see Figure 2. Where the fifth stage is focused on creating added value in relation to corporate strategy, the sixth stage adds a user-centered approach. Being both a business strategist and end user strategist, the CRE manager creates work environments that support work practices and encourage behavioral change, in alignment with both corporate goals and employee needs. In addition to the skills that are needed in stages 1–5, psychological knowledge is required to analyze, facilitate, and stimulate workers' differing and changing needs and behaviors. In an online interview with Joroff he supported this extension of the original model, emphasizing that a user-centered focus should be part of all previous stages as well. It is expected that the collaboration between CREM, HRM, and IT will be further extended by incorporating other disciplines such as labor psychologists, occupational health specialists, neurologists, data specialists, and artists.

Whereas the CREM maturity model has originally been developed by CREM-oriented people, the same trends are visible in FM. This supports the trend to a further integration of FM and CREM, which is clearly represented in current practice, where FM and CREM are often integrated in the same department or service center, with an internal distinction between strategy developers, project managers, and facility managers.

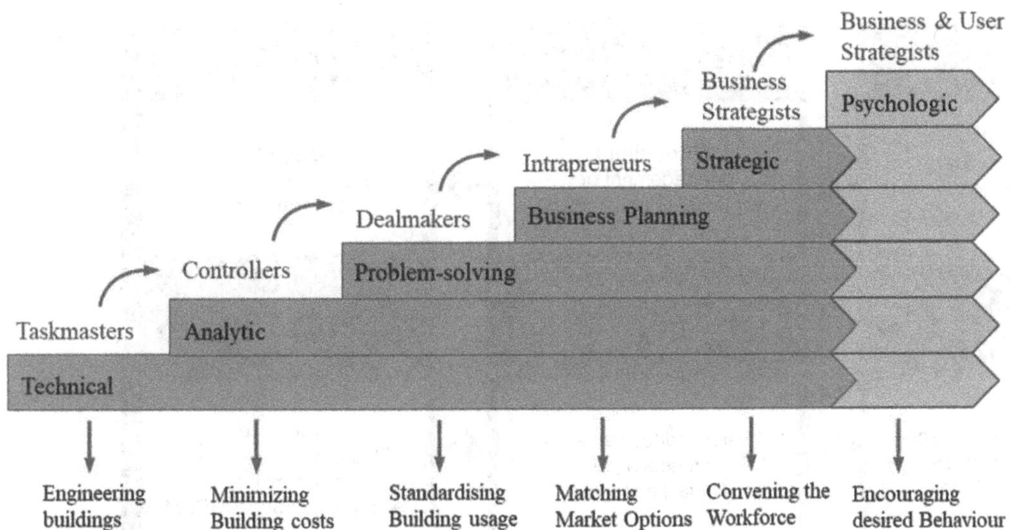

Figure 2 Five-stage real estate evolutionary model of Joroff et al (1993), extended with a sixth stage (Wijnja et al. 2021; Hoendervanger et al. 2022).

Adding Value through FM and CREM

The shift from a cost-oriented approach toward a more value-based approach inspired Per Anker Jensen, professor in FM at the technical university of Denmark (DTU), to start a EuroFM working group on the added value of FM. I was one of the participants from the start. This work culminated in a huge number of papers and conference presentations, and two anthologies: *The Added Value of FM: Concepts, Findings and Perspectives* (2012), and *Facilities Management and Corporate Real Estate Management as Value Drivers: How to manage and measure adding value* (2017).

Added value is defined as the extent to which the trade-off between benefits, costs, and risks of interventions in buildings, facilities, and services contributes to organizational goals and values, a better fit between people and buildings, and an optimal match with societal needs such as sustainability and corporate social responsibility. The books tried to open the black box of input -> throughput -> output -> outcome -> impact/added value by discussing a taxonomy of six types of interventions and different value parameters. The first book presents six value categories: (1) Use value: quality in relation to the needs and preferences of the end users; (2) Customer value: trade-off between benefits and costs for the customers or consumers; (3) Economic, financial or exchange value: the economic trade-off between costs and benefits; (4) Social value: connecting people by supporting social interaction, identity and civic pride; (5) Environmental value: environmental impact of FM, Green FM; (6) Relationship value e.g. getting high-quality services or experiencing a special treatment. Building on this work, the second book elaborated 12 value parameters by presenting state-of-the art research for each parameter and ways to manage: (1) people-related values (satisfaction, image, culture, health, and safety), (2) process and product (productivity, adaptability, innovation and creativity, risk), (3) economy (cost, value of assets), and (4) societal (sustainability, corporate social responsibility).

In order to support decision-makers in value adding FM and CREM, Hoendervanger, Bergsma, Van der Voordt and Jensen (2016) developed a Value Adding Management process model with four steps, see Figure 3. The VAM model is action oriented and follows the same steps in the renowned Deming cycle. The PDCA cycle is widely

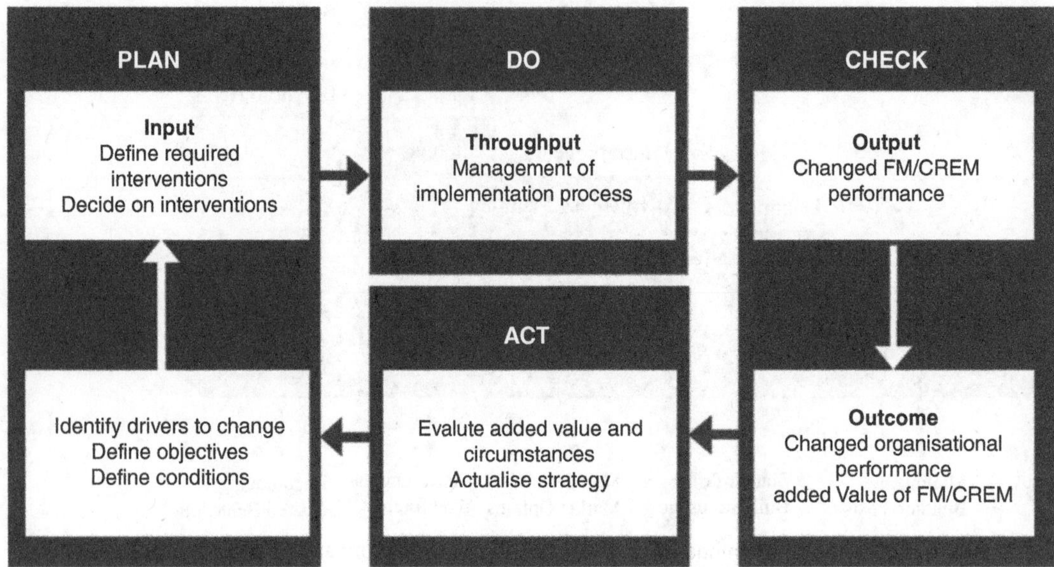

Figure 3 Value Adding Management process model (Hoendervanger et al., in Jensen and Van der Voordt 2017).

applied to support total quality management and is familiar to many practitioners. The principles of input-throughput-output-outcome/added value correspond with what to do and why, how to implement, and how to measure its impact. The link with FM and CREM is key in the VAM model.

The VAM model guides decision-makers through the process of adding value in four steps, from identification of performance gaps, objectives for improvement and selection of appropriate interventions to its implementation and a check on whether the objectives have been attained, what value has been added to whom, and which Key Performance Indicators are most appropriate to measure the added value by FM and CREM.

The main actions in the *Plan-phase* are to identify the drivers to change i.e., to define if there is a gap between the desired and actual performance of the organization and the accommodation, facilities, and services, and to define which interventions may result in improved performance. The Plan-phase ends with clear decisions about which interventions should be implemented and how to implement them. In order to support this first step, Jensen and Van der Voordt (2020b) developed a typology of Value Adding FM/CREM interventions. Analyzing the context of value adding management may start with exploring the different roles, interests and power of stakeholders involved, using stakeholder analysis. A SWOT analysis can help to identify the need and direction for change, concerning both the organization and the FM/CREM processes and products.

The *Do-phase* encompasses the implementation of the proposed interventions and management of the change process. Decisions to be made include who should be involved in the process and how, time schedules, how to cope with resistance to change, and how to cope with the different needs of different stakeholders. A major challenge is to keep focus on the initial goals. Implementation processes tend to develop their own dynamics, which can easily shift the focus from long-term strategic organizational goals to short-term tactical and operational goals of the participants.

In the *Check-phase* the costs and benefits of the intervention(s) and its impact on the performance of the organization and its facilities has to be measured, both during the change and ex-post, after the implementation of the intervention(s) has been realized. To be able to measure whether the performance has been improved, a baseline measurement, i.e., an ex ante measurement, before the intervention is implemented, is needed as well. It is also necessary to evaluate if the changed performance fits with the organizational strategy, mission, vison, and objectives and as such adds value to the organization.

The *Act-phase* is quite similar to the Plan-phase. However, whereas the Plan-phase may start with an analysis of changing internal or external circumstances or a strategic analysis of the strengths and weaknesses of the organization and FM/CREM products and processes, these factors are already considered before the Act-phase. When all objectives have been attained and maximum value has been added, the Act-phase may be limited to consolidation of the new situation, until new drivers to change come to the fore. If the objectives are not sufficiently attained or not optimally, or if too many negative side effects come to the fore, new interventions or broadening of earlier interventions should be considered. Another option is to reconsider the objectives. It may happen that the aimed performance was not realistic and feasible within the current conditions. Moreover, the context or conditions of the original objectives may have changed, which might force the organization to change its organizational or FM/CREM strategy. If new or revised interventions have to be implemented, the Plan- and Do-phases start again.

The cyclic character emphasizes that value adding management is or should be a continuous process. Evaluation of realized output/outcome/added value may be a starting point for new interventions.

Concluding Remarks

This contribution shows the growing maturity of our field. In the last decades, FM emerged from a rather tactical and operational discipline toward a discipline with a much wider and more strategic scope. Whereas traditionally FM focused on facilities and services in buildings-in-use, to support clients, customers, and end users, and corporate real estate was perceived as more strategic and important during the whole life cycle of buildings, considering the needs of clients and shareholders, nowadays both fields get more integrated, due to its many similarities, joint goals, and objectives. Both disciplines aim to optimize buildings, facilities, and services in order to get the best possible match between demand and supply, from a business point of view as well as with respect to end user needs and societal values. Both FM and CREM tend to become more data-driven, supported by modern technology, and shift from a cost reduction focus toward a wider scope and adding value. New terms such as workplace manager, accommodation manager, and sustainability coordinator who covers the whole range of buildings, places, facilities, services, and human behavior (also summarized as "bricks, bytes and behavior") come to the fore. Whereas the boundaries of FM become less clear, we should not call all building related activities FM. It might be worth rethinking the definition of FM. For the future, a further strengthening of the connection between FM and CREM and other disciplines may be expected, such as HR and IT, financial control, and social sciences, supported by new technologies like artificial intelligence, digital twins, and smart tools to manage and measure the added value of FM and CREM. Hopefully, this will help to provide smart, sustainable, and healthy buildings, facilities and services that enable businesses and people to act in a comfortable, efficient, and effective way and supports the quality of life.

Bibliography (selection)

Appel-Meulenbroek, R., Aussems, R., Van der Voordt, T. et al. (2020). Impact of activity-based workplaces on burnout and engagement. *Journal of Corporate Real Estate* 22 (4): 279–296. https://doi.org/10.1108/JCRE-09-2019-0041.

Arkesteijn, M.A., Van der Voordt, T., Remøy, H., and Chen, Y. (eds.) (2016). *Dear Is Durable*. Liber Amicorum for Hans de Jonge. TU Delft.

Bakker, I. and Van der Voordt, D.J.M. (2010). The influence of plants on productivity. A critical assessment of research findings and test methods. *Facilities* 28 (9/10): 416–439. https://doi.org/10.1108/02632771011057170.

Bakker, I., Van der Voordt, T., de Boon, J., and Vink, P. (2014). Pleasure, arousal, dominance: Mehrabian and Russell revisited. *Current Psycholog* 33 (3): 405–421. https://doi.org/10.1007/s12144-014-9219-4.

Beckers, R., Dewulf, G., and Van der Voordt, T. (2015). Aligning corporate real estate with the corporate strategies of higher education institutions. *Facilities* 33 (13/14): 775–793. https://doi.org/10.1108/F-04-2014-0035.

Beckers, R., Van der Voordt, T., and Dewulf, G. (2015). Management strategies for aligning higher education accommodation with the user needs. *Journal of Corporate Real Estate* 17 (2): 80–97. https://doi.org/10.1108/JCRE-10-2014-0025.

Beckers, R., Van der Voordt, T., and Dewulf, G. (2016). Why do they study there? Diary research into students' learning space choices in higher education. *Higher Education Research & Development* 35 (1): 142–157. https://doi.org/10.1080/07294360.2015.1123230.

Beckers, R., Van der Voordt, T., and Dewulf, G. (2017). Learning space preferences of higher education students. *Building and Environment* 104: 243–252. https://doi.org/10.1016/j.buildenv.2016.05.013.

Beckers, R., Dewulf, G., and Van der Voordt, T. (2015). A conceptual framework to identify spatial implications of new ways of learning in higher education. *Facilities* 33 (1/2): 2–19. https://doi.org/10.1108/F-02-2013-0013.

Brunia, S., De Been, I., and Van der Voordt, T. (2016). Accommodating New Ways of Working: lessons from best practices and worst cases. *Journal of Corporate Real Estate* 18 (1): 30–47. https://doi.org/10.1108/JCRE-10-2015-0028.

De Jong, T.M. and Van der Voordt, D.J.M. (eds.) (2002). *Ways to Study and Research Architectural, Urban and Technical Design*. Delft: Delft University Press/IOS Press.

De Vries, J.C., De Jonge, H., and Van der Voordt, T.J.M. (2008). Impact of real estate interventions on organisational performance. *Journal of Corporate Real Estate* 10 (3): 208–223. https://doi.org/10.1108/14630010810922094.

Geraedts, R.P., van der Voordt, T., and Remøy, H. (2017). Conversion potential assessment tools. In: *Building Resilience in Urban Settlements through Sustainable Change of Use* (ed. H. Remøy and S. Wilkinson). Chapter 7, 121–151. Wiley-Blackwell.

Gorgievski, M.G., Van der Voordt, Th., van Herpen, G.A., and Van Akkeren, S. (2010). After the fire. New ways of working in an academic setting. *Facilities* 28 (3/4): 206–224. https://doi.org/10.1108/02632771011023159.

Groen, B., Van der Voordt, T., Ypma, L., and Hoekstra, B. (2019). Impact of employee satisfaction with facilities on self-assessed productivity support. *Journal of Facilities Management* 17 (5): 442–462.

Hoendervanger, J. G., Bergsma, F., van der Voordt, T., Jensen, P. A., (2016 /10/4). 17 Tools to manage and measure adding value by FM and CREM. *Facilities Management and Corporate Real Estate Management Journal* 299: Taylor & Francis.

Hoendervanger, J.G., Van der Voordt, T., and Wijnja, J. (2022). *Huisvestingsmanagement: Van Strategie to Exploitatie*, 3e. Groningen: Noordhoff Uitgevers.

Houben, P.P.J. and Van der Voordt, D.J.M. (1993). New combinations of housing and care for the elderly in the Netherlands. *Netherlands Journal of Housing and the Built Environment* 8 (3): 301–325.

Jensen, P.A. and Van der Voordt, T. (2015). *The Added Value of FM: How Can FM Create Value for Organisations. A Critical Review of Papers from EuroFM Research Symposia 2013–2015*. Baarn: EuroFM publication. https://doi.org/10.13140/RG.2.1.2361.8089.

Jensen, P.A. and Van der Voordt, T. (eds.) (2016). *Facilities Management and Corporate Real Estate Management as Value Drivers: How to Manage and Measure Adding Value*. Oxfordshire: Routledge.

Jensen, P.A. and Van der Voordt, T. (2020a). Healthy workplaces: what we know and what we should know. *Journal of Corporate Real Estate* 22 (2): 95–112. https://doi.org/10.1108/JCRE-11-2018-0045.

Jensen, P.A. and Van der Voordt, T. (2020b). Typology of Value Adding FM and CREM interventions. *Journal of Corporate Real Estate* 22 (3): 197–214. https://doi.org/10.1108/JCRE-09-2019-0042.

Jensen, P.A. and Van der Voordt, T. (2021). Productivity as a value parameter. *Facilities* 39 (5/6): 305–320. https://doi.org/10.1108/F-04-2020-0038.

Jensen, P. and Van der Voordt, T. (2016 10/15). *Facilities Management and Corporate Real Estate Management as Value Drivers. How to Manage and Measure Adding Value*. Routledge/Taylor & Francis group. ISBN - 978-1-138-24387-3.

Jensen, P.A., van der Voordt, T., and Coenen, C. (eds.) (2012). *The Added Value of Facilities Management: Concepts, Findings and Perspectives*. Lyngby, Denmark: Centre for Facilities Management & Polyteknisk Forlag.

Khanna, C., Van der Voordt, D.J.M., and Koppels, P. (2013). Real Estate mirrors Brands. Conceptual framework and practical applications. *Journal of Corporate Real Estate* 15 (3/4): 213–203. https://doi.org/10.1108/JCRE-01-2013-0003.

Lynch, K. (1960). *Urban Planning, Architecture*. The MIT Press. 1e. 194. ISBN0-262-62001-4.

Maarleveld, M., Volker, L., and Van der Voordt, T.J.M. (2009). Measuring employee satisfaction in new offices – the WODI toolkit. *Journal of Facilities Management* 7 (3): 181–197. https://doi.org/10.1108/14725960910971469.

Palvalin, M., Van der Voordt, T., and Jylhä, T. (2017). The impact of workplaces and self-management practices on the productivity of knowledge workers. *Journal of Facilities Management* 15 (4): 423–438. https://doi.org/10.1108/JFM-03-2017-0010.

Plijter, E., Van der Voordt, D.J.M., and Rocco, R. (2014). Managing the workplace in a globalized world. The role of national culture in workplace management. *Facilities* 32 (13/14): 744–760. https://doi.org/10.1108/F-11-2012-0093.

Remøy, H.T. and Van der Voordt, D.J.M. (2007). A new life: conversion of vacant office buildings into housing. *Facilities* 25 (3–4): 88–103. https://doi.org/10.1108/02632770710729683.

Remøy, H.T. and Vander Voordt, D.J.M. (2014a). Adaptive reuse of office buildings: opportunities and risks of conversion into housing. *Building Research & Information* 42 (3): 381–390. https://doi.org/10.1080/09613218.2014.865922.

Remøy, H.T. and Vander Voordt, D.J.M. (2014b). Priorities in Accommodating office user preferences: impact on office users' decision to stay or go. *Journal of Corporate Real Estate* 16 (2): 140–154. https://doi.org/10.1108/JCRE-09-2013-0029.

Riratanaphong, C. and Van der Voordt, D.J.M. (2015). Measuring the added value of workplace change. Performance measurement in theory and practice. *Facilities* 33 (11/12): 773–792. https://doi.org/10.1108/F-12-2014-0095.

Van der Voordt, D.J.M. (1990). Building adaptable housing – from theory to practice. Current developments in the Netherlands. *Architecture and Behaviour* 6 (1): 17–37.

Van der Voordt, D.J.M. (1997a). Design for all: towards a barrierfree environment for everyone. *Cadernos Técnicos Aut* 3: 51–75. Faculdade de Architectura e Urbanismo, Universidade de São Paulo.

Van der Voordt, D.J.M. (1997b). Housing and care variants for older people with dementia. Current trends in the Netherlands. *American Journal of Alzheimer's Disease* 12 (2): 84–92. https://doi.org/10.1177/153331759701200206.

Van der Voordt, D.J.M. (2001). Lost in a nursing home. *IAPS Bulletin for People-Environment Studies* 18: 19–21. Spring 2001. Special issue on Environmental Cognition.

Van der Voordt, D.J.M. (2003). *Costs and benefits of innovative workplace design*. Center for People and Buildings, Delft & Centrum Facility Management, Naarden.

Van der Voordt, D.J.M. (2009). Quality of design and usability: a Vitruvian twin. *Ambiento Construido* 9 (2): 17–29.

Van der Voordt, D.J.M., Ikiz-Koppejan, Y.M.D., and Gosselink, A. (2012a). Evidence-based decision-making on office accommodation: accommodation choice model. In: *Enhancing Building Performance* (ed. S. Mallory-Hill, W.F.E. Preiser, and C. Watson). Chapter 18, 213–222. Chichester, West Sussex, UK: Wiley-Blackwell.

Van der Voordt, D.J.M. and Maarleveld, M. (2006). Performance of Office Buildings from a User's Perspective. *Ambiente Construido* 6 (3): 07–20.

Van der Voordt, D.J.M. and Van Wegen, H.B.R. (1983). Underpasses for pedestrians and cyclists. User requirements and implications for design. *Transportation Planning and Technology* 8 (1): 1–14.

Van der Voordt, D.J.M. and Van Wegen, H.B.R. (1990a). *Sociaal veilig ontwerpen. Checklist ten behoeve van het ontwikkelen en toetsen van (plannen voor) de gebouwde omgeving*. Publikatieburo Bouwkunde, Technische Universiteit Delft.

Van der Voordt, D.J.M. and Van Wegen, H.B.R. (1990b). Testing building plans on public safety: usefulness of the Delft checklist. *Netherlands Journal of Housing and Environmental Research* 5 (2): 129–154.

Van der Voordt, D.J.M. and Van Wegen, H.B.R.(1991). *Sociale veiligheid en gebouwde omgeving. Theorie, empirie en instrumentontwikkeling*. PhD thesis. Publikatieburo Bouwkunde, Technische Universiteit Delft.

Van der Voordt, D.J.M. and Van Wegen, H.B.R. (1993). The Delft checklist on safe neighborhoods. *Journal of Architectural and Planning Research* 10 (4): 341–356.

Van der Voordt, D.J.M. and Van Wegen, H.B.R. (2005). *Architecture in Use. An Introduction to the Programming, Design and Evaluation of Buildings*. Oxford: Elsevier, Architectural Press/Routledge.

van der Voordt, D.J.M. and van Wegen, H.B.R. (2005 March 8). *Architecture In Use*, 1e. Routledge. ISBN 9780750664578, 252.

Van der Voordt, T. (2004). Productivity and employee satisfaction in flexible offices. *Journal of Corporate Real Estate* 6 (2): 133–148. https://doi.org/10.1108/14630010410812306.

Van der Voordt, T. (2017). Facilities management and corporate real estate management: FM/CREM or FREM? *Journal of Facilities Management* 15 (3): 244–261. https://doi.org/10.1108/JFM-05-2016-0018.

Van der Voordt, T. (2021). Designing for health and wellbeing: various concepts, similar goals. *Gestão & Tecnologia de Projetos*. São Carlos, 16 (4): 13–31. https://doi.org/10.11606/gtp.v16i4.178190.

Van der Voordt, T. (2022). Value-sensitive design & management of buildings and facilities. In: *Teaching Designing for Values. Concepts, Tools, & Practices* (ed. R. Rocco, A. Thomas, and M. Novas Ferradás). Chapter 9, 220–243. Delft: Delft University of Technology. https://doi.org/10.34641/mg.54.

Van der Voordt, T. and Jensen, P.A. (2018). Measurement and benchmarking of workplace performance: key issues in value adding management. *Journal of Corporate Real Estate* 20 (3): 177–195. https://doi.org/10.1108/JCRE-10-2017-0032.

Van der Voordt, T. and Jensen, P.A. (2021). Value adding management of buildings, facilities and services. In: *A Handbook of Management Theories and Models for Office Environments and Services* (ed. V. Danivska and R. Appel Meulenbroek). Chapter 12, 140–151. New York: Routledge. https://doi.org/10.4324/9781003128786-12.

Van der Voordt, T. and Jensen, P.A. (2022). The impact of healthy workplaces on employee satisfaction, productivity and costs. *Journal of Corporate Real Estate* 25 (1): 29–49. https://doi.org/10.1108/JCRE-03-2021-0012.

Van der Voordt, T., Jensen, P.A., Hoendervanger, J.G., and Bergsma, F. (2016). Value Adding Management (VAM) of buildings and facility services in four steps. *Corporate Real Estate Journal* 6 (1): 42–56.

Van der Voordt, T., Van Meel, J., Smulders, F., and Teurlings, S. (2003). Corporate culture and design. Theoretical reflections on case-studies in the web design industry. *Environments by Design* 4 (2): 23–43.

Van der Voordt, T.J.M. (1999). Space requirements for accessibility. cross-cultural comparisons. In: *Measuring Enabling Environments* (ed. E.D. Steinfeld and S. Danford). New York: Plenum Press.

Van der Voordt, T.J.M., Dick, V., and Van Wegen, H. (1997). Comparative floorplan-analysis in programming and architectural design. *Design Studies* 18 (1): 67–88. https://doi.org/10.1016/S0142-694X(96)00016-6.

Van der Voordt, Th.J.M., De Been, I., and Maarleveld, M. (2012). Post-occupancy evaluation of facilities change. In: *Facilities Change Management* (ed. E. Finch), 137–154. Chichester, West Sussex: Wiley-Blackwell.

Van der Zwart, J. and Van der Voordt, T. (2013). Value adding management of hospital real estate. Balancing different stakeholders' perspectives. *(E)Hospital* 15 (3): 13, 15–17.

Van der Zwart, J. and Van der Voordt, T. (2015a). Adding value by hospital real estate: an exploration of Dutch practice. *HERD, Health Environments Research & Design Journal* 9 (2): 52–68. https://doi.org/10.1177/1937586715592649.

Van der Zwart, J. and Van der Voordt, T. (2015b). Pre-occupancy evaluation of patient satisfaction in hospitals. *HERD, Health Environments Research & Design Journal* 9 (1): 110–124. https://doi.org/10.1177/1937586715595506.

Van Hoogdalem, H., Van der Voordt, D.J.M., and Van Wegen, H.B.R. (1985). Comparative floorplan-analysis as a means to develop design guidelines. *Journal of Environmental Psychology* 5 (2): 153–179. https://doi.org/10.1016/S0272-4944(85)80015-3.

Van Sprang, H., Groen, B., and Van der Voordt, T. (2013). Spatial support of knowledge production in higher education. *Corporate Real Estate Journal* 3 (1): 75–88.

Wijnja, J., Van der Voordt, T., and Hoendervanger, J.G. (2021). Corporate real estate management maturity model: Joroff et al. one step ahead. In: *A Handbook of Management Theories and Models for Office Environments and Services* (ed. V. Danivska and R. Appel Meulenbroek). Chapter 2, 13–24. New York: Routledge. https://doi.org/10.4324/9781003128786-2.

Theo J.M. van der Voordt, PhD
Emeritus Associate Professor
Corporate Real Estate and Facilities Management
Department of Management in the Built Environment
Faculty of Architecture and the Built Environment
Delft University of Technology
Delft, Netherlands

Theo J.M. van der Voordt (MSc, PhD) is emeritus associate professor in Corporate Real Estate and Facilities Management at the Department of Management in the Built Environment, Faculty of Architecture and the Built Environment, Delft University of Technology. His research includes new ways of working, workplace management, performance measurement, transformation, and adaptive reuse as a means to cope with structurally vacant office buildings, successful real estate strategies, and adding value by real estate and related facilities and services.

He is the author of numerous journal papers, book chapters, and books, including *Costs and Benefits of Innovative Workplace Design* (2003), *Architecture in Use* (2005), *Transformation and Adaptive Re-use of Vacant Office Buildings* (2005, in Dutch), *The Added Value of Facilities Management* (2012), *Corporate Real Estate and Facilities Management Strategies* (2012, 2017, 2022, in Dutch), *Facilities Management and Corporate Real Estate Management as Value Drivers: How to Manage and Measure Adding Value* (2016), and *Dear Is Durable* (2016).

His work on offices is conducted in close co-operation with the Center for People and Buildings in Delft, a knowledge center that specializes in the relationship between people, working processes, and the working environment (www.cfpb.nl).

W: https://www.researchgate.net/profile/Theo-Van-Der-Voordt
https://www.linkedin.com/in/djmvandervoordt

3.12

My FM Path and Spelling and Defining Facilities Management and Facility Manager Holistically

Audrey L. Schultz

Introduction

My path to the field of facilities management is significantly different from others'. In the late 1970s, I was an art assistant for two art teachers at Eleanor Roosevelt Senior High School in Greenbelt, MD, USA. They both articulated that I should stay away from the arts since they never knew if they had a job from year to year. Back then, the school system frequently reduced its spending for the arts and music. At the same time, the school systems were starting to let female students register for shop classes and I decided to enroll in a shop class so that I could work on car engines. To be more specific my plan was to work on our 1967 Chevrolet Bel Air station wagon 327 v-8 Turbo-Fire 275 hp engine. Nonetheless none of the team captains chose me to be on their team. The reason I'm telling you this is because something amazing happened. Subsequently since I didn't get selected to be on a team, the teacher handed me a 4-inch-thick book on HVAC and refrigeration cycles and told me to read the entire book and develop a final exam that I would take at the end of the semester for a grade. He would prep me on the exam and grade it as well. This is where my appreciation and knowledge for mechanical systems originated and my initial exposure to facilities operations and maintenance (O&M) began. Since then, my immense fondness for facilities has always been in O&M and commissioning (Cx).

Shortly after I turned 16, a friend asked if I wanted to take over her job at Stride Rite Shoes, selling children's shoes at the local Mall. She was hired to work at the new Optometrist that opened in the neighborhood. My parents approved and shortly after I received my work permit and started selling Stride Right Shoes. About 1.5 years later I started working for Sears as a revolving sales associate working in various women's clothing, and the shoe department on busy Saturdays plus running the registers. I also enrolled in the work study program at high school, where I went to school half a day and worked in the afternoons. I'm mentioning this because I started working at 16 years of age and never stopped even through all four university degrees; I've always worked at least part-time.

When deciding what to do with my life after high school, and the fact that the art teachers discouraged me from going into the arts, I decided to major in Fashion Merchandising and enrolled at St. Petersburg Junior College in St. Petersburg, Florida. I worked for Maas Brothers department store in the women's clothing department and was recommended by one of my teachers to manage his friends' gift and import store on the weekends. Two years later I received my Associate of Science degree in Fashion Merchandise and moved back home to Maryland. Within two weeks I had an Assistant Store Management position and started a career in retail management.

Five years later I found myself exhausted and started contemplating going back to school. Though my retail experience was amazing, I mean who wouldn't want to work nights, weekends, holidays, and six days a week for

Facilities @ Management, Concept - Realization - Vision, A Global Perspective, First Edition. Edited by Edmond Rondeau and Michaela Hellerforth.

little pay. I managed women's better sportswear at Lord & Taylor, opened up a Bradlees store managing all women's clothing departments including lingerie, I was a sales manager for over 50 associates at Woodward and Lothrop: and became a store manager running women's discount stores. Many of my retail management positions were recommended to me by colleagues. I started taking courses at the University of Maryland's Adult Education Center in personnel management and labor relations, and marketing. I found the courses boring and not exactly what I was looking for. It occurred to me to go back into the arts, just not the fine arts. A career in architecture would be more lucrative than fine arts that I had once been discouraged from.

When researching local universities for architecture school, the DC area only had two choices: Catholic University and the University of Maryland. At the University of Maryland, you had to go to school and take general education core requirements for two years prior to even applying for architecture school. That meant that you could go to school for two years and not even know if you would get into architecture school. I decided to enroll in pre-design courses and loved them. My friend's brother who was an architect and had graduated from the University of Maryland's architecture school told me about the interior design program. I enrolled in the interior design program and then they lost their accreditation. Meaning, if I wanted to become a certified interior designer, similar to a registered architecture, I needed to graduate from an accredited degree program. The teacher told the class that Marymount University in Arlington, Virginia had an accredited interior design program. Therefore, I transferred to Marymount University and graduated 2.5 years later (December of 1991) with a Bachelor of Arts in Interior Design. This is where I was introduced to IFMA and facilities management and then became a member of the National Trust for Historic Preservation, AIA, ASID, and IIDA.

I was still working in retail management for Sassafras women's clothing stores and became a roving manager in the DC region. I had keys to almost all of their stores in the DC metro area, working 30 hours a week while going to school. I had two interior design internships, one at the Kennedy Center for Performing Arts. There I was going to work on designing the new bookstore and gift shop, but funding fell through, therefore I ended up working with the gift store manager in merchandising, sales and administration for the Texas Festival, Phantom of the Opera, and the other key events.

Through a colleague who was again, a friend of a friend I was offered a summer internship at the Architect of the Capitol. That was one of the best summers working on Capitol Hill and going to the Washington Design Center. I worked with the interior design staff and did architectural drafting, design, space planning, selecting interior materials, finishes and FF&E (furniture, fixtures, & equipment). Worked with people from the Library of Congress and Supreme Court. Projects were located in the Capitol, Rayburn and Longworth House Office Buildings and Russel and Hart Office Buildings. A sample of projects completed were the carpet and paint for Chairman Robert Roe's Committee Room; designing millwork and cabinets for Congressman Joseph McDade; the Committee on Energy and Commerce Lounge Room I chose the carpet, paint, and FF&E; the design and renovation of Senator Rockefeller's office; and Senator Carl Levin's Reception room which included FF&E and designing a wall display. It was a very successful and fun summer internship. For years I kept looking to see if any positions opened up that I could apply for. It would have been a wonderful career on Capitol Hill working for the Architect of the Capitol.

In fact, that summer is where I found my first full-time interior design position at the ASID job bank in the Design Center. I was in my last semester at Marymount and answered an ad for an interior designer at the Holland Lessard Group. It was a small husband and wife architecture firm; they taught me everything about running an architectural firm. I answered phones, managed the reception desk, architecture, and material library, did marketing and business development, and designed new marketing brochures for the company and used AutoCAD Release 9 and AutoCAD's first digitizers. We had arrangements with developers, and I coordinated space planning for tenant leases. I did a lot of architectural design and drawings, construction documents, plumbing diagrams, site plans, even designed and managed a new septic field installation for a small private school. I worked on commercial interiors, car dealerships, medical suites, dialysis centers and wrote most of the project specification manuals.

About four years later I was done with car dealerships, small architectural projects, and working two jobs. I felt that as an interior designer, and the only interior designer in a small architecture firm, that in order to grow professionally, I needed more facilities and space planning experience. Therefore, I became the facility planner of the northeast region at EDS's northern Virginia headquarters. At EDS I space planned over one million square feet of commercial and tech office space myself, managed the headquarters as-built drawings, designed a spec office building, and brought all EDS interior designers together for bi-weekly meetings. Worked with the facility manager in Camp Hill, PA, on their building renovation, helped spec interior finishes and designed the basements lobby from the loading dock to the server room and added an art installation to jazz it up. My design using tedlar textured wallpaper, and a specific rubber base and chair rail was pictured on the front cover of Koroseal's Wall Protection Systems brochure. I worked on Camp Hill's new call centers and server rooms. Became a member of northern Virginia's Art League and managed the headquarters lobby rotating art exhibit. They sent me down to Charlotte, NC, to design and manage a building renovation. We also met at national headquarters in Plano, TX, several times a year. At EDS we used Intergraph for design drawings.

After two years I felt that I had gained invaluable experience in facilities, design, and space planning and was prepared to move on. Once more, through suggestions from colleagues', I ended up at Cooper Lecky as head of the interior design department. Cooper Lecky was an iconic Washington, DC architecture firm, they collaborated with Maya Lin on the Vietnam Memorial developing the construction documents and seeing the project through fruition. They designed the Korean Memorial, as well. I was given the opportunity to manage the construction administration on Henrico Cultural Arts Center during the construction phase. At Copper Lecky, I developed FF&E specifications for clients, redesigned the architectural and material library. Worked on the Department of State's (DOS) historic Delegates Lounge renovation. My biggest and most exciting project was the renovation of seven floors in the nine-story historic office building for the Federal Election Commission (FEC). I did the initial planning, stacking redesign, schematic design, design development, managed the construction documents and specifications and construction administration. Working closely with the facilities officer, FEC, GSA, the developer, the contractor, subcontractors, and the property manager. This is when I realized that I truly enjoyed working on a construction site. It was perfect timing, I found out through a colleague that Virginia Polytechnic Institute and State University (Virginia Tech) was developing a graduate center in northern Virginia with University of Virginia (UVA). We were the first cohort for the Master of Science in Architecture with a concentration in Construction Management. Four and a half years later I graduated.

The FEC project was almost complete therefore I crossed over to the construction side of the A/E/C/FM/RE industry. I ended up at a small family-owned construction company out of Gaithersburg, MD, called Glenn Construction. There I was promoted 4 times in four months; started in the drawing and bid room, became project administrator, then project executive on a strip center and then assistant project manager for the executive office buildings. Our biggest project was building the new Kings Farm Community. I worked closely with the developer on special projects and did tenant fit out on the office building, working with the real estate project managers, tenants, architects, and designers. Since nobody wanted to finish their projects, I also developed a reputation as the punch list queen.

About a year earlier a colleague had recommended that I interview at CH2M Hill, an international environmental engineering firm. The team called me in for a conversation; however, they didn't have the budget for the position though we kept in touch. A year later CH2M Hill (CH) called me to see if I was still interested in working for them and shortly thereafter, I joined as a project manager. CH was purchased some years ago by Jacobs, another global environmental engineering firm. At CH I served as project manager for the US General Services Administration (GSA) Project Manager's Web Guide. I was the designer, project manager and construction administrator for the strategic planning and renovation of a prominent law firm. CH also had a building commissioning (BCx) department that I worked with. I had the opportunity to take a week-long BCx workshop at the University of Wisconsin, Madison. CH also trained me in project management and marketing and business development. I still use and teach many of the skills and process and procedures that I learned at CH to this day. And I held secret security clearance.

They offered me the first ever construction services marketing and business development global coordinator position. That was the first time I was 100% non-billable and did business development, marketing, public relations, and resource management functions for growing construction management services internally and internationally. Then I was selected to be the first program analyst as a third-party on-site staff for DOS, Overseas Building Operations (OBO) division, five-year A/E Support Services IDIQ Program. At OBO I supported acquisition initiatives for US Embassy's globally. Conducted design-build and A/E pre-qualifications, design coordination, technical A/E evaluation reviews; value engineering studies, developed statements of work for contracts and budgets. Reorganized and updated OBO's Building Code Guidelines, Manuals, and Standard Embassy Guidelines. The IDIQ contract was shortened due to new State Department leadership, and I went back to the office as a project manager.

My workload was lighter than usual, but they kept me busy working on construction litigation research on a wastewater project that went awry and updating the construction administration manual. I kept trying to market myself to get more work and colleagues kept telling me that I didn't have the experience they needed. It was mostly because I was female, an interior designer and project manager with no engineering degree. To top it off, I was told that I didn't have any Department of Defense (DOD) experience. Therefore, I looked around DC for the largest DOD project which was the Pentagon renovation and construction program. Months later I was hired as a design and construction area coordinator on the $1.5 billion, 6.5 million sq. ft. Pentagon Renovation and Construction Program (PENREN) working for DMJM H&N, who was bought by AECOM.

We were the joint venture who worked under Washington Headquarter Services and managed the PENREN DOD design-build program. There I served as the primary source of day-to-day design and construction status and program requirements for tenant agency space and had construction over-site of 134 tenant agency suites. I ensured that Hensel Phelps (HP) the design-build contractor constructed and provided appropriate program and tenant agency requirements as defined in tenant approved design intent documents (DIDs). Performed design and construction inspections; participated in solving design and construction issues; communicated daily insight and coordination in constructability; security; health and safety; monitored furniture moves; coordinated special events; commissioned building systems to include HVAC/Elec/IT/Data cabling, and warranty issues; established and maintained liaison with general contractor, sub-contractors, supply chain and DOD tenant agencies, keeping all informed of construction progress. Managed construction and inspections of sensitive compartmented information facilities (SCIFs). Worked with US military agencies to include Army, Navy, Marines, Air Force, Joint Chiefs of Staff, DIA (Defense Intelligence Agency), and Air National Guard. Managed special projects directed by the former US Secretary of Defense, Donald Rumsfeld. And held secret and top-secret security clearance.

One day I received a call from a professor I knew, she worked with DBIA (Design Build Institute of America) and taught design build workshops across the country and had provided workshops for us at PENREN. She asked me If I would leave Virginia to come out to California and teach construction management at California Polytechnic State University (CalPoly). I applied and hence, my academic career started, 16.5 years ago as I picked up everything and moved cross country to Los Osos, CA. While in California, I was recommended, of course, to finish up the construction on the $5.2 million, 8,500 sq. ft. San Luis Obispo Children's Museum. It was a non-paid volunteer construction management position that reported to the Executive Director. Duties consisted of design, construction, and subcontractor oversight; health and safety management; coordinating interior tenant fit-out construction; sprinkler system re-design, building inspections and permits. Worked closely with the City of San Luis Obispo Community Development Department, Fire Marshall and San Luis Obispo County Department of Planning and Buildings for all inspections and permits. I also took part in all the fund-raising events and integrated the project into CM curriculum at CalPoly.

I helped develop an integrated curriculum across the CM program, worked with a retired lawyer on a construction administration course, created an introduction to CM and managed a construction firm business course. After teaching for three years, I realized that if I wanted to stay in academia, I needed to get my PhD. The wheels started turning again and through the ASC (Associated Schools of Construction) academic networking I was introduced to Professor Eddy Finch at the University of Salford, Salford England. He was in charge of the facilities management department. Since I already had degrees in (fashion) interior design, architecture, and construction and none of the schools offered facilities management education, I decided that my PhD would concentrate on facilities management. At the time it was going to be BIM's integration to FM systems. I couldn't imagine what my parents were thinking when I quit my CalPoly job, sold everything and asked if I could move in with them for the summer because I'm going to England to get my PhD in September and I don't know when I'll be home again.

At the University of Salford, I was offered three research assistantships working on the Value *creation models in real estate business*, ARVO project with a colleague from Alto University, Espoo, Finland. ARVO is the Finnish word for "value." Funded by the Finnish Funding Agency of Technology and Innovation (TEKES). The principal objectives were to understand the effectiveness of applying lean principles to the context of facility management organizations. My PhD was changed forever, and my supervisor instructed me to use part of this research for my own PhD, I started backward in doing the literature research first before deciding on a topic and concentrated on process and procedures and visual management. My research was titled, Integrating Lean Visual Management in Facilities Management Systems. The aims and objectives explored the impact of integrating lean principles and visual management technologies in facilities management practices. The design science action research case study utilized the transformation and lean journey of the University of Salford's Estate and Property Services department. The university served as both my case study, and a source of employment; I spent more than a year as a part-time employee in the facilities management department.

Three years later and my research was complete, it was time to get back into the workforce. Hence the perfect tenure track faculty position opened up at Pratt Institute in New York City. It was serendipity, I was moving back to my roots. Not only was I born in Brooklyn, NY, I'm a third generation Pratt. My grandfather taught at Pratt and my uncle received his degree from Pratt. Three years later I received my PhD and have mainly taught in the master of science in facilities management program. I've held various academic leadership positions on committees, to include being an executive officer on the academic senate. I was also part of the team that developed the first ever Pratt Center for Teaching and Learning. On July 1, 2022, I became chairperson for the construction management, facilities management, and real estate practice programs (CM/FM/REP). We are bridging the gap between academia and industry concentrating on workforce development, student success and wellness and faculty development. Moving forward I intend to implement a cross-disciplinary A/C/FM/RE pedagogy that encompasses digital twins, energy management, lean construction, smart cities and buildings, sensor data sets linked with digitalization, and integrated project delivery systems. There won't be any silos while I'm chair because we'll be working together across disciplines to remove the obstacles. The sector is moving toward innovation and technology most of all. I sincerely hope you enjoyed following my journey to facilities management.

Spelling and Defining Facilities Management and Facility Manager Holistically

Every semester, while reviewing Master of Science in Facilities Management (FM) student's thesis I'm reminded yet again about the various spellings and definitions for facilities management and facility manager. It's the inconsistencies in how the various spellings of FM and the meanings can be confusing. Especially when teaching a room full of graduate students and having reviewed over 54 graduate thesis compositions. The questions that come to mind are as follows:

How does a teacher, faculty, professor, and FM professional make that distinction of how to use the various spellings and definitions of facilities management vs facility management? Is there any difference?

Another question would be: What is the difference in the spelling (and meaning) of facilities management vs facility management? Is there a difference?

This chapter will answer the questions above and explore the various spellings and definitions of facilities management; connections to the spellings and definitions will also be considered. Some of the references used in this chapter were provided over the past 32 years from 1990 to present. Which is apropos to this book in celebrating the past 40 years of this wondrous profession we call Facilities Management. Take note of how the phrase Facilities Management was capitalized and spelled in the preceding paragraph before the meaning and spelling of the term is mentioned.

The key to this conversation is "consistency"; uniformity in spelling, capitalization, and terminology. An author must maintain consistency in spelling, chaptalizing words, abbreviations and defining their subject while discussing their topic. In other words, if we use the first sentence of this chapter as an example, it is written in most part as: "while reviewing a Master of Science in Facilities Management"; Facilities and Management are capitalized in the first sentence, then the words "Facilities Management" should always be capitalized throughout the manuscript. What usually occurs is that the author writes it both capitalized and uncapitalized throughout the document. This can be confusing as the capitalized version of Facilities Management might mean something different than the lower-case version or it might have more significance. Facilities Management might differ from facilities management, the reader can become confused; always wondering does the chaptalized Facilities Management denote something different then the lower-case facilities management. In this scenario the capitalized Facilities Management described a formal name of a program in higher education. Therefore, every time a formal program in Higher Ed, or possibly a name of a company is written, the first letter of each word will be capitalized.

Furthermore, this raises the complication of facilities management's acronym. If a term is to be abbreviated, it must first be spelled out completely. Let's go back to the first sentence again: "Upon reading a Master of Science in Facilities Management (FM) student's thesis"; In this sentence the acronym for Facilities Management is illustrated as (FM). Meaning that the author will use FM throughout the written document occasionally to signify Facilities Management. Abbreviations should always be described in the document prior to using the acronym, next to the actual word and written in parentheses. In more formal documents such as a graduate level thesis or technical report a legend should be provided such as an Abbreviation list. While referencing the facilities management industry in the next portion of this chapter, lower case letters will be used unless otherwise noted. Here the author introduced the reader to the style of writing and meaning of facilities management as an industry, facilities spelled with an i-e-s.

The next big question is what is the difference between facilities management and facility management. Or Facilities Manager and Facility Manager. Is there a difference of the word definition based on the spelling? Not exactly. As a professor and long-term industry professional with 32 years of proficiency in the built environment; "facilities management" would be the correct spelling of word choice when referring to the facilities management industry. When using the spelling of facilities management, the author is indicating the broad FM industry, therefore the proper use of "i-e-s" after the "t" would be the correct way of writing the word "facilities." The entire facilities industry is being discussed as a whole, which covers many organizations and sectors. The proper formula here is for the author to be consistent in their spelling and that they do not move back and forth writing the word in one paragraph as facilities and in another paragraph or sentence as facility when describing the industry. Unless they are quoting authors that might be spelling facilities management as facility management differently. This will be addressed further in this chapter as we define FM from various authors. Is this making sense so far?

Let's look at the spelling for the facilities manager or is it facility manager. Here again one must be consistent in spelling the words. If the industry is spelled "f-a-c-i-l-i-t-i-e-s," with an "ies" after the "t," then the person or position should be in the singular form of facility; spelled with a "y" after the "t." Hence, describing the manager and or the person working in the business unit as a "facility" manager. If we in the industry keep this rule of thumb consistent, then we have set a high standard for the industry and correct spelling of facilities management and facility manager vs facility management and facilities manager. Nonetheless, this is not always the case and even in industry documents written by facilities associations and organizations, there are always inconsistencies in the spellings. The author recommends that when writing about the industry use facility(ies) management just like someone might do when describing many facilities that are owned or operated by an organization. When writing about a specific person or position of leadership in the facilities industry, one writes it as facility (ty) manager, or facility managers, when describing more than one manager. Uniformity is the rule of thumb when it comes to writing and spelling facilities management and facility manager.

The next topic is attempting a uniform definition and descriptions of FM, which has its own complexities. The European Committee for Standardization, also known as CEN, adopts the phrase: facilities management as "the integration of processes within an organization to maintain and develop the agreed services which support and improve the effectiveness of its primary activities," which is used by the British Institute of Facilities Management (BIFM) and the British Standard, also known as BSI (Wiggins 2010). The facility manager is viewed as a service provider that supports the client in the European concept of facilities management. Although the CEN definition is a little hazy, it clearly distinguishes between integrating processes and providing effective services. It doesn't specify the working environment or the workplace. As a result, in the UK and Europe, FM tends to refer to the organization's successful processes and services.

On the other hand, the Royal Institution of Chartered Surveyors (RICS), defines facilities management (FM) as: encompassing the overall management of all services that support the core business of an organization. According to Wiggins (2010) the connection between the core business functions, supporting business operations, and managing the real estate portfolio is the focus of FM across all sectors of business, industries, and services (Wiggins 2010). The RICS facilities management definition encompasses all services that holistically serve the parent company across all sectors of business, as well.

While back in 2010–2013, IFMA posted three definitions of the term *facility management* on their website under their FM Glossary resources. IFMA had defined the wording, *facilities management*, similar to BIFM and CEN. IFMA further defined *Facility Management* (capital F and M) as: "The practice of coordinating the physical workplace with the people and work of the organization; integrates the principles of business administration, architecture and the behavioral and engineering sciences" (IFMA 2010). Additionally, IFMA *described Facility Management 1* (capital F and M with a "1" after it) as "A profession that encompasses multiple disciplines to ensure functionality of the built environment by integrating people, place, process, and technology" (IFMA 2013 -https://www.ifma.org/about/what-is-fm/). This is the most prevalent term used when conversing about the facilities management profession. It includes the workplace and the built environment while also integrating people, place, processes and technology (Hodges 2004; Rondeau et al. 2006; Cotts et al. 2010).

According to Theriault (2010), facilities management is a multidisciplinary field that integrates the cooperation of several industries, including IT, HR, building, architecture, finance, real estate, and design. Therefore, we might infer that the term facility manager refers to the person who is in charge of operating facilities. This individual is also known by other industry terms such as building manager, building superintendent, facility operations manager, or a maintenance supervisor. According to Boonwisut (2022) a facility manager's primary responsibility is to oversee the functioning of facilities in order to maximize an organization's growth by lowering operating costs and raising profits (Boonwisut 2022). This theory can also be traced back to Alexander (2003), Becker (1990), and Nutt's (2000) perspectives on the facilities management sector. Alexander (2003) believed that a way to motivate FM employees is by simplifying work processes and procedures, which in turn creates a positive work environment and added value to the organization. Employees will enthusiastically embrace their job responsibilities as a consequence, creating a more profitable and productive organization. Linking employee satisfaction to the triple bottom line.

Alternatively, FM is described as "an integrated approach" of the physical environment by Barrett and Baldry (2003, p. xiii), who claim that it "creates" and supports the goals and values of the organization. In addition to capturing the relationship between FM and the parent company, Barrett and Baldry's 2003 study captures the essence of facilities management as a discipline inside the workplace environment. RICS accepts the notion of a cross-industry service role and links FM to the core business, as well.

The field of facilities management has traditionally been viewed as one in which administrative tasks are supported by systems, processes, and procedures. Facilities management functions are value adding services that support and improve organizational performance and serve as the organization's gatekeepers for the health and safety of both internal and external end users. Facility managers must correctly carry out work activities to improve and safeguard the workplace, tying workplace productivity to the main objectives, missions, and aspirations of the entire company while keeping in mind the health, safety, and welfare of its employees (Wiggins 2010; Schultz 2016). FM must be viewed as a system process approach of ecologies that influence both internal and external environments in order to provide value to the company as a whole.

Furthermore, Price and Akhlaghi (1999) contend that FM has evolved from managing facilities to managing people and place which they label as third generation facilities management. That was back in 1999 when they stated in their paper that the new way of thinking about facilities management was in terms of strategic development and organizational best practices. Price and Akhlaghi (1999) allege that originally there was a first generation of FM, referred to as the traditional facilities management which grew out of managing the real estate portfolio or building assets. The second generation of facilities management comprised of IFMA's definition that FM is all about managing people, process, and places. What then became the third generation of facilities management "might be seen as more concerned with the creation of space which enables different levels and forms of performance" (Price and Akhlaghi 1999, p. 164).

The theories of Price and Akhlaghi (1999) and Becker and Steele (1995), who adapted facilities management as organizational ecologies of the workplace, are consistent. The networks and spaces that facility managers oversee make up a portion of the organization's ecosystem. This interplay of ecologies within the work environment fosters a more chaotic business system that results in a range of end-user performance levels (Price and Akhlaghi 1999). Price and Akhlaghi's third, fourth, and fifth generations of facilities management may then take advantage of creating new ecologies and artefacts that will encourage FM to strategically integrate more firmly with the business's strategic plan and initiatives (Boonwisut 2022).

The idea that facility managers must manage a safe and healthy workplace has not received enough attention in the literature throughout the years. Up until about 2013, when the International Facility Management Association's IFMA Foundation's (IFMA Foundation) Facility Management Accreditation Commission, or FMAC, an organization that accredits FM education on a global scale, benchmarked the 12 areas of expertise as a foundation for accreditation of university FM programs. The IFMA Foundation updated the FMAC accredited degree program (ADP) standards in February 2013. The addition of leadership, sustainability, and stewardship as well as a safe, secure environment through contingency planning and emergency readiness was introduced (IFMA Foundation, 2013).

Through the previous 40 years, there have been many disasters such as the continuous California wildfires; hurricanes with names such as Katrina, Super Storm Sandy and Ida; 9–11; and the COVID-19 global pandemic. Subsequently, facility managers are responsible for managing the workplace and keeping the organization operations running. Therefore, a comprehensive knowledge of building codes, both nationally and locally should be emphasized. Furthermore, the facility manager needs to have a thorough toolbox of business continuity emergency preparedness planning and disaster recovery plans in place. The facilities management profession has evolved, and as facility managers become leaders of the workplace, health and safety become a priority. This statement can be illustrated in the fact that IFMA World Workplace 2013 had a Deep Dive Session on Disasters are the New Normal. Additionally, more recently during the COVID-19 global pandemic, IFMA collaborated with IFMA Foundation and ABM to develop the Pandemic Manual, Planning and Responding to a Global Healthy Crisis for Facility Management Professionals published in 2020. Notice that IFMA and IFMA Foundation spell facilities management as facility management. Here lies the inconsistencies with the spelling of facilities management vs facility management.

To summarize the definition of FM, a Master of Science in Facilities Management student created a survey on defining facilities management. Student Boonwisut (2022), requested participants to define facilities management based on three various definitions that were provided. The 37 respondents were asked to also describe facilities management in one word. These are the words that the survey participants used to describe facilities management: "people, built environment, organization, mission, well-being, productivity, construction, design, operation & maintenance, finance, safety, health, workplace, quality of life, multifaceted career, process, resources, tools, integrated, seamless service, scope, condition monitoring, life cycle, sustainability, technology, human talent, human resources, physical assets, and digital" (Boonwisut 2022, p. 44).

The survey provided three various definitions of facilities management and asked the participants to rate the definitions, they are as follows (Boonwisut 2022, p. 43):

> Facilities Management is a profession that integrates a variety of disciplines and industries in order to manage, maintain, operate and ensure functionality of a sustainable built environment by integrating people, place, processes, and technology to encompass a healthy, safe and efficient, value driven workplace aligning the FM function to enterprise initiatives holistically.
>
> *Dr. Audrey Schultz (2016)*

> Facility Management is an organizational function which integrates people, place and process within the built environment with the purpose of improving the quality of life of people and the productivity of the core business.
>
> *IFMA (2022)*

> FM is responsible for co-ordinating all efforts related to planning, designing and managing buildings and their systems, equipment and furniture to enhance the organisation's ability to compete successfully in a rapidly changing world.
>
> *Becker (1990), (Boonwisut 2022, p. 43)*

The participants were told to rank each term, as described above. The majority of participants agreed with Schultz's 2016 definition 1. Six participants thought that definition 1 was the least appealing. The third and oldest definition of facilities management was retrieved from Becker's 1990 research. Becker's FM definition was developed 32 years ago; the poll participants did not agree with it. The definition from IFMA (2022) came in second. The majority of participants thought that Schultz's (2016) definition 1 was more pertinent to the current facilities management industry.

The relevance of facilities management as a field of science and technology, as well as the makeup of its workforce, have seen a remarkable transition over the previous 40 years since it was supposedly formalized in the United States. The article "The Evolution of Facilities Management, From Backend to Impacting Business Outcomes" on the Commercial Design India website quotes Sandeep Sethi (2020), a Managing Director at JLL, West Asia, as saying, "We have seen tremendous transformation in the Facilities Management function in recent years, facility management today is hardly just a back-end function anymore, having gradually transitioned to a strategic lever for driving business outcomes" (Jampana 2022). Sethi (2020) spells facilities management, the industry with an ies; yet he changes the spelling when he talks about FM today and spells it with a ty.

There has been an industrial and technological shift in the workplace. The facilities management industry is still devoting efforts to gain recognition as a distinct profession. According to Ankerstjerne (2019), the Chief Strategy Officer of Planon Group, on the website *Proptech Zone*, "FM was originally formed as a technical discipline, but as it developed, the technical competence required to deliver FM did not disappear; it changed and adapted to the new requirements of the market" (Ankerstjerne 2019, para. 3; Jampana 2022).

According to Jampana (2022) every new market growth has necessitated an equivalent FM development, and much like markets, each new phase in the development of the facilities management industry has built on the previous cycle. This is similar to Price and Akhlaghi (1999) generational theory of facilities management recognized in the beginning of this chapter. FM is currently experiencing a technological "growth cycle" or another generation of evolution. The industry is being affected by significant global events, pandemics, wars, political upheaval, climate change and advancement in technology. Sustainability, stewardship, the human experience, and workplace strategies are all new professional services and activities offered by FM service providers. As Jampana (2022) states that "simply" maintaining the operations of building systems and facilities is old school and the need for employee and customer satisfaction and engagement is key to becoming a value add to any organization. Given the rising importance of user-centricity, customer experience and biophilic design in the workplace, it is vital to reevaluate current distribution strategies, definitions, and spellings of facilities management and facility managers.

References

Alexander, K. (2003). A strategy for facilities management. *Facilities* 21 (11/12): 269–274.

Ankerstjerne, P. (2019). *The industry shifts that define Facility Management today*. PropTech Zone. Retrieved from https://proptech.zone/the-industry-shifts-that-define-facility-management-today.

Barrett, P. and Baldry, D. (2003). *Facilities Management Towards Best Practice*, 2e. Oxford: Blackwell Science Ltd.

Becker, F. (1990). *The Total Workplace: Facilities Management and the Elastic Organisation*. New York: Van Nostrand Reinhold.

Becker, F. D., & Steele, F. (1995). *Workplace by design: Mapping the high-performance workscape*. Jossey-Bass, Database: APA PsycInfo.

Boonwisut, P. (2022). *Benefit of education and credentials in facilities management for facility managers*. Master of Science in Facilities Management Thesis. Pratt Institute, New York.

Cotts, D., Roper, K., and Payant, R. (2010). *The Facility Management Handbook*, 3e. New York: AMACOM.

Hodges, C.P. (2004). *Operations & Maintenance An IFMA Competency-Based Course*. Houston: IFMA International Facility Management Association.

International Facilities Management Association (IFMA) (2010). Retrieved 11 November, 2010, and 6 December, 2013, from http://www.ifma.org.

IFMA (2013). Website. Retrieved from https://www.ifma.org/about/what-is-fm/.

IFMA (2022). Website. Retrieved from https://www.ifma.org.

Jampana, P. (2022). *The impact of covid-19 and its influence on women's work-life balance in the facilities management workforce*. Master of Science in Facilities Management Thesis. Pratt Institute, New York.

Nutt, B. (2000 March). Four competing futures for facility management. *Facilities* 18 (3/4): 124–132. https://doi.org/10.1108/02632770010315670.

Price, I. and Akhlaghi, F. (1999). New patterns in facilities management: industry best practice and new organisational theory. *Facilities* 17 (5/6): 159–166.

Rondeau, E.P., Brown, R.K., and Lapides, P.D. (2006). *Facility Management*, 2e. Hoboken: John Wiley & Sons, Inc.

Schultz, A. (2016). *Integrating lean visual management in facilities management systems*. Ph.D. Thesis. The University of Salford, Salford, UK.

Sethi, S. (2020). The evolution of Facilities Management–from backend to impacting business outcomes. *Commercial Design*. Retrieved from https://www.commercialdesignindia.com/people/5517-the-evolution-of-facilities-management-from-backend-to-impacting-business-outcomes.

Theriault, M. (2010). *Managing Facilities Management & Real Estate*. Toronto, Canada: Wood Stone Press.

Wiggins, J. (2010). *Facilities Manager's Desk Reference*. Oxford: Wiley-Blackwell, A John Wiley & Sons, Ltd., Publication.

Audrey Schultz, Ph.D. | Chairperson, Professor
Construction Management, Facilities Management,
and Real Estate Practice Programs
PRATT INSTITUTE
School of Architecture
Brooklyn, New York USA

Dr. Schultz has been active in the whole building life cycle for 32 years, retaining positions in interior architecture/design, construction and facilities management and has overseen the planning, design, construction, inspection, and commissioning services on government and military programs.

Professor Schultz is the Chairperson for the Construction Management, Facilities Management and Real Estate Practice Program department in the School of Architecture at Pratt Institute, previously coordinated the Minor in Entrepreneurship. She supervises and mentors' administrative staff, faculty, and students; engages and advises undergraduate and graduate students throughout their academic career. She has taught courses in facilities condition assessment and life cycle cost analysis, strategic planning, business continuity and emergency management, project management and leadership, lean principles and visual management, and integrated project delivery systems.

Dr. Schultz retained leadership positions on the Pratt Academic Senate, Co-Chairs the Campus Safety Committee and Pratt Institutes MSCHE reaccreditation Self Study Standard VI Working Group and is involved in various other Institute and School of Architecture initiatives. She is a program evaluator, PEV for ABET accreditation organization and has had additional universities requesting her expertise to peer review their program. Her international research and publishing are in design science and action research, transformational strategies and change management, lean visual management systems, BIM and PPP's. A member of IFMA, WIFM and WE (Workplace Evolutionaries) Community, CoreNet, and currently Vice President on the Executive Board for CMAA Metro NY/NJ Chapter, leading student outreach and scholarships.

3.13

The Genesis of MIT's Creation of the World's First Comprehensive Facilities Management System and Later Creation of the Nonprofit INSITE® Consortium

Kreon L. Cyros

This contribution records the genesis of the Massachusetts Institute of Technology (MIT) senior management's decision to develop and deploy the same decision support approach for the management of MIT's portfolio of real property as used to manage its financial and human resources.

Development began in late 1966 with the hiring of the author by MIT's Planning Office. MIT invested close to US$5 million over the next 10 years of research, development, real time field testing, addressing MIT's administrative and academic departmental needs, and further testing via a pilot Consortium of 3 other institutions. This culminated in 1976 with MIT's creation of the Office of Facilities Management Systems dba INSITE® (OFMS), and appointment of the author and INSITE's initial developer as its founding Director. Led by the author, its mission was to independently carry on with the development and support of INSITE's web-based database and linked floorplan suite for the Consortium Members (CMs) – all having learned about the INSITE Consortium via word-of-mouth.

The author provides MIT's rationale and approach to developing an enterprise-wide system for managing their real property portfolio. The concept of "space accounting" is introduced and explained in the context of responding to three strategic questions about one's physical assets – buildings, land, equipment (movable and fixed), and the financial information associated with each. Those key questions are:

1) What do we have and where is it? (Inventory)
2) How well are we using what we have? (Utilization)
3) When do we need more or less of those resources, and what are the costs? (Planning)

The capability is then present to provide effective decision support to any aspect of those resources to reveal important options to those critical questions.

The report concludes with two topics. The first is an overview of the design and application of MIT's multimillion-dollar, enterprise-wide system (INSITE™) developed specifically for managing their large building portfolio. The second is a review of the international consortium established within MIT in 1973 as a pilot study, and its evolution in 1996 as an independent nonprofit corporation created by MIT lawyers to:

- Further transfer the INSITE™ decision-support technology to other nonprofit universities, medical centers, and governmental agencies.
- Provide the training and expertise to apply the technology appropriately and efficiently from a business process perspective.
- Gather and share ideas and knowledge from others with similar environments and concerns.

Facilities @ Management, Concept - Realization - Vision, A Global Perspective, First Edition. Edited by Edmond Rondeau and Michaela Hellerforth.

The Beginnings of a Concept

In 1966, MIT's late Provost, Dr. Jerome Wiesner (Nobel Laureate; former Science Advisor to President John F. Kennedy's seeking to have a US astronaut landing on the moon; and later President of MIT) asked his Director of MIT's Planning Office to eliminate the increasing dilemma facing him when carrying out his space allocation duties as Chair of MIT's Space Committee. Meeting with a faculty or staff member requesting additional space, the Provost's inevitable question was, "What space did you have in mind?" It usually resulted in a finger pointing exercise between the requestor and the often unaware "Donator" of the space. Dr. Wiesner would invite both parties to a second meeting that he described to the author this way: "Most of these second 1-hour meetings resulted in 50 minutes questioning the data used by each side, pro and con, and only 10 minutes discussing the strategic space and program implications for each party as well as the Institute who "owned" the space." Dr. Wiesner's solution for reversing that trend was his authorization to fund a new staff position for MIT's Planning Office, led by its distinguished Director O. Robert Simha and past President of the Society for College and University Planning (SCUP). The budget was initially perceived as "temporary" support for a MIS-trained staff member to "develop a computer-based, decision-support, space system on MIT's main frame computer, and then move on."

While the concept of developing software within a non-IT department was novel at that time, Dr. Wiesner's reasoning was simple. MIS staff were seen then as taking but a few days to "understand" a client's needs, while weeks could be spent developing the end product, too frequently with little or no further contact with the client. The results often caused some clients unhappiness at best, or seriously flawed functionality in meeting the intended goals at worst. Dr. Wiesner insisted that this pattern could be minimized, if not eliminated, by forcing the IT designer/developer and client to work in the same office. (That modus operandi is now more common.)

From late 1966 to late 1970, several new iterations of MIT's space inventory system were developed, providing and testing working models and data maintenance process fully in place. Each iteration improved upon the lessons learned from the last. Those primary lessons, while simple in concept, yet oftentimes proving to be time-consuming to implement and test.

Some key takeaways were:

- organizations in constant change require a system architecture that can be changed as rapidly as the environment in which it has to accommodate, without significant and costly re-programming.
- the typical data analyses methods of the planning and management world must be duplicated in the digital world. This allows the answer of one's first query's result to smoothly lead to a more focused next question, and so on, until a more clarified response is found within the data at hand – if there is one. This presents a better result than from a database's hard coded "standard" reports, and
- having a database filled with raw data has no value unless that data can be converted into strategic information, and that the business processes in place allow that information to be transformed into knowledge.

The Birth of a Consortium

From the early MIT-wide use of the INSITE's suite, an equipment inventory function linked to individual spaces was added in 1969. In 1973 MIT's floorplans were seamlessly integrated with INSITE's space database. The MIT Planning Office was able to build its knowledge of the MIT campus using their INSITE system suite as an important planning and decision-support tool. The system's functionalities, designed for the MIT planners to aid their upcoming master planning efforts for a major growth of the MIT campus, drew attention from several sister universities via word-of-mouth. As such, MIT asked the author to attend several University-focused US conferences to present MIT's findings of its multi-million-dollar development effort to gather feedback, other application suggestions, and ideas.

This led to visits and calls from several universities seeking to purchase the INSITE system. MIT's then Chancellor (and later President), Dr. Paul Grey, was not in favor of individual sales. Two ideas then arose. Either license the opportunity to an MIT-known venture capital group to actively market it for their profit generation and MIT's licensing royalty income, or form an in-house Consortium, as done by many MIT academic groups for sharing their activities and research. Dr. Grey's final decision was to authorize the 1973 creation of a two-year pilot Consortium with Brown University, Syracuse University, Harvard University's Medical School, and MIT.

While Dr. Grey maintained a cautious approach to being involved in any activity that had even the slightest chance of mediocrity in its application, he agreed to these terms for CMs (Consortium Members):

1) MIT would freely install the system's object code, i.e., the binary interpretation of the code itself, on each participant's mainframe, with the commitment that if OFMS could duplicate a user's system anomaly, OFMS would fix it with a new code patch.
2) In turn, each CM would reimburse MIT the author's time and travel expense, without overhead costs, to train the CM's staff in the ongoing use and effective application of the system, and provide application support and system support should system anomalies arise,
3) Additionally, each CM agreed to share non-sensitive data and information informally and privately with each other and MIT to achieve new ideas and knowledge more rapidly for best managing their real property portfolio. (CMs reasoned that this shared knowledge would allow for more informed decision-making about their facilities than they individually might achieve in the same amount of time.)

After two years, the pilot study results were re-visited by President Dr. Grey with input from each CM. While positive overall for the continuation of the Consortium, the participants asked MIT for a reasonable assurance that it would not abandon the continued development and support of the system without cause. Rejecting that proposal, Dr. Grey countered with a US$1,900 budget to have the author report on the pilot study findings at several regional conferences. The purpose was to listen to the questions and comments of the audiences to determine if the idea of freely licensing a multimillion-dollar system with an included site license for the sake of sharing knowledge of one's building management practices, made sense to others. The results were that nine other universities and medical centers asked to join the Consortium while presenting at the two conferences. With this, MIT officially recognized the existence of the Consortium, and pledged its support as long as the endeavor was self-funding.

The Birth of MIT's Office of Facilities Management Systems (OFMS)

As MIT pledged its support of the Consortium with its caveat, it also recognized the need to have an MIT office focused on its own internal facilities management efforts as well adequate support of the INSITE CMs. Thus, in mid-1976, MIT created the Office of Facilities Management Systems (OFMS), appointing the author as the department's founding Director, reporting initially to MIT's Vice President of Operations, and finally to MIT's Senior Vice President. Within OFMS's mission statement, these three groups were included:

1) A self-funding system development group to continue to enhance, improve, and maintain a technology edge in MIT's space management suite.
2) A self-funding group to provide continuous application support to Consortium Members.
3) A group for updating the INSITE system's database and linked floorplans, producing reports, institutional analyses, participate in internal and external audits of the data, audit support, and allied service to the MIT community at large,

In 1979, OFMS was asked to create and manage two other capabilities for MIT. The first was to create a Property Management Office to barcode all of MIT's movable equipment, enhance barcode scanner software to auto-update the INSITE space database after field surveys, store critical data of official nomenclature to assure consistent data entry of new equipment, and each item's previous location if different than found in the current field survey.

The second was a broader definition of INSITE's "Space Accounting" capability. A welcomed US federal government program provided reimbursement of operations and maintenance costs of spaces, and equipment depreciation, based on each space's use or partial use for doing federal-sponsored research and contracts. It was in recognition of those costs otherwise borne by the institution for working on Federal grants and contracts. (Some CMs use the same feature for calculating auditable space use charges for one department using the space assigned to another, or charges by the institution to any departmental use of institutional space.)

This Federal reimbursement program is not without pages of rules and regulations. While also heavily audited, these efforts ultimately led to the annual recovery of millions of US dollars to individual Consortium Members and other institutions, coining the term "Space Accounting" as another of INSITE's capabilities. It was the first comprehensive indirect cost recovery application within a CAFM system, and the first to provide equipment depreciation.

Facilities Management (FM)

Shortly after the energy crisis in the United States, a new management concept was introduced, called Facilities Management, or FM, for short. Some suggest that this was a result of the energy crisis between April 1979 and April 1980. Whatever the case, FM was introduced by the Facilities Management Institute (FMI) created by Herman Miller, one of the US prominent furniture manufacturers. Of the many concepts developed by FMI's talented staff, the key was their focus on corporate physical assets as being equally important in meeting corporate business plans as were the corporation's financial and human resources. Heretofore, the physical assets of a corporation were long excluded as being no more than a necessary item on one's balance sheet, and worthy of little Board Room attention.

Facilities Management (FM) – Continued

FM is a bone fide management concept that needed corporate attention that coincided with the advent of the Personal Computer. The result was the rapid appearance of "for-profit" corporations offering the newly recognized breed of Facilities Managers (FM'ers) the idea of CAFM (Computer-Aided Facilities Management). These CAFM developers offered a plethora of COTS tools (Commercial-Off-The-Shelf), that included Project Management; Computerized Maintenance Management; Building Control, Wire and Cable Management; Space Inventory; and Computer-Aided Design (CAD) systems, to name a few. Later, many of which became stand-alone and more robust applications.

Real Property Portfolio Management

Real Property Portfolio Management can be defined as the systematic activity to plan, manage, and utilize an organization's physical assets and workplace environments from a strategic management and capital preservation viewpoint. It is a concept far more strategic in nature than the well-touted business of FM and, surprisingly to some, has a far greater impact on an organization's financial well-being.

Treating a collection, or portfolio, of buildings as a strategic management issue versus managing individual buildings as entities unto themselves was accepted by MIT's Senior Management in the late 1980s. At that time, MIT adopted a real property portfolio management perspective for managing MIT's large campus after enlisting the assistance of Dr. Ranko Bon of MIT's School of Architecture and Planning (later a distinguished faculty member at Reading University). His seminal research work on building economics, coupled with supportive research developed by his colleague, Mr. Michael Joroff, Senior Researcher and Lecturer, also of MIT's School of Architecture and Planning, convinced MIT's Senior Vice President William R Dickson, MIT SB '56, responsible for all of the Institute's physical assets, to adopt this new approach.

The Need for Enterprise Solutions to Effectively Manage Real Property Portfolios

When MIT adopted the concept of strategically managing their physical assets based upon an economic portfolio perspective, it was quickly recognized that the application of a Commercial Off The Shelf (COTS) package to respond to their space management and accounting needs would not suffice. With the emerging technology of minicomputers in the 1980s and later advances in Client/Server technology of the 1990s and beyond, MIT's need already existed in the INSITE suite with minimal further development required. Initially designed as an enterprise-wide, audit-ready system was more important than ever, e.g., "who changed what, when, and what was the previous value." The INSITE system had that capability based on several CM's earlier suggestions and advanced knowledge of MIT's newly adopted HR and student record enterprise systems. All a tribute to the concept of a Consortium of like institution's situations willing to share. With MIT's then annual Indirect Cost Reimbursement (ICR) request from the Government (approximately US$70 million at that time), new hurdles had to be considered. The system had to be prepared for government audits of indirect cost recovery of their space and equipment operating and maintenance costs deployed to carrying out government-sponsored research and contracts; the data had to support detailed scrutiny by government auditors and one relying solely on a timely, accurate, and auditable space accounting capability, an enterprise-wide, flexible, and easily used solution by administrative individuals was imperative. had no reasonable alternatives.

The Foundation of a Real Property Portfolio Management System

While simple in concept and mundane in some aspects of its execution, the foundation of a real property portfolio management system is a space accounting approach. This approach asks the followers to put in place the capability to respond to these three basic issues:

1) *Inventory*: This requires an accurate and timely response to the space-related questions of, "What do we have?" and "Where is it?"
2) *Utilization*: Next is the thoughtful and analytical approach to developing the answer to the question, "How well is it being used?"
3) *Planning and Costing*: This should be the end game of all space-related inquiries, the answers to which are both strategic in their development, and financially devastating in the erroneous resolution. The key questions to be answered by this effort are, "When do we need more or less?" and "What are the hidden, or indirect, costs, of these decisions?"

It is from these space accounting techniques that MIT eventually came to name their space accounting system as INSITE™ (**IN**stitutional **S**pace **I**nventory **TE**chniques).

The Basic Steps to be Taken

There are seven basic steps needed to be taken to manage any portfolio of real property. They are:

1) Inventory all facilities in one's portfolio with just these four basic elements, at the outset:
   a) A unique space identification for every space encountered on a floor, e.g., building-floor-room number. This includes both the assignable areas such as offices, labs, and the like, as well as the non-assignable areas. These include building service areas (e.g., janitorial supply closets), circulation areas (e.g., stairwells, corridors and elevator shafts at each floor penetration), and mechanical areas (e.g., utility rooms/closets, inaccessible utility shaft penetration on each floor).
   b) Area. Record both the inside and the centerline of wall areas. The former is for space allocation purposes, the latter, for space reconfiguration purposes.
   c) Architectural room use. Avoid the temptation to record a room's use by the architectural function for which it was designed rather than the function to which it is being used. Such anomalies should be recorded elsewhere.
   d) Organizational assignment. The task here is to record the specific organizational unit to which the occupant of the room is assigned, or in the case of multiple occupants, the organizational unit of each.
2) Become "database-centric" versus "CAD-centric" in your approach. That is, do not cloud the vision of your goal with the colorful pictures presented by CAD-oriented COTS (Commercial Off The Shelf) vendors. It sells systems, but it is the traditional "cart leading the horse" approach to FM. While linking floor plans to a database is a capability that will provide added rewards on the output side of one's IT endeavors in space management, the data identification, collection, input, and analyses activities – all database focused activities – must be one's primary focus.
3) Identify and measure each building's indirect costs of their operation, maintenance, repair, and refurbishment. Pay close attention to the quality of this data, and the specific building areas to which it is related.
4) Distribute these indirect costs to their applicable building, first; then distribute those "per-square-foot costs to the occupying tenants of the building, or "cost centers."
5) Measure each building's utilization. While each industry may have their own unique measures, it is imperative to understand two things about this exercise. The first is that there are no magic utilization ratios that are guaranteed to work for your purpose. The second is that this process is simply one of ratio analyses. That is, an approach to measuring utilization is to analyze a number of ratios of which the numerator denotes some measure of the organization's resource, the denominator is some measure of productivity from the use of that resource. While "square meters (or feet) per person" is often the measure of choice, there are numerous cases in which that has little or no meaning. One example is a measure of square meters per rank of the person occupying the space – a far more realistic understanding of the space required by the differing ranks of occupants. Recognizing that each rank may well have a different productivity mission (teaching, research, doctoral candidate mentoring, etc.), and therefore different rates of space utilization are needed for a full understanding.
6) Factor into each space both the indirect costs of occupancy as identified and distributed in steps #3 and #4 above, and the percentage of that space's use by the occupant/occupants as described in step #5 above. A simple reminder of the significance of this step is this: if the indirect costs of operating and maintaining a laboratory equals "X," and if that laboratory is in use only 50% of the time, the actual cost to the organization for having that laboratory is "2X."
7) One can now transform data into information. The institutional database of real property used to meet the business goals of that institution is now able to be used as a Decision Support System providing senior managers with the ability to make more informed decisions about their portfolio, based upon accurate and timely data, complete space costs, and true space use data.

Summary

A strategic approach to effectively managing a large building portfolio requires the same level of expertise, commitment, and focus as given the institution's financial and human resources. The benefits reaped from such an approach are enormous, as proven by some of the global corporate giants that use this technique. On the other hand, the implementation costs for such an approach is surprisingly small, but requiring these often-difficult commitments:

- Adoption and maintenance of a solid business process that assures a consistent and timely data capture process.
- Implementation of a rigorous process of quality control in obtaining that data.
- A commitment to creativeness in converting the harvested data into information.
- A willingness to challenge yesterday's paradigms to enhance the knowledge we need to prepare successfully to meet the inevitable changes that occur constantly in our real property portfolios, whether those change are to the portfolio's value, quantity, quality, or any combination thereof.

Epilogue

As an epilogue of this report, MIT spun-off the international Consortium of INSITE™ users in 1996 as a nonprofit entity to share more effectively the technology and the expertise for its application. Naming this new entity "OFMS, Inc. dba INSITE" for its MIT benefactor, the MIT's Office of Facilities Management Systems, the company was provided a royalty-free. world-wide license with the mission to:

- Continue to support the existing Consortium Members – now located in 6 countries.
- Allow for the growth of Consortium members to better serve all involved through a wider base of benchmark data and portfolio management experience.
- Utilize the broad base of users to continue with the enhancement of the existing INSITE™ system, and the further development of new, related tools and systems.

The Consortium is leading the way to sharing the richness of benchmarking and planning data from the base of University and Medical Center participants, reap the benefits of economies of scale in development of tomorrow's latest IT technology, and link the brightest academic and administrative minds of university colleagues and research hospitals to manage the world's leading institutional portfolios of academic and medical buildings.

In closing, the author would do well to recall the prophetic words of Dr. John Hinks in his presentation at the CIB W70 Symposium of 2000; "*We need facilities management... we don't need facilities.*" The author believes that this might serve well as our mantra for this century, and perhaps even beyond.

Kreon L. Cyros, MIT SB '56, SM '62
President and Chairman
OFMS, Inc. dba INSITE®
Peabody, MA USA

Kreon L. Cyros is an MIT graduate with both a bachelor's degree in Civil Engineering and a master's degree in Construction Management, with a minor in Management Information Systems. He is a retired US Army Captain as a Combat Engineering Officer, serving 2 years with nearly 18 months in Korea. He was jump certified and a graduate of the military's Escape & Evasion school in Okinawa.

His civilian professional life started as a Civil Engineer in heavy construction in both the US and abroad, and later was the program liaison with NASA for the MIT Draper Lab's Apollo Guidance and Navigation system project.

In 1966, Kreon was hired by MIT's Planning Office to design and develop a facilities management Decision-Support System (DSS) for managing a rapidly expanding campus. It was named INSITE®, an acronym for INstitutional Space Inventory TEchniques. As word was spread that MIT was willing to share INSITE with other Institutions, MIT created a new department, "Office of Facilities Management" (OFMS). Kreon was appointed as its founding Director and tasked with a number of Facilities-related campus responsibilities and to create and support a Consortium of INSITE users from higher education, healthcare, and Government institutions to share the technology, provide application and systems support, and collectively gather ideas for new applications and uses to then implement the best.

In 1996, given the Consortium's growth, MIT lawyers created its spin-off as a nonprofit, independent, corporation, "OFMS, Inc. dba INSITE." Named as its "Sole Incorporator," Kreon was appointed as President & Chairman of OFMS, Inc., responsible for the continued development and support of the INSITE system, and protection of the exclusive, royalty free, worldwide license from MIT.

3.14

Evolution of Facility Management and Its Merging in a Pattern of Industrial Sustainability and Resilience Perspectives – Experiences – Views

Henning Balck

Overview of Contributions

Introduction

Technological Paradigm Shift "Industry 4.0" Provides the Basis for Comprehensive Sustainability

One of the original ideas of facility management from the 1980s was the IT-based integration of data from the project phases of construction with subsequent data from operation and management. This has remained a vision to this day, a solution is only now becoming recognizable through the industrial paradigm shift of "Industry 4.0."

Part 1

IT Technology and Cost Effectiveness – Steps of Development in German Facility Management – Commented with the Author's Experience

The historical starting points in German Facility Management was in the mid-1980s. The development was initiated with the emerging CAD technology. This made it possible to transfer CAD data from the construction phases to operation after completion. Examples from the USA provided the impetus (in particular through initiatives by Dr. Mack).

Part 2

Challenge Industry 4.0 in Industrial Sites. The Importance of Local Infrastructure Is Changing from a Cost-Only-Status to a Factor of Competition and Sustainability

In a series of pilot projects, the author had the task of investigating real estate portfolios in industrial locations with mixed teams. The aim was to provide evidence of deficiencies in use and operation and, as a result, to prepare plannable modernizations.

Facilities @ Management, Concept - Realization - Vision, A Global Perspective, First Edition. Edited by Edmond Rondeau and Michaela Hellerforth.

Case studies:

- Industrial building of Siemens AG (Operator "Betriebsunterhalt"(=Facility Management), Siemens Erlangen, 1988–1992)
- Industrial Real Estate Roche Diagnostics (Operator Roche Diagnostics Mannheim, 2000–2004)
- Passenger stations of Deutsche Bahn (operator DB Station & Service, 2009)
- Administrative and traffic structures at Frankfurt am Main Airport (operator Fraport, 2010–2013)
- Industrial building in the InfraServe Chemical Park (Operator: InfraServ Gendorf, 2016–2018)

In these projects, it became obvious to the author that progress in the modernization, renewal or planning of new investments had not only to do with the clarification and elimination of structural defects. It has always been about a complex bundle of dimensions of inventory valuation. Above all, it is about improving user processes, maintaining functional capabilities in systems and components and, increasingly, improving energy efficiency. However, the decision-making process as to what had to be done at what time was always the key question. Therefore, time priorities were decisive, in each case depending on the critical need.

The focus on "sustainability" has not yet been pronounced in the projects mentioned. However, the criteria that apply today already played a role to some extent. However, they were mostly subordinated to economic considerations. Today, on the other hand, a fundamental change toward sustainability strategies can be observed in the companies mentioned.

Part 3

The Success of the Built Environment Must be Measured in Its Sustainability during Operating Phases and Life Cycles

If the methodological bracket of the integration of knowledge and data between the phases of construction projects and the subsequent operational phases is to be mastered, a fundamental reorientation of this value chain is required. The explanations are intended to make it clear that a deeply rooted but outdated economic project model must also be reconsidered. The target triangle of cost-deadlines-qualities that comes from operation research must be detached from its previous narrow framework and transferred into a larger context of sustainability. The approach, presented in methodological simplification, is a combination of industrial value creation with overarching models of the sustainability of a "built environment" to be realized.

Part 4

A Renaissance of Serial Construction Enables Sustainability Goals in Realigned Value Chains – Merging Design and Operation of the "Built Environment"

The adjustments in the building stock described and described above are a starting point. Ultimately, however, they only have a perspective if the imperative of sustainability is made possible at the same time by the introduction of technological innovations. The new approach of "Industry 4.0" described at the beginning offers the core potential for this. Construction projects that have been carried out with serial prefabrication in recent years show astonishing leaps in efficiency in terms of costs and throughput times with high quality levels of sustainability at the same time.

At the time of this writing, it is also evident that the original vision of facility management of integrated value chains of construction and operation is now becoming a reality.

The individual chapters described are independent in terms of content and can be read on their own. However, they are factually and methodologically related. The author has therefore initiated a "Learning Network" with interested operators, architects, engineers and industrial partners, in which the following mission statement applies:

> ***Industrial progress is only possible in networks and thus Future-proof through networking***
> *(H.Balck)*

Introduction

The Technological Paradigm Shift "Industry 4.0" Plus "Green Deal" Provides the Basis for Comprehensive Sustainability

At the now legendary UN conference in Rio de Janeiro in 1992, the world public was confronted with a program of global change in response to climate change. Undoubtedly, it is only now, three decades later, that the insight into compliance with sustainable living conditions is recognizable worldwide. Politically, in business and even in international finance. Building is an essential part of this, and thus especially the subsequent "Built Environments."

End-to-end digitized process chains are in demand. They include manufacturing to use processes and further to the recycling of building products in their end-of-use phases. This is made possible for the first time in a technological paradigm shift that is just beginning – through Industry 4.0.

The term "Industry 4.0" is a German neologism, Hannovermesse (2011). Subsequently, it has spread as a term in other language areas. It is about the historical transition in the development chain from Industry 1.0 (eighteenth century) to Industry 2.0 (mechanization and electrification in the nineteenth century). Then to Industry 3.0 (computer-aided technologies in the twentieth century) and currently to an integrated pattern of complex automation combined with artificial intelligence that goes far beyond this. In Germany industrial plant engineering and the chemical-pharmaceutical process industry are pioneers of this development. In the USA, the initiative originated under the name "Industrial Internet Consortium" (IIC for short), one focus is the Industrial Internet of Things (IIoT).

One of the most important features of this still very new industry paradigm is the consistent focus on sustainability – i.e., on future viability! This now gives rise to a wide range of transfer and spin-off potentials for the construction of buildings and technical infrastructure. For sustainable methods, processes and innovations, their importance should not be underestimated. This is because such transfers are currently taking place in rapid disruptions. This is different from the previous historical phases of the industrialization of construction. It was typical for them to follow the development stages of technological progress late and also at a slower pace. The main reason for the now surprising acceleration of the mechanization of buildings and built environments is the connection to the 4.0 mainstream of industrial change.

We are not very prepared for this in the construction industry and are undoubtedly still at the very beginning of a multi-layered reorientation. Therefore, both exploratory and experimental action is required in the sectors of built infrastructure. This will not be easy, but it will be possible if proven success factors come together: Committed entrepreneurial will, economic competence and innovative technologies. However, this also requires a new avant-garde of future-oriented investors, builders, architects, engineers and companies.

Part 1

IT Technology and Cost Effectiveness – Steps of Development in German Facility Management – Commented with the Author's Experience

IT-Innovations and IT-Problems – Beginnings of German Facility Management (mid-1980s)

The historical starting points in German Facility Management was in the mid-1980s. I heard a lecture by Walther Moslener (founder of the German section IFMA) in 1985. He reported on a complex challenge in the operation of the Berlin AMK exhibition center (the largest building of its kind in Germany at the time). In an interview with an AMK operator, he explained that his main problems were insufficient data on maintenance, interference eliminations, and repairs. He emphasized an urgent need for a software solution to manage services for numerous and complex technical equipment. And such kind of software, specialized in buildings, did not exist at that time. The idea of computer aided Facility management (CAFM) was still almost unknown in Germany.

With the same problem I was confronted in 1989 at the Siemens site in Erlangen (about 17,000 employees separated in an area like a small city). This time with the clear task of finding complex IT solutions. The aim was to realign and optimize existing support processes. Two different IT projects were pursued in-house. First, Computer Aided Facility Management (CAFM), which has just become known in Germany, coming from the USA. Second, special software for order processing, developed by Siemens experts. Both IT- initiatives were integrated in an organizational development for the technical internal Services. The whole project, we called "Introducing Computerized Facility Management."

But we had to realize that the available CAFM software, which at that time was still largely oriented toward floor plan data, could not map any processes. And vice versa the order processing software that we had to develop was well suited to commercial requirements but was not sufficiently linked to building data. In short, both approaches were only partially usable and could not be integrated at that time. CAFM solutions, which are both object-oriented and suitable for commercial processes, emerged in Germany only in the mid-1990s.

The Conflict between Insourcing and Outsourcing (1990s up to the Present Day)

To this day, facility management is inconceivable without consistent IT support. But the real success factors lie elsewhere. With the groundbreaking publications by Michael Porter on the distinction between processes of direct value creation (primary processes in the core business) and the complementary support processes (secondary processes), the paradigm for all service industries (around 2000) was established. Since then, it has also been used in maintenance, cleaning, catering, and other infrastructural services. Porter's value chain model became fundamental to consulting practice and academic work. Constitutive was a distinction that resulted from the experiences of numerous organizational projects. The focus was on the conflict between traditional internal services, established over decades, and the newly emerging external service market. It was simply a question of outsourcing or insourcing.

But in organizational projects which I have led myself, I had to realize that sourcing decisions are difficult. It always depended on which "service culture" was given priority. Especially in an FM project for the National Bank Vienna, it became clear to me that internal security and the associated high level of trust requirements for service employees can be indispensable. Then the decision for insourcing is clear – at least for safety-critical services.

The Distinction between Core Business and Support Processes Devaluates Facility Services with a Blemish (1990s up to the Present Day)

Outsourcing has been a global success story since the 1990s. Conversely, internal services, which have been able to assert themselves especially in large companies, are under increasing pressure to succeed. I have often experienced outsourcing threats by management in operating organizations. A two-pronged strategy proved to be a solution: becoming more effective and efficient in one's own company at the same time. In the end, comparative analyses were decisive, which made clear advantages over external providers. Proof of internal competitiveness was therefore indispensable. However, the separation of processes with and without added value soon showed a dark side. Those who belonged to departments of non-value creation were left behind. Demonstrable performance and qualification did not help to overcome the fact that those affected were now afflicted with an obvious flaw. This has only changed recently, where services are experiencing upgrades that are critical to the success of buildings for their sustainability in our entire natural and social environment.

Facility Management (FM) and Corporate Real Estate Management (CREM) Developed in Parallel – Swaying between Synergies and Conflicts

The emergence of specialized service industries and management models such as insourcing or outsourcing were not sufficient to manage fundamental changes in the world of physical assets and the associated service world. Almost simultaneously with the beginning of facility services, real estate management developed in parallel. The scientific pacesetter in Germany was the Institute for Real Estate Economics at the EBS (European Business School), founded by Prof. Karl-Werner Schulte in 1994. Discussions with him and his research associates showed a fundamentally similar approach to FM. A jointly offered CREM-FM-course was attended by participants who came from both in-house services and the real estate sector. A central theme was the transfer of real estate management methods into in-house services. This means "rents" were introduced instead of internal company settlements for the use of building space. The most important consequence was that services for users were now billable costs for ongoing real estate operations.

Organizationally results an established differentiation between property management and facility management both within the company and in the service market. This change is now firmly established, but it cannot be taken for granted, nor is it conflict-free. Real estate orientation always means primarily bringing about economic success by renting or selling real estate objects. Facility management, on the other hand, is originally anchored in own corporate culture and is therefore primarily quality oriented. I remember Siemens service employees who explained that they like to maintain company-owned buildings and facilities, just like their own houses, but not to achieve financial results. In connection with real estate management, however, this quality orientation is inevitably restricted. The responsibility for real estate assets, dominated exclusively by financial goals, transforms Facility services into pure cost factors.

In the incipient worldwide transformation of the entire economy the main objective is sustainability. For this reason, socially important quality goals must take precedence over overpowering economic goals. However, this can only be achieved if technological change becomes a driver. Economic actions are in the role of finding ways and means.

Part 2

Challenge Industry 4.0 in Industrial Sites. The Importance of Local Infrastructure is Changing from a Cost-Only-Status to a Factor of Competition and Sustainability

The contribution is based on the publication "Industry 4.0 in Industriestandorten" in the German journal Facility Management, Bauverlag (2018) – © Prof. H. Balck

Industry 4.0 does not only include the digitization of business processes. It is about much more – about mastering fundamental transformations of the technological infrastructure in widely ramified industries. At the same time dramatic acceleration of automation through robotics, Internet of things and artificial intelligence, can be observed. All this takes place in assets of industrial infrastructure, such as real estate, technical networks, and their operation. However, this brings together fundamentally different, opposing patterns of investing – with questions about how they can be sustainably shaped at the same time.

Industry 4.0 Is Transforming the Embedding of User Equipment's in Building Infrastructure

The following explanations are an attempt to derive the emerging need for structural changes in industrial locations embedded in the global change to Industry 4.0.

Technology trends such as robotics, artificial intelligence and the Internet of Things are fundamental changes, considered a given. But they remain incomprehensible if you don't consider their locations and local peculiarities. Technological innovations end in producing and operating industrial equipment. And this has its place in buildings and built environments. This form of embedding is an old phenomenon in the history of technology. User equipment and built environment have always had fundamentally different lifespans. Mobile machines and equipment with short life cycles must fit in long-lasting buildings. But this established relationship is currently in crisis. The growing difference in the two speeds in the development of user technologies and immobile infrastructure is becoming a structural problem. The drivers of Industry 4.0 affect both worlds and require synchronization (see Figure 2.1).

Future-proofing renewals of user equipment induce structural and technical adaptation needs in local infrastructure. However, part of the everyday experience of CREM/FM is that mobile equipment can be replaced more easily and quickly than replacing immovable objects such as buildings and their components. The ever-shorter historical cycle in which mobile industrial goods become obsolete will therefore not lead to synchronous restructuring in the more permanent structures of industrial sites. On the other hand, growing adjustment backlogs in the portfolio of industrial properties are to be expected. This applies in the same way to infrastructure of real estate and services.

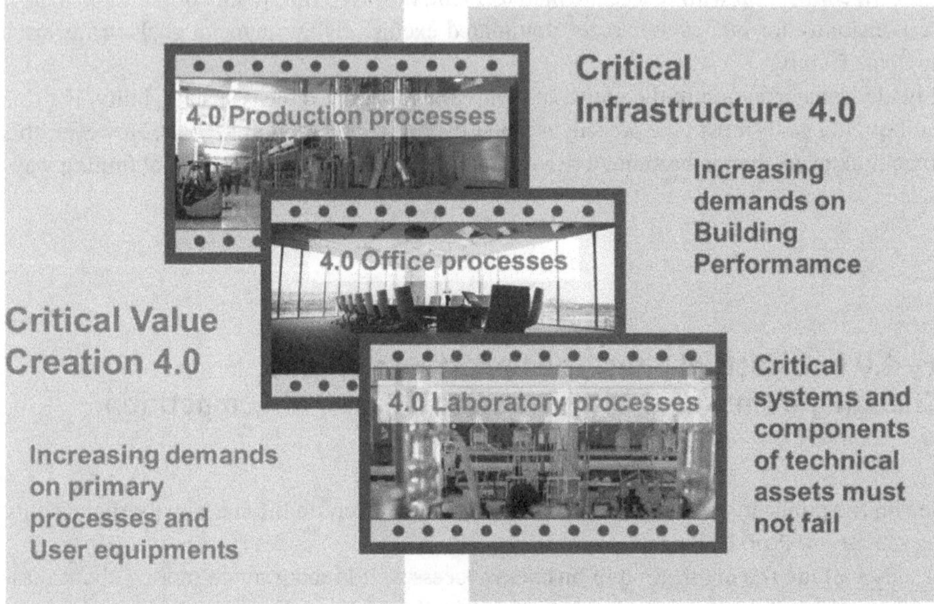

Figure 2.1 Industry 4.0 is transforming the embedding of user equipment's in building infrastructure. © Prof. H. Balck.

Increase of Support Requirements in Industrial Sites

Industry 4.0 replacements in equipment of the user processes needs reengineering of local infrastructure as well. In addition, there are transformations of the support organization. That means nothing less than a new paradigm shift, both in corporate real estate management and in the associated facility management. According to Michael Porter, companies are focused on value chains. The separation between primary activities (actual value creation) and secondary activities (non-value-adding activities) is critical to success. With the advent of Industry 4.0 in production, logistics and research, however, this distinction is becoming less important. The combined responsibility of CREM (Corporate Real Estate Management) and FM (Facility Management) is thus becoming more important, associated with an ascent in the ranking of company divisions.

Due to the often-sudden increase in support requirements (especially for predictive maintenance and security equipment), supporting processes also receive added value. For example, applications of the Internet of Things are tangible. IoT Systems enable the reduction of cyclic control capacities during inspections, maintenance and testing services. "Maintenance 4.0" enables significant cost reductions plus much higher reliability. Examples are fire alarm systems in which fire detectors report data about their contamination status to a service center. The processes initiated are carried out "on demand" with replacements of faulty components.

There is no doubt that this path is fit for the future. However, it should not be underestimated that structural change is not an easy path. For example, trouble-free operation of technical systems facilitated by Industry 4.0 technologies is an extreme requirement. It can only be reached if the operating staff is familiar with the associated IT Systems. In the service teams, this is a demanding learning process. But that's not enough. It is necessary to link investment processes and operational processes end-to-end. But in many cases, investment processes in construction departments are still sealed off from operators and users. This results in a variety of interface problems. Particularly critical are torn data chains between construction projects and subsequent operations.

Structural Break between Building Investment and Operation

The structural break between investment and operation is exacerbated by increasing requirements for data transparency and data proof. There is hope for this. Building information modeling (BIM) renews the much-invoked vision that as-built data will be handed over to the operator-organization at the end of construction projects (see Part 4). When processes are digitized in this way, demands on knowledge and skills shift. IT competence is becoming a necessary condition for successful service teams – and this is a problem of great importance, exacerbated not only by a shortage of skilled workers, but above all by professional channeling. Overcoming this includes the establishment of learning organizations in companies, interprofessional and cooperatively in mixed teams. I have had best experience in quality circles.

Introduction of the "Pareto Magnifying Glass" to Manage Inventory Data

The possibilities of technological and organizational change in industrial sites go hand in hand with answers to the following key questions:

- Which 4.0 value-added processes occur in which objects of the built infrastructure?
- How can built infrastructure be upgraded together with related services?
- How do data and software contribute to this?

Numerous attempts to introduce CAFM got stuck at that time, because the data depth alone failed "down to the last screw" at the limits of feasibility. But such attempts with invisibly recorded data volumes had to fail, because

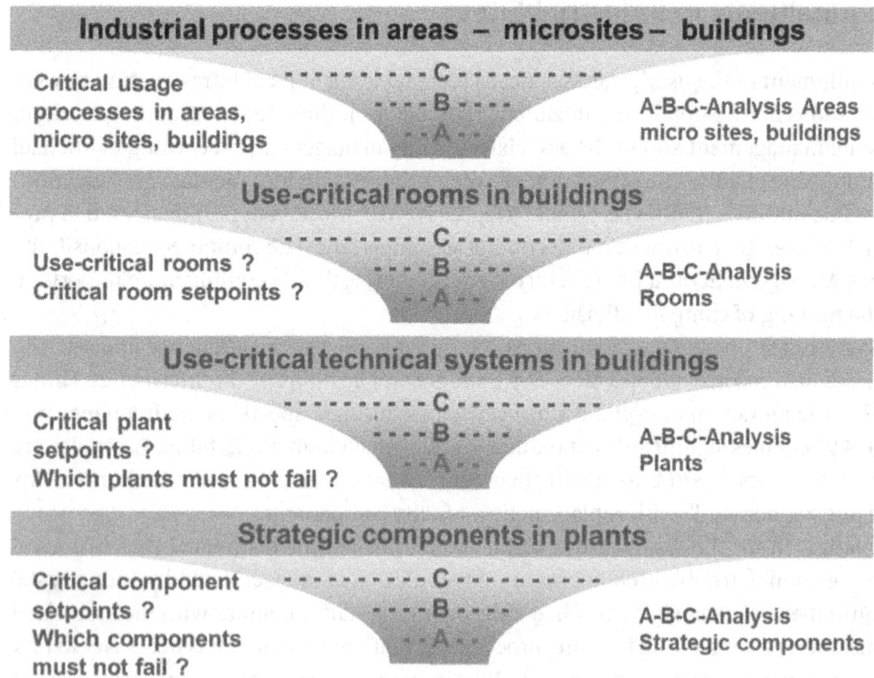

Figure 2.2 Hierarchical step sequence to identify individual objects with relevance for re-engineering. © Prof. H. Balck.

the subsequent processes of data maintenance could not be mastered. This phenomenon has been known since the introduction of CAFM systems in the 1980s. The digitalization of today's form must not repeat this mistake. That's why we take a different approach in our reengineering projects. I have developed a "Pareto magnifying glass" for the recording of existing properties. As a result, we were able to control the amount of information in our teams and, above all, keep it manageable. The classic commandment of "reduction of complexity" was decisive. At all levels of an object hierarchy (area / buildings / technical plants / components), the unfolding complexity was gradually reduced with the help of A-B-C analyses. The 80:20 Pareto rule serves as a guideline. In a hierarchical step sequence, we were able to identify individual objects that were relevant for reengineering (Figure 2.2).

The Pareto-focus makes it possible to limit different degrees of effectiveness and efficiency. As a result, increased requirements for Industry 4.0 transformations could be focused on and implemented more realistically. A further advantage is the manageability of the amount of data to be handled. In contrast to unsurpassable widespread big data analysis, this procedure involves a reversal: a small data digitization strategy.

Aging and Obsolescence – Management of Renewal

A frequently asked question by those responsible in corporate real estate management concerns the degree of ageing and obsolescence in buildings and facilities. When and to what extent are modernizations, renovations, extensions, or new buildings necessary? This often involves a heterogeneous building stock between 100 and 300 properties. These are usually many old buildings and various newer buildings. If you try to estimate the overall need for renewal costs at a given time, the result is often frightening. Such a cumulative need for renewal is hardly compatible within the framework of feasible financing. Then previously serious renewal plans are quietly dropped, and the previous regime of spontaneous investments is returning.

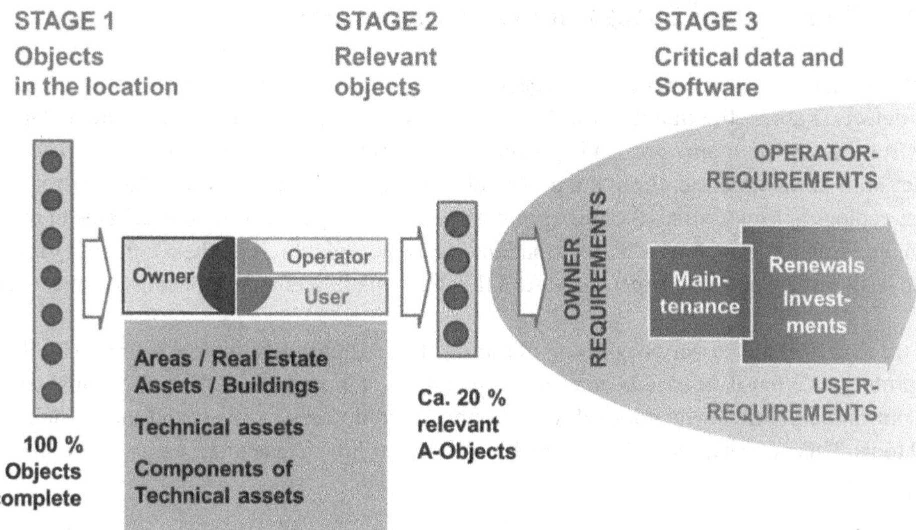

Figure 2.3 Pareto focus in projects of site-transformation with assignment of actors. © Prof. H. Balck.

Our experience in industrial projects has taught us that applications of the "Pareto magnifying glass" can help to make pragmatic limitations on the basis of the Pareto 20–80 rules. Figure 2.3 provides the model which can simplify change management in industrial asset portfolios. After a comprehensive survey of the entire building stock approximately 20% of those buildings are selected according to a joint set of development criteria, which are relevant both in terms of usage aspects as well as in terms of operator processes and real estate aspects: for optimizations and inventive changes. Subsequently, targeted measures for improvements or renewals of structural and technical facilities are determined.

Results of transformation projects carried out by the author show a double effect: The development of a fitness program "Industry 4.0" that is selectively geared to structural and technical facilities opens rationalization potential in steps. At the same time, potentials for sustainable location development are tapped.

PART 3

The Success of Built Environment Must be Measured in Its Sustainability during Operating Phases and Life Cycles

The contribution is based on the publication "Transformationen des Bauens" in the German journal Facility Management, Bauverlag (2022) – © Prof. H. Balck

In the now legendary UN conference in Rio de Janeiro in 1992, the world public was confronted with a program of global change in response to climate change. In fact, only now, three decades later, is the will to sustainability recognizable worldwide. Politically, in business and even in international finance. Built environments should be integrated into overcoming obstacles in long-standing established industry patterns of the construction and real estate industry. What is needed are end-to-end process chains of sustainable value creation – digitized in the BIM paradigm.

The creation of buildings begins with mental anticipation and ends – like all production – after plan-based realization. Unlike typical products in trade and industry, buildings have extreme spatial dimensions and long service lifespan at the same time. Since the beginning of the history of construction, this has had corresponding consequences: high material resource consumption, combined with irreversibly harmful natural interventions. In addition, with increasing mechanization, energy-dependent building operation has become a major cause of climate change.

Critique of the Usual Definition of Success for Construction Projects

At present, in critical reflection, it is understood that the success of a building no longer consists only in handing over a "work" without defects. Rather, the quality of building completion must be measured by the extent to which a sustainable environment is maintained or made possible. This in turn requires different goal definitions and success criteria. They shift in the timeline. From the project phases (years) to the operating phases (decades) and beyond to ecological-geological time courses (centuries and more). Thus, the beginning-end scheme of manufacturing is in question. Real success is no longer a precision landing as "completion" but the confluence of suitability and testing of the built during phases of operation. And this happens in a changing social, technical and natural environment.

We are not prepared for this. Still at the very beginning of a multi-layered paradigm shift, confronted with a variety of time-critical problems. What is needed is determined social action, in which many things must come together synergically. Committed political will, coupled with economic willingness to invest and the availability of suitable methods and tools. This means nothing less than a broad-based innovative change, activated in rapidly shorter time windows!

Critique of Traditional Economic Rationality

One reason for decades of refusal to take action to avert climate change is the long-established success patterns of economic action. Although energy efficiency and even comprehensive sustainability have long been recognized goals in public works, their pursuit has almost always failed due to rigid budget limits. The focus is on cost ceilings of the investment. Follow-up costs in subsequent operations are usually ignored. Rigid cost orientation and time pressure are standard. "Additional costs" for better energy or resource efficiency are rarely accepted.

Missed Sustainability in Facility Management

With the introduction of facility management in the 1980s, a fundamental reorientation in the value chain of construction was already required. The aim was an integration of project planning and operational planning. A claim that comes very close to current approaches to sustainability. But it was premature and could not succeed. One of the main reasons is the one-sidedness of economic principles of project management. Costs and deadlines are in the foreground. Qualities are very often only boundary conditions. Priority is given to costs and time deadlines (Figure 3.1).

This model was introduced in construction projects in the 1960s, following the operations research of the 1940s/1950s. In the 1980s, it was seamlessly adopted into the new practice of facility management – and even tightened. To this day, the following applies almost unrestrictedly: Costs down for such services that are necessary but have no direct contribution to business success or the usual standard of use. Conversely, qualities in service only play a significant role if they are due to special, especially legal requirements. The established facility management and services coordinated within it are therefore trapped in the same target system of operations research as the preceding project management. Both with a formally identical structure in the economically controlled triangular relationship of costs – qualities – dates/times. Worse, they are separated by a threshold of non-communication (Figure 3.2).

With regard to time-critical sustainability goals, this radical separation of the project and operational worlds has fatal consequences. Let's take a look at new workflows of creating sustainable buildings. They start with identifications of relevant parts in the building system: choice of sustainable building materials for building constructions (e.g., recyclability), energetically relevant technical systems and components. Different qualitative aspects of sustainability must be considered, which, like follow-up costs, only have an impact on operation, maintenance, and management. Due to the communicative barrier between the project world and the operating world, however, object aspects of sustainability are hardly noticed in the planning phase or simply ignored for cost reasons.

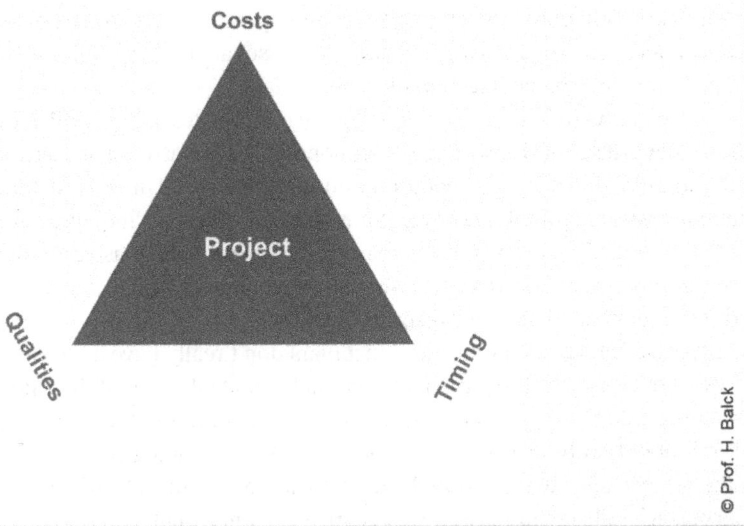

Figure 3.1 The simplified triangle cost-time-quality – first defined in Operation Research in the 1950s. (© Prof. H. Balck).

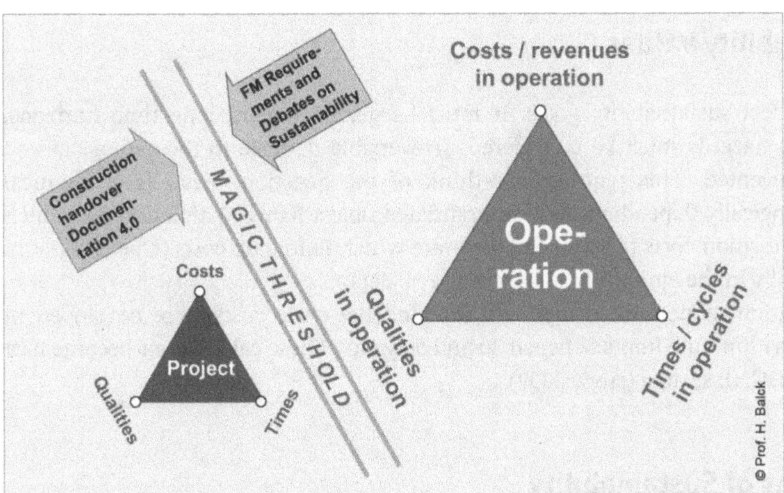

Figure 3.2 Radical separation in the management between project world and operating world. (© Prof. H. Balck).

In other words, whether decisions in the construction project are sustainable – or not – is thus not recognizable as success or failure in operation. There is a lack of feedback from users and operators on building owners, planners and even more so on product manufacturers.

Critical Phase Transitions

Let's take a look at the process chains in the project down to the data level. From the design phases to the tenders and awards, from construction management to commissioning and at the end from the (rarely organized) construction handover to subsequent operation. If these phases are modeled in actual process maps, a torn overall picture emerges organizational structural breaks, media breaks, interface, and data problems. The most difficult is the phase transition from project activities to operator services. Project activities end with planners and installers leaving the stage. Subsequently, users come and are "left alone." After that, users and operators have to start over. As a rule, with the elimination of construction defects. Instructions for operating processes are missing. Submitted documentation is incomplete, partly incorrect. There can be no question of "as-built data."

For this reason, a glaring problem arises with the onset of maintenance and legally required tests: In insufficient construction handovers, sufficiently accurate "plant lists" and associated "component lists" in as-built status are missing. They are indispensable as an alphanumeric database for operation and maintenance. This means that time-consuming subsequent as-built documentation is required. However, for reasons of cost and time, these are usually passed on by the operators to the commissioned maintenance companies. The effort to be mastered now is then contractually "owed" again in a new way. But not demanded as follow-up services from planners and companies. This results in unnecessary, sometimes high risks in operator responsibility.

Taken together, a strangely established anti-logic emerges. Planners and installers don't really have a problem at all. And neither do service companies! However, as soon as the pursuit of sustainability at all stages of the value chain is required by builders and investors, this outdated fragile industry pattern cannot continue. It needs to be corrected, better still completely reshaped. It is about nothing less than methodically and technologically modelling the critical phase transition between the project and the operational world. However, significant difficulties due to insufficient inventory data remain an obstacle. This requires new ways with end-to-end digitization. In part 4, approaches are shown on the basis of a renaissance of industrial value chains.

Life Cycle Costs and Sustainability Values

Economic action can no longer neglect sustainability goals. It must be geared toward long time horizons. Foreseeable as well as unforeseeable hazards must be considered. Irreversible damage to the geological and biological environment must be prevented. This requires a rethink of the question: What is economical construction? Answers are methodologically dependent on a differentiated object focus. Which designs, which systems, which components have acquisition costs (Capex) and generate which follow-up costs (Opex)? First in the project, then in operation and finally in the end-of-life phases of the objects.

In order to combine both, calculation models for component-oriented life cycle costs were developed in particular with component lifespans within time limits between 20 and 50 years. These calculations became part in rating systems first in the German DGNB system (since 2009).

Geo-ecological Time Horizons of Sustainability

But how can follow-up costs and long-term qualities be modeled if longer time horizons in geological-ecological dimensions are considered? In principle, this is the case when the effects of greenhouse gas emissions, such as CO_2, have to be modeled. Recently, a dramatic phenomenon has been added when natural disasters, such as floods, raise the question of the suitability of our buildings. If the unsuitability of existing buildings becomes obvious, even locations are questioned. And in the worst case with reduced market values of endangered real estate assets.

Therefore, construction requirements must change. What is required now are other construction methods, technical concepts and materials which have been proven to have sustainable performance in the use and operation phases.

In addition to the basic concerns of environmental protection, it is also specifically about requirements for "building resilience," i.e., preventive measures against possible extreme loads. Precursors for this are proven security solutions, e.g., fire protection, fire brigade, devices against water damage. So complex rethinking is necessary.

First of all, what we need is an extended, dual economic logic

- Classic criteria for investments according to economic principles, such as scarcity and efficiency as a basis
- Inclusion of costs for additional measures that are considered "uneconomical" compared to the previous standard but are considered necessary for the required sustainability in operational phases and beyond in long-term scenarios.

Securing the Future through Networking – Race against Time

Our social and economic future stands or falls with the success of major transformations toward sustainability. This includes the development process of sustainable construction. This is accompanied by fundamentally changed content in goals and methods. An indispensable prerequisite is the use of modern information technology tools, i.e., integration into the overarching transformation of digitization.

Chances of advancing depend on the time available. In the race against time, we need more speed through networking, which has not been sufficiently practiced so far. "Learning networks" must be created. Partnerships of technology partners, public and private investors, real estate companies, builders, operators, and research institutions. It is about ecologically, socially, and economically sustainable ways of integrated development by projects (Figure 3.3). This is a model for teams. Solutions for a livable and nature-friendly future can only be found cooperatively.

Sustainability in the Real Estate Industry

In the real estate industry, the willingness to sustainability is currently present to an astonishing extent. The reason is the EU Disclosure Regulation, which came into force in 2021, with its ESG taxonomy (Environmental – Social – Governance). It is a universal valuation system that also covers all areas of financing in the real estate industry. At the European MIPIM real estate fair 2021 in Cannes ESG was a much-noticed focus. However, this realignment of investment is not only about expanding responsibility by adding ethical obligations. Rather, a dual economic pattern can also be seen here:

- Real estate economical basics and income models remain the basis.
- In addition, there are forecasting models for value enhancements resulting from targeted sustainability measures. They are based on demonstrable energy and resource efficiency and proven social qualities of real estate. To the extent that they can be advantageously represented in ESG scorings, there are marketing advantages.

The EU taxonomy changes the value status of certifiable real estate if consistently observed. Properties without certifications according to EU taxonomy are simply less attractive and therefore have a lower market value, with an impact on financing conditions for banks or investors.

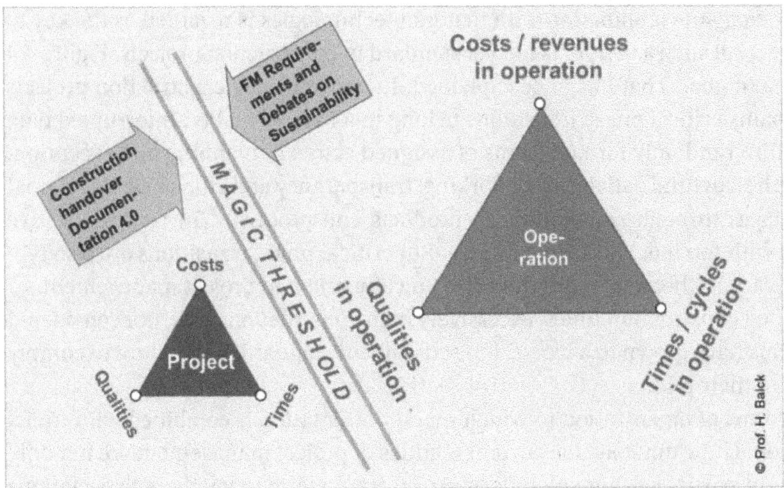

Figure 3.3 Synthesis of goals in project and goals in operation – a formula of sustainability. (© Prof. H. Balck).

Part 4

A Renaissance of Serial Construction Enables Sustainability Goals in Realigned Value Chains – Merging Design and Operation of Built Environment

The contribution is based on the publication "Industrielle Wertschöpfung" in the German journal Facility Management, Bauverlag (2022) – © Prof. H. Balck

Since the 1930s, the industrialization of construction has been a program of technological orientation in architecture and construction. All implementations, most recently the large-scale experiment with building systems in the 1960s, have not succeeded. And today? Far-reaching new approaches can be seen during the new Industrial revolution. Sustainability becomes the guiding mode.

Industrial Plan of the Green Deal – Tailwind of Technological Change

The turning point has begun. Action against climate change is both social and political will. At the world economic forum 2023 in Davos, Ms. von der Leyen (President of the European Commission) presented the "Industrial Plan of Green Deal":

> *In less than three decades, we want to be carbon neutral. Everyone knows that the key to managing this transformation, or better yet, leading the way, lies in the clean technology industry.*

In the temporal race against growing dangers from natural disasters, a key question arises for all construction actors: How is accelerated industrialization possible? With about 35% of CO_2 emissions and 55% of waste generated (Germany 2023), the construction sector is in a difficult starting position. In contrast to mechanical engineering, the construction industry, which is anchored in the skilled trades, has always been the stepchild of mechanization. To this day, handicraft activities dominate. Only their transfer by far-reaching industrialized processes can enable significant CO_2 reductions and materiel resource efficiency. This is currently happening in a fundamental change in the construction and real estate industry.

Paradigm Shift in Construction and Building Technology (See First Chapter Introduction Industry 4.0)

It is undisputed that information technology in combination with material technologies is regarded as the key to industrial progress. Figure 4.1 illustrates that such a merger is not yet standard in construction projects. Figure 4.4 illustrates that it is a defined development goal. That has to be explained. In conventional construction projects and even in their overarching value chains critical phase transitions belong to a bad normality. Interrupted data chains complicate communication within randomly formed teams of assigned actors in planning and execution. This results in project organizations with structural deficits in co-working, transparency and efficiency. Its original purpose was to realize market advantages through competition in products and processes. In fact, innovative results are rarely achieved. But always with too much time and money. But critical phase transitions are also typical in construction projects that take place in the strict regime of economically oriented project management.

Negative consequences dominate. Long construction times, excessively high construction costs, not considered life cycle costs and even quality problems have grown to a worse, but self-evident standard. The failure to comply with sustainability goals are currently in their midst.

What is needed, therefore, are new forms of organization in which market orientation is combined with transparency and unhindered communication. Unfortunately, the current realities of project management are not only backward, but also resistant to reform. In particular, bureaucratic approval processes destroy these expectations again and again. But there is a way to reshape the outdated project paradigm: industrialization of all project phases from the beginnings of component production to the handover of the buildings and on to their operational phases.

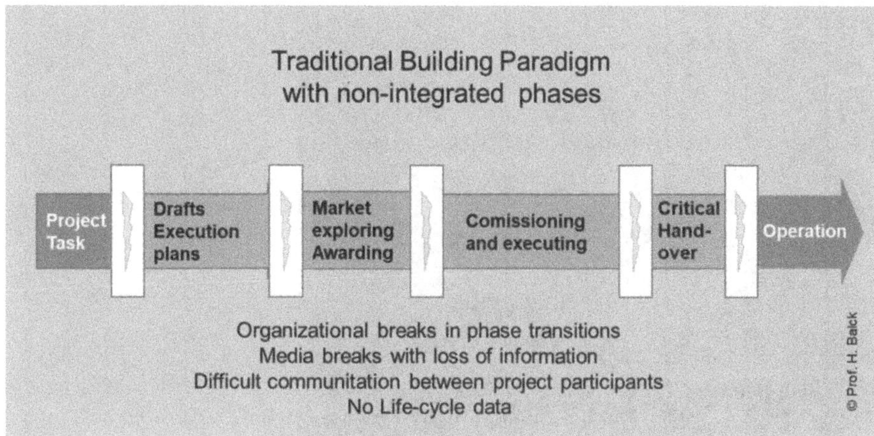

Figure 4.1 Critical phase transitions and structural breaks in the traditional building paradigm. (© Prof. H. Balck).

Sustainable Production – Computer-based Planning Plus Robotics

Achieving sustainable buildings cannot succeed if outdated methods of serial production are repeated. Instead of rigid standardized components, as was common in modular systems in the 1960s, components can now be manufactured with a high degree of flexibility. Today's level of computer-controlled production enables variable standard parts. Noteworthy is a related change in the series principle. By linking IT-supported design and the subsequent use of robots in prefabrication, non-identical parts can also be produced. This is followed by high-precision assemblies with significantly less effort and thus shorter construction times (average time savings of 25%, as of 2020) than with conventional construction methods.

A pioneer of this methodology is the architect Prof. Achim Menges from the Institute for Computer-based Design and Construction Manufacturing, University of Stuttgart. In cooperation with industrial partners, he has shown in a series of research and pilot projects that ecological orientation and extreme material efficiency can go hand in hand. And that along with constructive aesthetics (Figures 4.2.–4.3). Achim Menges and his team are not only interested in specialized research. He sees his work as prime examples of overcoming outdated industry patterns.

Figure 4.2 Prof. Achim Menges – Buga 2019 Pavilion Heilbronn – Serially produced wooden cassettes, where no two are the same. Foto © ICD / ITKE University Stuttgart.

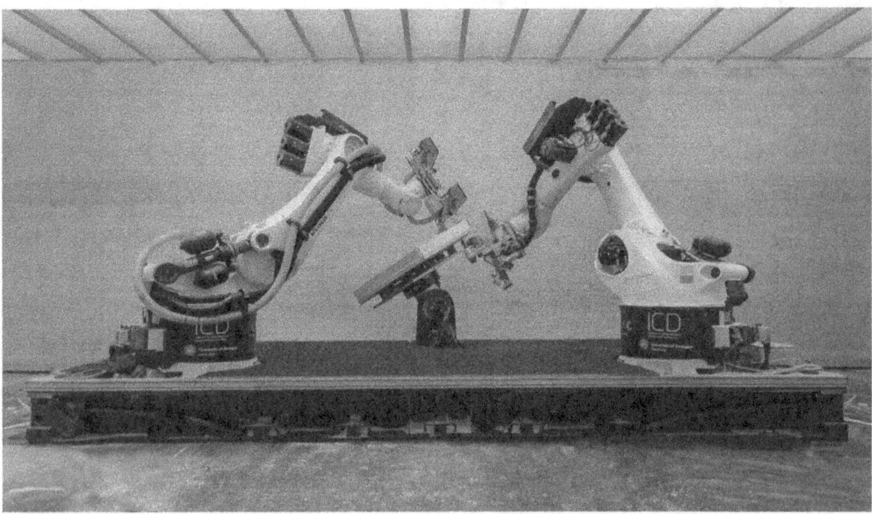

Figure 4.3 Prof. Achim Menges – Component production by robots in combination with computer-aided design. Foto © ICD / ITKE University Stuttgart.

Planning with Elemental Systems – BIM Enables Integrated Construction Planning in Industrial Production

Design and planning are the strategic anchors of any construction project. This has been true since the historical beginnings of architecture in antiquity and remains unchanged in the planning of industrialized construction processes. Even in the upheaval of digitization, nothing changes in this constant. It's just the media that's changing. In the traditional planning process, architects and engineers have a high degree of freedom from preliminary design to execution planning. The "creative search space" for concepts, materials and applicable products is ultimately the entire spectrum of available technologies and products. Unfortunately, this possible diversity is also an approved degree of freedom for commissioned architects and engineers. The great freedom of design that is usually permitted causes a major source of problems. This shows again and again the failure of large-scale buildings, especially for public clients.

A reorientation is necessary. Freedom of design has to be defined in an industrial approach. A fundamental success factor of industrial production is the limitation of the variety of components through standardization. Design with system elements necessarily has degrees of freedom, which are inherent in their combinatorics.

In the 1960s, this was done with rigidly defined elements and correspondingly monotonous constructions (largely failed early industrialization). In the technological restart that is now beginning construction tasks are carried out by designing and planning configurations of flexible components. The equally innovative IT medium is Building Information Modeling BIM (Figure 4.4).

In innovative prefabrication BIM is already indispensable. In the subsequent project phases planners use this information basis in specified BIM applications. Indeed, this is still revolutionary. At the time of this writing (2023), attempts to introduce BIM in conventional projects are exceptions. This technology is still in the early stages of innovation and too complicated and expensive for most architects and engineers.

The introduction of BIM is completely different when you see its application in industrial construction. Everything runs in parallel here. At the beginning there is the production of series components and at the same time the use of BIM. The same happens in the subsequent production steps. Each step has its BIM counterpart. BIM-based architectural and engineering models can be set up much more easily and, above all, faster on this basis. It is obvious that the achievement of sustainability goals is also made possible in this approach. Simply because the focus is on components, the selection of which is crucial for sustainable resource efficiency.

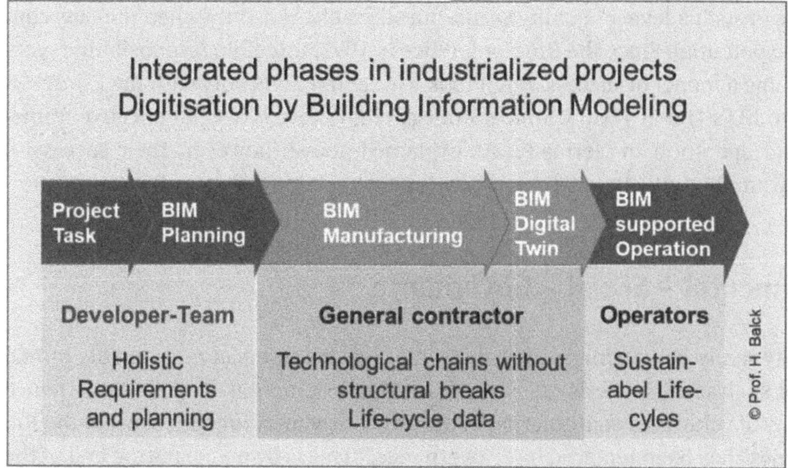

Figure 4.4 Integrated phases in industrialized projects – Digitisation by Building Information Modeling. (© Prof. H. Balck).

Developers and Construction Companies Reinvent Themselves

Construction companies have become system providers. They commission planning offices, operate in cooperation with project developers or combine product development and project-related construction planning in their own departments. Of particular note are the integration models of a new generation of general contractors (GCs) and their sustainability goals. The greater their technological flexibility, the more diverse the possible designs. GC planners regularly decide in consultation with production which components are suitable and required. The decisive factor for success is always the resulting ecological footprint, which is determined in accompanying computer simulations.

This, in turn, is not only decisive for the achieved customer benefit concerning sustainability. Therefore, proof of quality in the form of international certifications (types of certifications above) have an important effect on all partners involved: developers, architects, planners, construction firms, manufacturers. It must be emphasized that such quality assurance is unusual in the previous practice of GCs. This has never happened before. The introduction of a completely new standard of values is a historic turning point.

A new complex of automated industries is emerging, which is committed to the principles of sustainability. The Sustainable Development Goals (defined in the EU in 2015) will become a recognized guideline with key variables:

- Reduction of energy and materials in production (CO2 reduction and resource economy)
- Use of ecologically flawless materials (e.g., renewable raw materials) in constructions / plants
- CO_2 reduction in transport (supply chains, final assembly)
- Energy efficiency in the operation of devices, systems, and components
- Unmixed dismantling of installed materials (recycling) through reversible constructions.
- Construction companies can record installed parts with a return guarantee (circular economy)
- Use of recycling agents (e.g., recycled concrete from old concrete (R-concrete))

On national and international markets an avant-garde enters the stage of competition. Developers and construction companies offer IT-integrated buildings as a whole. Flexibly manufacturable prefabricated parts of the shell construction are supplemented by adaptable systems of structural finishing elements. Combined with alternative energy supply systems, such as photovoltaics, solar thermal energy, heat pumps, first-class buildings are created. This is also proven by measurements during the commissioning of prototypes and sample solutions.

At first glance, the cost-effectiveness of such a level of quality seems questionable. It is only when the new context is considered that one sees realistic potential. Since the Rio Conference in 1992, threefold responsibility – ecological + economic + social – has become a model of success. But it took a long time. It is only with the European "Green New Deal" (see above) and the EU's "Long Term Climate Strategy" that there is also an effective framework for progressive construction and operation in Germany. As explained above, however, their successful implementation cannot be separated from the Industry 4.0 transformation that is taking place at the same time.

ESG-value Principle: Environmental – Social – Governance

Especially in Europe, there are already many convincing and at the same time economical investments in real estate that comply with EU regulations, such as ESG criteria of sustainability: **E**nvironmental – **S**ocial – **G**overnance (entered into force in 2019). This trilogy of values is a continuation of the previously mentioned scheme of the Rio Conference. The position for "economics" has been replaced by "Governance." This is by no means the end of the economic aspect, but quite the opposite. It has now been confronted with the three ESG value dimensions as a whole. Thus, the Rio value scheme is now even sharper.

In this context economic goals no longer remain the only valid ones. They are extended to these extended value principles for investment and operation. Its effectiveness is obvious. Builders and construction companies combine sustainable financial models with demonstrably sustainable construction quality. Today, the success of real estate investments is the total product of sustainable construction and operation. The overall quality achieved is proven by certifications such as DGNB, (Germany), LEED (USA), BREEAM (UK). Accordingly, patterns of project development are changing.

If monetary results are no longer the only criterion for success, this means a considerable rethinking of business models. If we succeed in combining sustainability data with financial performance in the new ESG-driven real estate business, we can argue: Sustainable quality pays off! Nevertheless, it should not be overlooked that this approach is not an easy act and must also be enforced against a variety of resistance.

Managing the Phase Transition: From Planning and Construction up to Operation – the Original Vision of Facility Management Becomes Feasible

With an increasing understanding of the need for a sustainable built environment, the criterion for success shifts from merely flawless project execution to a definition of success that only becomes apparent during operation. A prime example is the optimization of energy efficiency in computer simulations and subsequent verification of reduced consumption values. Measurable usability and availability of the operated systems and components become criteria for measuring success. It is about nothing less than data and knowledge about "life cycle objects" throughout their lifetimes. And beyond that, it is about the consequences that have been triggered by realizing buildings in the long-term perspectives of climate change. Suppliers of industrial building systems must meet these requirements. A new generation of architects, planners, and companies is faced with the challenge of both technologically renewing their value chains and consistently aligning them with sustainability goals.

Ongoing Paradigm Shift

Toward a Pattern of Industrial Sustainability and Resilience

Business as usual has been around for decades for established markets simpler services, called facility services. It's different with facility management. With its introduction in the 1980s, this term was associated with a message.

It was about fundamental changes in the understanding of buildings and their use. The focus was on overcoming technical boundaries in construction planning. Combined with innovative IT technology, especially CAD, process chains were to be integrated into planning, implementation, and operation. Architects and engineers, as well as all other participants "around construction," should understand as co-actors that their works can only succeed if they define the beginning of a construction project as a joint and thus interdisciplinary task. Their result, the "performance" in the company, should be measurable with IT support. The aim was to build a bridge to users and operators. Above all, technically complex buildings should be handed over to the operational stakeholders in a user-friendly manner after completion with data transparency. In addition, users and operators should exchange experiences with project stakeholders on an equal footing.

As of today, this has not come close to succeeding. But by no means hopeless. On the contrary, the potential of an incipient industrial revolution opens up new perspectives. A fusion of holistic approaches of original facility approaches (1980s) with currently cross-industry IT-driven "Industry 4.0" is recognizable.

In addition, however, there is another socio-ecological megatrend that goes beyond this, which is reshaping both approaches. The preservation of our (still) habitable earth by establishing, putting into use, or adapting the built environment is on the brink. It is therefore essential for all construction measures to meet comprehensive sustainability requirements. But not only to prevent the consequences of climate change. In the future, these will increasingly become a reality. In a "rewilding earth" (J.Rifkin – The Age of Resilience, 2022) with devastation and catastrophes determines everyday life, then we need not only strategies of prevention but also those that offer us protection in the event of unavoidable natural attacks. These include, in particular, "resilient buildings" and robust geo-facilities.

Sustainable construction and operation is thus becoming a key issue for today's and tomorrow's living environment. The industry of the past has put us in this position. The industry of the future must be used to ensure our survival. And all this in a race against time.

Prof. Henning Balck
Heidelberg, Germany

Studied mathematics at the University of Marburg
Studied architecture at the University of Karlsruhe
Studied architecture at the University of Stuttgart (1973, graduate engineer in architecture)
Since 1970 Freelance consulting and cooperation in project management and business consulting
2003–2021 Managing Partner in the BALCK+PARTNER Group for Facility Management and Facility Engineering, Heidelberg (Germany)
Teaching and further training
1990–2023 Numerous lectures and training events on project management and facility management (Germany, Austria, Switzerland)
Since 1994 member of the advisory board of the journal Facility Management, Bauverlag (Germany)
Extensive Publications, see Knowledge in www.ips-institut.de
1994–2000 EBS European Business School, Oestrich-Winkel (Germany)
Course Corporate Real Estate Management + Facility Management – Participation in the development of the course together with Prof. W. Schulte and implementation of lectures in Facility Management
1996–2010 University of Applied Sciences in Mittweida (Germany)
Honorary Professor of Integrated Facility Management and Real Estate Management
Participation in the development of the study program and implementation of lectures and exercises
1999–2009 Krems Danube University (Austria)
University Course Facility Management and Real Estate Management (Master of Science) – Scientific Advisory Board – Lectures and supervision of Master's theses

1999–2011 IFZ Institute for Financial Services in Zug (Switzerland)
Lectures on Facility Management in the Real Estate Industry
1990–2023 IPS Institute for Project Methodology and Systemic Services (Germany). Consulting – Research and Development, www.ips-institut.de.

The IPS Institute was founded in Mannheim in 1990 by Henning Balck. He is the owner and scientific director of the institute. The institute is a project-oriented organization. In projects, consulting services are combined with innovative developments. The IPS cooperates with scientists and experts from a wide range of disciplines and industries. The institute's tasks are broadly based. In addition to research and development, this includes innovative impacts in consulting projects.

3.15

The Robert L. Preger Intelligent Workplace™ a Transformative Living Laboratory at Carnegie Mellon University

Joint Contribution By Volker Hartkopf and Vivian Loftness

Climate change, rapid advances in technology, and the global pandemic have significantly changed the nature of work and the workplaces that are best suited to ensuring a healthy, productive workforce. Over 25 years, the Advanced Building Systems Integration Consortium (ABSIC), an NSF Industry-University-Government collaborative based in the Center for Building Performance & Diagnostics (CBPD), has researched and advanced the definition of high-performance, intelligent workplaces and intelligent facilities management to support:

1) *Individual Productivity and Comfort*
   High-performance workplaces ensure individual comfort, health, and productivity through quality enclosure, HVAC, lighting, and interior systems that deliver thermal, acoustic, visual, and air quality, as well as spatial and ergonomic support.
2) *Organizational Flexibility*
   High-performance workplaces ensure sustained individual productivity and collaborative creativity in the face of ongoing organizational and technological change through user-centric, customizable plug-and-play infrastructures that support spatial change.
3) *Technological Adaptability*
   High-performance workplaces support advances in technology and connectivity for individual and collaborative work, through accessible and open interior systems and engineering infrastructures that support changing technological demands.
4) *Environmental Sustainability*
   High-performance workplaces demonstrate the highest level of environmental sustainability in energy, water, material use, and assembly design, for health, resource efficacy, just-in-time delivery, flexibility, and maintainability, with a focus on life cycle design and natural conditioning.

The ABSIC consortium of university, industry, and government leaders researched advanced buildings around the world (Hartkopf et al. 1997), debated assemblies and integrated systems that promote more sustainable, flexible, and user-based workplaces for the future, and built a "Living Laboratory" at Carnegie Mellon University – the Robert L. Preger Intelligent Workplace (IW). This living laboratory supports integrated testing of advanced building technologies and building science research, towards linking the design, construction, and management of high-performance workplaces to measurable gains in environmental and human outcomes. These advances are critically needed around the world, but especially in the US where the built environment:

1) creates 40% of land-fill waste by weight, and 30% by volume;
2) consumes almost 40% of the US's primary energy for operation and an additional 10–20% in embodied energy for building materials production;

Facilities @ Management, Concept - Realization - Vision, A Global Perspective, First Edition. Edited by Edmond Rondeau and Michaela Hellerforth.

3) consumes 70% of the US electricity production; and is the largest contributor to destabilizing peak load demands;
4) contributes over 500 million tons of CO_2 into the atmosphere per year from the generation of the electricity used for building operations;
5) offers potential for health and productivity savings in the range of $20–200 billion annually through improved practices and systems integration.

The Robert L. Preger Intelligent Workplace

A Living and Lived-in Laboratory

Every university's Architecture, Building Science, and Building Engineering program should have an Intelligent Workplace – a living laboratory of component and subsystem innovations and a testing ground for the next generation of building systems integration for indoor environmental quality and resource sustainability.

The Robert L. Preger Intelligent Workplace™ (IW) is a 700 m² (7000 sq. ft.) living laboratory of component and subsystem innovations in an occupied, lived-in laboratory with integrated passive-active systems for sustainability (Figures 1 and 2). Developed with leading practitioners and manufacturers of high-performance building systems, the Intelligent Workplace is an ongoing test bed for advances in building enclosure, mechanical, lighting,

Figures 1 and 2 The Intelligent Workplace™, a 700 m² rooftop extension at Carnegie Mellon University has flexible, floor-based infrastructures to support research and organizational dynamics. (*Source:* authors).

telecommunications and interior systems as well as the next generation of building controls to support human health and productivity as well as the highest level of environmental sustainability (Hartkopf and Loftness 2004; Hartkopf et al. 2017).

Led by Dr. Volker Hartkopf, the IW living lab is the result of an unprecedented collaboration between the eight faculty and several generations of graduate students in the Center for Building Performance and Diagnostics (the first National Science Foundation Industry/University Cooperative Research Center in the building industry) and the Advanced Building Systems Integration Consortium (ABSIC), a collaborative body of industry and government leaders in the built environment (see acknowledgements). A major gift from Robert L Preger, a Carnegie Mellon electrical engineering alumnus and leader at Oracle, triggered university and industry commitments to building this $4 million laboratory.

Over more than 25 years, the Intelligent Workplace has provided a forum for education, demonstration and research with a global constituency focused on:

1) *innovations in enclosure, lighting, HVAC, power and data networking, as well as interior systems to achieve the highest level of thermal, air, visual, acoustic, and spatial quality;*
2) *innovations in spatial flexibility and collaborative work environments to meet the changing nature of work and organizations;*
3) *innovative product assemblies for ensuring performance in an integrated setting;*
4) *"state of the art" IT systems as well as sensing, actuating, and controls technologies to enable effective performance of systems to achieve sustainability in building operations;*
5) *training on material, component, and systems choices and their integration for performance;*
6) *hands-on training with instrumentation and metrics for evaluating performance and occupancy comfort;*
7) *and field validated development of computational tools for design, simulation and management.*

The IW enables the interchangeability and side-by-side demonstrations of innovations in HVAC, enclosure, lighting, interior, networking and control components and assemblies. Most importantly, as a "lived-in" office, research, and educational environment, the IW provides a testing ground to assess the performance of new products in an integrated, occupied setting.

Designing for Sustainability and Organizational Change

Flexible Grid – Flexible Density – Flexible Closure Systems

To avoid failures in indoor environmental quality and early obsolescence of building systems, it is critical to invest in user-based infrastructures that are modular, reconfigurable, and expandable for all key services – ventilation air, thermal conditioning, lighting, data/voice, and power networks, as well as interior systems. The dynamic reconfigurations of space and technology typical in buildings today cannot be accommodated through the existing embedded service infrastructures – either the large zone "blanket systems" for uniform open-plan configurations or the idiosyncratic systems for pre-occupancy floor plans that are obsolete even before move in (Figure 3). Instead, what are needed are flexible infrastructures capable of changing both location and density of services:

Flexible Grid – Flexible Density – Flexible Closure Systems are a constellation of building subsystems that permit each individual to set the location and density of HVAC, lighting, telecommunications, and furniture, as well as the level of workspace enclosure.

In contrast to the embedded and blanket services of the past, advanced buildings demonstrate modular, floor-based infrastructures, in well organized plenums for reconfigurable service, to more effectively support the dynamic workplace. Floor-based "nodes of service" such as HVAC diffusers or plug terminals can be continuously relocated, added, or subtracted, to meet changing organizational or individual needs. With modular floor-based services, the ceiling can become more playful and elegant – as a light and acoustic diffuser – defining working groups and working neighborhoods as well as re-exposing the beautiful ceilings of landmark buildings (Hartkopf et al. 1999).

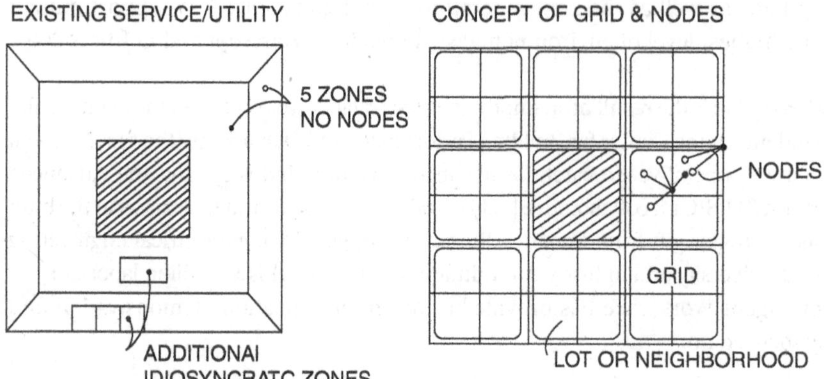

Figure 3 The future of high-performance offices are flexible infrastructures with micro-zoning and user reconfigure-ability of heat, light, air, networks and spatial settings for individual and collaborative work. (*Source:* authors).

Building services and their infrastructures need to be designed to be as dynamic as the spaces they are to support. Given ongoing changes in occupant density and work hours, alongside the evolution of office technologies and web-based work tools, the shift away from cast-in-place or embedded infrastructures to more flexible work environments are more critical than ever (Loftness 2009).

Designing, Constructing and Operating Buildings to Eliminate Waste

The Intelligent Workplace demonstrated a 70–90% reduction of waste during the production and assembly of materials and systems for the structure, façade, roof, and floor plenum, when compared to a conventional building. The IW's modular design supported deep material efficiencies and its off-site prefabrication led to no on-site waste during the construction phase – from the steel structure to the facade erection to interior fit-out. This also ensured significant time-savings during construction, with measurable capital savings through shortened delivery time. Accessible plenums, bolted and pinned assemblies (rather than welded and caulked) and plug-and-play technologies allow for component-by-component, or system-by-system changeout of technology without destruction or waste. Flexible, plug-and-play infrastructures also insure that the building is a renewable asset for its investors and will not become a "straightjacket" that eventually has to be discarded in whole or in part. In the operational phase, material waste is reduced through reconfigurable and relocatable interior systems with modular interfaces to the HVAC, lighting, and networking systems.

The integrated, modular, and demountable systems in the IW reflect the fact that buildings are made from components with different life cycles. Investing in quality not quantity, in plug-and-play enhancements over redundancy, allows for easy change-out and advancement of technology as the need or opportunity arises. Three system diagrams have been included at the end of this article for clarity (see Figures 16, 17 and 18) As a result, the waste management and environmental benefits of the IW, given site project management by Professor Steve Lee, include:

1) *During the manufacturing phase, the materials, components and systems as designed and specified require a fraction of the energies and produce a fraction of the emissions of comparable systems.*
2) *During the construction phase, waste is eliminated through prefabrication and modular design.*
3) *During the operational phase, the greatest reduction in energy and carbon waste is through natural conditioning by passive solar heating in winter and night cooling in summer, by natural ventilation for six or more months, and by daylighting 100% of the work places during daylight hours.*
4) *During the operational phase, relocatable and expandable grids and nodes of service for HVAC, lighting, power, communication and interior systems manage obsolescence and support organizational and technological changes on-demand.*

5) *The design for disassembly anticipates a complete decommissioning of the building and its constituent parts. The "long life systems" such as structure and enclosure and high quality lighting and interior systems can be redeployed elsewhere. In all other cases, the materials, components, and subsystems are designed for disassembly (not fused) and can be recycled into primary materials.*

Leading with Efficiency and Passive Conditioning, Supported by Active Low-Energy Systems

The global race to net zero buildings often relies on extensive use of renewable energy to offset demands. The Intelligent Workplace is designed to maximize energy conservation and passive energies before using active low-energy systems or renewables. Insulated and airtight enclosure assemblies, high-performance glazing for daylight with appropriate thermal resistance, external shading, natural ventilation and building mass, maximize the number of hours for which no electric lighting is needed and the number of months for which no cooling, heating or ventilation is needed. Passive strategies such as cross ventilation, stack ventilation, fan-assisted ventilation, and night ventilation are an ongoing area of research, alongside passive solar heating and daylighting through time of day and seasonal window management.

In the IW Living Lab, mechanical systems are engineered to maximize passive conditioning through tiered responses. When wind or thermally induced ventilation is no longer viable naturally, economizer ventilation with 70% heat recovery is provided by an enthalpy recovery wheel air handling unit. When outdoor humidity levels exceed the effective comfort zone for economizers, desiccant cooling is the next level of mechanical assistance. Only when all else fails, during the hottest days, does refrigerant cooling provide conditioned breathing air and locally controlled comfort through radiant ceilings and chilled beams. Energy efficiency and indoor environmental quality are both met strategically through the separation of ventilation with "fresh" air from thermal conditioning with innovative water-based thermal systems. Independent heating, cooling and ventilation innovations that support passive conditioning is an ongoing area of research in the "living lab," engaging both nature (passive conditioning) and humans in the control loop.

Electric lighting is also designed to maximize passive conditioning. Daylighting is pervasive in the IW, providing 100% of the workspaces light for 80% of the work hours throughout the year. Daylight redirection by season is achieved through dynamic external louvers and manual and automated interior blinds ensure effective daylighting even on low light days as well as shading on hot days. Every light fixture is an IP addressable point of control, enabling occupancy, daylight levels, and user preferences to support highly individualized lighting control for visual quality and energy efficiency. Research on façade assemblies for effective views and connection with nature, for daylighting without overheating and for electric lighting innovations and controls continue to be a long-term research focus.

Over 25 years of innovations with passive and active design and operational strategies, the energy demands of the Intelligent Workplace are 20% of a standard US office building. Achieving this level of performance is a critical pre-condition for creating net positive energy buildings, with demands so low they can be met through rooftops or site photovoltaics.

Creating a Generation of Building Industry and Facility Leaders through Building Science Research and Education in a Living Laboratory

Carnegie Mellon University's Intelligent Workplace has supported over 25 years of graduate and undergraduate thesis projects, with path-breaking results in multiple research areas:

1) Learning from the Field – Post Occupancy Evaluation with Measurement (POE+M).
2) Advancing and Quantifying the Performance of Integrated Systems

3) Advancing Innovative Controls with the Internet of Things (IoT) to bring humans and nature “in the loop.”
4) Advancing Lab and Field-tested algorithms and software tools for designing and operating buildings – moving from BIM to BAS to BEM.
5) Ensuring Net Zero Carbon Futures

Learning from the Field – *expanding beyond post occupancy evaluation surveys to evaluate and correlate the technical attributes of building systems and quantitative measurements of indoor environmental quality (POE+M).*

Every student of architecture, building engineering and facilities management should spend substantial time in the field studying the integrated technical systems and resulting indoor environmental quality in occupied buildings. Without on-site exposure to the installed enclosure, mechanical, lighting, networking, and interior systems that define our workplaces, and their modification and performance over time, it is not possible to advance the design and operation of complex buildings for sustainability, organizational change, or productivity and health. In addition to surveys on user satisfaction and perceived health and productivity, post occupancy evaluation (POE) must include actual measurements of indoor environmental quality (POE+M) as well assembling up-to-date records of the technical attributes of building systems (Loftness 2017).

In collaboration with the US General Services Administration (GSA), the CBPD faculty, and students have refined toolkits in the IW for the side-by-side and before-and-after POE+M evaluation of office, school, lab and assembly facilities worldwide. Led by Professors Azizan Aziz and Vivian Loftness, the National Environmental Assessment Toolkit (NEAT) combines on-site physical measurements of thermal, air, lighting and acoustic conditions, user satisfaction questionnaires, and records of the technical attributes of building enclosure, lighting, interior, and HVAC systems, to generate statistically significant guidelines for sustaining or improving the quality of the indoor environment. (Aziz et al. 2012; Loftness et al. 2020).

Dr. Ying Hua (2011 PhD BPD) used field data from POE+M in 27 federal offices to statistically reveal that distributed dedicated meeting and copy rooms both improve collaboration and reduce distraction as compared to centralized shared amenities or open plan locations (Hua et al. 2010, 2011). Dr. Jihyun Park (2015 PhD BPD) gathered the CBPD POE+M data for over 1,600 workstations in 65 office buildings, to statistically demonstrate that IEQ assessment methods and procedures centered on the occupant are critical to identifying the significant discrepancy between major IEQ standards and actual human perception (Park 2015). Through multivariate regression, multiple correlation coefficients, and Pearson correlation statistical analysis, this thesis demonstrated that humans are effective direct and indirect sensors for certain IEQ metrics, but not for others, with findings comparable to complex instrumentation. Moreover, the thesis identified that present environmental thresholds are not adequate for capturing acceptable thermal, acoustic, visual, and air quality conditions for occupant satisfaction, missing critical human factors and physical workspace characteristics (Park et al. 2013). Figures 4 and 5 reveals two of these findings relative to the occupants of office buildings in the US: first that summer temperature setpoints should be raised from 22C (72F) to 24 C (75F), and that mechanical zone sizes need to be substantially smaller than one zone per 10 occupants, if 80% satisfaction – the US standard for facilities – is to be ensured.

Advancing and Quantifying the Performance of Integrated Systems – developing and testing innovations in integrated mechanical, lighting, enclosure and interior systems for energy effectiveness, productivity, and health.

Buildings designed and engineered as tightly fit but poorly integrated components and systems often do not ensure IEQ or energy performance over time. In advanced building systems research, the faculty and students in the CBPD are actively researching innovations in the integration of mechanical, lighting, enclosure and interior systems, using the IW living lab to advance the highest level of sustained thermal, air, acoustic, and visual quality critical for physiological, psychological and social health, comfort, and well-being.

There are at least 12 facade design decisions that critically impact visual, thermal, air quality, and energy performance of buildings, designs that in turn demand operational integration with lighting, mechanical and even power generation systems (Hartkopf et al. 2020). Transoms are critical to effective daylighting and night ventilation,

Coling season user satisfaction with air temperature at 0.6 m
CBPD, CMU, Cooling Season (n = 446)

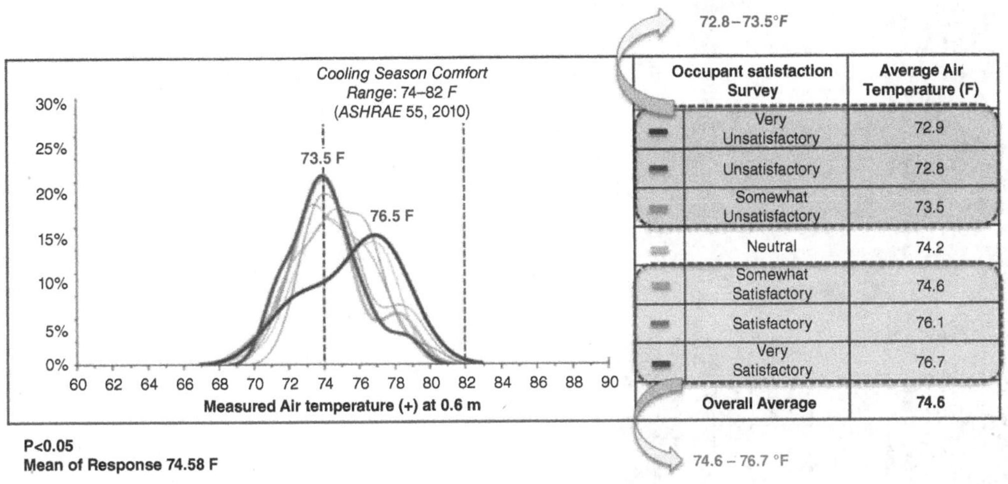

Occupant satisfaction Survey	Average Air Temperature (F)
Very Unsatisfactory	72.9
Unsatisfactory	72.8
Somewhat Unsatisfactory	73.5
Neutral	74.2
Somewhat Satisfactory	74.6
Satisfactory	76.1
Very Satisfactory	76.7
Overall Average	**74.6**

Temperature Satisfaction by "Size of Zone"

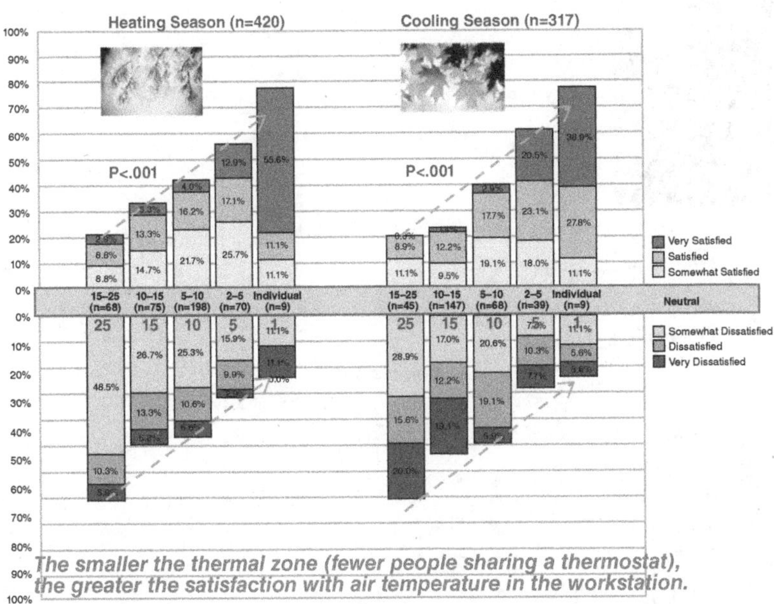

Figures 4 and 5 US office occupants are much more comfortable at 74.6–76.7oF in summer (left), and only achieve ASHRAE's mandate for 80% satisfied in individually controlled zones (44 buildings, n=737) (right). (*Source:* Park 2015).

while the vertical fields below are critical to views and natural ventilation, each to be balanced against shading, heat loss, heat gain and infiltration. While US buildings attempt to solve these competing forces in a "single plane" of façade construction, often with universal solutions without respect to climate, high-performance buildings recognize the power of three horizontal layers of design – external to the façade, integral and interior. In both new and existing construction, these layers become especially impactful for energy and indoor environmental quality (IEQ) when they are *dynamic*, responding to time of day and seasonal climate changes (Figures 6, 7, and 8).

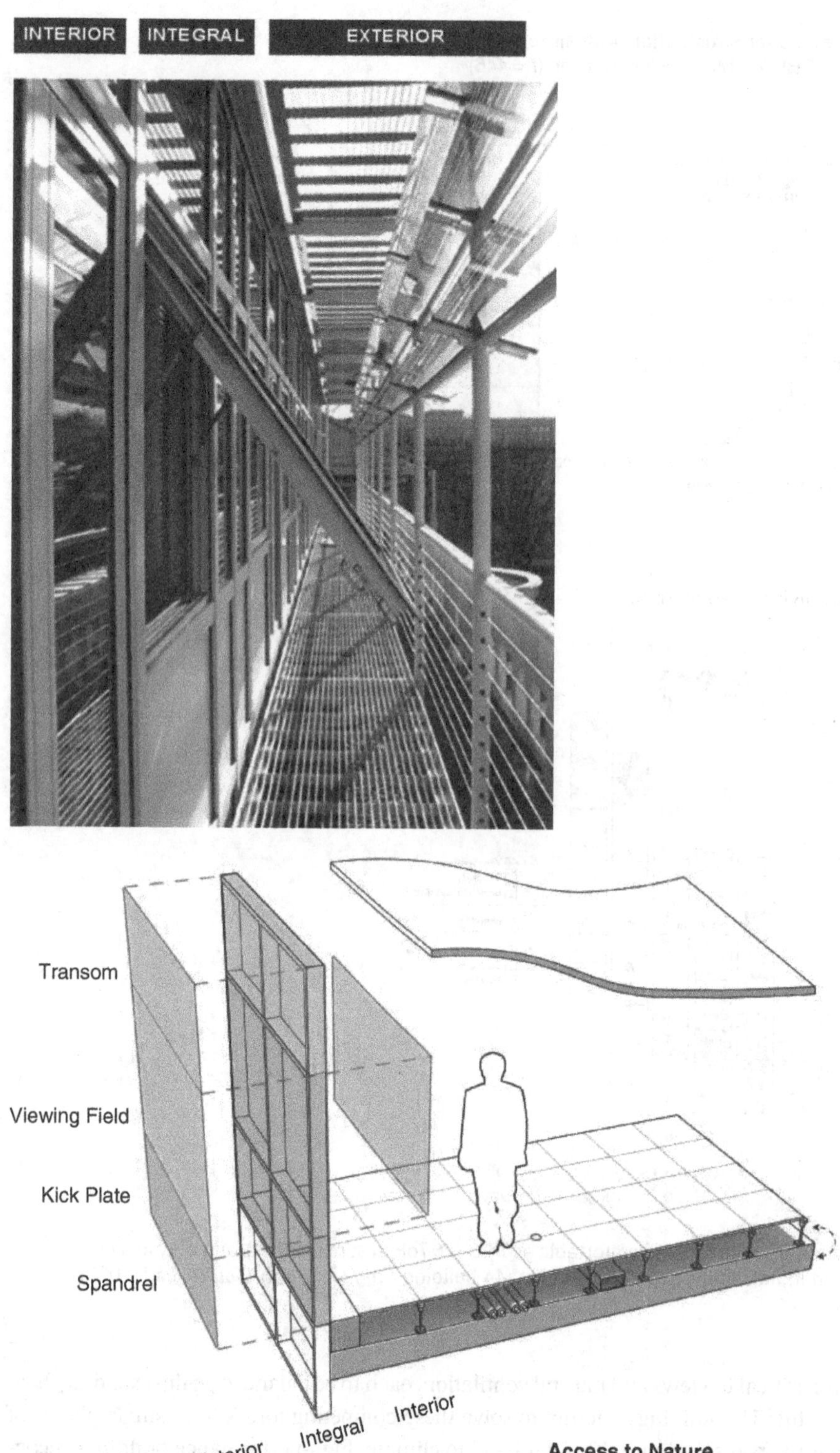

Figures 6, 7 and 8 There are at least 12 facade design decisions that critically impact visual, thermal, air quality, and energy performance of buildings, including the integration of PV shading. (*Source:* authors).

Figures 6, 7 and 8 (Cont'd)

In the IW Living Lab, Dr. Yun Gu (PhD BPD 2011) researched the impact of integrating personalized lighting, ambient lighting and daylighting control with real time knowledge and expert feedback, for improving user satisfaction, task performance, and energy savings in the workplace. This thesis demonstrated that real-time energy feedback reduces demand by 40%, and that expert advice on the lighting and daylighting choices, with an index of the carbon benefits, further reduce energy by 20%, for a total of 60% energy savings over conventional ceiling luminaires with fixed light output, in addition to higher levels of user satisfaction for paper-based tasks (Figure 8) (Gu 2011).

In the IW Living Laboratory, Dr. Omer Karaguzel (PhD BPD 2013) developed computational methods and procedures to extend the performance of multifunctional solar PV shading systems through a multi-domain parametric framework. The thesis engaged dynamic PV louvers exterior to the south facade achieve reductions in heating, cooling, and electric lighting energy consumption and maximize generation of renewable solar power while maintaining acceptable levels of thermal and visual comfort conditions for the occupants. This dissertation identified: that the area of PV integrated glazing systems with respect to opaque and clear glazing area is the most influential design factor for all climate types; that solar and visible transmittance of semi-transparent PV systems can be influential to net energy balance under high solar radiation conditions; and that geometrically and electrically optimized BSiPV (building system integrated pv) in high-performance south-facing facades of typical medium-sized office buildings in the US can offset around 25.4% of the heating and 26.1% of the cooling in whole building energy balance with respect to national benchmark models (Karaguzel 2013).

Advancing Innovative Controls with the Internet of Things (IoT) to Bring Humans and Nature "In the Loop"

Living Laboratories are ideal settings to advance environmental controls research – demonstrating the potential for building occupants to act as sensors and controllers through innovations in wirelessly addressable building systems for IEQ and energy savings. This is especially critical since most current research on smart controls for buildings is focused on optimizing active systems to deliver uniform conditions at the lowest energy costs (the yacht). Control innovation is critically needed to integrate active system controls (mechanical/electrical) and passive system controls (façade) into low carbon, dynamic systems (the sailboat). In addition, future building control systems must critically put "humans in the loop" including bio-signal controls and "nature in the loop" including mixed mode or hybrid conditioning – for thermal, air, and lighting control (Loftness 2019a).

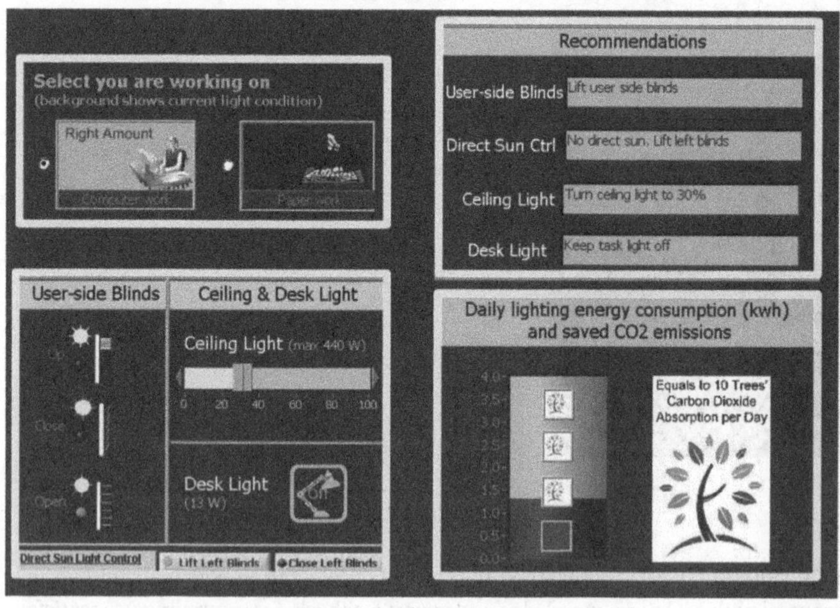

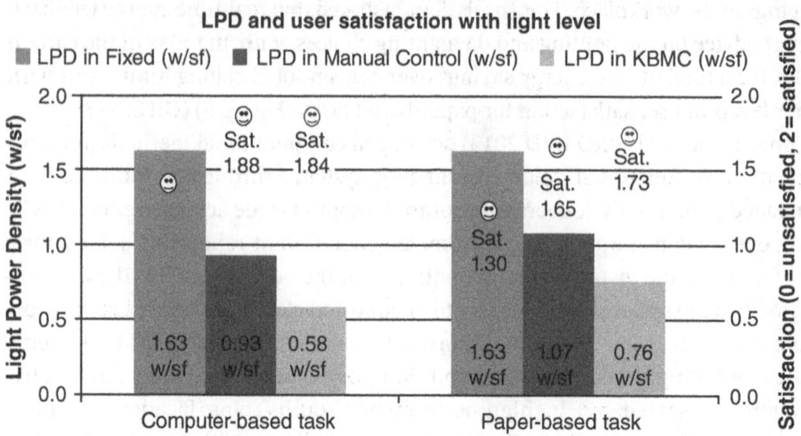

Figures 9 and 10 Individual control of ambient, task, and daylight yield 40% energy savings and high satisfaction, rising to 60% savings with dashboards offering real-time, knowledge-based individual control. (*Source:* Gu 2011).

The Intelligent Workplace is one of the most sensored and controllable workplaces worldwide, shifting from traditional settings with one control for every 20 occupants to 20 controls for every occupant. With the emergence of wireless sensors and controllers and the Internet of Things (IoT), the IW is a testbed for the engagement of occupants as both sensors and controllers for the improvement of environmental quality and energy conservation.

The emergence of IoT has major implications for the design and management of buildings. With wirelessly addressable environmental control components, every ceiling light, task light, and connected venetian blind, every radiant panel, cooling chilled beam, motorized window, fan, and desktop technology can be a "user or nature" point of control for energy savings, indoor environmental quality and organizational effectiveness. In addition, every sensor in the building for local temperature, humidity, particulates, CO2, daylight, and electric light levels, and the energy use of all of our desktop technologies are now IoT nodes that can inform the "user and nature" points of control.

In a controlled experiment in the IW, Dr. Joonho Choi (BPD PhD 2010) demonstrated the potential of human bio-signals – such as heart rate and skin temperatures – to act as thermal comfort controllers for the zone control of HVAC systems in summer. The resulting 93% satisfaction with thermal conditions – as compared to less than 60% in thermostat-controlled offices – and 5.9% in cooling energy savings is accompanied by the potential for wrist bands to modulate thermal conditions for the infirmed (Choi 2010, 2012). The power of bio-signal controllers is that they recognize the integrated thermal conditions created by HVAC, enclosure, interior systems and individual health, age, gender, activity and more – in-situ conditions which will affect physiological responses to thermal, air and visual conditions.

In 2010, the CBPD faculty began a long-term collaboration with the students of Dr. Bernd Bruegge in the Institut for Informatik at the Technical University of Munich (TUM). Through this collaboration, the IW has been the laboratory for TUM Bachelor's, Master's, and PhD thesis projects related to the use of smart phones to provide communication, expert feedback, and intuitive control of building systems. The true power of the Internet of Things in buildings is the ability to customize our environment to individual needs at the lowest energy demand, potentially with the lowest effort if that is desired (Figures 11–14). For those who prefer lower lighting for computer work, smart phones can let you dim the light with a gesture, using the smart phone's integral compass and gyroscope (Peters 2011). For those who want less air conditioning, a smart phone controller could allow us to lower the fan speed or partially close the air diffuser. For those who want all of this with less effort, calendaring and geo-fencing can automatically turn off selected technologies when we leave the office, so even reminders are unnecessary. The mutually supportive opportunity for saving energy and increasing indoor environmental quality is long overdue (Peters 2016).

The power of user interfaces and the internet of things is the subject of another thesis focused on the development of Intelligent Dashboards for Occupants (ID-O). Dr. Ray Yun (PhD CD/BPD 2014) developed innovative energy dashboards for office workers focused on communication/ feedback, expert consulting, and multiple levels of control (C^3 see Figure 15). The impacts of nine critical interventions for behavioral change were studied in his dissertation, structured in three sets: Instructional interventions – education, advice and self-monitoring; Motivational interventions – goal setting, comparison and engagement; and Supportive interventions – communication, control and reward. With a focus on controlling plug loads, the fastest growing energy end use in commercial buildings, the nine month controlled field experiment with 80 office workers revealed that occupant dashboards for controlling desktop technology, with ongoing energy communication and expert consulting generated by the occupant's own data set, can generate up to 40% energy savings in plug loads (Yun et al. 2014a, 2014b).

Advancing Lab and Field-tested Algorithms and Software Tools for Designing and Operating Buildings – Moving from BIM and BAS to BEM

For the highest level of energy conservation and environmental quality, building information modeling (BIM) needs to fully engage building automation (BAS) and environmental life cycles into a future that that might be reconceived as building environmental management (BEM).

Through the leadership of professors Dr. Ardeshir Mahdavi, Dr. Khee Poh Lam, and Dr. Omer Karaguzel, the CBPD has been developing dynamic life cycle building information models (DLC-BIM) focused on Total Building Performance to ensure best practices in sustainable and green architecture. The DLC-BIM is designed to support multidisciplinary and concurrent multi-domain building design and operational decision making, developed and refined through the power of testbeds such as the IW. The model structure is primarily based on the Industry Foundation Class (IFC) schema that captures the "static" building information generated during the design and construction process to: support building performance predictions using building simulation tools

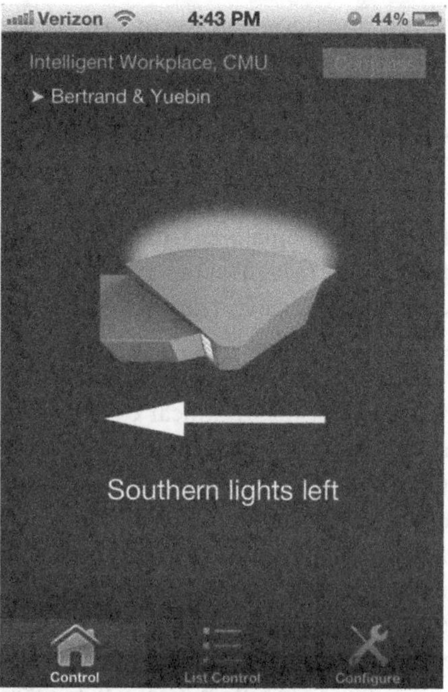

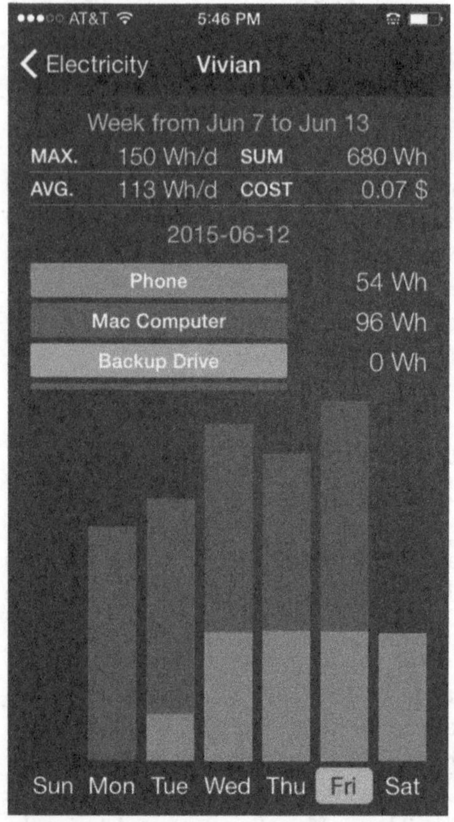

Figures 11, 12, 13 and 14 Smart phones can support intuitive gesture control of every fixture (left 2), provide incentivizing energy use information (center) and offer personal readings of individual sensors and set points (right), (Peters 2011; Mazza 2015).

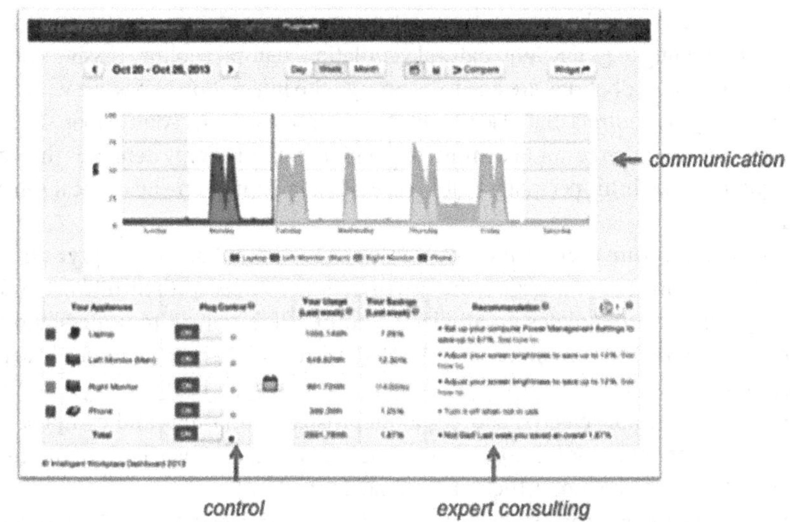

Figure 15 Desktop energy use "feedback and control" dashboards for occupants yielded as much as 40% sustained savings from an already efficient workstation (Yun, 2014b).

(energy, CFD, lighting, acoustics, etc.) and benchmark evaluations (e.g., LEED) with advanced design optimization; and capture "dynamic" (operational) building information generated from large-scale occupancy detection and environmental sensing networks for on-going commissioning, whole building performance monitoring and advanced adaptive controls based on occupant behavioral studies. Specific applications include Occupancy Behavior-based Predicted HVAC Control; Passive, Active and Hybrid HVAC and Controls; Dynamic 3-D Architectural Model and Sensor Information Visualization; and Evaluating Embodied Energy and Carbon Content in Primary Building Materials.

In an IW partnership with the Phipps Center for Sustainable Landscapes offices (with Living Building platinum designation), Dr. Jie Zhao (PhD BPD 2015) developed and demonstrated a design-build-operate Energy Information Modeling infrastructure (DBO-EIM) that can be deployed in each stage of the building life cycle to improve energy and IEQ performance. An occupant-oriented mixed-mode EnergyPlus predictive control system (OME+ PC) fully engages "nature and humans in the loop" through mixed mode operation and user control of thermal conditions through dashboards. The advances in algorithms and software in the OME+ PC field research resulted in a 29.37% reduction in annual HVAC energy consumption for a building that was already deeply energy efficient, alongside measurable gains in user satisfaction (Zhao 2015).

The emerging capabilities of data-driven model predictive control (MPC) and machine learning for data analytics offer the building community unprecedented opportunity to advance the performance of existing buildings in portfolio wide analysis. With two years of advanced energy data and HVAC fault detection data for 47 federal buildings managed by the US General Services Administration, the CBPD's data analytics with LEAN and LGBM machine learning tools revealed priority actions for portfolio-wide management programs to correct the faults (FDD) captured in GSALink Sparks with the highest electric and gas energy costs and the greatest frequency of portfolio-wide hardware or software faults. In aggregate, the research confirmed that consistent response to BAS faults would save a total of 65 million kWh of electricity and 220 million cubic feet of gas for the 47 buildings, and an average of $160,000 in operating cost per building per year (Loftness et al. 2019b).

Advancing a Carbon Positive Future

Carbon Neutral and even Carbon Positive buildings are longstanding goals of the Center for Building Performance and Diagnostics at Carnegie Mellon University. The deepest contributions to carbon neutrality that will simultaneously improve indoor environmental quality must be achieved through energy efficiency (e.g., air tightness, thermal insulation, shading) and passive conditioning (e.g., daylight, natural ventilation, natural cooling, passive solar heating with dynamic facades). The next level of contributions are ensured through the integration of passive with deeply efficient active conditioning systems, alongside controls that place "nature and humans" in the control loop. These advances in building design, engineering, and operation result in commercial buildings with energy demands that are so low (site total EUI under 25 kbtu/sq.ft. or 75 kwh/m^2 per year) that their loads can be met on-site or local renewable energy sources for carbon neutrality.

Carbon positive goals, however, require even further innovation in the production of energy at the building and campus level, recognizing the importance of energy – the use of energy sources at their highest and most appropriate level. As the building world contemplates an all-electric future, it is critical to understand that building thermal demands can be significantly met by more energy effective thermal energy sources – with waste heat from power generation or air conditioning as well as with solar and geothermal heat and cooling. Carbon neutral buildings cannot be defined by renewable electric sources alone. Instead, buildings and campuses need to generate energy and cascade the power generation energy waste (such as steam and heat) to ensure the highest generation efficiency in combination with the most efficient demand controls.

Toward cascading energy management, the IW has been the home to a generation of "Intelligent Workplace Energy Supply Systems (IWESS)" projects under the leadership of Dr. David Archer. A biodiesel CCHP generator, concentrating solar collectors with absorption cooling, and enthalpy recovery desiccant conditioning thesis projects demonstrated the ability to simultaneously address building power and thermal energy demands, including the use of high temperature reject heat to drive desiccant and absorption systems. Dr. Ming Qu (PhD BPD 2010b) demonstrated that two bays of solar thermal concentrators (refer back to Figure 1) could supply 39% of the cooling (through absorption chillers) and 20% of heating energy for the IW (Qu et al. 2010a, 2010b). Dr. Fred Betz (PhD BPD 2009a) proved the viability of Combined Cooling, Heating, Power, and Ventilation systems (CHPV) with a small biodiesel power and steam generating system integrated with the desiccant AHU in the IW to deliver an average annual power generation efficiency of about 68% and a peak of 78% (Betz 2009a; Betz and Archer 2009b).

To demonstrate a net positive carbon future for buildings, the faculty of the CBPD have proposed the next generation Living Laboratory as an addition to the Margaret Morrison building on which the IW sits (Figure 16). This Building-as-Power-Plant (BAPP) would integrate advanced enclosure, heating, ventilation, air-conditioning and lighting technologies with innovative distributed energy generation systems with energy cascades (from power to

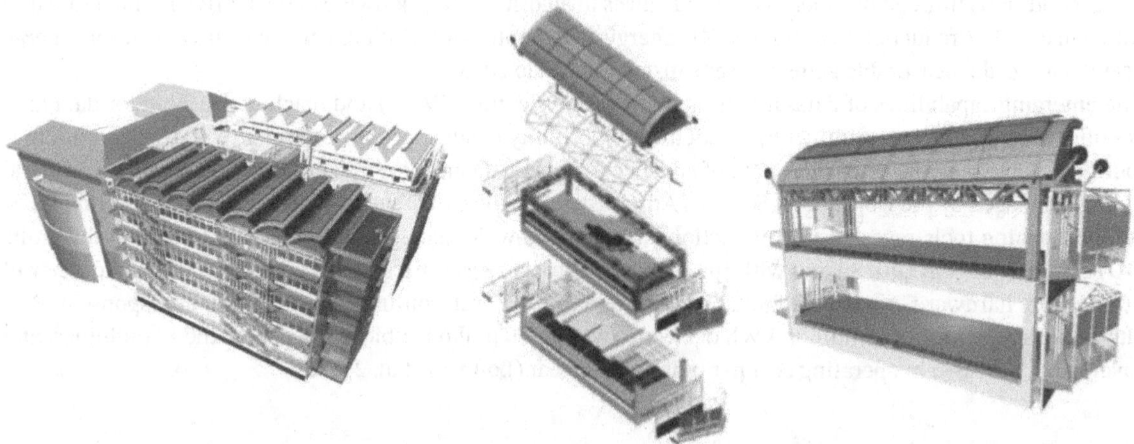

Figure 16 Column-free design, floor plans designed for change with plug and play infrastructures and modular interiors can be combined with CCHP and solar (PV and thermal) to make the next Living Laboratory a Building as Power Plant. (*Source:* author).

thermal demands), such that all the building's energy needs for heating, cooling, ventilating and lighting are met on-site, with excess energy exported to the campus – each new building or wing as a power plant.

Every Campus Should Have a Living Lab and be a Living Lab

The Robert L Preger Intelligent Workplace has been the proving ground for over $12 million in research over 25 years, the hub for the Advanced Building Systems Integration Consortium, the setting for monthly design charrettes with public and private building clients, the learning and research instrument for hundreds of undergraduate and graduate students, and the destination showcase for international clients, practitioners, and the building industry. The IW Living Lab gave rise to multiple living and lived-in labs around the world, including the GSA Adaptable Workplace Lab in Washington DC, the EDF Center for Cognition lab in Paris, and Tsinghua's Low-Energy Demonstration building in Beijing (AWL Hartkopf et al. 1999).

Every Architecture, Building Science, and Building Engineering program should have an Intelligent Workplace, a living laboratory of innovative component and subsystem, and a testing ground for the next generation of building systems integration for indoor environmental quality and resource sustainability. These "living laboratories" should be seen as scientific instruments to rival engineering and science labs, support ongoing collaboration with

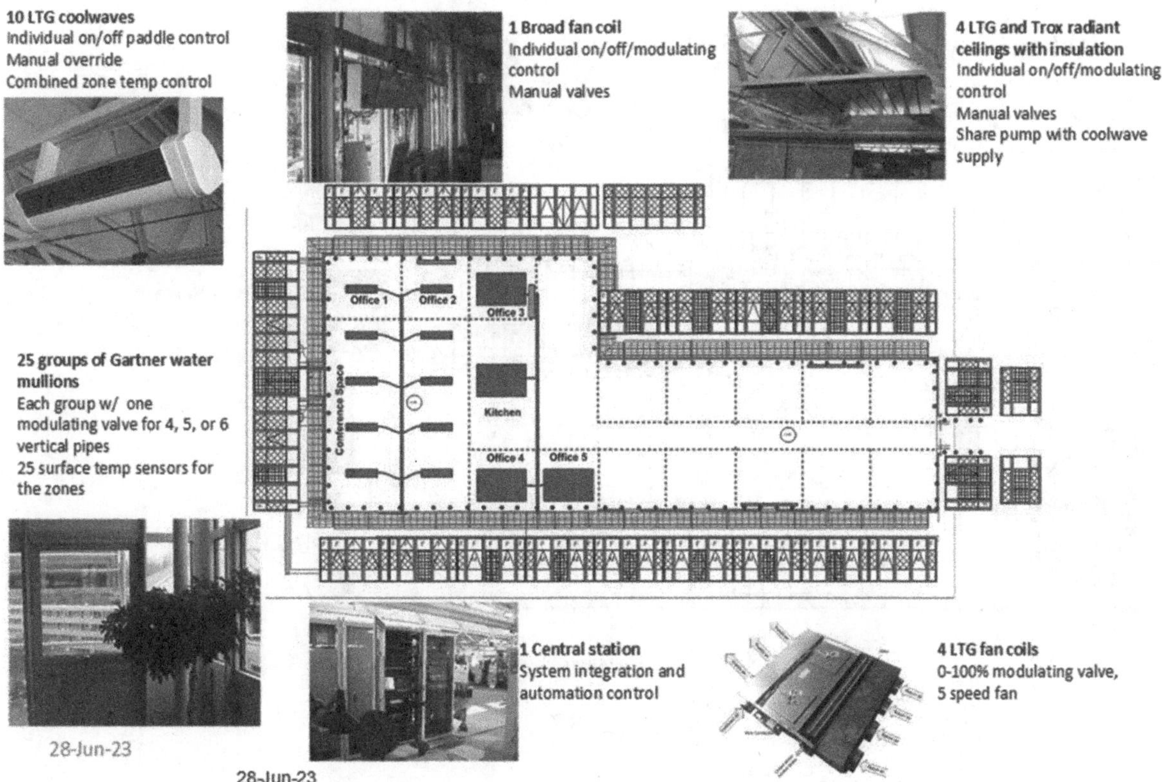

Figures 17, 18 and 19 Enclosure, Thermal, Ventilation and Lighting Components, Systems and Controls in Carnegie Mellon's Intelligent Workplace. (*Source:* Authors).

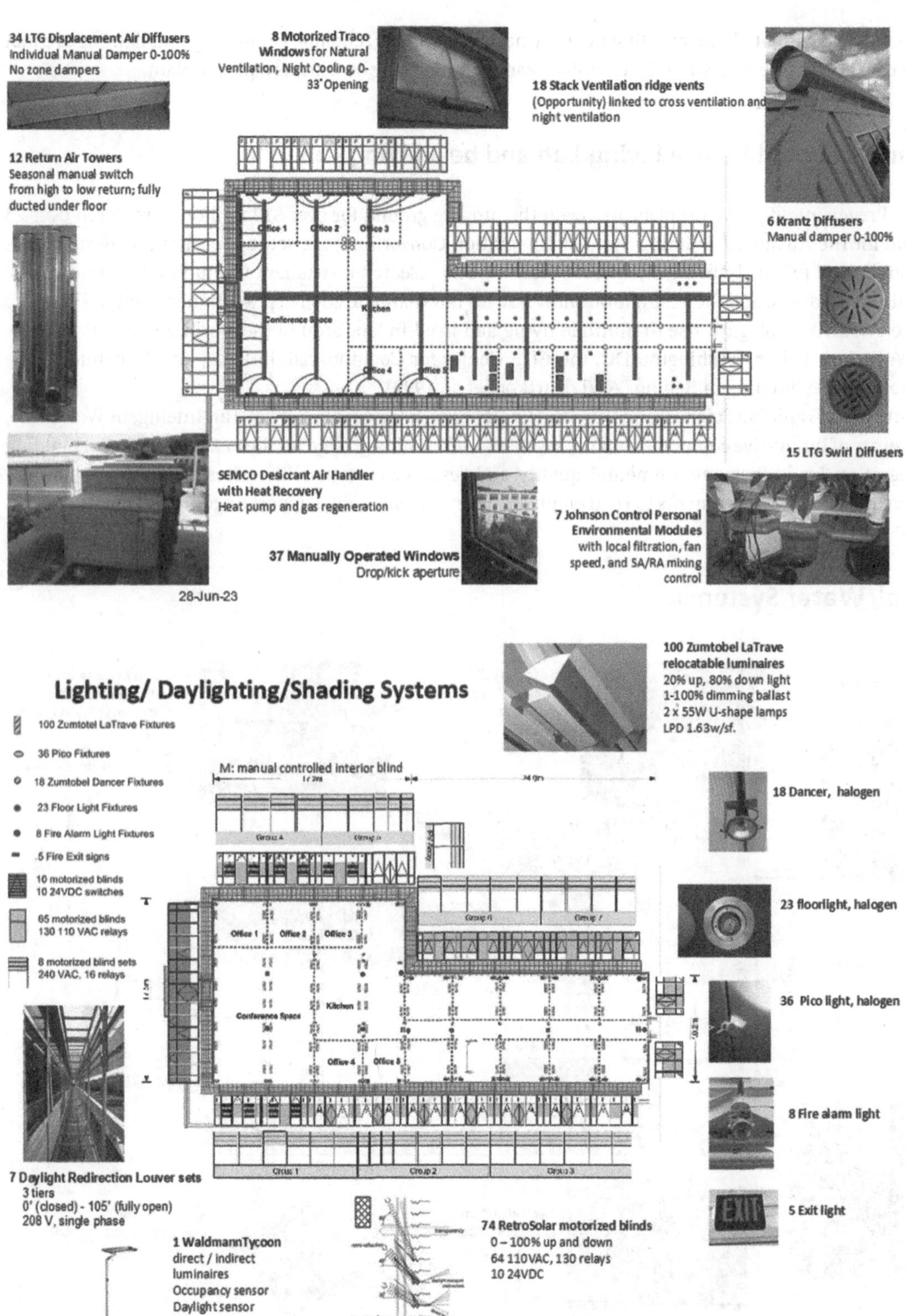

Ventilation/ Thermal Air Systems
34 LTG Displacement Air Diffusers
Individual Manual Damper 0-100%
No zone dampers
8 Motorized Traco Windows for Natural Ventilation, Night Cooling, 0-33° Opening
18 Stack Ventilation ridge vents
(Opportunity) linked to cross ventilation and night ventilation
12 Return Air Towers
Seasonal manual switch from high to low return; fully ducted under floor
6 Krantz Diffusers
Manual damper 0-100%
Office 1
Office 2
Office 3
Kitchen
Conference Space
Office 4
Office 5
15 LTG Swirl Diffusers
SEMCO Desiccant Air Handler with Heat Recovery
Heat pump and gas regeneration
7 Johnson Control Personal Environmental Modules
with local filtration, fan speed, and SA/RA mixing control
37 Manually Operated Windows
Drop/kick aperture
28-Jun-23
100 Zumtobel LaTrave relocatable luminaires
20% up, 80% down light
1-100% dimming ballast
2 x 55W U-shape lamps
LPD 1.63w/sf.
Lighting/ Daylighting/Shading Systems
100 Zumtotel LaTrave Fixtures
36 Pico Fixtures
18 Zumtobel Dancer Fixtures
23 Floor Light Fixtures
8 Fire Alarm Light Fixtures
5 Fire Exit signs
10 motorized blinds
10 24VDC switches
65 motorized blinds
130 110 VAC relays
8 motorized blind sets
240 VAC, 16 relays
M: manual controlled interior blind
18 Dancer, halogen
23 floorlight, halogen
36 Pico light, halogen
8 Fire alarm light
5 Exit light
EXIT
7 Daylight Redirection Louver sets
3 tiers
0° (closed) - 105° (fully open)
208 V, single phase
1 WaldmannTycoon
direct / indirect
luminaires
Occupancy sensor
Daylight sensor
Interior
74 RetroSolar motorized blinds
0 – 100% up and down
64 110VAC, 130 relays
10 24VDC

Figure 17, 18 and 19(Cont'd)

the building industry for research and education of a new generation of building scientists and FM practitioners. These "living labs" should be the testing and proving grounds of innovative and integrated building systems, controls, and operations, alongside the development of life cycle cost-justifications (Loftness et al. 2014). They should be showcases and collaborative project development meeting places for the building industry, supported by government and industry alike. Beyond the individual living laboratory buildings, however, the campuses themselves should be advanced as "living laboratories" to advance energy, water, materials, and environmental quality – ensuring that the next generation of graduates lead in practice and innovation for our shared future.

Acknowledgements

CBPD and IW Research Faculty:
Volker Hartkopf, PhD; Stephen Lee, AIA; Vivian Loftness, FAIA; Ardeshir Mahdavi, PhD; Azizan Aziz; Khee Poh Lam, PhD; David Archer, PhD; Bertrand Lasternas; Erica Cochran, PhD; Omer Karaguzel, PhD; Nina Baird, PhD; Bernd Bruegge (TUM), PhD; Azadeh Sawyer, PhD.

ABSIC Industrial Research Partners
Air Advice; Aircuity; ALCOA; American Bridge Company/Continental Engineering Corporation; AMP: Architect of the Capitol (AOC); Armstrong World Industries Inc.; Bank of America; Bayer USA; Bechtel Corporation; Bell of Pennsylvania; Bosch GmbH; Bosse Design; BP Solar; Bricsnet; BROAD Air-conditioning, China; Consolidated Edison of New York; Duquesne Light Company; Echelon; Edo Rocha, Architects, Brazil; Electricité de France; Grahl Industries, Inc; Hüppe Form; ICC Technologies; Interface Inc.; Johnson Controls, Inc.; Josef Gartner & Company; Kimball International, Inc.; LG; Honeywell Co., Ltd.; LTG Aktiengesellschaft; Mahle GmbH (now Linder); Mori Biru, Japan; North West Energy Efficiency Alliance; Nucor; OsiSoft; Osram/Sylvania; PPG Industries; RetroSolar; Siemens Energy & Automation, Inc.; Siemens Building Technologies; Somfy; Steelcase, Inc.; Teknion Inc.; Thyssen Krupp AG; Tyco Electronics; United Technologies/Carrier; Vanadium; Westinghouse Electronic/The Knoll Group; Zumtobel Staff Lighting, Inc.

ABSIC Government Research Partners
General Services Administration; US National Science Foundation; US Department of Defense; US Department of Energy; US Department of State; US Environmental Protection Agency

University Research Partners
Robert L. Preger and Carnegie Mellon University leadership
Gale Foundation; Heinz Foundation; Wege Foundation
Technical University of Munich; Texas A&M ESL; University of Maryland, CHP; Darmstadt University; University of Braunschweig; Sierra-Nevada/University California Davis; Milwaukee School of Engineering; Tsinghua University

Design team of the IW
CBPD Faculty – Hartkopf, Lee, Loftness, Aziz, Lam, Mahdavi with a team of graduate students; Architects – Pierre Zoelly, Zoelly Associates; Peter Bohlin, BCJ; Rob Pfaffmann, BCJ; Greg Mottola, BCJ; Jon Jackson, BCJ; Engineers – Rick Yates, Mechanical; Tom Brzuz, Hornfeck Electric; R.M. Gensert, Structural; Construction – Louis Romano, Tedco; Consultants – Peter Mill, PWC; Fred Dubin, Dubin-Bloome.

References

Aziz, A., Park, J., Loftness, V., and Cochran, E., (2012). *Field measurement protocols for evaluating indoor environmental quality and user satisfaction in relation to energy efficiency*. U.S. DOE The Energy Efficient Buildings Hub, Dept. of Energy.

Betz, F. (2009a). *Combined cooling, heating, power, and ventilation*. PhD dissertation, Carnegie Mellon University.

Betz, F. and Archer, D. (2009b). Biodiesel fueled engine generator with heat recovery: comparing biodiesel to diesel performance. *ASME 2009 3rd International Conference on Energy Sustainability*, 2, San Francisco, CA, USA, July 19–23, 2009.

Choi, J. (2010). *CoBi: bio-sensing building mechanical system controls for sustainably enhancing individual thermal comfort*. PhD Thesis, Carnegie Mellon University.

Choi, J. and Loftness, V. (2012). Investigation of human body skin temperatures as a bio-signal to indicate overall thermal sensations. *Building and Environment* 58: 258–269.

Gu, Y. (2011). *The impacts of real-time knowledge based personal lighting control on energy consumption, user satisfaction and task performance in offices*. PhD dissertation, Carnegie Mellon University.

Hartkopf, V. et al. (1997 September). An integrated approach to design and engineering of intelligent buildings—The Intelligent Workplace at Carnegie Mellon University, Center for Building Performance and Diagnostics, Carnegie Mellon University, Pittsburgh, PA 15213, USA. *Automation in Construction* 6 (5–6): 401–415. https://doi.org/10.1016/S0926-5805(97)00019-8.

Hartkopf, V. et al. (1999). The GSA adaptable workplace laboratory. In: *Cooperative Buildings: Integrated Information, Organizations and Architecture* (ed. N.A. Streitz, J. Siegel, V. Hartkopf, and S. Konomi). Germany: Springer.

Hartkopf, V., Aziz, A., and Loftness, V. (2020). Facades and enclosures: building for sustainability. In: *Sustainable Built Environments. Encyclopedia of Sustainability Science and Technology Series* (ed. V. Loftness). NY: Springer. https://doi.org/10.1007/978-1-0716-0684-1_873.

Hartkopf, V. and Loftness, V. (2004). Architecture, the workplace, and environmental policy. In: *The Innovative University* (ed. D.P. Resnick and D.S. Scott), 181–194. Pittsburgh: Carnegie Mellon University Press.

Hartkopf, V., Loftness, V., and Aziz, A. et al. (2017). The Robert L Preger intelligent workplace: the living and lived-in laboratory. In: *Creating the Productive Workplace*, 3e (ed. D. Clements-Croome). Routledge, an imprint of Taylor and Francis.

Hua, Y., Loftness, V., Heerwagen, J., and Powell, K.M. (2011). Relationship between workplace spatial settings and occupant-perceived support for collaboration. *Environment and Behavior* 43 (6): 807–826.

Hua, Y., Loftness, V., Kraut, R., and Powell, K.M. (2010). Workplace collaborative space layout typology and occupant perception of collaboration environment. *Environment and Planning B: Planning and Design* 37 (3): 429–448.

Karaguzel, O. (2013). *Simulation-based parametric analysis of building systems integrative solar photovoltaics*. PhD Dissertation, Carnegie Mellon University.

Loftness, V., Aziz, A., Heerwagen, J. (2009 1 October). The value of post-occupancy evaluation for building occupants and facility managers. *Engineering, Intelligent Buildings International* 1 (4).

Loftness, V., Aziz, A., and Hartkopf, V. et al. (2019a). Humans and Nature in the Loop: integrating occupants & natural conditioning into advanced controls for high performance buildings. *International Building Performance Conference IBPC 2018*: Healthy, Intelligent, Resilient Syracuse, NY September 24–26, 2018.

Loftness, V., Aziz, A., Zhang, C., and Xu, Y. (2019b). *Evaluation of GSALink total estimated cost impacts (TECI) of building system maintenance and management*; 2018–2019 study, presentations and final report (115 pages). GSA Contract #, submitted October 2019.

Loftness, V., Hartkopf, V., and Aziz, A. et al. (2017). Critical frameworks for building evaluation: user satisfaction, environmental measurements and the technical attrutes of building systems (POE+M). In: *Building Performance Evaluation: From Delivery Process to Life Cycle Phases*. Springer Publishing.

Loftness, V., Aziz, A., Son, Y.J. (2020), National Environmental Assessment Toolkit (NEAT) POE+M Study of Interface, Inc. *Work Environments*; 2019 study, presentation, and report (70 pages).

Loftness, V., Srivastava, R., Dadia, D. et al. (2014 December). The triple bottom line benefits of climate responsive dynamic facades. *Proceedings of PLEA 2014: Sustainable Habitat for Developing Societies*, Ahmedabad, India.

Mazza, F. (2015). Nonlinear incremental analysis of fire-damaged r.c. base-isolated structures subjected to near-fault ground motions. *Soil Dynamics and Earthquake Engineering* 77: 192–202. Springer.

Park, J. (2015). *Are humans good sensors? Using occupants as sensors for indoor environmental quality assessment and for developing thresholds that matter.* PhD Thesis, Carnegie Mellon University.

Park, J., Aziz, A., and Loftness, V. (2013). Post occupancy evaluation for energy conservation, superior IEQ, and increased occupant satisfaction. *IFMA's World Workplace Conference and Expo 2013*, Philadelphia, PA.

Peters, S. (2011). *A framework for the intuitive control of smart home and office environments.* Masters Thesis in Informatics, TUM.

Peters, S. (2016). *MIBO – a framework for the integration of multimodal intuitive controls in smart buildings.* PhD Thesis in Informatics, TUM.

Qu, M., Yin, H., and Archer, D. (2010a). A solar thermal cooling and heating system for a building: experimental and model based performance analysis and design. *Solar Energy* 84 (2): 166–182.

Qu, M., Yin, H., and Archer, D. (2010b). Experimental and model based performance analysis of a linear parabolic trough solar collector in a high temperature solar cooling and heating system. *Journal of Solar Energy Engineering* 13: 021004-1-12.

Yun, R., Aziz, A., Lasternas, B. et al. (2014a). The design and evaluation of intelligent energy dashboard for sustainability in the workplace. *Proceedings, HCI International 2014*, Creta Maris, Heraklion, Crete, Greece. June 22–27, 2014. Springer Verlag.

Yun, R., Lasternas, B., Aziz, A. et al. (2014b). Toward the design of a dashboard to promote environmentally sustainable behavior among office workers. In: *Persuasive Technology*, 253–265. Berlin: Springer.

Zhao, J. (2015). *Design-build-operate energy information modeling for occupant-oriented predictive building control.* PhD Dissertation, Carnegie Mellon University.

Volker Hartkopf, PhD., Dr. h.c., Dipl. Ing.
Professor of Architecture, Emeritus Director
The Center for Building Performance & Diagnostics (CBPD)
Carnegie Mellon University, Pittsburgh, PA, USA

Volker Hartkopf is Professor Emeritus in the Carnegie Mellon School of Architecture, founder and Director Emeritus of the Center for Building Performance and Diagnostics (CBPD). For 50 years, Professor Hartkopf has been teaching and leading research in advanced building systems for performance at Carnegie Mellon University. This research and his international consulting have spanned the full range of global design challenges, from third-world housing and disaster prevention to advanced technologies, building systems integration, deep energy conservation, sustainability, and urban revitalization. As an architect, Dr. Hartkopf has realized building projects in Germany, Bangladesh, Peru, China and the United States, including the first inner-city passive/active house followed by an inner-city neighborhood of low energy new and retrofit homes. With his colleagues, Dr. Hartkopf has also led sustainable master planning efforts for Volkswagen and the City of Wolfsburg, EXPO 2000 Hanover, and Berlin-Lichtenberg in Germany, as well as master planning for the CERL Labs in Champagne, Illinois.

In 1975, Prof. Hartkopf co-initiated and subsequently directed the first multidisciplinary program in Architecture, Engineering and Planning in the USA with grants from the National Science Foundation and the building industry. In 1981, he co-founded the Center for Building Performance and Diagnostics (CBPD) at Carnegie Mellon. Between 1981 and 1985, Prof. Hartkopf developed jointly with Vivian Loftness and Peter A.D. Mill, the Total Building Performance Evaluation Method at Public Works Canada whilst on an Executive Interchange Program.

To address the long-standing research and development needs in building performance, Prof. Hartkopf created and directed the Advanced Building Systems Integration Consortium (ABSIC) of over 50 leading building industries from around the world, six US and foreign governmental agencies, hosted at Carnegie Mellon. In operation since 1988, the ABSIC consortium's research and demonstration effort focuses on the impact of advanced technology on the physical, environmental, and social settings in office buildings, towards creating high-performance work environments. This leadership forum, in cooperation with Carnegie Mellon, designed and constructed a living laboratory – the Robert L Preger Intelligent Workplace – which officially opened in the winter of 1997. This living laboratory has been transformative for education, research, and the building industry, and is the home of the Center for Building Performance and Diagnostics (CBPD), a National Science Foundation Industry/University Cooperative Research Center, the first building-focused NSF/IUCRC.

Under the leadership of Professor Dr. Hartkopf, the CBPD and ABSIC have been instrumental in advancing the importance of systems integration and performance-based decision-making in design, engineering, and operation of buildings. In collaboration with clients around the world, the USA to Germany, China, Korea, Africa, and France, the CBPD team has received prestigious awards for research and practice.

An award winning teacher and a frequent keynote speaker in Australia, Europe, Asia and the Americas, Professor Hartkopf has authored books, book chapters, journal articles, and over 100 technical publications. He continues his consulting with such organizations as DaimlerChrysler, Volkswagen, Thyssen Krupp, Electricite de France, the US Department of State, US Department of Energy, and Siemens.

Prof. Hartkopf's next vision for research and education is a "Building as Power Plant" living laboratory to demonstrate net positive energy solutions for a more sustainable built environment. With support from the US Congress, DOE and DOD, a National Test-bed for Advanced Energy Technology in Building was initiated to integrate advanced energy-effective building technologies with innovative energy generation systems, such that all of the buildings energy needs for heating, cooling, ventilation, lighting, as well as plug loads are met on-site, delivering excess renewable energies back to the campus. Dr. Hartkopf's career continues to be focused on living laboratories as the future of education, research, and innovation in the built environment.

EDUCATION:
Pre-Diploma (Vordiplom) in Architecture, University of Stuttgart (1964)
Graduate Engineer (Dipl.-Ing)., Architect, University of Stuttgart (1969)
M. Arch., University of Texas, Austin (Fulbright Scholar, 1970–1972)
Dr. Ing. (PhD), University of Stuttgart (1989)
Dr. h.c., Sierra Nevada College (2004)

Vivian Loftness, FAIA
University Professor & Paul Mellon Chair in Architecture
Co-Director, Center for Building Performance & Diagnostics
Leed AP, Certified Passive House Consultant
Fellow of Design Futures & New Buildings Institute
Carnegie Mellon University, Pittsburgh, PA, USA

Vivian Loftness is Paul Mellon Chair and University Professor at Carnegie Mellon University and former Head of the Carnegie Mellon School of Architecture. She is an internationally renowned researcher, author, and educator with over forty years of experience in building science research for industry and government. In addition to editing the 2013 and 2020 Springer Encyclopedia on Sustainable Built Environments, she has authored books, research reports, and book chapters on climate and regionalism in architecture, environmental design and sustainability, advanced building systems integration, and design for performance in the workplace of the future to enhance productivity, health, and the triple bottom line.

Vivian has served on over 25 board of directors, including the US Environmental Protection Agency's (EPA)'s National Advisory Council for Environmental Policy and Technology (NACEPT), the Department of Energy's (DOE) Federal Energy Management Advisory Committee (FEMAC), the National American Institute of Architects (AIA), International Living Future Institute (ILFI), and US Green Building Council (USGBC) boards, as well as on 15 National Academy of Science (NAS) panels on sustainable built environments.

Vivian has been recognized as one of 13 Stars of Building Science by the Building Research Establishment in the UK, received the Award of Distinction from AIA Pennsylvania and from the Northeast Sustainable Energy Association (NESEA), holds a National Educator Honor Award from the American Institute of Architecture Students (AIAS), and a USGBC "Sacred Tree" Award. Vivian has a Bachelor of Science and a Master of Architecture from MIT.

4

Summary and FM Outlook

We trust that you, the reader, have enjoyed reading and considering the many contributions in Chapters 1, 2, and 3 that span the over 40 years of international FM. You have seen that each contribution provided a unique and focused perspective on FM that was relative to their career, country, and when FM became a focus in their life.

The co-editors will now provide a Summary of what has been provided and some thoughts as an FM Outlook as to what the reader may consider and need to address in their FM future.

Overview

In this anthology, the different perspectives one can have on FM, different points of view, approaches and ideas were presented. Of these many experiences, concepts and visions, I would like to highlight only a few that seem particularly significant to me and can be seen as representative of other experiences around FM.

It turns out that FM is still a topic of the future from a symbiotic point of view among the smaller market participants. It's not just about integrated technical FM; it's about integrating FM as a process. Even though this has been theoretically clear for many years, it has not reached all companies. The perspective on FM as a process and as a management discipline, in non-property companies in the sense of CREM, integrates the entire entrepreneurial context as a cross-lifecycle concept. This is the only way to achieve a holistic implementation in the sense of the company's real estate as a value-added factor and a possibility for value leverage for the entire company. In fact, the holistic FM seems to disappear into the system in many companies, it is lost in relation to the core business. The code for Corporate Real Estate Management of the ZIA (Zentraler Immobilie Ausschuss, Central Real Estate Committee) speaks of CREM as an equal partner of the management and encourages the people working in the field of CREM to act on an equal eye level with the CEO. It should be pointed out that working together brings advantages for the whole company.

The process has also changed – and not just since COVID. Reading a technical system, management and evaluation processes have taken on a different face. While working remotely remains uncommon for many older employees, the younger generation sees working, meeting, discussing, and conducting online as normal. This goes hand in hand with BIM, which is not only a computer system but enables a new way of collaboration, supported by artificial intelligence. This changes the whole of life, and also FM.

But nevertheless, and especially in FM, the human factor remains decisive on every level, whether personally or only on the screen. This New Work has changed our leadership images, styles and cultures, which have been shaped over decades, and the question of vision: how will FM change in the future and where do we want to go with FM has been answered in various ways in the book. Whether this also changes the mindset with regard to

Facilities @ Management, Concept - Realization - Vision, A Global Perspective, First Edition. Edited by Edmond Rondeau and Michaela Hellerforth.

attitudes toward FM and neighboring disciplines will become one of the critical future success factors, especially with the different approaches to the discipline of FM, which can be clearly seen both country-specific and in connection with different academic and professional backgrounds. The job description of facility services, i.e., on-site activities, is also facing new challenges as a result of technical developments.

An important focus of the anthology is the role of women in FM. When I was younger, I was especially confronted with some of the well-known prejudices against women working in a typical field for men, which seemed particular to be a theme in teaching and research. On the other hand, and not least due to my superiors, but also to my self-employment, I never felt such prejudice or simply ignored it. But let us look at the facts: at least the proportion of women in management positions around FM speaks against equal opportunities or is it because FM is still technically dominated. And of course even in this book we have not reached a 50–50-percentage.

Although the idea of the process was already part of the first FM definitions, it remains a success factor, and in many companies the middle ground between "living the processes" and "letting breathe" still have to be found here, not least a leadership issue.

Germany in particular is characterized by small and medium-sized enterprises (SMEs), which are even considered the backbone of the German economy. And their development potential toward FM is enormous and will become increasingly important, not least when buildings are considered as "energy guzzlers" and associated building changes, the application of new key figures to measure progress and the achievement of ESG goals. This opens up enormous scope for action through the management process around FM, combined with all disciplines, such as CREM, REIM, asset and property management, but also a great responsibility, in shaping the future of our buildings, the built environment and sustainability.

Sustainability will be the theme and the task of the future because it decides our future, so that FM and all the management disciplines around FM will have to work on solutions. In spite of the fact that approaches on FM around the globe are different this also means a supranational collaboration in this field. This anthology might help people in FM to meet and find solutions for our Future Challenges.
Michaela Hellerforth, Co-editor

Summary

The reader has seen in this anthology of contributions that FM began over 40 years ago as a thought, then a concept and finally as a profession. The contributors have shared their work and evolution as they met the challenges to improve the work environment, manage organizational space and real estate, their staff, and the budgets they developed. All of this was designed to provide their organizations with the best possible workplace solutions within the time, funds and resources available to support the success of their organization in the products and/or services they sold, researched, provided.

FM has developed as a profession over time in countries around the world in the last 40 years. In some countries, FM was quickly embraced while in other countries the history and realization of the benefits of FM were less well understood and/or accepted. FM education in many countries has evolved where academics now provide university degree programs, private organizations provide FM training, and FM associations now provide FM certification or certificate programs which supports the FM profession and those who support their FM and business requirements.

The reader has seen that the careers of many of the Contributors started in many professions such as architecture, engineering, interior design, consulting, sales, marketing, business administration, finance, accounting, construction, etc. The road to FM has been for many FMs a series of career opportunities and challenges before there were formal associate, bachelor, masters or PhD degree programs. High school and college students now see and recognize that an FM career path can provide them with an interesting, challenging, and rewarding lifetime of working with "people, process, place, and technology" to them and to their organizations.
Edmond P. Rondeau, Co-editor

FM Outlook

In Chapter 3, Vision, the reader has seen that many Contributors focused on the FM outlook as to what FM will look like in the future. As FM has evolved, we have seen how issues such as COVID, sustainability, climate change, energy, material availability, and customer expectations affect the FM profession and the new solutions that FMs have and/or will need to develop.

Cultural Considerations

Cultures around the world have a history, and a long list of successes and failures including within the built environment. Each culture translates FM into how people look at materials, services, expectations, and their built environments. As these cultures developed from different beginnings and personal points of evolution, their role in FM exists in various stages of maturity. Their FM education, FM standards. services and expectations have evolved at various rates. Some for financial reasons, some for political reasons, some for purely timing issues, and some for specific regional and/or country issues.

Sustainability[1]

In 2015, the United Nations (UN) set an ambitious 15-year plan to address some of the most pressing issues that were faced by the world.

By supporting ISO members to maximize the benefits of international standardization and ensure the uptake of ISO standards, this UN committee is helping to meet the United Nations Sustainable Development Goals (SDGs). Economic, environmental and societal dimensions are all directly addressed by ISO standards. Organizations and companies looking to contribute to the SDGs will find that International Standards provide effective tools to help them rise to the challenge.

Sustainable Development Goals to Transform Our World[2]

The UN Sustainable Development Goals are a call for action by all countries – poor, rich, and middle-income – to promote prosperity while protecting the planet. They recognize that ending poverty must go hand-in-hand with strategies that build economic growth and address a range of social needs including education, health, social protection, and job opportunities, while tackling climate change and environmental protection. More important than ever, the goals provide a critical framework for COVID-19 recovery.

Goal 1: End poverty in all its forms
Goal 2: Zero Hunger
Goal 3: Health
Goal 4: Education
Goal 5: Gender equality and women's empowerment
Goal 6: Water and Sanitation
Goal 7: Energy
Goal 8: Economic Growth
Goal 9: Infrastructure, industrialization
Goal 10: Inequality
Goal 11: Cities

1 https://www.iso.org/sdgs.html.

2 https://sdgs.un.org/goals.

Goal 12: Sustainable consumption and production
Goal 13: Climate Action
Goal 14: Oceans
Goal 15: Biodiversity, forests, desertification
Goal 16: Peace, justice and strong institutions
Goal 17: Partnerships

Some of these Goals are directly related to FM and others are related to supporting people and the built environment,

ISO Standards[3]

The ISO (International Organization for Standardization) is an independent, non-governmental international organization with a membership of 168 national standards bodies.

Through its members, it brings together experts to share knowledge and develop voluntary, consensus-based, market relevant International Standards that support innovation and provide solutions to global challenges.

ISO Standards on Facility Management[4]

ISO Standards on Facility Management[4]The International Standards on Facility Management (FM) developed by ISO/TC 267 describe the characteristics of facility management and are intended for use in both the private and public sectors. The standards are periodically reviewed and revised as the members of the committees see the need for revisions and/or clarifications.

See also Appendix D, Glossary on ISO FM details and definitions.

Information Technology (IT) and Artificial Intelligence (AI)

During the past 40 years we have seen IT in the workplace move from an item of interest to working tools available to most all FMs. The cost of IT has dropped dramatically from large main frame computer systems to networked workstations and printers, cell phones, digital pads, virtual communications and laptops that can access vast amounts of information on the Internet and provide real time information, video, streaming services, etc.

Software and hardware have become part of the background of our daily FM work life as we seek to find and use these tools to support our organization's FM data base and graphic requirements. We see that software/data base and graphic programs such as:

- CADD (Computer-Aided Design and Drafting)
- CAFM (Computer-Aided Facility Management)
- CMMS (Computer Maintenance Management System)
- BIM (Building Information Modeling) and
- other related programs for Corporate Real Estate Management, Space Management, Budgeting, Ordering, Inventories, Maintenance Management, Finance and Accounting, etc. that may be simplified and improved in the future by AI development for FM.

FM technology will continue to require advanced protection from the hacking and theft of organizational computers, databases, computer networks, Wi-Fi accesses to networks, e-mails, web site(s), cell phones, laptops, notebooks, printers, databases, clouds, and software and devices that will be developed in the future. In the early years

3 https://www.iso.org/about-us.html.
4 https://www.iso.org/committee/652901/x/catalogue.

of FM, many FM departments had the leftovers of IT hardware and software from other departments. Today and in the future FMs have and will need to have the latest hardware and software equal to any department in their organization.

Artificial intelligence (AI) is intelligence demonstrated by machines, as opposed to intelligence of humans and other animals. Example tasks in which this is done include speech recognition, computer vision, translation between (natural) languages, as well as other mappings of inputs.[5]

AI applications include advanced web search engines (e.g., Google Search), recommendation systems (used by YouTube, Amazon, and Netflix), understanding human speech (such as Siri and Alexa), self-driving cars (e.g., Waymo), generative or creative tools (ChatGPT and AI art), automated decision-making, and competing at the highest level in strategic game systems (such as chess and Go).[6]

As machines become increasingly capable, tasks considered to require "intelligence" are often removed from the definition of AI, a phenomenon known as the AI effect. For instance, optical character recognition is often excluded from things considered to be AI, having become a routine technology.[7]

Most organizations today have real estate databases and CAFM digital information for their leased and owned properties, know their monthly lease and operating cost payments, when their leases expire, know how many options they have, know where these properties are located, what the operating budgets and expenses for each building are by year, the strategic importance of each property, the services provided to each property, the maintenance and operations services provided, staff based there, and the future expectations for these locations.

AI is still evolving. Where will AI take FM in the future? Will FM professionals be prepared to understand and use AI tools and information to help them and their organizations to succeed beyond their current uses of IT?

FM Networking and People Skills

In our work at home discussions, we see that FM networking will become even more important as FMs are asked to provide solutions for many customers who only occasionally or seldom go to the office. FMs will need to develop many unique and organizational focused ways to meet and provide services to their customers and their staff, whether working remotely and/or working in the office, but not every day. Our phone, e-mail, virtual, media, writing, and other communication skills will be tested. Our meeting and sharing of information may require us to make extra efforts to ensure our customers and their staff are fully aware of their requirements to receive the FM services they need.

People skills will become more important for remote communications, The FMs hard and soft skills may be more critical in the FMs success when networking and working with more remote customers.

During and after COVID the FMs career networking needs have become more difficult as remote meeting with FM peers is very difficult to achieve the understanding and sharing that one-on-one meetings provide. Special effort will be needed for FMs to maintain and expand their networking group where conferences and educational programs can be used to keep growing the FMs networking base.

Leadership

As Facility Departments evolve in the future, it is very important for FMs to continue to recognize, encourage, and develop leaders for various services and for leaders to hire, train, manage, and understand their staff. Leadership must become familiar with the many senior managers in their organization who can help them and who they can

5 https://en.wikipedia.org/wiki/Artificial_intelligence and

6 https://web.archive.org/web/20160310191926/https://wildoftech.com/alphago-google-deepmind (2016).

7 Schank, Roger C. (1991). "Where's the AI." *AI* magazine. Vol. 12, no. 4, page 38.

help with their FM requirements. It is important to develop leadership skills and to expand their FM knowledge and responsibilities as opportunities arise.

This leadership requires that the FM supports the various local, regional and national FM educational programs for the growth of the profession and to bring high school and college students to become familiar with FM and consider FM as a profession. With the projected shortfall of FMs available in the near future, attracting new FMs into the profession is a leadership challenge.

Energy

FMs will continue to be in the forefront of energy management, managing reduced energy programs, and using alternative energy sources. Local, regional and country energy programs are seeking various way to provide customers with the energy they need with less carbon emissions, providing energy from alternative sources such as solar, wind, water, and being able to store energy more efficiently for later use.

The development of zero-based energy buildings will continue to grow as technology improves and as material and labor costs become more cost effective. Alternative energy vehicles should become more standard such as electric, hybrid, natural gas, hydrogen and other alternative fuel vehicles will continue to evolve and FMs will need to be aware of the energy site and off-site services that may be needed to support their organizations business and FM services.

Climate Change

Some of today's weather, weather changes and its impact on our ability to exist in specific areas of the Earth will effect on how we live and conduct our FM work. Air temperature swings, highs and lows, rain fall and lack of rain, snow fall or lack of snow, winds, flooding, water temperature, rising seas, tornados, hurricanes/typhoons, water tables, the growth of deserts, air pollution, CO_2 emissions, etc. will continue to and possibly increase in negative effects that will impact on how we live and accomplish our FM work.

These challenges and solutions will become or have become issues that will affect how we do business, where we do business, how we can work with cities and municipalities, and how our FM customers need to change where they can work in these changing times of climate change. Our organizations will be looking for carbon neutral solutions, and how the business can support the climate goals that their country has set whether in sustainability, CO_2, energy efficiency, zero waste, recycling, new production processes, plastic recycling, degradable plastics, reduced landfills, etc.

Health, Safety, and Security

With our COVID-19 experience, we are now more aware of how a pandemic can affect our lives, our families, our FM work, our co-workers, our customers, our companies, our cities, our regions and our countries. Working at home and/or at remote locations provided other health, safety and security issues. Now that the pandemic appears to be a smaller risk, working at home or remotely may become the norm. Only time will tell the direction working for the organization will take.

Safety continues to be an issue that impacts on our workforce and workplaces. We continue to see an increasing number of cameras, card access entrances, secure parking facilities, safe rooms, panic buttons, keypads, networked databases, direct access to police and fire departments, etc. to ensure that our customers can feel safe in their work environments.

Our organizations will continue to review our security issues, increasing our security services or outsourcing security services to a lot more than just a sign-in service. These security services have additional training for all kinds of emergencies, working with local police and fire departments, and providing a safe environment for our workforce. They will also consult with FMs and other customers to ensure the safety requirements have been addressed and have reviewed the installed safety requirements.

Politics

Our local, regional, and national politics will provide a safe or unsecure environment for our customers and their workforce. Country politics that are stable and supportive are the norm that organizations look to invest their business capital and people. Countries whose politics are not stable or subject to sudden changes will continue to be risky places to do business. This also applies to local and regional political situations whose change

Also, the FM must be aware of internal organizational politics in the future, and how changes in leadership can change and affect your FM program. Tied to politics is the availability of funds, the organizational strategic plans, and a growth or no growth direction for the organization.

Strategic and Long-Range Planning

We expect that the FM's role in their organization will continue to develop to work with senior management and to be recognized for their leadership and solutions for their work for the organization. The FM is interested in the development of the organization and their customers' plans based on the strategic and long-range planning that they participated with senior management.

FMs will continue to be considered as the go to service to find real estate, develop properties, set and manage realistic budgets, manage workable designs, set expectations, manage construction and move-ins, maintain and operate the facilities, and dispose of the property when no longer required by the organization.

FM Research

The future of FM must have more ongoing research that can be shared with FMs for simple and complex problems that need to be addressed and investigated. Some areas and countries have more history of FM research by FM associations, from university programs and some research is sponsored by FM vendors and services. Some results are published, some results are presented at conferences and some results remain the property of the vendor or service.

For some of the above tasks and considerations in this FM Outlook, some research will become available and usable by FMs either through training or from the application of reading and testing of published research. FM Research provides an ongoing need to help the maturing of the profession. With the complexity of working globally, FM research with need to work with city requirements, codes, permit requirements, CO2 issues, sustainability, energy, the built environments laws, AI, and other issues that have not yet become a problem at this time.

FM Outsourcing

The formal outsourcing of FM management and services is almost as old as the profession. Some of this history has been very smooth and some have been embarrassing for the outsourcing FM vendor and those with the organization that hired the vendor. Almost 40 years later FM outsourcing has matured, consolidation of companies has taken place internationally, and organizations have become more aware of the pitfalls and benefits of the outsourcing of some non-critical and non-core FM services.

Where outsourcing was initially an arm's length relationship, many organizations and FM outsourcing vendors have chosen to trust and share information. This provided opportunities to sit together to discuss and solve issues and provide some common information that would not previously be shared. Outsourcing will continue to grow in some countries and services, and these vendors will continue to use their expertise, technology and management to support their outsourcing customers and FM departments.

Real Estate and the Built Environment

Real estate refers to real, or physical, property, and can include land, buildings, air rights above the land, and underground rights below the land. As a business term, real estate also refers to producing, buying, and selling property.[8] This is one of today's definitions that may be changing as the industry resettles from COVID and the fallout from many workers having to work at home for 2+ years. What will corporations and organizations do with the office, industrial, manufacturing, retail, etc. space in the coming years? The jury is still out as to where the real estate business is going in the future.

The Built Environment is a term used to refer to human-made conditions and is often used in architecture, landscape architecture, urban planning, public health, sociology, and anthropology, among others. These curated spaces provide the setting for human activity and were created to fulfill human desires and needs. The term can refer to a plethora of components including the traditionally associated buildings, cities, public infrastructure, transportation, open space, as well as more conceptual components like farmlands, dammed rivers, wildlife management, and even domesticated animals.[9]

As we go through the many changes that must take place in the future, the built environment will evolve to meet the climate, energy, sustainability, and business needs of those who invest and manage this changing world. The FMs that have these responsibilities will have challenges to face to help their organizations succeed in the economies and working environments that will be faced.

Conclusion

So, who are the organizations, associations, and people that will manage and succeed in the FM future? This question and the answer will provide an opportunity for FMs to review in another 40 years from now. Whatever the outcome, the FM profession will be more mature, will be better able to address their issues, be more aware of what their senior management's strategic plans are, and will have many new issues to face that those FMs who came before did not face.

Technology will be a key to the solutions that FMs will use and their ability to use the data this technology can provide. FM education will continue to provide research and training for those who seek to understand and be able to manage this newly built environment. FM associations and related associations will continue to help their members to find and suggest some of the many solutions that will be necessary in this changing world. As we continue to network with our peers around the world, shared issues and their solutions can be available to help in the FMs daily and on-going work requirements, perhaps in outer space, on the Moon or even Mars. The FM will be there.

Michaela Hellerforth, Co-editor
Edmond P. Rondeau, Co-editor

---

8 https://www.thebalancemoney.com/real-estate-what-it-is-and-how-it-works-3305882#:~:text=Real%20estate%20refers%20to%20real%2C%20or%20physical%2C%20property%2C,also%20refers%20to%20producing%2C%20buying%2C%20and%20selling%20property.

9 McClure, Bartuska, Wendy, Tom (2007). The Built Environment: A Collaborative Inquiry into Design and Planning (2nd ed.). Canada and Hoboken, New Jersey: John Wiley & Sons. pp. 5–6.

Appendix A

Facility Management (FM) Associations

Facility Management

FM in the USA started as an academic idea and concept in the late 1970s that gained exposure from a private commercial furniture company, Herman Miller, Company, then through their research arm, Herman Miller Research Corporation and its Facility Management Institute (FMI) based in Ann Arbor MI USA. By 1978/1979, FMI had been providing FM education seminars to corporate end users (later known as FM professionals) throughout the USA.

A number of these corporate end users who attended FMI seminars met in 1979 with FMI assistance to consider forming a FM association. On May 29, 1980, at a meeting in Houston, TX, USA, sponsored by an end user held at his corporate headquarters was attended by 16 facility management professionals (corporate end users) who agreed to form the National Facility Management Association (NFMA), an independent nonprofit association. In 1982 with the chartering of the Toronto, Canada Chapter, the name was changed to the International Facility Management Association (IFMA).

This FM association chose to grow based on the development of local chapters with FM members using a bottoms-up approach that influenced the development and growth of association and chapter management, membership, networking, programs, services, education, conferences, scholarships, etc. IFMA moved its headquarters operations from Ann Harbor, Michigan to Houston, Texas, USA, in 1984.

As of December 2022, IFMA had over 130 chapters with over 20,000 members located in 100 countries around the world.[1] Members included facility professionals, architects, engineers, interior designers, planners, financial administrators, manufacturers, consultants, property and real estate managers, landlords, and service and product providers to corporation facilities.

2021 IFMA **Vision:**[2]

> Lead the future of the built environment to make the world a better place.

---

1 https://www.ifma.org/membership/networks/local-chapter-map.

2 https://www.ifma.org/news/whats-new-at-ifma-new/ifma-announces-new-vision-mission-and-launches-refreshed-brand-website.

Facilities @ Management, Concept - Realization - Vision, A Global Perspective, First Edition. Edited by Edmond Rondeau and Michaela Hellerforth.

2021 IFMA **Mission**:[3]

> We advance our collective knowledge, value and growth for facility management professionals to perform at the highest level.

IFMA also works with architectural, engineering, interior design, and other associations to share information and provide educational opportunities. The skills that Facility Management Professionals use are defined by IFMA as the following 11 core competencies as of 2022:[4]

1) Communications
2) Quality
3) Technology
4) Operations & maintenance
5) Human factors
6) Finance & business
7) Emergency planning & business continuity
8) Real estate & property management
9) Project management
10) Environmental stewardship & sustainability

Since 1980 other FM associations have been formed in countries around the world in Asia; Australia/New Zealand; North, Central, and South America; Europe; and Africa.

FM Associations

FM Associations in many countries may have developed and provide some of the following for their members:

- assist in marketing, and membership growth, chapter formation and chartering in the country.
- support the country association Board of Directors and its Presidents and Chairs
- develop finances to support their programs, events, staff, and financial reserves.
- define and adopt the FM Core Competencies
- hold FM conferences for members and guests with exhibits
- identify FM professionals who are recognized in award programs
- support the special focused councils
- support the membership with the adoption of a code of ethics
- support the development and publication of their definition of facility management
- develop alliances with FM associations in other countries and regions.
- publish an association newsletter and a magazine
- support the membership in the establishment of an association foundation for FM research, and raise funds for and award FM scholarships to university students
- develop a FM certification test and award
- develop FM education and testing programs in the country language
- define and adopt revised FM core competencies in the association country
- work with Global FM as an FM association member

3 https://www.ifma.org/news/whats-new-at-ifma-new/ifma-announces-new-vision-mission-and-launches-refreshed-brand-website.
4 http://cdn.ifma.org/sfcdn/knowledge-base/ifmas-11-core-competencies.pdf?sfvrsn=0.

- support the association foundation in the development of FM awareness programs for high schools and community colleges. This also includes obtaining grant programs from state/providence/countries to provide FM training programs.
- address the challenges that COVID brought to members and their association

We know that FM association members and their work organizations in the 2020s have faced new FM challenges as the result of changing work environments that resulted from the 2+ years of COVID. These challenges and their possible solutions are still a moving target as organizations seek to find workplace solutions in the coming years.

Country associations in the 2020s will rely on their members and their sharing of these solutions with their chapters, with members in articles, and as presenters at FM and other conferences.

FM and Related Associations

In the past 40 years we have seen a number of FM and related associations that have supported FM in their countries. The related associations shown are also found as country centric associations internationally:

FM Associations (not an all-encompassing list): found on the Internet:

1) Alliance of Infrastructure and Facility Managers of India (AIFMI)
2) Association of Physical Plant Administrators – now APPA
3) Association des Responsible Services Généraux. (France) (ARSEG)
4) Association of Property and Facility Managers (Singapore) (APFM)
5) British Institute of Facilities Management (BIFM)
6) Bulgarian Facility Management Association (BGFMA)
7) Canadian Recreation Facilities Council (Canadian) (CRFC)
8) Conference for Catholic Facility Management (US) (CCFM)
9) Denmark Facilities Management (DFM)
10) European Facility Management Network (EuroFM)
11) Facility Management Association of Australia (FMA)
12) Facility Management Austria (FMA)
13) Facility Management Netherlands (FMN)
14) FM-ARENA (Switzerland)
15) German Facility Management Association (GEFMA)
16) Global Facility Management Association (Global FM – has grown from 8 to 14 current member-centric FM associations).
    - ABRAFAC – Brasil
    - FMA – Australia
    - FMANZ – New Zealand
    - HFMS – Hungary
    - IFMA – Headquarters in Texas, US
    - IWFM – Great Britain
    - MEFMA – United Arab Emirates
    - SAFMA – South Africa
    - TRFMA – Turkey
    - APAFAM – Panama
    - ACFM – Catalonia, Spain

- EGYFMA – Egypt
- AFMPN – Nigeria
- SFMA – Saudi Arabia

17) Health Estates & Facilities (UK) (HEFMA)
18) Higher Education Facility Management Association of Southern Africa (HEFMA)
19) Hong Kong Institute of Facility Management (HKIFM)
20) Institute of Workplace and Facilities Management (IWFM)
21) International Facility Management Association (IFMA)
22) Irish Property & Facility Management Association (IPFMA)
23) Japan Facility Management Association (JFMA)
24) Maintenance and Facility Management Society of Switzerland (MFS)
25) National Association of Church Facilities Managers (NACFM)
26) National Association of Industrial & Office Properties (US) (NAIOP)
27) Restaurant Facility Management Association (RFMA)
28) Romanian Facility Management Association (ROFMA)
29) Royal Institute of Chartered Surveyors (RICS) – Chartered in Facility Management
30) Tertiary Education Facilities Management Association (Australasian) (TEFMA)

Related Associations (not an all-encompassing list – some are international) found on the Internet:

1) American Academy of Environmental Engineers and Scientists (AAEES)
2) American Institute of Architects (AIA)
3) American Planning Association (APA)
4) American Society of Civil Engineers (ASCE)
5) American Society of Heating, Refrigerating and Air-Conditioning Engineers (ASHRAE)
6) American Society of Interior Designers (ASID)
7) American Society of Landscape Architects (ASLA)
8) American Society of Mechanical Engineers (ASME)
9) American Society of Plumbing Engineers (ASPE)
10) Association for Facilities Engineering (AFE)
11) Association for Facilities Engineering (AFE)
12) Association of Architecture Organizations (AAO)
13) Association of Asset Management Professionals (AAMP)
14) Association of Collegiate Schools of Architecture (ACSA)
15) Association of Physical Plant Administrators (APPA)
16) Building Owners and Managers Association (BOMA)
17) Building Owners and Managers Institute International (BOMI)
18) Campus FM Technology Association (CFTA)
19) Commercial Real Estate Women (North America) (CREW Network)
20) CoreNet Global – International Corporate Real Estate Association
21) Facilities Management Institute (FMI)
22) Institute of Electrical and Electronic Engineers (IEEE)
23) Institute of Real Estate Management (US) (IREM)
24) Interior Design Society (IDS)
25) International Association of Museum Facility Administrators (IAMFA)
26) International Furnishings and Design Association (IFDA)
27) International Interior Design Association (IIDA)

28) National Fire Protection Association (NFPA)
29) National Society of Professional Engineers (NSPE)
30) Royal Institution of Chartered Surveyors (RICS)
31) Society of American Military Engineers (SAME)
32) The Society for Maintenance & Reliability Professionals (SMRP)
33) Urban Land Institute (ULI)
34) Women in Construction (WC)

Appendix B

Facility Management (FM) Education

FM Education

Many FM associations provide some form of education for their members, whether through formal programs, at conferences, through chapter meetings, online, virtually, etc. The education programs may lead to a certificate, certification, or professional recognition. Some FM educational programs in many countries have a number of offerings in a number of languages, while some have a few and are growing. There are also a number of educational companies, consultants, and continuing education departments in universities that offer FM educational programs, some with testing, and some without testing.

IFMA Education

In October of 1980 the facility management profession was fortunate to have 16 founding members who met in Houston, Texas, USA, to begin and charter the National Facility Management Association (NFMA). In 1982 the NFMA changed its name to the International Facility Management Association (IFMA) to accommodate new members and the chapter from Toronto, Ontario, Canada.

One of IFMA's early goals was to develop one-, two-, or three-day FM education courses based on the eight FM core competencies adopted in 1984:

- Long Range and Annual Planning
- Financial Forecasting & Management
- Telephone, Security & Administration
- Interior Space Planning & Space Mgmt.
- Architecture & Engineering
- New Construction & Renovations
- Maintenance & Operations
- Real Estate

Initially many of these early IFMA core courses were developed and presented by IFMA members. In the early 1990s IFMA began to develop and own a number of certification programs and testing for these certifications for the above core competencies.

Facilities @ Management, Concept - Realization - Vision, A Global Perspective, First Edition. Edited by Edmond Rondeau and Michaela Hellerforth.

Certified Facility Manager® (CFM[1])

In 1990 IFMA started the development of the Certified Facility Manager® (CFM) credential program and related testing. This sets the industry standard for ensuring the knowledge and competence of practicing facility managers. This certification program and testing have been translated into a number of languages for FM in many countries. In 1993 the first 32 IFMA credential "Certified Facility Manager" (CFM) designations were awarded as shown to the left.

Essentials of Facility Management™ (EoFM®)

IFMA's facility management introductory course for those that want to learn more about the field of facility management and are new to the profession.

Facility Management Professional™ (FMP®)

IFMA's FMP is a knowledge-based credential for FM professionals looking to increase their depth-of-knowledge in the core FM topics deemed critical by employers.

Sustainability Facility Professional®

IFMA's Sustainability Facility Professional (SFP) is an assessment-based certificate program delivering a specialty credential in sustainability

1 Source: www.ifma.org.

IFMA also works with architectural, engineering, interior design, and other associations such as the AIA, SME, RICS, and EUROFM to share information and education opportunities.

The skills that Facility Management Professionals use are defined by IFMA as the following 11 core competencies as of 2005:[2]

1) Communications
2) Quality
3) Technology
4) Operations & Maintenance
5) Human Factors
6) Finance & Business
7) Emergency Planning & Business Continuity
8) Leadership & Strategy
9) Real Estate & Property Management
10) Project Management
11) Environmental Stewardship & Sustainability

IFMA Foundation

Established in 1990 as a nonprofit, 501(c) (3) corporation, and a separate entity from IFMA, the IFMA Foundation, based in Houston, TX, USA, works for the public good to promote priority research and educational opportunities for the advancement of facility management. It is dedicated to the mission of promoting education for the facility management profession around the world and making FM a Career of Choice.[3]

The IFMA Foundation's purpose is to provide greater resources and educational opportunities to facility management professionals and those who support the FM industry.

Mission

Position facility management as a career of choice by promoting and supporting educational opportunities, related scholarships, and research initiatives worldwide.

Purpose

Foster FM workforce development by collaborating with higher education institutions worldwide on FM degree programs, FM student scholarships, and FM research.

IFMA Foundation Programs

Education – The foundation promotes higher FM education through its Accredited Degree Program (ADP).

Scholarships – The foundation supports aspiring and practicing FMs in their efforts to reach higher, go further and make a difference.

Research – The foundation conducts research significant to the advancement of the profession.

The IFMA Foundation is supported by the generosity of the FM community including IFMA members, chapters, councils, corporate sponsors, and private contributors who share the belief that education and research improve the FM profession. FM students from around the world may apply for IFMA Foundation scholarships which are awarded annually.

---

2 IBID.

3 Source: https://www.ifma.org/about/about-ifma/ifma-foundation.

The Foundation has developed standards for excellence in FM education that enables universities and colleges worldwide to apply for accreditation for Associate, Bachelor, and Master's FM programs globally. At the present time through the Facility Management Accreditation Commission (FMAC[4]) and the ABET[5] 38 schools are accredited around the world, including:

- USA: 25
- Netherlands: 4
- Singapore: 3
- Hong Kong: 1
- Korea: 1
- Sri Lanka: 1
- Germany: 1
- Ireland: 1
- Turkey: 1

The Foundation encourages higher education schools around the globe to start FM programs and has initiated a The Global Workforce Initiative that addresses the triple bottom line of Economy, Equity, and Environment. Meeting this demand is critical to the energy efficiency of the built environment, reducing greenhouse gas emissions, increasing economic productivity of these buildings, and providing education and employment opportunities in the profession of facility management.

GWI programs are offered through the IFMA Foundation's

FM Accredited Degree Programs (ADP)
FM Registered Degree Programs (RDP)
FM Talent Development Pipeline (TDP)

Students enrolled in GWI (Global Workplace Initative) programs (see Figure 1) have the opportunity to participate in scholarship programs, internships, job shadowing, and our signature student competition called IgniteFM! The Student Challenge.

Many of the FMs who started FM and started with a FM association in their country are aging, and many have or are considering retirement. The IFMA Foundation Board of Directors have approved the FM Career Ambassador Speaker's Kit and Program to make it easy for the current aging and retiring group of FM professionals to speak on the value of an FM career to different groups – grade and high school students, high school counselors, college students, community colleges counselors, community organizations, and the general public.

Each kit includes a Speakers Guide (see Figure 2) in the note section to walk the FM professional through the process and their presentation, a PowerPoint presentation that can be modified for audience, a script, and videos. This program is also being delivered directly to high schools, community colleges, and soon-to-be retirees and discharged members of the US armed forces who may want to consider a future in FM.

As of 2022, FM is a $1.2 trillion global industry, 5% of the economy is Facility Services, and there are 25 million FM practitioners worldwide.[6]

In 2022 the IFMA Foundation was Granted Special Consultative Status with United Nations Economic and Social Council as Nongovernmental Organization (see Figure 3).

4 FMAC is made up of FM academic, professional members of IFMA and FM consultants who set the academic standards for all degree programs and visit candidates, virtually or in person, to evaluate their FM programs.

5 ABET (Accreditation Board for Engineering and Technology) is a nonprofit, ISO 9001 certified organization that accredits college and university programs in applied and natural science, computing, engineering and engineering technology in 41 countries.

6 Source: https://www.ifma.org/about/about-ifma/ifma-foundation.

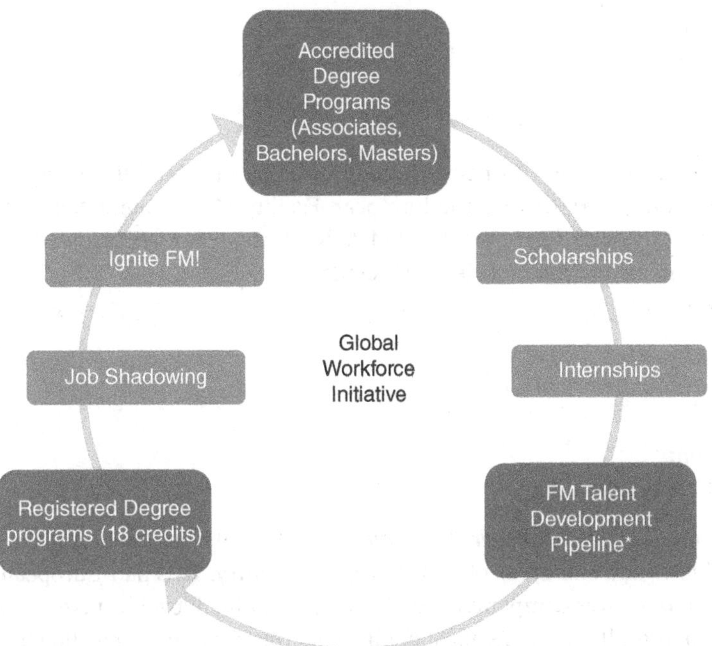

Figure 1 IFMA Foundation – IgniteFM! (*Source:* IFMA Foundation / https://foundation.ifma.org/wp-content/uploads/2019/07/IFMA-Foundation-Brochure-FINAL-Sept-2019.pdf).

Figure 2 IFMA Foundation – Building FM Careers. (*Source:* IFMA Foundation).

Figure 3 IFMA Foundation – UN Granted Special Consultative Status (*Source:* IFMA Foundation).

EuroFM Education[7]

In 1987 the first exploratory meeting to create a European FM network was hosted by Mr. Bart Bleker in the Netherlands. He helped develop the association till 1990. In 1993 the European Facility Management Network was officially registered by NEFMA, the Dutch FM association now called FMN, the Danish FM association DFM, and the British Centre for Facilities Management led by Professor Keith Alexander.

As of this publication, the association focuses on:

- Promoting FM across Europe;
- Adding value to our members;
- Financial Stability;
- Dissemination of knowledge and information;
- To facilitate networking opportunities to share best practice

In collaboration with REUG, EuroFM has developed a *Site Manager Certificate* which sets international standards and at the same time is localized to the local laws and standards of each European country. This pan-European Certificate supports the provision of services and assures comparable quality at international level. It takes into account transparent bidding processes and supports all requirements of public tendering procedures by allowing to measure and improve on the quality of service and the skills of people delivering it.

Site managers are the link between tenants, real estate, and facility services. They understand the tenant's core business and are thus able to define the tenant's requirements with regard to FM, to organize hard and soft services, to implement FM processes and services at site level, and to adapt real estate use accordingly.

The Site Manager Certification is designed for people who already work in this role in FM departments or at service companies, either as internal site managers on the tenant's premises or as external site managers working for service providers.

The competencies required to achieve the Certificate are based on the European Standard EN 15221-4 (technical and infrastructural knowledge, legal knowledge and compliance, organizational, business and commercial skills, knowledge about workplace and related strategies, construction and renovation, social and methodological skills) The questions used to assess the knowledge of the candidates are localized to the national context (legislation, norms, and standards as well as the language) by EuroFM in collaboration with the local FM organizations performing the examination.

EuroFM Education Group[8]

The focus of the Education Group is on students and their lecturers, specifically universities and their Facility Management programs. In addition, this group includes members of EuroFM who are not affiliated to universities but who have a strong interest in the educational field and who provide training related to the FM profession.

Through personal relations and organizing exchange and international teamwork the Education Group hopes to inspire students and their lecturers to actively join in networking through both physical and online platforms. Not only for social networking where members can exchange knowledge and receive commentary and input on new ideas. Indeed sharing knowledge, exchanging experiences, helping each other to become part of the FM profession will aid in its future development. The Education Group is aiming for a powerful platform where open, transparent, and sustainable partnerships are established, based on mutual trust and shared goals.

The role of the Education Group is summarized below:

- To facilitate an active education network in Europe reflecting the integrated approach to FM education, research, and business;

7 Source: https://eurofm.org/about-fm/site-manager-certification/.

8 Source: https://eurofm.org/education.

- To assist educational institutions in drafting their FM curricula and to set standards of FM education in Europe;
- To encourage and facilitate student exchange between EuroFM member universities;
- To encourage and facilitate an active knowledge sharing and staff exchange culture among the EuroFM member universities;
- To enhance internships within the FM business.

The intention of the Education Group is to reach young professionals and teach them about the added value of the EuroFM platform. Through student and staff exchange, students see the differences in not only FM knowledge, but also in how to work together. This will not only benefit students, but also EuroFM members who can learn a great deal from the next FM generation. Various initiatives developed within EuroFM provide students access to the FM world through digital stay connected activities, where business, young talents, educators, and researchers meet.

There are different types of activities and network opportunities organized during the year:

Summer- and winter schools: during the summer and winter a member university will host a one-week project where students and their lecturers work together on interesting and present-day projects. The aim is to deliver interesting data to the Research Group within EuroFM:

- The EuroFM Student Competition: students are asked to participate in a video competition where they present internship, thesis, or project results during the EuroFM International Conference.
- Collaborative Online International Learning (COIL) projects: students and staff from member universities work together through digital collaboration for a 6- to 12-week project.
- Student and staff exchange: students can study for 15–30 credits at FM programs at member universities. Staff can teach as guest lecturers within the different FM programs.
- Internship platform: the EuroFM network gives opportunities for students to find an internship placement in the international FM market;
- Evidence-based research: bachelor FM students participate in different projects to add value to the EuroFM network and its members.
- Current and/or future activities:
  - To coach students from the bachelor to master to researcher role by facilitating opportunities within the EuroFM network;
  - To set up an EuroFM alumni community where participants of summer and winter schools and student competitions can exchange their knowledge and experience;
  - To develop an inventory of present FM curricula in EuroFM member universities which makes the present study offering visible to all interested parties;
  - To develop Internship Programs within EuroFM Universities: researching the possibility to professionalize internships with an aim to build a “EuroFM internships program.” The main objective of this project is to benefit EuroFM business;
  - To develop a EuroFM Honors/Excellence Program: an inventory of present programs in order to create an overview of all special programs. The aim is to develop a specific Honors/Excellence Program for the highly motivated and skilled students within the EuroFM member universities.

BOMI Education[9]

BOMI (Building Owners and Managers Institute International) is based in Annapolis, MD, USA. BOMI is the international provider of educational products and services to the property and facility management industries.

9 Source: https://www.bomi.org/Students/Educational-Offerings/Designations-and-Certificates/Certificate-Programs.aspx.

As the Independent Institute for Property and Facility Management Education, BOMI, based in the USA, is dedicated to improving the skills of professionals with property, facilities, and systems responsibilities.

Founded in 1970, BOMI International is a 501(c)(3) not-for-profit educational institute that has earned a reputation as the trusted property and facility educational resource for top corporations, government agencies, property management firms, educational institutions, unions, and trade associations.

Core Purpose:

To add value through learning.

Core Values:

- Client-centered value and service
- Integrity and ethics throughout
- Relevant and continuous lifelong learning
- Flexibility and nimbleness for a changing marketplace
- Operational excellence and continuous improvement
- Embracing diversity and inclusion
- Collaborating across boundaries
- Vision:

To be the career-long learning and development partner of choice to the property and facility management professions and related disciplines.

Certification

Building Know-How

Whether you are new to the commercial real estate industry or a professional looking to enhance your skills and gain recognition, a BOMI certificate program is right for you. All certificate programs provide the fundamental knowledge you need to better understand your job responsibilities and are a great stepping stone to a BOMI designation.

To earn a BOMI certificate, learners must complete three courses and pass an exam at the end of each course to show their understanding of program content. Certificates may be earned until a designation is achieved.

Plan Your Path

Property Administrator Certificate

Want to keep your property in top operational condition, inside and out? Learn how to manage the ongoing operation and maintenance of building systems and to maximize building efficiency and cost-effectiveness. For those who manage the overall operations of a building or a portfolio of buildings.

Property Management Financial Proficiency Certificate
Take control of your assets and improve overall efficiencies by understanding all areas of investment decision-making, how to analyze financial statements, and how to construct productive property facility budgets. For those responsible for analyzing, managing, and investing in real estate assets.

Certified Manager of Commercial Properties™
Designed for early-career property professionals who are looking to validate their industry knowledge, this new certification will take your career to the next level. The CMCP Exam Prep course offers 30 hours of coursework, fulfilling the requirements of the CMCP certification. For those looking to take the first step in building a successful career in commercial real estate

Facilities Management Certificate
Teaches the information necessary to manage the ongoing operation and maintenance of building systems and to maximize building efficiency and their cost-effectiveness.

For those who manage the ongoing operation and maintenance of facilities.

Building Systems Maintenance Certificate
Obtain the in-depth information you need on key building principles, including efficient energy management and water treatment. In addition, you'll gain a better understanding of HVAC, plumbing, and other building systems that work together to provide a comfortable indoor environment. For those who operate and maintain multiple building systems.

Building Energy Certificate
Learn the necessary foundational concepts that relate to reducing energy consumption from a technical perspective. Gain the skills necessary to implement sustainability efforts in the built environment for all building operators.

For those in operational and system maintenance roles.

Learners must complete their certificate enrollment within a three-year period. Learners selecting not to complete their certificate enrollment within the term will be considered inactive. All active certificate learners enrolled prior to January 1, 2020, will be given three years from January 1, 2020, to complete their current certificate program. Enrollees on or after January 1, 2020, will have three years from their enrollment date to complete their certificate program.

Royal Institute of Chartered Surveyors (RICS)[10]

The Royal Institution of Chartered Surveyors based in London, UK, promotes and enforces the highest professional qualifications and standards in the development and management of land, real estate, construction, and infrastructure. With over 100,000 highly qualified trainees and professionals, and offices in many significant financial markets, RICS is placed to influence policy and embed their standards within local marketplaces in order to protect consumers and businesses. In doing so, RICS seeks to innovate and progress the development of spaces and places so they are fit for future generations, in addition to the challenges faced in the present.

RICS FM Training:
In a marketplace with a complex workforce, of varying levels of knowledge and skills; ensuring staff are working toward international best practice and consistent standards is a growing priority for all organizations. Drawing from the RICS breadth of knowledge across disciplines, RICS is uniquely positioned to work within the professional FM community to address industry trends and development gaps.

To meet the demand of the RICS evolving market, RICS ensures professionals will have the skills and resources necessary to anticipate and prepare for the challenges and opportunities presented by the FM industry.

All RICS training programs have been developed with their global network of professionals, incorporating their expertise, knowledge, relevance, and practical insight. RICS training professionals work with the local teams to discuss, identify, and develop appropriate training solutions tailored to local needs and aligned to local business goals. RICS training also is designed and delivered to allow significant interaction to maximize the learning experience, supplemented by course materials that support the application of learning in the workplace. RICS' operates globally, working with multinational clients, while balancing the need for local relevance and applicability.

Some RICS FM Training Courses:

Procurement of facility management
This code of practice provides guidance on the various factors that need to be considered throughout a FM procurement process, including activities and key decisions during planning, procurement, and post-procurement. It aims to help the professional choose an appropriate procurement route and the various factors in delivering an effective FM procurement process that results in a successful contract with benefits for both the client organization and the supplier.

The document is aimed at public and private sector FM property professionals involved in a facility management (FM) procurement process either within their territory or region, or globally. This includes property managers, directors of estates, heads of FM, consultants, RICS-regulated firms acting for a landlord, and FM suppliers

10 Source: https://www.rics.org/north-america/about-rics/.

procuring services from subcontractors. Those managing in-house teams delivering FM services may also find some of the content helpful.

New code brings global consistency to facility management procurement

RICS has launched a new global industry code of practice which sets out high-level steps to bring transparency and consistency to the facility management procurement process.

It comes at a time when the expertise of facility management professionals is playing a crucial role in supporting occupiers and building owners to manage challenges through the COVID-19 pandemic.

Clients and service providers benefit from new clear guidance within the code outlining the key factors that need to be considered through the planning, procurement, and postprocurement of facility management services.

Designed to be applicable globally, the new code will support public and private sector property professionals to help clients understand the value of facility management expertise.

RICS Training courses deliver global skills with local knowledge and RICS promotes and enforces the highest level of professionalism and standards in the valuation, development of real estate, construction, infrastructure and its management through property and facility management.

With a global footprint and network of professionals practicing in the major political and financial centers of the world, RICS' market presence means they are placed to drive thought leadership, influence policy, embed standards, and gather market insights.

With over 100,000 global professionals qualified with RICS credentials, industry stakeholders who work with RICS registered professionals have confidence in the quality and ethics of the services they receive.

APPA (formerly the Association of Physical Plant Administrators)[11]

APPA's CEFP certification is the only facilities credential that focuses on the FMs' professional development, while encompassing the full, multidisciplinary range of educational facilities management principles and practices – from planning, design, and construction to daily operations and general management.

As the facilities profession becomes more complex, it's critical for FMs to demonstrate their well-rounded knowledge of all the systems that make their campus an efficient place to learn, live, and work.

Today's APPA facilities professionals face a variety of challenges, such as:

- Aging infrastructure of buildings and utilities
- Expanding workload and expectations
- Reduced resources and skills gaps
- Utilizing space more efficiently

Professional certification ensures that FMs are well equipped to meet these challenges and demonstrates their commitment to lifelong learning.

11 https://www.appa.org/certification/cefp-credential/.

The Certified Educational Facilities Professional credential (CEFP) is a certification designed for both aspiring and existing educational facilities professionals with eight years of combined education and professional facilities management experience. Earning the CEFP demonstrates that the FM has a mastery of professional expertise and is a mark of superior proficiency in the core competencies for education facilities professionals.

The Certified Educational Facilities Professional (CEFP) designation provides a professional credentialing continuum for the educational facilities professional.

To maintain the CEFP credential, the FM must demonstrate their commitment to ongoing professional development and continuing leadership by means of a recertification process every four (4) years. The process utilizes a point system that assigns recertification credits for participation in professional development programs, as well as APPA and non-APPA volunteer leadership and service in support of the educational facilities profession.

Reasons for FM Education

Many other FM associations not listed have educational programs and courses to expand their members FM and business knowledge to further their careers and their ability to help their organizations succeed and provide a safe, secure, and efficient work environment.

Some FMs have gotten into FM through a career change or challenge, and they needed to increase their basic FM knowledge to help them manage their responsibilities and projects, to understand the hard and soft skills needed for their FM career and to provide outstanding service to their customers and their work organizations.

FM education is continuing to grow and provide FMs the information and skills they need and/or will need in the organizations and countries they serve.

Appendix C

Co-Editors and Co-Initiators

Edmond P. Rondeau

AIA Emeritus, RCFM, IFMA Fellow, Atlanta, Georgia USA

Ed Rondeau is the stateside initiator, organizer, and co-editor for this publication, has, as an educated architect from the Georgia Institute of Technology (GA Tech), Atlanta, GA, USA, and a member of the American Institute of Architects (AIA), had very early contact with FM development. He served the IFMA Association under different functions since 1983 nationally and has experience as an instructor for over three decades as teacher and consultant in the Americas, Asia, Australia, and Europe.

His involvement with FM began in 1981 when he was a Construction Manager with the Coca-Cola Company in Atlanta, GA, USA, and obtained a Master of Business Administration in Real Estate from Georgia State University, Atlanta, GA, in 1982. In the same year he became one of the co-founders of the IFMA Atlanta, GA, chapter. He was President of the Chapter in 1983 and was asked to be an IFMA Regional Vice President/Member of the IFMA Board of Directors.

He was elected as the 1987/1988 IFMA President, and in this leadership position he initiated meetings with other groups forming and met FM associations, including the Japan Facility Management Association (JFMA) in the USA and later in Japan in 1987.

A FM trip in 1988 to Australia and New Zealand with Prof. Franklin D. Becker of the FM degree program at Cornell University was organized to meet with architects, engineers, interior designers, and corporate end users. This trip was sponsored by an Australian dealer of Herman Miller to share FM with various audiences.

As the Past President of IFMA in 1989, he worked with the IFMA Executive Director Melvin R. Schlitt to develop the basis for establishing the IFMA Foundation which was approved and founded in 1990 by the IFMA Board of Directors.

The IFMA Foundation set up the nonprofit education, research, and scholarship organization for IFMA and for FM. He was the first Chair of the IFMA Foundation Board of Directors and began the raising of donations from companies and IFMA chapters to fund student scholarships for FM degree related university programs.

As part of the IFMA Foundation, the Facility Management Accreditation Commission (FMAC) developed the academic FM standards for Associate, Bachelor and Master degrees FM programs in the USA and internationally. He served as a review team member to two FM universities in the USA and to two candidate FM universities in Singapore.

Mr. Rondeau has worked in all phases of FM and corporate real estate for many large and some small organizations within the USA and internationally. This included training and managing staff and budgets, developing FM and Real Estate strategies, database, and IT systems, hiring and managing architects, engineers, interior designers,

Facilities @ Management, Concept - Realization - Vision, A Global Perspective, First Edition. Edited by Edmond Rondeau and Michaela Hellerforth.

consultants, vendors, and contractors, and working with senior management, customers, attorneys, finance and accounting, cash flow, internal audit, insurance, IT, FM maintenance and operations, work orders, furniture, moving, parking, food service, landscaping, janitorial, mail, telephone, security, local Police and Fire Departments, energy conservation, environmental issues, sustainability and safety, permits, and with local and state officials.

He has been invited as a speaker or guest FM lecturer at many FM conferences, programs and universities in the USA, Canada, Mexico, Cuba, Trinidad & Tobago, Venezuela, Brazil, Argentina, Singapore, Malaysia, India, Hong Kong, China, Japan, the Philippines, Australia, the UK, Northern Ireland, the Netherlands, Germany, Austria, Denmark, Sweden, Spain, and Italy.

He has spoken and written for other associations, including the American Institute of Architects (AIA), BIFM, the Building Owners and Managers Institute International (BOMI), EUROFM, Construction Owners Association of America (COAA), CoreNet Global (International Corporate Real Estate), Industrial Asset Management Council (IAMC), and the International Development Research Council (IDRC).

Ed has received the following professional recognition:

- 2017/2023 – IFMA Qualified Trainer: CFM Instructor and FMP Instructor
- 2015 – Atlanta IFMA Chapter, Anchor Award for service to the Chapter and the FM Profession.
- 2009/2010/2019/2020 – IFMA Education Department, selected as a Subject Matter Expert (SME) for revisions to the IFMA CFM and FMP Learning System Curriculum programs;
- 2004 – IFMA Distinguished Author Award for the book The "Facility Manager's Guide to Finance & Budgeting"
- 2001 – Accepted for the IDRC the American Society of Association Executives (ASAE) 2001 Gold Circle Award Trophy for International Association Management for the IDRC Global Operations and Learning Department
- 1999 – IFMA Foundation Award – Outstanding Support
- 1999 – BCCR Award – Award for Outstanding Contributions to BCCR – Distinguished Service and Dedication
- 1997 – IFMA Distinguished Author Award for the book "Facility Management".
- 1993 – Awarded Certified Facility Manager (CFM) certification from the IFMA
- 1992 – Elected a Fellow in the first class of the College of Fellows of the IFMA
- 1989 – IFMA Distinguished Author Award
- 1988 – IFMA Distinguished Member Award
- 1988 – IFMA Facility Management Achievement Award

His related academic experience has included co-authoring and/or as a lead author for seven books on real estate and facility management which have been used as textbooks for numerous FM programs in universities:

- Co-editor of this Wiley book with co-editor Michaela Hellerforth (Germany), 2024
- Lead co-author, 2nd Edition, *Facility Management*, John Wiley & Sons, New York, NY, 550 pages, Jan. 2006. Edmond P. Rondeau, Robert Kevin Brown & Paul D. Lapides
- Co-author, *The Facility Manager's Guide to Finance & Budgeting*, AMACOM Books (now published by CollinsHarper), published in September 2003, David G. Cotts and Edmond P. Rondeau
- Co-editor, *Facility Management 2*, Springer Publishing, Berlin, Germany (published in German), January 2001, Walther F. Moslener and Edmond P. Rondeau
- Lead co-author, 1st Edition, *Facility Management*, John Wiley & Sons, New York, NY, 613 pages, September 1995, Edmond P. Rondeau, Robert Kevin Brown & Paul D. Lapides
- Lead co-author: *Managing Corporate Real Estate: Forms and Procedures*, John Wiley & Sons, 450 pp, April 1994, Edmond P. Rondeau, Robert Kevin Brown & Paul D. Lapides
- Co-author: *Managing Corporate Real Estate*, John Wiley & Sons, 650 pages, April 1993, Robert Kevin Brown, Alvin L. Arnold, Joseph S. Rabianski, Neil G. Carn, Paul D. Lapides, Scott D. Blanchard, Edmond P. Rondeau
- Author, *Principles of Corporate Real Estate* manual, self-published, 250+ pages, 2019, used at a number of universities in the USA.

Professor, Dr. Michaela Hellerforth
Luedenscheid, Germany

Michaela Hellerforth studied Economics and Law at the renowned University of Muenster, where she specialized in Financing and Taxation. Hellerforth left the faculty in 1990 with a diploma thesis about investment and rentability of investment in real estate, at a time when real estate aspects were usually just a matter of macroeconomic studies at universities.

Following that, she worked as a scientific consultant for the German Federal Association of Independent Housing Companies (Bundesverband Freier Wohnungsunternehmen, BFW), a real estate association, settled in Bonn and Berlin, Germany, where she soon after became a member of the board of directors. Besides her work on the board, she became a Managing Director at one of the largest real estate service providers in South Westphalia, Germany.

Hellerforth started her doctorate at the University of Cologne in 1994. The thesis, published in 1996, was investigating the relationships of the subject matter experts (SMEs) of the real estate industry in between the single European domestic market and awarded with the Hugo-Thienhaus-Prize in 1997.

Hellerforth moved to London in 1998, where she worked in real estate and financing business for the Euro Hypo, subsidiary of Deutsche Bank. Back in Germany, she started her work as a Professor for Facility Management at the Westphalian University of Applied Sciences in Gelsenkirchen in 2000, where she teaches in different fields around Facility Management today.

During this time, she started extensively publicizing and got in touch with a relatively new field in connection with real estate: Facility Management, which was a subject that did not matter for property companies at that time, while it developed more and more in companies from non-property fields. During and after her doctorate, she became a pioneer in the area of Facility Management and published more than 130 articles and 14 books, in an around this field, today. Besides, she also continues to teach Facility Management at the South Westphalian University of Applied Sciences.

Since turn of the millennium, Michaela Hellerforth lectured at different institutions, for example at the University of Cologne, at the Frankfurt School of Finance and Management, at the University of Applied Sciences in Dortmund and the Hamburger Distance Learning University as well as for different continuing education providers, such as the Technical Academy Wuppertal in connection with RWTH Aachen.

Over the years, Hellerforth served as speaker for different companies and institutions. She is also active in consulting public and private institutions in matters of Facility Management, especially, restructuring the organizations and parts of organizations to working in CREM and Facility Management divisions, strategic aspects in Facility Management and Service provider Management as well as contract management and service provider control. Even so she is active in implementation advice and control.

Michaela Hellerforth published – as said – a number of books and articles: At Springer she published *Outsourcing in Real Estate* and *Handbuch Facility Management für Immobilienunternehmen* in 2006, the follower of a former published book by Haufe-Verlag in Freiburg. This book aimed to talk about the basics and instruments for professional management of properties, buildings, and equipment to guarantee a sustainable and persistent preservation of value, savings while managing buildings and cost controlling.

In 2022 she published – after 11 years – the second edition of a book called *Gebäudemanagement*, at Cornelson publishing house, which is aimed at professionals who want to specialize in Facility Management during their professional training. The second edition included an important development around the field of Facility Management including new trends such as building information modelling, new technical opportunities, and innovative market players who changed a lot of the FM issues. Furthermore, the second edition of *Immobilienwirtschaft kompakt* will be published in 2024.

Co-Initiator

Dr. Dr. Josef R. G. Mack

Frankfurt/Main, Germany

Josef Mack, the European initiator and organizer of this anthology is a trained social-economic researcher, scientist, and consultant. He holds PhD degrees and two graduate diplomas from the HfW Hochschule für Welthandel, respectively University of Economics in Vienna, Austria and passed the doctoral exams of HSG Hochschule St. Gallen für Wirtschafts- und Sozialwissenschaften, now University of St. Gallen, Switzerland. He also received a diploma in Cybernetics from the Technical University in Vienna. While still postgraduate, he started his research and consulting career in 1974, focusing as a freelancer and independent contract researcher in co-operation with university departments in transfer of management know-how, social technologies for hierarchical institutions, and innovative concepts and technologies.

An assignment from the middle of the 1970s stretching into the early 1980s under the auspices of the Federal Minister of Finance and the Federal Ministry of Health and Environmental Protection of the Austrian Government (Cabinet Kreisky II) with the aim of a basic reorganization of the Health Care and Hospital System of Austria. This brought many international and interdisciplinary contacts to premier universities in Western and Eastern Europe and North America, including cooperative research with relevant think tanks (Battelle, RAND) and enterprises (IBM, Siemens, Unidata).

This initial research project included problem solving tasks connected to specialized facilities, housing, laboratories, and hospitals, in combination with care, training, educational and administrative units. To evaluate their very particular cost structure in order to optimize processes required data then mostly not available nor even measured. Despite the availability and use of computers in mathematical calculation, in finance, statistics, and other processes requiring large calculations, neither adequate hardware nor software tools existed then to handle the unmanageable amount of partially non-digital design and architectural layout information, essential for planning, building, FM and inventory. This experience helped him to later become involved and embrace FM.

In autumn 2021 Edmond P. Rondeau suggested to use the upcoming anniversary in 2022 of the change from "National" to "INTERNATIONAL" Facility Management Association (IFMA) in 1982, as a chance to fill a gap which not only he had registered: despite the many publications existing about FM by then already, no objective description of why and how FM developed, who did it, who sponsored it, who got involved and why, to create out of an obvious need not only an industry of quite some size, but a profession, which now has an international academic acceptance, titles, professors, curricula, and an industry with a multi-billion turnover worldwide.

Together Ed Rondeau and Josef Mack decided to join in an initiative to organize 20 years after their last mutual publication on FM to trace, find, and renew old contacts, establish if necessary new ones, and gather protagonists of the early days, and from today, to have them contribute their own personal history to an anthology about their experience, work, and career, complemented by their own views and expectations on the future of FM. After some visits in the USA earlier last year, the UK, the Netherlands, Austria, Switzerland, and extended travels to many leading academic and corporate protagonists of FM also in Germany, the interest in supporting this very special project was overwhelming.

As gender equality in the industry and in professional education in the USA seemed to be more advanced compared to other parts of the world, we were pleased to win early last year a German female co-editor, Prof. Dr. Michaela Hellerforth, with relevant European and US experience of her own, to direct and underline due attention to the increasingly important role of women in the working world. We furthermore worked to include a number of female contributors, to receive insightful articles about their careers in FM practice in the past and the present.

This was important, as for the future a balance at least in the German-speaking countries is predictable, as at least in academia and especially in the universities of applied sciences a demographical balance of students already can be stated, while companies in Europe appear to be still lagging behind in gender distribution compared to the USA.

The realization of this unique anthology as a private, non-sponsored, not for profit activity, with an unparalleled variety of participants and performers from academia, business, and associations, all having contributed to the development of Facility Management, was only possible by a joint waiver of reimbursement by all contributors, and the editors/initiators alike made this possible.

Last but not least thanks to our research partner, the late Dr. Dr. John H. Smalley, who supported our contacts in the UK and in the Netherlands in the late 1980s, and by forming with Dr. Mack the IMI - Innovation Management Institute EEIG (now UKEIG) in 1998, networking with universities in England (Reading) and Scotland (Glasgow) and consulting organizations like DEGW in London, and TwynstraGudde in the Netherlands.

Appendix D

Glossary

FM associations, standards organizations, and educational organizations have developed and adopted a glossary of FM terms and definitions for use by association members, organizations, and educators and students.

The following are glossary locations and sites from a number of organizations for the readers use.

IFMA Knowledge Library (https://knowledgelibrary.ifma.org/glossary-search):
The IFMA Knowledge Library provides access to hundreds of facility management key terms and definitions.[1]

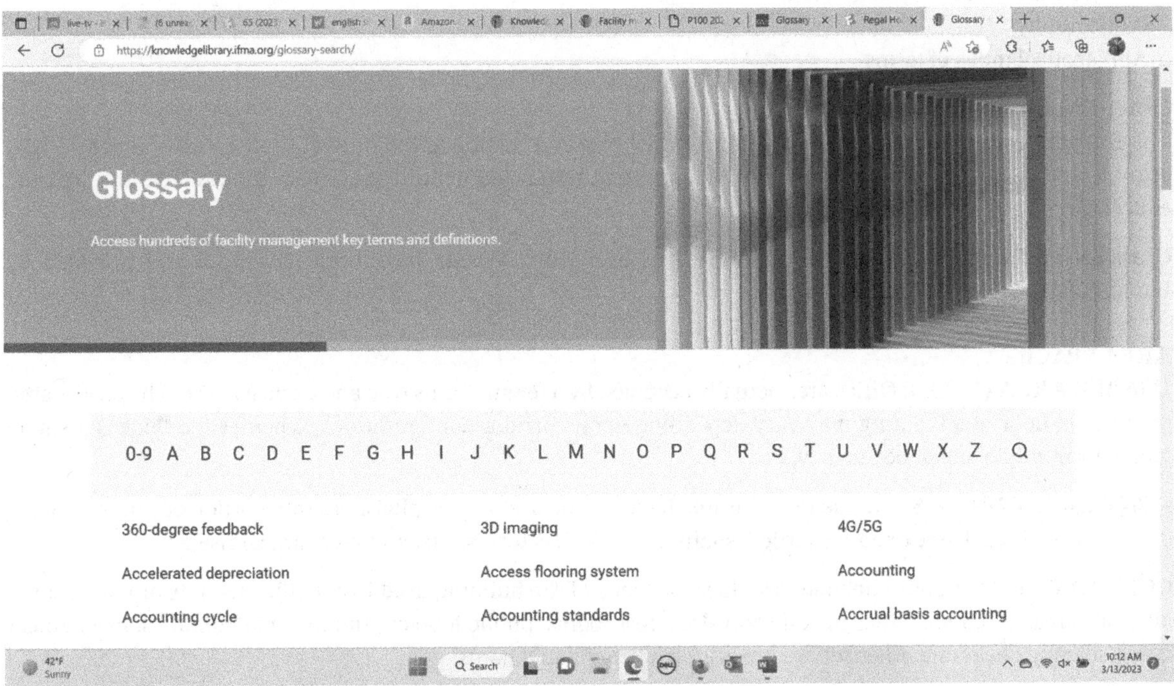

1 https://knowledgelibrary.ifma.org/glossary-search.

Facilities @ Management, Concept - Realization - Vision, A Global Perspective, First Edition. Edited by Edmond Rondeau and Michaela Hellerforth.

International Standards Organization (ISO): https://www.iso.org/obp/ui/#iso:std:iso:41011:ed-1:v1:en
The International Standards on facility management (FM) developed by ISO/TC 267 describe the characteristics of facility management and are intended for use in both the private and public sectors.[2]

ISO 41011:2017(en)
Facility management – Vocabulary
Table of Contents

Other Professional Associations:
The Building Owner and Managers Institute (BOMI.org) and the Royal Institute of Chartered Surveyors (RICS.org) have developed Glossary's of Facility Management terms and definitions. These are available through the education courses they offer.

Glossary of Facility Management Definitions and Buzz Words have been developed and provided by Edmond P. Rondeau, Co-Editor:

AREA SPACE -
USABLE AREA (U.S.F.): Office area actually occupied by a tenant for its sole and exclusive use. The usable area on a single floor of a building may vary depending upon corridor configurations, whether the floor is a single tenant or multiple tenant occupancy, etc.

RENTABLE AREA (R.S.F.): The rentable area includes the usable area plus a pro rata portion of common areas of the entire office floor excluding vertical shafts, such as elevators, stairs, and mechanical risers.

COMMON AREAS: Common areas are those portions of the building used by all office tenants or which serve all office areas. Common areas include corridors, rest rooms, public lobbies, and in some buildings, mechanical space, loading docks and other service areas which benefit all tenants.

LOAD FACTOR/COMMON AREA ALLOCATION: That percentage of the building in which common area is allocated to the tenants to increase their usable area to rentable area. A Load Factor of 10% means that 900 usable square feet (U.S.F.) would be 990 rentable square feet (R.S.F.).

2 https://www.iso.org/obp/ui/#iso:std:iso:41011:ed-1:v1:en.

B.O.M.A. STANDARD MEASUREMENT: A defined way of measuring space by the Building Owners and Managers Association (B.O.M.A.). Landlords may choose their own method to measure space, normally by increasing the amount of common areas added to the usable area.

EFFICIENCY (Note: Efficiency and load factor are not the same): The percent of rentable area which is usable area, i.e., a 90% efficient building offers 900 usable square feet for every 1000 rentable square feet.

LAYOUT EFFICIENCY: Efficiency of the usable area to meet tenant's workflow requirements, office design, personnel, etc. Efficiency of usable area is dictated by building shape, core location, floor size, leasing depth, corridors, etc.

RENTAL & EXPENSES -
RENTAL: The cost charged per rentable square foot (R.S.F.) on a monthly or annual basis for a leased area.

BASE RENTAL: The initial rental rate, normally identified as the annual rent in a gross lease.

EXPENSE STOP: An identified dollar amount, either on a dollar per square foot per year basis or a pro rata share basis of the total operating expense cost, that the landlord is responsible to pay. Any increase over the expense stop will be allocated to the tenant.

GROSS RENTAL (Gross Lease): A rental rate which includes normal building standard services as provided by the landlord within the base year rental.

NET OR SEMI-GROSS RENT (Net Lease): A rental rate which includes some services to be provided by the landlord; normally the tenant is responsible for cost of janitorial and utilities services.

NET NET NET RENT (Triple Net Lease): A lease in which the tenant is responsible for every and all expenses associated with their proportionate share of occupancy of the building.

OPERATING EXPENSES: Those normal expenses associated with the operation of a standard office building included, but not limited to: management fees, utilities, janitorial service, consumable supplies, taxes, insurance, refurbishing of building common areas, maintenance and repair, etc.

BASE YEAR: The year of building operation, normally a calendar year, in which the landlord fixes or identifies the operating costs which are included in a gross or semi-net lease. Any increase in operating expenses over the base year are "passed through" to the tenant on a pro rata of rentable

PRO RATA SHARE: The ratio between the tenant's percent of occupancy of the rentable square footage of the building and the entire building rentable area.

EXPENSE ALLOCATION: The allocation of all expenses (Gross or Semi-Net Lease) or a proportionate share of increased expenses to a tenant based upon tenant's pro rata share.

DIRECT CHARGE OF OPERATING EXPENSES: On a Gross or Semi-Net lease where the tenant does not pay costs directly, the landlord directly bills tenants for pro rata share of occupancy costs. In most instances, this will be done on a "good faith", "best estimate" advanced payment basis, wherein the landlord bills the tenant for estimated operating expense costs during the lease term.

RECAPTURE: The billing to tenants of their pro rata share of increased operating expenses after those expenses have been incurred and paid for by the landlord.

PROJECTED OPERATING EXPENSE INCREASE: A "good faith" estimate by the landlord as to current operating expenses increases over a base year, which are billed to the tenant as additional rent.

RENTAL COMPONENTS -
Three basic rental components are:

OWNERSHIP COSTS: The cost to the owner to own the building, service existing debt, or receive a return on their equity. Also included would be costs of capital improvements, repair and upkeep which would not be considered standard operating costs.

OPERATING EXPENSES: Those expenses necessary for the day-to-day operation of the building which have been outlined above. Included can also be amortization of capital improvements which are necessary or required for the more efficient operation of the building or to meet certain code costs.

AMORTIZATION OF TENANT IMPROVEMENTS OR TENANT ALLOWANCES: The return to the landlord over the term of the lease those costs included in the landlord's building standard work letter and any other costs which landlord has agreed to assume or amortize.

RENTAL "GAMES" -
FREE RENT: Period of time in which the tenant occupies the premises under the lease but does not pay rent.

STAIR-STEPPED RENT: A rental rate which increases by fixed amounts during the period of the lease term.

RENTAL INCREASES/RENTAL REVIEW: Changes in the base rent during the term of the lease. Not to be confused with operating expense recapture increase or expense billings, usually tied to "fair market" or C.P.I. reviews.

EFFICIENT RENT: The dollar amount per square foot per year figure which the tenant pays on an average over the term of the lease. This would be the average of specified rents in a stair stepped lease as well as the average of a lease with substantial free rent period. Example: A five-year lease with six months' free rent offers a 10% discount from the face rate.

FACE RATE (Contract Rate): The identified rental rate in a lease which is subsequently discounted by concessions offered by the landlord, also known as contract rate.

CONCESSIONS: Those inducements offered by a landlord to a tenant to sign a lease. Common concessions are free rent, extra tenant improvement allowances, payment of moving costs and lease pick-ups.

LEASE PICKUP: Landlord's commitment to pay the costs associated with assuming the financial obligations of paying a tenant's rent in premises to be vacated which are still under lease.

LEASE BUYOUT: A cash inducement offered by a landlord to a tenant's previous landlord or by the tenant to their current landlord to cancel the remaining term of tenant's lease.

SUBLEASE: Leasing of premises by the current tenant to another party for the remaining balance of an existing lease term.

MOVING ALLOWANCE: An offer by a landlord to pay all or part of tenant's moving costs.

FAIR MARKET VALUE: The rental value of space similar to the leased premises for comparison purposes in rental adjustments.

PRE-LEASE: Leasing of premises in a building under construction which is not yet ready for occupancy.

IMPROVEMENTS & ALLOWANCES -
TENANT IMPROVEMENTS BUILDING STANDARDS: Standard building materials and quantities as identified by the landlord which are to be provided at no cost to the tenant to improve tenant premises. Normally

included are partitioning, doors, walls, hardware, ceiling, lighting, window and floor coverings, telephone and electrical outlets and HVAC (Heating, Ventilation & Air Conditioning systems).

BUILDING STANDARD WORK LETTER: A document which delineates the type and quality of materials and quantities to be furnished by the landlord as building standard.

WORK LETTER: A document which would include the above plus any additional items to be paid for by the landlord or by the tenant with an indication as to who is responsible for each item.

IMPROVEMENT ALLOWANCE: The estimated or dollar value of the building standard work letter being offered by the landlord.

ABOVE BUILDING STANDARD: Materials not included in the work letter which are subject to negotiation between the landlord and tenant.

AMORTIZATION OF TENANT IMPROVEMENTS: An agreement on the part of the landlord to pay for above-building standard improvements and amortize those improvements at a defined interest rate over a fixed term as additional rental.

"TURN KEY": A complete build-out of tenant's premises to the tenant's specifications.

SUBSTITUTION AND CREDITS: The ability to substitute non-standard materials for landlord-supplied materials as specified in the work letter, or to receive dollar amount credits for those materials if not utilized.

LAY OUT (Space Plan): A plan created by a space planner/interior designer/architect showing locations of tenant improvements and the utilization of the space by the tenant.

WORKING DRAWINGS: Drawings necessary to fully price the work, to obtain a building permit and to construct the tenant improvements.

CODE REQUIREMENTS & COMPLIANCE CONDITIONS -

CODE REQUIREMENTS: Building code requirements which must be satisfied by either the tenant or the landlord in preparing space or building for tenant occupancy. Included are seismic (earthquake), life safety, energy, hazardous/toxic materials, and handicapped code requirements.

CODE COMPLIANCE TRIGGER: Certain costs, structural changes or other defined events which necessitate the compliance of a building to current codes.

ASBESTOS/HAZARDOUS MATERIALS: Materials which may exist in office structures which may pose a detriment to tenant's occupancy.

LIFE SAFETY: Government regulations and building code requirements for buildings relative to seismic, fire, handicapped and existing requirements.

RE CORRIDORS: Special corridors with partitioning designed to create escape routes in time of fire.

COMPARTMENTALIZATION: A code requirement to divide large floor plates into smaller units to meet fire code requirements.

SPRINKLERS: A fire suppression system, usually water, designed into many buildings to reduce compartmentalization and to provide additional fire protection.

HANDICAPPED REQUIREMENTS: Code required features designed to accommodate handicapped persons. Included are entry ramps, rest rooms, rest room fixtures, hardware, special doors, etc.

"FIRE RATED": Special building materials, such as partitioning and doors which have been constructed and have been tested to provide greater fire resistance that normal building standard materials.

BUILDING DEFINITIONS -

DEMISING WALLS: Those walls between one tenant's area and another as well as walls between tenant areas and public corridors.

PARTITIONING: The divisions between offices, separate office suites, tenant areas and corridors.

FLOOR PLATE: A broker's buzz work for rentable floor size.

BUILDING MODULE: Standard dimensions within leased areas dictated by spacing of window mullions or columns, i.e. a 5-foot module dictates offices in multiples of 5-feet dimensions.

LEASING DEPTH: The distance from the building window line to the building corridor.

BUILDING CORE: The "guts" of a building, which normally includes building elevators, rest rooms, smoke towers, fire stairs, mechanical shafts, janitorial, electrical and phone closets.

SHARED TENANT SERVICES: Services provided by a building to allow tenants to share the costs and benefits of sophisticated telecommunications and other technical services.

UNDER FLOOR DUCT SYSTEM: A system of ducts permanently located in floors to assist in the installation of telephone, computer and electrical wiring.

CROSS-OVER FLOOR: A floor in which one bank of elevators connects to another bank of elevators, allowing tenants to have access to floors in other elevator banks without returning to the lobby of the building.

ZONES: The identified portions of an office area served by the HVAC system which have separate thermostatic and temperature controls.

SMART BUILDING: A building which has additional technical capabilities to provide enhanced building management and operating efficiency.

AREA TAKE-OFFS -

DEFINITION OF TERMS – The following terms and definitions are supplied for purposes of explanation and understanding of the methods and uses for area take-offs. While these terms and definitions were taken from published documents as listed, they may not conform to methods or formulae that you may use. They are offered as <u>one</u> method for purposes of explanation and illustration.

LEASED AREAS -

SPACE: The generic definition of a particular enclosed area. A space, may be a building, floor, or any defined area. As a practical matter, a space is defined as that area defined by the drawing.

USER: The generic definition of the occupant of a space. This may be a tenant, a company, a department, etc. A given space may have more than one user for each tier of definition except Tier 1.

GROSS AREA: Building or floor footprint; the total area of the building or floor within the outline of the building extremities. This area is computed by measuring the inside finished perimeter of the dominant portion of the permanent outer building walls. The DOMINANT PORTION shall mean that portion of the inside, finished surface of the permanent outside building wall which is 50% or more of the vertical floor-to-ceiling dimension measured at the dominant portion. If the dominant portion is not vertical, the measurement for area shall be to the inside, finished surface of the permanent out building wall at the point of intersection of the vertical wall and the finished floor.

AREA OF PENETRATION: The sum of the area of those physical objects which vertically penetrate the space which serve more than one floor. E.g., elevator or HVAC shafts, stair wells, flues, and their enclosing walls.

COMMON AREA FACTOR (RENTABLE/USABLE RATIO): The factor used to determine a tenant's pro rata share of the common area.
Equation:

$$\mathbf{CAF(R / U\ RATIO) = RENTABLE\ AREA / USABLE\ AREA}$$

COMMON AREA: That area that has common access to all Users within a Gross space. E.g., Public corridors, rest rooms, mechanical or utility rooms, vestibules.
Equation:

$$\mathbf{COMMON\ AREA = RENTABLE\ AREA - USABLE\ AREA}$$

RENTABLE AREA: The area of a given Space defined as the Gross Area less the areas of penetration. NOTE: This area is fixed to the life of the building unless major modifications such as the addition of elevators occur.
Equation:

$$\mathbf{RENTABLE\ AREA = GROSS\ AREA - AREA\ OF\ PENETRATION}$$

And

$$\mathbf{RENTABLE\ AREA = USABLE\ AREA / COMMON\ AREA\ FACTOR}$$

EXAMPLE: Rentable area of floor = 20,000 S.F.
ABC Company usable = 17,500 S.F.
R = 20,000 = 1.144 COMMON AREA FAC.(CAF)
U = 17,500

1,500 Usable S.F. × 1.144 = 1,716 Rentable S.F.

1,716 S.F. Rentable = 1,500 Rentable S.F.
1.144 CAF
USABLE AREA: That area of a space that may actually be occupied by a User.

Equation:

$$\mathbf{USABLE\ AREA = RENTABLE\ AREA - COMMON\ AREA}$$

LEASED RENTABLE AREA: That area to which the base lease rate applies. This is defined as the tenant's usable area times the CAF. This area may then be multiplied by the lease rate to determine the lease amount.

Equation: **TENANT USABLE AREA X CAF**

OFFICE AREAS -
SPACE LIMITS: The area of a given space is generally determined by some physical limit. These are most often walls, partitions or panels.

WALLS: These are generally defined as outside walls, load-bearing walls, fire walls or those walls that are not considered as movable or removable.

PARTITIONS: These are generally defined as inside floor-to-ceiling structures not otherwise meeting the criteria of Walls. Partitions are movable or removable.

PANELS: Panels are generally defined as modular furniture sections used to define the limits of a workstation. Panels do not extend from floor to ceiling.

WORKSTATION: Any space for which a function is accomplished. This may be an enclosed space or a space in an open area. E.g., a Conference Room, an Executive Office, a Coffee Station, a Reception Area, a Data Input Station. A workstation does <u>not</u> necessarily require that a person or persons be assigned to that particular workstation.

WORKSTATION NET AREA -
ENCLOSED SPACE: The floor area inside the enclosed space, measured to the inside surface of walls, major protrusions or other surfaces which define the limit of functionally usable floor surface. This does not include the area of free-standing columns which inhibit functional use of the space and does not include circulation area outside the space.

OPEN SPACE: The floor area available inside a workstation for furniture, equipment and internal circulation. This area is measured from the inside face of furniture screens or panels, to the outside edge of furniture, equipment or carpet that bounds the intended work area. Does not include area of circulation outside the immediate work area.

CIRCULATION -
INTRA-OFFICE: The common area <u>between</u> departments, sections, etc. used for corridors, aisles, or walkways. These would generally be determined between users defined on the same TIER as compared to the previous tier.

INTER-OFFICE: The common area <u>within</u> departments, sections, etc. used for corridors, aisles or walkways. These would generally be determined between users defined on the same TIER as compared to the previous tier.

TIER 1: The overall working area of the space being considered. This may be the Gross Area of a floor or the Rentable Area of a floor. TIER 1 should use the space description as the USER of TIER 1. E.g. “Globe Building, 8th Floor.” TIER 1 may have only **<u>one</u>** user.

TIER 2: A spatial area definition under the TIER 1 definition. This may be a tenant, section, department, etc. TIER 2 may have multiple Users of similar definition. E.g., “XYZ and Associates,” “Acme Computers,” or “Accounting Department,” “Marketing Department.”.

TIER 3: A spatial area definition under the TIER 2 definition. This may be a department, unit, etc. TIER 3 may have multiple Users of similar definition.

TIER 4: A spatial area definition under the TIER 3 definition. This may be a department, unit, etc. TIER 4 may have multiple Users of similar definition.

TIER 5: The lowest level of spatial definition. This is generally used to define a work station, room or individual space.

A typical tier structure might be depicted as follows:

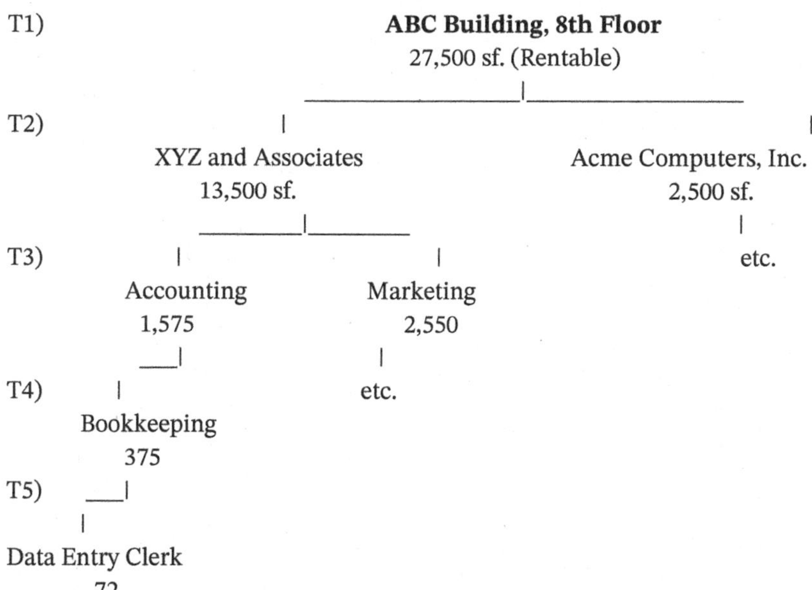

Index

Note: Page numbers followed by "*f*" and "*t*" refers to figures and tables, respectively; Page numbers followed by "n" refer to footnotes.

Facilities @ Management, Concept - Realization - Vision, A Global Perspective, First Edition. Edited by Edmond Rondeau and Michaela Hellerforth.

C

d

e

f

g

S